Net-Zero/Positive Energy Buildings and Districts

Net-Zero/Positive Energy Buildings and Districts

Editors

Ala Hasan
Francesco Reda

MDPI • Basel • Beijing • Wuhan • Barcelona • Belgrade • Manchester • Tokyo • Cluj • Tianjin

Editors

Ala Hasan
Smart Energy and Built
Environment
VTT Technical Research
Centre of Finland
Espoo
Finland

Francesco Reda
Industrial Energy and
Hydrogen
VTT Technical Research
Centre of Finland
Espoo
Finland

Editorial Office

MDPI
St. Alban-Anlage 66
4052 Basel, Switzerland

This is a reprint of articles from the Special Issue published online in the open access journal *Buildings* (ISSN 2075-5309) (available at: www.mdpi.com/journal/buildings/special_issues/Net_Zero_Positi_Energ_Build_District).

For citation purposes, cite each article independently as indicated on the article page online and as indicated below:

LastName, A.A.; LastName, B.B.; LastName, C.C. Article Title. *Journal Name* **Year**, *Volume Number*, Page Range.

ISBN 978-3-0365-4564-6 (Hbk)
ISBN 978-3-0365-4563-9 (PDF)

Contents

About the Editors

Ala Hasan

Dr. Ala Hasan is Principal Scientist in the Smart Energy and Built Environment Research Area at VTT, the Technical Research Centre of Finland. He has been working in the field of energy in buildings since 1982. He joined VTT in 2013. Prior to that, he worked at the Helsinki University of Technology and Aalto University since 1998. He held the position of Research Fellow of the Academy of Finland and has been PI of several Academy of Finland research projects. He was evaluated as University Full Professor (2015). Dr. Hasan has an excellent portfolio in research work, scientific publications, earning research funding, as well as offering service to the international scientific community. He has more than 100 publications in international scientific journals, conferences, book chapters, and reports. He has been listed within the top 2% most cited scientists in the world in his field as published by Elsevier BV and Stanford University in years 2020 and 2021. He is a member of the Editorial Board of the *Buildings* journal MDPI Switzerland and a member of the Editorial Board of the International Journal of Low Carbon Technologies–Oxford Journals UK. He acted as advisor and examiner of several Ph.D. and Licentiate degree students in Finland, Sweden, and Denmark. He was elected as Fellow of IBPSA (The International Building Performance Simulation Association) 2021. He was the first president of IBPSA-Nordic and was a member of IBPSA-World Board of Directors. He is a member of the Executive Committee of the International Energy Agency (IEA), Energy in Buildings and Communities (EBC) programme representing Finland.

Francesco Reda

Dr. Francesco Reda. Currently he is Research Manager of the Industrial Energy and Hydrogen Research Area at VTT, the Technical Research Centre of Finland. He has 10-year experience in R&D of renewable energy production system applications, conceiving concept solutions for Net Zero and Positive Energy communities, combining both technical and user-centred solutions. His research activities are devoted to developing socio-technical energy solutions as a trigger of a "renewable revolution"for preserving the fragile interaction between the Earth, as a finite ecosystem, and the human beings, as consumers of resources coming from such a fragile ecosystem. He has been and is involved in several Research and Innovation European and International cooperation projects focusing on smart city and building renewable energy solutions. He is also active in the international energy research and innovation programme in the buildings and communities field. Currently he is the Operating Agent of the IEA-EBC Annex 83 "Positive Energy Districts". He has authored more than 30 scientific publications.

Preface to "Net-Zero/Positive Energy Buildings and Districts"

Buildings account for more than one-third of the global final energy consumption and CO2 emissions. The building sector offers a significant potential in the transition towards the decarbonisation of societies. To achieve this goal, different concepts and implementations of Net-Zero/Positive Energy Buildings and Districts (NZPEBD) have emerged in recent years and are still in progress. Countries around the world have adapted different strategies and regulations for the decarbonisation of the building sector. An example is the EU Directive on Energy Performance of Buildings (EPBD) legislation that aims at achieving Nearly Zero Energy Buildings. Another example is the European Strategic Energy Technology (SET) Plan that supports the planning, deployment, and replication of 100 Positive Energy Neighbourhoods by 2025.

The concept of NZPEBD is not only about achieving a net or positive energy balance between the exported and imported energy from a building or a district within a specified time horizon. Research on NZPEBD encompasses all related aspects of energy in buildings and districts, starting from the definition of the basics of the concept, the boundary of the building/district, types of energy credits to be balanced, energy efficiency of the building envelope, integration of onsite renewable energy generation, technologies for short and long term storage, performance control and operational optimisation, onsite load-matching, offering flexibility to the grids, as well as people's acceptance and contribution and the economic feasibility.

Due to the importance of the NZPEBD topic, we opened a call for articles submission in a Special Issue "Net-Zero/Positive Energy Buildings and Districts"of the *Buildings* journal. This book is the collection of the articles published in that Special Issue. The book includes 17 research articles covering different aspects of NZPEBD planning, technologies, economics, building design, and retrofitting, citizen engagement, and collection of energy data.

We would like to thank the authors of the articles for their valuable contribution. We also wish to express our gratitude to the reviewers of the articles for their time and effort.

Ala Hasan and Francesco Reda
Editors

buildings

Editorial

Special Issue "Net-Zero/Positive Energy Buildings and Districts"

Ala Hasan * and Francesco Reda

VTT Technical Research Centre of Finland Ltd., P.O. Box 1000, FI-02044 Espoo, Finland; francesco.reda@vtt.fi
* Correspondence: ala.hasan@vtt.fi

The important goal of decarbonization of communities and cities has resulted in the emergence of new concepts and implementations of Net-Zero/Positive-Energy Buildings and Districts (NZPEBD) in recent years. Research on NZPEBD comprises all related aspects of energy in buildings and communities, from the basic definition of the concept, including the boundary and types of energy credits definitions, to the characteristics of the building envelope, onsite renewable energy system components and integration, interaction with the external grids, performance control and optimisation, etc., as well as social and economic aspects.

This Special Issue includes a total of 17 papers covering different aspects of NZPEBD planning, technologies and their economics, building design and retrofitting, citizen engagement and collection of energy data.

Under NZPEBD planning, six papers investigated different aspects of NPZEBD planning including definitions, replication methods, obstacles and international collaboration. Lindholm et al. [1] presented the essential factors that determine the planning of PEDs in the EU by studying different definitions of PEDs, features and availability of the renewable energy, consumption behaviours, populations, costs and regulations. Uspenskaia et al. [2] studied common trends in technologies and replication strategies for positive-energy buildings/districts in smart city projects. A case study was performed in Leipzig, one of the lighthouse cities in the SPARCS project, which emphasised the importance of formulation of replication modelling for the upscaling process. Tuerk et al. [3] considered the economic optimisation and market integration opportunities provided by the Clean Energy Package for Plus Energy Buildings (PEBs) and Plus Energy Districts (PEDs). They identified the regulatory limitations at the national level with regard to transposing the set of EU Clean Energy Package provisions. Different options for PEBs and PEDs were studied based on the H2020 EXCESS project. Makvandia and Safiuddin [4] studied the challenges for net-zero buildings in single-family homes in the Greater Toronto Area. The main challenges were technical obstacles, lack of governmental and institutional support, lack of standardisation and low public awareness measures. Recommendations were provided for governmental and academic support for technological uptake and financial incentives. Zhang et al. [5] analysed 60 PED projects in Europe by their main characteristics: geographical information, spatial–temporal scale, energy concepts, building archetypes, finance source, keywords, finance model and challenges/barriers. Many projects use an annual scale; about one-third of the projects have an area smaller than 0.2 km^2 and the most common renewable energy systems are solar, district heating/cooling, wind and geothermal energy. Hedman et al. [6] explained Annex 83 "Positive Energy Districts" of the International Energy Agency—Energy in Buildings and Communities Programme (IEA-EBC). The structure of Annex 83, including its four subtasks, and the working plan were described. The main topics of discussion were the definitions of PEDs, virtual and geographical boundaries, evaluation approaches, the role of different stakeholders, environmental, economic and societal implications, and learnings from realised PED projects.

Citation: Hasan, A.; Reda, F. Special Issue "Net-Zero/Positive Energy Buildings and Districts". *Buildings* **2022**, *12*, 382. https://doi.org/10.3390/buildings12030382

Received: 23 February 2022
Accepted: 17 March 2022
Published: 21 March 2022

Under technologies and their economics, five papers studied the technologies of fuel cell, PV, PVT and waste heat with borehole storage. Lindholm et al. [7] studied the use of electricity generated by solar photovoltaic (PV) panels to produce hydrogen gas and its seasonal storage for a reversible solid oxide cell (RSOC) operation. A case study showed that the system can achieve higher utilisation of the generated PV electricity, resulting in achieving a net-zero annual energy balance. Lovati et al. [8] proposed a peer-to-peer (P2P) business model for PV in a small community of 48 individual prosumer buildings in Sweden, considering the energy use behaviour, electricity/financial flows, ownerships and trading in a local electricity market. The results show different use of the common PV resource by the buildings and diverse self-sufficiency features. Lovati et al. [9] presented an agent-based modelling environment for shared urban photovoltaic (PV) systems between 48 households in a local grid of a positive energy district with optimised self-sufficiency. Various scenarios were explored by varying the number of owners of the PV systems and their pricing profiles. Penaka et al. [10] presented a techno-economic study of a typical PVT system for a single-family house to generate electricity and domestic hot water in 85 locations worldwide. The economic performance was assessed using net present value and payback period under two financial models. Hirvonen and Kosonen [11] studied the utilisation of excess heat from waste incineration together with borehole thermal energy storage as seasonal energy storage to supplement conventional district heating of a new residential area. A total of 36 different storage configurations were investigated to obtain the techno-economic performance. In case the district boundary is expanded to include the waste heat generation, the community as a whole can progress toward net-zero energy.

The first step towards achieving net NZPEBD is to minimise the energy demand of buildings by adapting higher energy conservation measures. There are four papers on building design and retrofitting. Hirvonen et al. [12] investigated the potential of Energy retrofitting of buildings for achieving CO_2 emissions neutrality in six Finnish building types by comparing the emissions reduction, investment and life cycle costs. The results indicate that it is possible to reduce the emissions cost-neutrally by 20 to 70% in buildings with district heating and by 70 to 95% with heat pumps. Switching single-family homes with oil or wood boilers to heat pumps produced the largest emission reduction potential. Albatayneh [13] used multi-objective optimisation with various design variables in the building's envelope to reduce the heating and cooling energy in residential buildings in the city of Ma'an, Jordan. The results indicate savings of 88.1%, 94.2% and 78.5% in the total energy consumption, cooling load and heating load, respectively, compared with a baseline building. Koke et al. [14] studied design strategies for suitable building concepts and energy systems to be used in Nearly Zero-Energy Container Buildings (NZECBs) for different climates. Container-based lightweight buildings have high ecological and economic potential. Three cases in Sweden, Germany and Ethiopia were demonstrated and compared, particularly regarding energy self-sufficiency. The influence of different climate zones on the energy efficiency of a single-family house was studied, as well as the influence of the insulation and battery size. Quintana et al. [15] presented a digital spatial map of both electricity use and district heating demand of a set of buildings in the city of Borlänge, Sweden. A toolkit for top-down data analysis was used based on an energy database of monthly consumption of the buildings, which consisted of 228 and 105 geocoded addresses. Digital mapping showed a spatial representation of hotspots of electricity use in high-occupancy/-density areas and for district heating needs. The results can provide an understanding of the existing energy distributions for stakeholders and energy advisors.

Citizen engagement is very important as it is essential to keep citizens informed and engaged with the increasing numbers of technologies and the large scale of urban development. Fatima et al. [16] examined citizen engagement in Espoo (Finland) and Leipzig (Germany) to assess readiness for developing and implementing positive energy districts (PEDs). They studied the cities' operations and methods for citizens' participation.

As lighthouse cities, findings from these two cities can be used to assist other cities in Europe and beyond to plan and operate PEDs together with citizens.

Assessing the energy performance of buildings for PEDs at the urban scale requires large amounts of data. However, these data can be challenging for communities to acquire. Han et al. [17] used a Gaussian mixture model (GMM) with an Expectation-Maximisation (EM) algorithm to produce synthetic building energy data. This method is tested on real datasets. The developed approach is useful for building simulations and optimisations with spatio-temporal mapping.

The heterogeneity of the involved research underlines the importance of following a holistic approach when undertaking NZPEBD planning and implementation. It also shows that further multi-disciplinary research is needed for the integration of advanced technology together with developed people's engagement practices and financing mechanisms on individual and community levels. The NZPEBD challenges will be greatly intensified when going from a small-scale community to a district or a city level.

We would like to thank the authors of the papers for their valuable contribution to this Special Issue. We also wish to express our gratitude to the reviewers of the papers for their time and effort.

Funding: This research received no external funding.

Conflicts of Interest: The authors declare no conflict of interest.

References

1. Lindholm, O.; Rehman, H.U.; Reda, F. Positioning positive energy districts in european cities. *Buildings* **2021**, *11*, 19. [CrossRef]
2. Uspenskaia, D.; Specht, K.; Kondziella, H.; Bruckner, T. Challenges and barriers for net-zero/positive energy buildings and districts—Empirical evidence from the smart city project SPARCS. *Buildings* **2021**, *11*, 78. [CrossRef]
3. Tuerk, A.; Frieden, D.; Neumann, C.; Latanis, K.; Tsitsanis, A.; Kousouris, S.; Llorente, J.; Heimonen, I.; Reda, F.; Ala-Juusela, M.; et al. Integrating plus energy buildings and districts with the EU energy community framework: Regulatory opportunities, barriers and technological solutions. *Buildings* **2021**, *11*, 468. [CrossRef]
4. Makvandia, G.; Safiuddin, M. Obstacles to developing net-zero energy (NZE) homes in greater Toronto area. *Buildings* **2021**, *11*, 95. [CrossRef]
5. Zhang, X.; Penaka, S.R.; Giriraj, S.; Sánchez, M.N.; Civiero, P.; Vandevyvere, H. Characterizing positive energy district (PED) through a preliminary review of 60 existing projects in europe. *Buildings* **2021**, *11*, 318. [CrossRef]
6. Hedman, Å.; Rehman, H.U.; Gabaldón, A.; Bisello, A.; Albert-Seifried, V.; Zhang, X.; Guarino, F.; Grynning, S.; Eicker, U.; Neumann, H.-M.; et al. IEA EBC annex83 positive energy districts. *Buildings* **2021**, *11*, 130. [CrossRef]
7. Lindholm, O.; Weiss, R.; Hasan, A.; Pettersson, F.; Shemeikka, J. A MILP optimization method for building seasonal energy storage: A case study for a reversible solid oxide cell and hydrogen storage system. *Buildings* **2020**, *10*, 123. [CrossRef]
8. Lovati, M.; Zhang, X.; Huang, P.; Olsmats, C.; Maturi, L. Optimal simulation of three peer to peer (P2P) business models for individual PV prosumers in a local electricity market using agent-based modelling. *Buildings* **2020**, *10*, 138. [CrossRef]
9. Lovati, M.; Huang, P.; Olsmats, C.; Yan, D.; Zhang, X. Agent based modelling of a local energy market: A study of the economic interactions between autonomous PV owners within a micro-grid. *Buildings* **2021**, *11*, 160. [CrossRef]
10. Reddy Penaka, S.; Kumar Saini, P.; Zhang, X.; Amo, A.d. Digital mapping of techno-economic performance of a water-based solar photovoltaic/thermal (PVT) system for buildings over large geographical cities. *Buildings* **2020**, *10*, 148. [CrossRef]
11. Hirvonen, J.; Kosonen, R. Waste incineration heat and seasonal thermal energy storage for promoting economically optimal net-zero energy districts in Finland. *Buildings* **2020**, *10*, 205. [CrossRef]
12. Hirvonen, J.; Jokisalo, J.; Sankelo, P.; Niemelä, T.; Kosonen, R. Emission reduction potential of different types of finnish buildings through energy retrofits. *Buildings* **2020**, *10*, 234. [CrossRef]
13. Albatayneh, A. Optimising the parameters of a building envelope in the east mediterranean saharan, cool climate zone. *Buildings* **2021**, *11*, 43. [CrossRef]
14. Koke, J.; Schippmann, A.; Shen, J.; Zhang, X.; Kaufmann, P.; Krause, S. Strategies of design concepts and energy systems for nearly zero-energy container buildings (NZECBs) in different climates. *Buildings* **2021**, *11*, 364. [CrossRef]
15. Quintana, S.; Huang, P.; Han, M.; Zhang, X. A top-down digital mapping of spatial-temporal energy use for municipality-owned buildings: A case study in Borlänge, Sweden. *Buildings* **2021**, *11*, 72. [CrossRef]
16. Fatima, Z.; Pollmer, U.; Santala, S.-S.; Kontu, K.; Ticklen, M. Citizens and positive energy districts: Are espoo and leipzig ready for PEDs? *Buildings* **2021**, *11*, 102. [CrossRef]
17. Han, M.; Wang, Z.; Zhang, X. An approach to data acquisition for urban building energy modeling using a gaussian mixture model and expectation-maximization algorithm. *Buildings* **2021**, *11*, 30. [CrossRef]

buildings

MDPI

Article

Positioning Positive Energy Districts in European Cities

Oscar Lindholm *, Hassam ur Rehman and Francesco Reda

VTT Technical Research Centre of Finland, P.O. Box 1000, FI-02044 VTT, Finland; hassam.rehman@vtt.fi (H.u.R.); francesco.reda@vtt.fi (F.R.)

* Correspondence: oscar.lindholm@vtt.fi

Abstract: There are many concepts for buildings with integrated renewable energy systems that have received increased attention during the last few years. However, these concepts only strive to streamline building-level renewable energy solutions. In order to improve the flexibility of decentralized energy generation, individual buildings and energy systems should be able to interact with each other. The positive energy district (PED) concept highlights the importance of active interaction between energy generation systems, energy consumers and energy storage within a district. This paper strives to inform the public, decision makers and fellow researchers about the aspects that should be accounted for when planning and implementing different types of PEDs in different regions throughout the European Union. The renewable energy environment varies between different EU regions, in terms of the available renewable energy sources, energy storage potential, population, energy consumption behaviour, costs and regulations, which affect the design and operation of PEDs, and hence, no PED is like the other. This paper provides clear definitions for different types of PEDs, a survey of the renewable energy market circumstances in the EU and a detailed analysis of factors that play an essential role in the PED planning process.

Keywords: PED; energy flexibility; socioeconomic analysis; techno-economic analysis; regions; regulation; renewable energy; energy storage; urban environment; climatic zones

Citation: Lindholm, O.; Rehman, H.u.; Reda, F. Positioning Positive Energy Districts in European Cities *Buildings* **2021**, *11*, 19. https://doi.org/buildings11010019

Received: 27 November 2020
Accepted: 27 December 2020
Published: 4 January 2021

1. Introduction

Various zero energy building (ZEB) concepts have been applied and used in the building sector all over the world. The overall ZEB definition states that "the building can be considered as ZEB after it shows through actual measurements that the energy delivered to the building is less than or equal to the onsite renewable exported energy" [1]. These concepts are, however, only applicable on individual buildings or groups of buildings and consider neither the impact on society at large nor the interaction with other energy consumers and producers. Most ZEBs even neglect the fact that the mobility sector is gradually connecting to the electricity grid. Nevertheless, ZEBs have recently received immense academic and political interest around the world [2–7].

The USA established the Energy Independence and Security Act of 2007 [8] to support the building sector to create zero energy commercial buildings by the year 2030. It also mentions converting 50% of American commercial buildings to ZEBs by 2040 and converting all commercial buildings into ZEBs by 2050 [9]. Similar legislation and regulations have been adopted by the EU in the form of Directive on Energy Performance of Buildings (EPBD), which aims to make all public buildings and new buildings nearly zero energy buildings by 2020 [7].

In Europe, the European Union (EU) has developed a framework that aims to reduce the emissions from buildings by improving the energy efficiency at the building level. The Directive on Energy Performance of Buildings (EPBD) initiated in May 2010 states that a nearly ZEB is a building with a high efficiency in terms of energy utilization and an energy demand that is mostly covered by on-site renewable energy generation [7]. The EPBD also mentions that the nearly ZEB definition can be flexible and adjusted to national or local requirements and targets. Nearly ZEB impact factors, such as the energy efficiency of the buildings and renewable energy sources, are also defined on a national or a regional level. Moreover, the primary energy factor is defined by the EU states based on national policies.

Cost-optimal buildings are also introduced in the EPBD in the Delegated Regulation No 244/2012 [10]. The cost-optimality is defined as the optimal ratio between the life cycle costs and the energy efficiency of the building [11]. This ratio varies from one EU region to another because the climate and the building standards are different in different EU regions [12–14].

According to the US Department of Energy (DOE), a ZEB is an "energy efficient building where, on a source energy basis, the actual annual delivered energy is less than or equal to the on-site renewable exported energy" [6]. Here, the source energy means the total life cycle energy of the building, including the building energy; the energy used for the extraction, transportation and processing of primary fuels; energy losses in the thermal and electrical plant; and energy losses in transport and energy distribution to the building site. Building energy refers to the on-site building energy consumption, including heating, cooling, ventilation, domestic hot water, indoor and outdoor use, lights, plug loads, process energy, elevators, conveying systems and intra-building transportation.

In addition to the ZEB concept, the DOE has defined three other concepts, the zero energy community (ZECo), zero energy campus (ZEC) and zero energy portfolio (ZEP), which consider a cluster of buildings that operates as a unit that shares the same renewable energy systems [6]. These concepts are, however, not clearly defined, and the differences between the concepts are quite indistinct. The main advantage of the ZEB, ZECo, ZEC and ZEP concepts is that they strive to cover the aggregated demand of the buildings, and thus, the generated renewable energy is used where it is needed the most.

The International Energy Agency (IEA) has proposed a concept called autonomous ZEB, which is an extension of the ZEB [15]. These buildings are self-sustaining buildings with no connection to the grid and are able to produce enough on-site energy to satisfy their own energy demand. In order to satisfy the energy demand day and night, summer and winter, energy must be stored. This differs from the net ZEB concept, as the net ZEBs are able to interact with the external grid as long as the annual energy export is equal to the annual energy import. The IEA does also bring up energy plus buildings (+ZEB), which export more energy than they import [15].

Table 1 shows a summary of the zero energy and positive energy definitions found in the literature. Here, it can be observed that most of the definitions address building-level applications and that only one definition addresses building energy efficiency, renewable energy, energy storage and as energy trading.

Table 1. Definition comparison in the literature (✓ = addressed in the definition, x = not addressed in the definition, ? = not defined/unclear).

Term	Definition	Building Energy Efficiency	Renewable Energy	Energy Storage	Energy Trading	Application	Reference
Nearly zero energy building	A high-energy-efficiency building that covers a large amount of its energy demand with on-site or nearby renewable energy generation	✓	✓	x	x	Building	[10]
Net-zero energy building	A building that exports an amount of energy to the grid equal to what it imports from the grid	x	✓	x	✓	Building	[16]
Zero energy building	A building that does not consume any energy	x	x	x	x	Building	[17]
Zero emission building	A building that does not release any emissions	x	x	x	x	Building	[17]
Net-zero source energy building	A building that generates all the energy it consumes, based on primary energy consumption	x	✓	?	x	Building	[18]
Net-zero site energy building	A building that generates all the energy it consumes, based on building energy consumption	x	✓	?	x	Building	[18]
Net-zero energy cost building	A building that covers the cost of imported energy by exporting on-site-generated renewable energy	x	✓	x	✓	Building	[18]
Autonomous zero energy building	A building that generates all the energy it consumes	x	✓	x	✓	Building	[15]
Photovoltaicor wind zero energy building	A building with a low energy demand and on-site PV panes and wind turbines	✓	✓	x	x	Building	[11]
Photovoltaic + solar thermal + heat pump zero energy building	A building that covers its energy demand via PV panels, solar thermal collectors, heat pumps and energy storage	x	✓	✓	x	Building	[11]
Wind + solar thermal + heat pump zero energy building	An energy efficient building that covers its energy demand via wind turbines, solar thermal collectors and heat pumps	✓	✓	✓	x	Building	[11]
Positive energy building	A building with a negative annual energy consumption	✓	✓	✓	✓	Building	[19]
Net-zero energy district	A building that exports an amount of energy to the grid equal to what it imports from the grid	x	✓	x	✓	District	[20]
Energy positive neighbourhood	A neighbourhood in which the energy demand is lower than the supply from local renewables	x	✓	x	✓	District	[21]

Most of the definitions found in the literature are different versions of the ZEB concept, and only a few definitions treat zero energy and energy positivity on a district or neighbourhood level. Especially, energy positivity on a district scale, so-called positive energy districts (PEDs), seems to be in an early conceptual phase. PEDs are gaining interest in the EU, as the European Clean Energy package is opening up a new opportunity for so-called energy communities, consisting of multiple small-scale energy consumers and providers, to share energy in local energy networks [22]. PEDs are hence becoming an increasingly interesting and a more competitive alternative to ZEBs, as they are able to exploit more of the available energy generation and energy storage potential of the community, i.e., the district. In PEDs, the renewable energy supply and demand can be unevenly distributed throughout the district, which allows a more strategical installation of renewable energy systems and energy storage.

As PEDs are starting to gain interest in various research projects, clear standardized definitions and frameworks need to be established. The objective of this paper is to provide clear definitions for different types of PEDs, a survey of the renewable energy market circumstances in the EU and a detailed analysis of factors that play an essential role in the PED planning process. The analysis discusses the available alternatives for constructing PEDs and networks of PEDs in European cities as well as the regulative aspects that are relevant for the implementation of PEDs in the EU. This analysis is also the novelty of the paper, as no other study of PEDs in the literature treats the preconditions and available options for PEDs on a general level.

2. Methods

The study was carried out in three steps. The first step was to produce a description of the PED concept and to develop a clear set of criteria that every PED must satisfy. This step was conducted by gathering information about the PED definition from the limited amount of literature about PEDs that is available.

The second step was to investigate the renewable energy environment in the EU. This step was carried out by collecting data and information about:

- The renewable energy generation and energy storage potential in different EU regions;
- The techno-economic properties of different renewable energy and energy storage technologies;
- The energy consumption and energy consumption trends in different EU regions;
- The electricity prices in different EU regions and factors affecting the electricity price.

In order to keep the collected data at a manageable size, the authors mainly collected data for four EU countries (Finland, the Netherlands, Germany and Italy) and their corresponding capital cities.

The first two steps of the study served as a base for the PED implementation analysis, which was conducted in the third and final step. The aim of the PED implementation analysis was to form a conception about the possibilities of implementing PEDs in the EU. The authors also used references from the scientific literature to gather ideas of how to implement different technologies. Additionally, the authors suggest a new idea of how networks of PEDs could be constructed and how PEDs could interact with each other. Finally, the analysis assesses how the current EU regulations and guidelines could affect the implementation of PEDs in the EU.

A schematic diagram of the research design is presented in Figure 1.

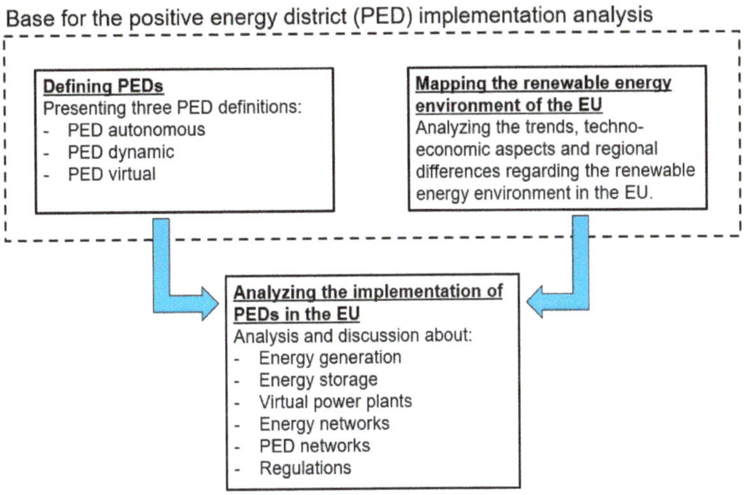

Figure 1. Schematic diagram of the research design.

3. Positive Energy District Definition

A positive energy district generates more renewable energy than it consumes on a yearly basis [23,24]. This is achieved by integrating renewable energy systems and energy storage as well as improving the energy efficiency of the district by optimizing the energy flows between the energy consumers, producers and storage. As a part of the European Strategic Energy Technology Plan (SET Plan), PEDs are considered as a building block for reducing the carbon emissions of cities. Three frameworks were developed in a PED definition workshop organized by the European Energy Research Alliance (EERA) Joint Programme Smart Cities [25]:

- PED autonomous—a district with clear geographical boundaries that is completely self-sufficient energy wise, meaning that the energy demand is covered by internally generated renewable energy. The district is thus not allowed to import any energy from the external electricity grid or district heating/gas network. The export of excess renewable energy is, however, allowed.
- PED dynamic—a district with clear geographical boundaries that has an annual on-site renewable energy generation that is higher than its annual energy demand. The district can openly interact with other PEDs as well as the external electricity grid and district heating/gas network.
- PED virtual—a district that allows the implementation of virtual renewable energy systems and energy storage outside its geographical boundaries. The combined annual energy generation of the virtual renewable energy systems and the on-site renewable energy systems must, however, be greater than the annual energy demand of the district.

Figures 2–4 show examples of how the three different PEDs could look. PED autonomous and PED dynamic are both constrained by geographical boundaries. PED autonomous is a completely self-sufficient energy system, which means that the energy demand is covered by internally generated renewable energy. PED dynamic allows the energy system to import externally generated energy, as long as the annual energy balance is positive. This means that PED dynamic must export more energy than it imports, on a yearly basis.

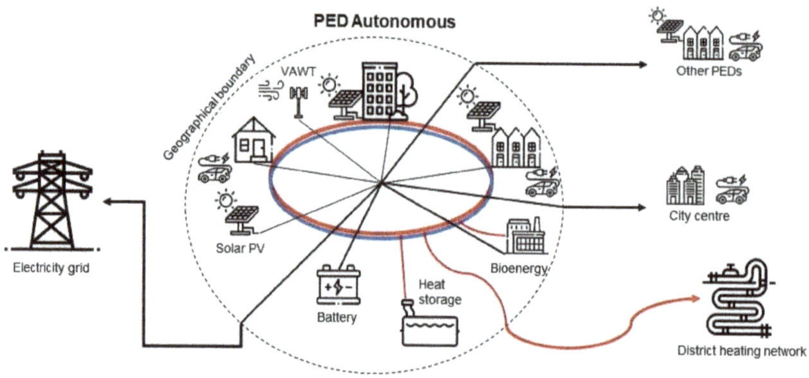

Figure 2. Graphical explanation of a PED autonomous. The PED autonomous is completely self-sufficient, which means that it covers the on-site energy demand with on-site renewable energy generation. It is, however, possible for the PED autonomous to export excess energy to other PEDs as well as the external electricity grid and district heating network. VAWT stands for vertical axis wind turbine.

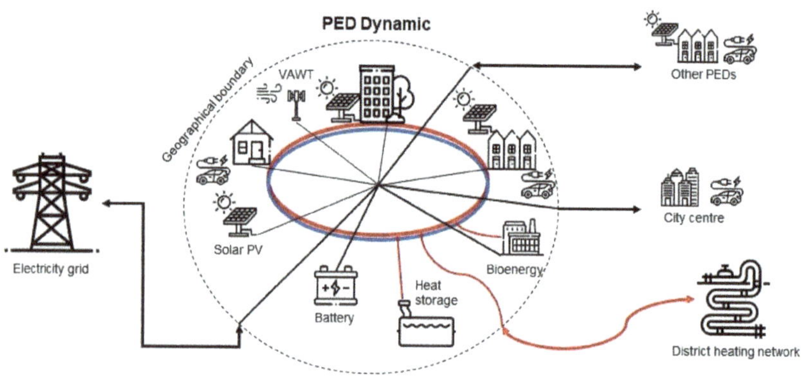

Figure 3. Graphical explanation of a PED dynamic. PED dynamic bidirectional energy trading with other PEDs as well as the external electricity grid and district heating network. VAWT stands for vertical axis wind turbine.

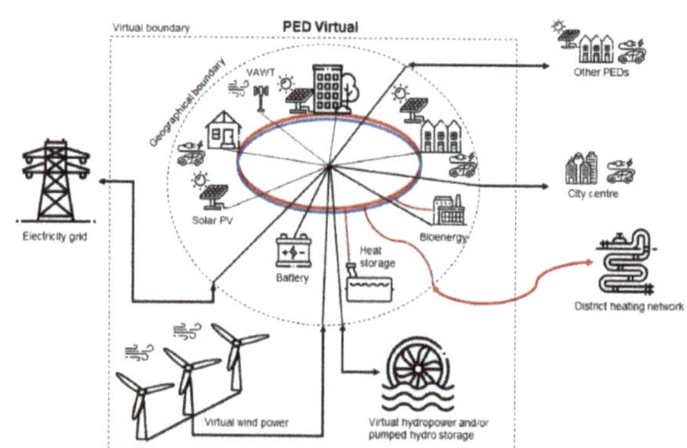

Figure 4. Graphical explanation of a PED virtual. PED virtual allows virtual renewable energy generation and energy storage operation outside the geographical boundaries of the district. VAWT stands for vertical axis wind turbine.

The third definition, PED virtual, operates within virtual boundaries, which means that the energy system can operate outside the geographical boundaries of the district and consequently utilize renewable energy sources or energy storage to greater extents. However, the part of the energy system that operates outside the borders of the district must be an asset of the district in order to be classified as a PED virtual.

In European cities, transportation is one of the greatest polluters. Today, the transport and mobility sector contributes to 27% of the emissions in Europe [26]. As emission reduction is one of the key objectives of PEDs, the use of electric vehicles (EVs) and other emission-free alternatives for transport and mobility in urban areas should be fostered. At the moment, it is expected that by the year 2030, EV usage will increase to 44 million cars globally [27]. Many cities are thus already including the electrification of mobility in their city plans [28]. Hence, all of the above-mentioned PED definitions should account for an increasing EV charging capacity and support for other emission-free transport and mobility.

4. Renewable Energy Market Circumstances in the European Union

Europe is divided into several countries and regions with different preconditions for PEDs. Renewable energy generation methods, such as solar, wind and hydro, are highly dependent on the geographical properties of the site. Even some energy storage methods, such as compressed air and pumped hydro storage, are only suitable for some geographical locations. The same applies to energy demand and electricity prices, although these are also highly dependent on the regulations and socioeconomic factors of the region.

This chapter addresses the techno-economic aspects of different renewable energy and energy storage technologies as well as their suitability for different geographical locations within the EU. The energy consumption and electricity prices for the different EU regions are also presented. Four EU countries with different climates and their corresponding capital cities are examined more thoroughly in this chapter in order to highlight regional differences in the renewable environment within the EU. The studied cities as well as their precipitation and temperatures are presented in Figure 5.

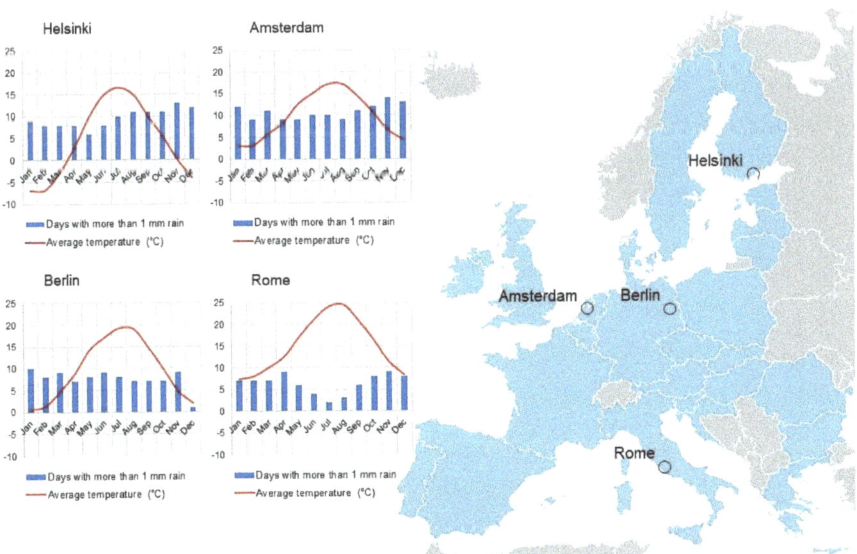

Figure 5. Monthly precipitation and temperatures for Helsinki, Amsterdam, Berlin and Rome.

4.1. *Renewable Energy Sources and Their Availability in the EU*

During the last 15 years, the amount of new investments in renewable energy has increased dramatically, especially for solar and wind energy technologies. The compound annual growth rate (CAGR) for the 2004–2018 period for solar and wind were 20% and 15%, respectively [29]. This investment boom in solar and wind energy has contributed to technological improvements and larger markets, and consequently cost reductions for technology [30]. The evolution of the renewable energy market has engendered better preconditions for PEDs and other energy transition projects. Table A1 in the Appendix A presents the average costs and capacity factors of different renewable energy technologies that can be utilized in PEDs.

Most renewable energy sources are somehow dependent on the geographical location of the site. Solar and wind energy are dependent on the climatic conditions of the site, while hydro power is dependent on the climatic as well as topographical and geological conditions of the site. Bioenergy, i.e., power generated by biomass, on the other hand, is not as dependent on the geographical conditions as the above-mentioned technologies.

The geographical differences in renewable energy environments are evident when the renewable energy mixes for different EU countries is studied. The renewable energy mixes of Finland, the Netherlands, Italy and Germany are presented in Table 2. It can be observed that there are some considerable differences in the renewable energy generation between these countries, which are partly caused by geographical factors. Figure 6 shows that the renewable energy share has increased significantly during the last ten years for all of the above-listed countries. The growth rate of the renewable energy share does, however, vary slightly between the countries.

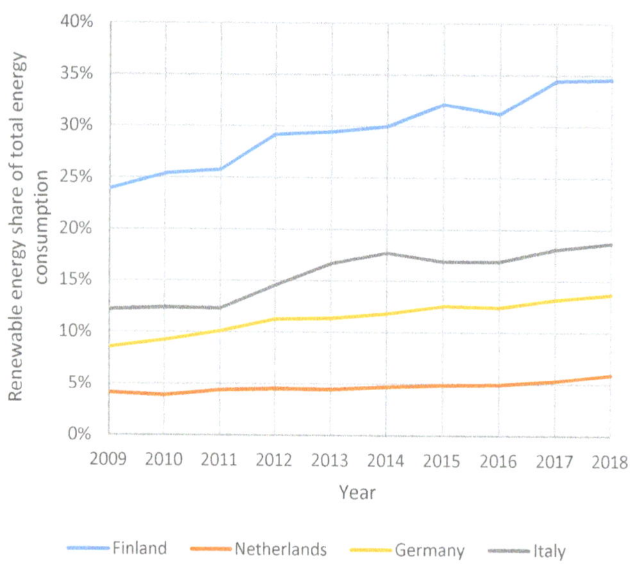

Figure 6. Renewable energy (heat and electricity) trends of Finland, the Netherlands, Germany and Italy [31].

Table 2. Gross renewable energy (heat and electricity) consumption mix of Finland, the Netherlands, Germany and Italy in 2018 [31].

Renewable Energy Source	Finland	Netherlands	Germany	Italy
Hydro	9.5%	0.1%	3.6%	14.3%
Wind	4.2%	19.6%	22.0%	5.2%
Solar PV	0.1%	6.9%	9.1%	6.7%
Solar thermal	0.0%	0.6%	1.8%	0.7%
Biofuels and renewable waste	81.6%	66.2%	60.2%	45.7%
Geothermal	0.0%	1.9%	0.7%	18.5%
Ambient heat (heat pumps)	4.7%	4.7%	2.7%	8.9%

4.1.1. Solar

The cost of solar electricity has decreased dramatically during the last decade. Particularly, the cost of solar photovoltaic (PV) installations has decreased significantly during the last few years. According to IRENA, the global weighted average total installed cost of utility-scale solar PV projects dropped from 4621 USD/kW in 2010 to 1210 USD/kW in 2018 [32]. The weighted average capacity factor for solar PV increased from 14% to 18% during the same time interval [32].

The installation cost of concentrating solar power (CSP) also fell from around 8829 USD/kW in 2010 to 5204 USD/kW in 2018, although the year-on-year variability has been relatively high due to the small scale of the market [32]. In 2010, the levelized cost of electricity (LCOE) for CSP was 0.19 USD/kWh, which is nearly twice as high as the LCOE for solar PV in the same year [32]. Hence, solar PV is still more profitable than CSP for electricity generating purposes. However, according to a cost reduction potential analysis by IRENA 2016, the gap between the LCOEs for CSP and solar PV will be reduced to 0.02 USD/kWh by 2025 [30].

Apart from electricity generated by PV and CSP, solar radiation can also be used for generating heat using so-called solar thermal collectors. These solar thermal collectors generate thermal energy in the form of hot water, which is usually used directly as domestic hot water [33]. The two most popular solar thermal collector technologies are flat plate collectors and evacuated tube collectors. Evacuated tube collectors cost 20% to 50% more than flat plate collectors, but they are more efficient and easier to apply in cold climates and regions without a substantial amount of sunlight. Due to the excellent performance of evacuated tube collectors in cold climates, it is also possible to use them as a heat source for building heating systems.

Despite the great potential of solar thermal collectors, the growth rate of solar heat in Europe has, according to Madsen and Hansen (2019), decreased during the last five years [34]. Only a few large projects have been executed in EU countries such as Poland and Denmark during the last few years [35].

The electrical power generated by solar PV and CSP as well as the heat produced by solar thermal collectors is dependent on solar irradiation [36,37]. Since the EU region covers over 30 degrees of latitude, there are considerable differences in solar radiation between some EU regions. Figure 7 presents the monthly global irradiation data for the four European example cities. The data were collected from the Photovoltaic Geographic Information System (PVGIS), which is a geographical information system developed by the Joint Research Centre of the European Commission [38,39].

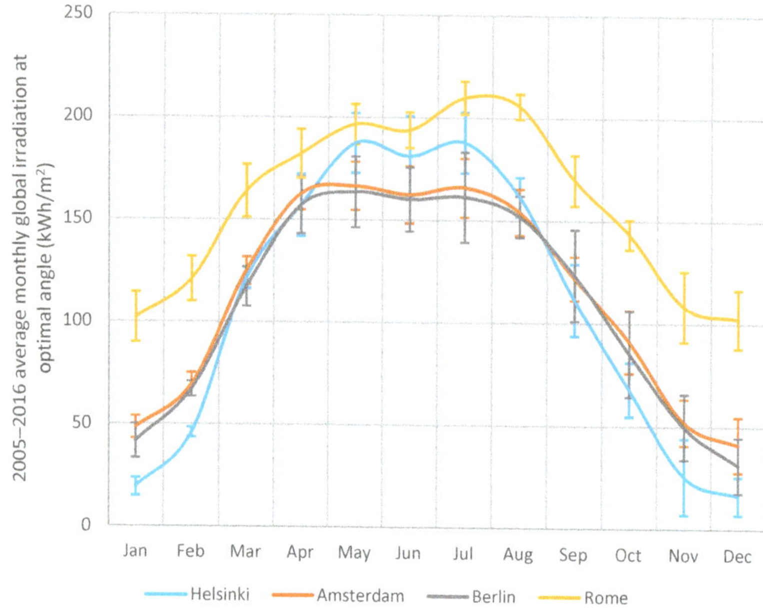

Figure 7. Monthly average solar irradiation of Helsinki, Amsterdam, Berlin and Rome [38,39].

Ambient temperature and wind speed are other geography-dependent factors that affect the photovoltaic capacity. The efficiency of photovoltaic cells is negatively affected by increasing cell temperature [40]. A high ambient temperature and still air thus have an unfavourable effect on PV power generation. According to Adeh et al. (2019), a 10 °C increase in ambient temperature reduces the efficiency by about 0.5 percentage points [40]. A wind speed reduction from 1.5 to 0.5 m/s also entails a 0.5 percentage point decrease in efficiency [40]. Since typical photovoltaic efficiency is below 20% [41,42], these seemingly small variations in efficiency actually have a significant impact on photovoltaic power generation. CSP, on the other hand, benefits from high temperatures and low wind speeds, and is thus suitable for the hottest regions within the EU [43].

4.1.2. Wind

Despite the huge increase in wind power investments, the decrease in wind power installation costs has not been as remarkable as for solar energy. For onshore wind power installations, the weighted average total installed cost dropped from 1915 to 1499 USD/kW in the period between 2010 and 2018. The weighted average total installed cost for offshore wind power installations has, however, only experienced a modest decrease during the last few years. In 2018, the weighted average total installed cost for offshore wind turbines was 4353 USD/kW [32].

Nevertheless, there have been significant technological improvements in wind turbine technology. These technology improvements, however, mainly benefit wind power installations in regions with low annual wind speeds [44]. In regions with high wind speeds, the benefits of the increased power generation are smaller than the cost difference between the old and new turbine technology. Hence, the benefits of the technological improvements of wind turbines are primarily observed for less windy sites. The improvements in wind power resulted in an increase in rotor diameter and turbine size between 2010 and 2017. In France, the rotor diameter of newly commissioned projects increased by 25% between 2010 and 2017 [32].

The wind power potential of a region is primarily determined by two physical qualities: wind speed and air density. These qualities are, however, dependent on the meteorological conditions, topography and surface roughness of the site [45]. Due to the sensitivity to different variables, it is challenging to accurately estimate the exploitable wind energy [45]. Estimating local extreme wind speeds and turbulence is also considered a challenge; these have an impact on the wind power potential.

When the wind speed and air density are known, the exploitable wind power can be calculated as follows:

$$P = \frac{1}{2}\rho A v^3. \quad (1)$$

where ρ is the air density, A is the cross-sectional area of the wind turbine and v is the wind speed [46]. Since the generated power is expressed as a cubic function of the wind speed, even a small variation in wind speed entails a significant difference in generated electrical power. The average wind speeds of the four European cities are presented in Table 3.

Table 3. Onshore and offshore wind speeds at 100 m height in Helsinki, Amsterdam, Berlin and Rome [47].

City	Average Wind Speed for Onshore Wind Turbines (m/s)	Average Wind Speed for Offshore Wind Turbines (m/s)
Helsinki	8.4	9.1
Amsterdam	8.5	9.4
Berlin	7.4	-
Rome	5.7	6.7

4.1.3. Hydro

According to the International Renewable Energy Agency, about half of the European hydropower potential is already utilized [48]. There are two types of hydro power—run-of-river (ROR) hydropower and reservoir hydropower—both of which are highly dependent on the geography of the region [49]. ROR hydropower uses the natural flow of rivers to generate electricity through a turbine and is thus dependent on seasonal variations in precipitation in the upstream catchment area [48,49]. Reservoir hydropower can store large volumes of water and is therefore less dependent on a continuous water flow. Reservoir hydropower is, however, highly dependent on the topography of the site [49]. The EU countries with the highest amounts of hydropower generation per capita are listed in Table 4.

Table 4. EU countries with the highest average annual hydropower generation [18].

Country	Hydropower Generation, 2009–2018 Yearly Average (kWh/Capita)
Sweden	6.52
Austria	4.44
Finland	2.57
Slovenia	2.13
Croatia	1.72
Latvia	1.55
Portugal	1.11
France	0.88
Romania	0.84
Slovakia	0.78

4.1.4. Biomass

Biomass is a controversial renewable energy source, which is sometimes not considered to be as "green" as the above-mentioned renewable energy technologies. Unlike other renewable energy technologies, bioenergy produces emissions that might be problematic in densely populated areas. Biomass is, however, by far, the most flexible of these renewable energy sources since it is possible to both transport and store biomass in large volumes. Biomass is therefore a great alternative to fossil fuels in energy system balancing applications [50].

Today, 75% of the energy generated by biomass is used for heating and cooling purposes, 13% for electricity generation and 12% for transport [50]. In 2017, biomass represented 60% of the total production of primary renewable energy [51].

Biomass is often used as a fuel in combined heat and power (CHP) plants. These plants often consist of a biomass-fired boiler where heat is transferred to a steam cycle [52]. The steam runs true turbines to generate electricity and through heat exchangers to acquire heat. Biomass contains significantly more moisture than fossil fuel. Hence, there is a significant amount of latent heat that can be recovered from the exhaust gases of a biomass-based CHP plant. This means that the exhaust gas heat recovery has a crucial impact on the overall efficiency of the plant. For wood-chip-fired CHP plants, the latent heat recovery is the highest when the exhaust gas temperature is 30–65 °C [53].

Biomass is competing with natural gas, which is considered as a relatively clean fossil fuel. The investment cost for biomass power plants is, in general, more expensive than that for natural gas power plants, although the fuel costs are on the same level. Unrefined biomass contains high levels of potassium and other alkalis, which cause deposit formation and corrosion in the boilers of combustion power plants [54,55]. This costly problem can be minimized by pretreating the biomass and/or by adding sulfur in the combustion process [55,56]. These countermeasures are also reflected in the cost of bioenergy technology. Another challenge with biomass is the limited availability of reliable, affordable and sustainable biomass [50].

Bioenergy generation is not as dependent on the geographical location of the generation site as the earlier-mentioned renewable energy sources, since biomass can be transported and utilized far away from the extraction site. Most of the bio-based energy generation within the EU is descended from forestry. Transporting wood-based biomass is expensive due to its low energy density compared to fossil fuels (the lower heating value (LHV) of wood pellets is 17.5 MJ/kg, which is less than half of the LHV of most fossil fuels) [50,57,58]. According to IEA-ETSAP and IRENA 2015, the cost of locally collected biomass ranges from USD 4 to 8 per GJ, while the cost of globally traded biomass ranges from USD 8 to 12 per GJ [52]. Biomass energy generation is therefore more lucrative in countries with domestic forest resources. The biggest bioenergy consumers per capita in Europe are the Nordic and Baltic countries as well as Austria [50]. These are all countries with great forest resources. Other EU countries with a forest land cover over 40% are Bulgaria, Croatia, Slovenia and Slovakia.

4.1.5. Geothermal

Geothermal power plants use thermal energy generated and stored in the Earth to generate power through steam turbines [59]. These plants usually require deep wells to reach sufficient temperatures (often higher than 180 °C). A large share of the installation costs are comprised of costs related to the construction of the wells. The weighted average total installation cost of geothermal power plants is 3976 USD/kW, which is quite high compared to other renewable energy technologies [32]. Although the installation costs of geothermal power plants are usually high, the operating costs are relatively low and predictable [32,59].

In theory, geothermal energy can be accessed everywhere as long as there are not any restrictions on the deepness of boreholes. In reality, it is not economically and technically possible to drill deep enough to harness geothermal energy. The vast majority of geothermal energy systems that are suitable for power generation are located in areas with volcanic activity, usually close to tectonic plate boundaries [59]. In Europe, these areas are mainly found in southern Italy and Iceland.

4.2. Energy Storage Systems and Their Application in the EU

Energy storage can provide sufficient flexibility to the energy system by evening out peaks in energy generation and demand. Hence, the importance of energy storage methods is growing as the amount of intermittent renewable energy is increasing.

The potential of some energy storage methods is highly dependent on the geographical characteristics of the site. Especially, large-scale energy storage methods such as pumped hydro and compressed air energy storage (CAES) are only suitable for some geographical regions, partly because of their low energy density and partly because of their dependency on suitable geographical features, such as geology, topography and precipitation. Even small-scale energy storage methods, such as different kinds of batteries, are somewhat dependent on the environment.

4.2.1. Pumped Hydro

Pumped hydro storage (PHS) is the most widely deployed technology for large-scale energy storage worldwide [60]. According to the International Hydropower Association, PHS accounts for 94% of the global installed energy storage capacity [61]. PHS systems store energy by pumping water from a reservoir to another reservoir at a higher altitude. The potential power of the water in the higher water storage reservoir is then extracted by releasing the water through a turbine to the lower reservoir [60].

The main advantages of PHS systems are their low installation cost per storage capacity and long life spans. The PHS potential and investment cost are heavily dependent on the geographical properties of the site. The performance of a PHS system depends on the water volume involved as well as the height difference between the higher and lower water storage reservoirs [62]. These characteristics are defined by the availability of water as well as the topography and geology of the site [62]. Due to the very site-specific nature of PHS systems, the installation cost varies between 5 and 100 USD/kWh [60].

The geographical limitations of PHS systems can definitely be considered as a major drawback for the storage method [61]. Another downside with PHS technology is its relatively slow reaction time compared to batteries [61]. This prevents PHS systems from being used as short-erm storage for balancing frequent variations in energy demand and supply.

A study by Gimeno-Gutiérrez and Lacal-Arántegui (2014) presents estimates of the pumped hydro storage potential in different European countries, based on a geographical information system (GIS) developed by a team from the Joint Research Centre and University College Cork [63]. The study used the GIS to identify potential pumped hydro storage sites where two existing reservoirs at different altitudes can be connected. The report presents both an unconstrained theoretical potential and a constrained realizable potential. Table 5 presents a part of the results that were obtained from the study.

Table 5. Pumped hydro storage potential in Finland, the Netherlands, Germany and Italy [63].

Country	Pumped Hydro Storage Potential									
	Theoretical Potential [1]					Realizable Potential				
	20 km	10 km	5 km	2 km	1 km	20 km	10 km	5 km	2 km	1 km
Finland	12	0	0	0	0	12	0	0	0	0
Netherlands	0	0	0	0	0	0	0	0	0	0
Germany	168	28	0	0	0	0	0	0	0	0
Italy	1867	661	218	85	11	3	670	99	5.5	4.6

[1] The theoretical potential includes the following constraints: a minimum head of 150 m, a minimum reservoir capacity of 100,000 m^3, a minimum distance of 500 m to inhabited sites, a minimum distance of 200 m to existing transportation infrastructure, a maximum distance of 20 km to electricity transmission, and location outside Natura 2000 conservation areas and UNESCO sites.

According to the results of the study by Gimeno-Gutiérrez and Lacal-Arántegui (2014), countries with mountain ranges, such as Austria, Switzerland, France, Italy, Spain, the UK and Norway, have the highest potential for pumped hydro storage [63]. Naturally flowing water and evaporation are, however, not taken into account in the report by the JRC, which will affect the PHS potential.

4.2.2. Compressed Air Storage

The operating principle of compressed air is to use electrical power to compress air and thereby store potential energy in the form of compressed air in an airtight storage cavern or vessel. The potential energy is released by letting the compressed air run through a turbine that generates electricity. Compressed air storage is an energy storage method that is particularly suitable in areas with already-available storage caverns, such as empty salt caverns and depleted oil and gas fields. Storing the compressed air in storage vessels is also possible, although it is significantly more expensive and not competitive with other energy storage methods [60].

This method does have a few major drawbacks. One problem is the low energy density of the storage method, and another problem is the low roundtrip efficiency. The low roundtrip efficiency is mainly caused by the compression process, which generates a significant amount of waste heat, and the air expansion process, which needs external heating to improve the power quality of the turbine [60].

Salt caverns and depleted oil and natural gas fields are by far the cheapest storage space for compressed air storage, but suitable storage caverns are unevenly distributed around Europe [60,64]. In Europe, eligible onshore salt caverns are located in northern and central Germany, Poland, parts of the UK, Denmark, eastern and northern parts of the Netherlands, northeast Spain, eastern France, western Portugal, central Romania, northeast Ukraine, eastern Bosnia and Herzegovina, western Greece and central Albania [64].

4.2.3. Batteries

Batteries are excellent for short-term residential energy storage, due to their short reaction time and high discharge rate. The residential energy storage systems market has also surged as a result of enhancements in residential solar PV regulations, subsidies and tax incentives as well as rapid price reductions for battery technologies [65]. Residential energy storage system shipments are, hence, expected to grow with a CAGR of 30% during the coming years [65].

Li-ion batteries are expected to lead the residential energy storage market in the future, due to their high efficiency and the declining cost of lithium [65,66]. In 2018, Li-ion batteries accounted for 49.3% of the residential energy storage market, followed by lead-acid batteries, with 40.7%.

The Li-ion battery market has grown rapidly during recent years as a consequence of the increase in electrical vehicle manufacturing [67]. There has also been a significant increase in the market for Li-ion batteries in residential energy storage applications. The market of residential Li-ion batteries is expected to continue to grow at a CAGR of 33.2% [65]. Large-scale production facilities and significant investments in R&D also drove the Li-ion battery price down from 1000 USD/kWh in 2010 to 254 USD/kWh in 2017. The price is expected to reach 100 USD/kWh by 2025, if not sooner [65].

Lead-acid batteries are also popular on the residential energy storage market. These batteries are cheaper than Li-ion batteries, but their performance is not as good [65]. The market for lead-acid batteries in residential energy storage applications is also expected to grow significantly in the near future [65].

Both Li-ion and lead-acid batteries degrade over time. The degradation is additionally accelerated by chemical side reactions caused by the environmental and functional conditions [68]. The more the battery is charged and discharged, the shorter the lifespan is. Today, the average lifespan of Li-ion and lead-acid batteries is 5–20 years [60].

As mentioned earlier, batteries are electrochemical devices that degrade over time, partly because of the functional conditions. High and low temperatures are driving factors that accelerate the degradation of Li-ion batteries [68,69]. Li-ion battery operation is also negatively affected by low temperatures. Battery capacity is significantly lower for sub-zero operating temperatures [69]. In cold conditions, it is possible to partly heat the batteries with internally generated heat, but at temperatures as cold as −20 °C, internally generated heat is not enough to maintain the operating temperature at a sufficient level [70]. Temperature-related problems can, however, be overcome by installing auxiliary cooling and heating units.

4.2.4. Thermal Energy Storage

Thermal energy can be stored in three forms: latent energy, chemical energy and sensible energy [71,72]. Latent thermal energy storage (TES) uses phase change materials to store thermal energy, while chemical TES stores thermal energy through chemical reactions. Sensible TES stores thermal energy by varying the temperature of the storage medium. Of these three TES methods, sensible TES is the most developed and cost-effective method.

Sensible TES can be used in both short-term and long-term storage applications. Long-term or seasonal sensible TES systems are typically placed underground and use liquids, usually water, and solids, e.g., soil or rock, as storage media. The most common seasonal sensible TES systems are hot water tank storage, water–gravel pit storage and borehole TES.

Seasonal sensible TES combined with solar heat collectors has become a popular method for storing "green" heat. This technology is widely used in Germany, especially on a district level, to provide heat for space heating and domestic hot water preparation [73]. In Germany, central solar heating plants with integrated seasonal sensible TES can reduce the district heating fossil fuel demand by more than 50%.

The main concern with seasonal sensible TES today is heat losses [71]. The heat losses are determined by the temperature gradient, storage volume and storage medium. Low heat losses are, hence, accomplished by minimizing the temperature difference between the storage medium and the surroundings, maximizing the storage volume and using a storage medium with a high specific heat capacity. Short-term TES systems are usually not affected by heat losses and can therefore operate at high temperatures and with relatively low storage volumes. Seasonal TES systems, on the other hand, are more exposed to heat losses and therefore usually operate at low temperatures. These types of TES systems need auxiliary heating systems such as preheaters or heat pumps to increase the temperature of the stored thermal energy before so that it can be used for space heating and domestic hot water [71].

According to cost data for sensible TES systems in Germany, there is a strong decrease in investment costs per storage capacity with increasing storage volume [73]. Despite the strong volume dependency, sensible TES installation costs are relatively low compared to electricity storage. The cost range for sensible TES storage is 0.1–10 USD/kWh [72].

Thermal Energy Storage Potential in Different EU Regions

Sensible TES systems, especially for seasonal storage applications, are negatively affected by low ambient temperatures. Low ambient temperatures entail greater heat losses from the TES to the surroundings. Particularly, regions with sub-zero-degree temperatures might cause problems for sensible TES systems that use water as the storage medium [71].

Borehole TES systems are extremely sensitive to the underground conditions of the site. The thermal conductivity and the heat capacity of the soil as well as the ground water level and flow affect the performance of borehole TES systems, and the stress distribution within a geologic medium might also affect the drilling [71,74,75]. These soil and ground water properties are highly dependent on the geographical location of the storage site.

4.3. Energy Demand in Different EU Regions

The energy demand of household consumers in Europe is highly dependent on climatic conditions [76]. An analysis by Tzeiranaki et al. (2019) shows a strong positive correlation between heating degree days and household energy consumption [76]. It is, however, hard to show the direct impact of climatic conditions on total household energy consumption, as the energy consumption is also affected by other regional factors, e.g., building regulations, consumer behaviour, economic conditions and population density.

The U-value requirement, i.e., the highest allowable coefficient of heat transfer to the surroundings, is an essential part of the building regulations that affect the heating demand of buildings in a particular region. The U-value requirements define the maximum allowed heat loss rate of a building. Table 6 shows the U-value regulations of Helsinki, Amsterdam, Rome and Berlin.

Table 6. U-value requirements for new buildings in Helsinki, Amsterdam, Berlin and Rome [77].

City	U-Value Requirements (W/m^2 K)		
	Walls	Roof	Floor
Helsinki	0.25	0.16	0.25
Amsterdam	0.37	0.37	0.37
Berlin	0.30–0.38	0.24–0.30	0.30–0.45
Rome	0.50	0.46	0.46

Energy consumption behaviour is also an essential factor that affects the energy demand of a region. Consumption behaviour can be divided into several categories or sectors, such as household, transport, industry and commercial and public services. The total household energy consumption of Finland, the Netherlands, Germany and Italy is depicted in Figure 8. It can be observed that Finland, which is the northernmost of the selected countries, has the highest annual energy consumption. The latitude of the region is, however, not the only factor that affects energy consumption. The fact that energy consumption in the Netherlands is lower than in Italy supports this statement.

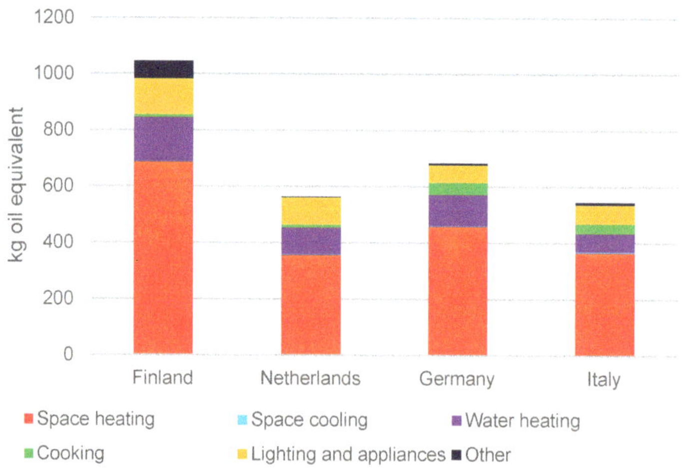

Figure 8. Total household energy consumption per capita, 2017 [78,79].

Countries with high GDPs tend to consume more energy than countries with low GDPs. According to Tsemekidi-Tzeiranaki et al. (2018), the countries with the highest energy consumption per capita in the residential sector also have GDPs that are above the European average [76]. The same trend can also be observed by studying the European countries with the lowest energy consumption per capita; these countries have GDPs that are lower than average.

The transport sector is currently in a transition phase, since petrol and diesel cars are slowly being replaced by plug-in hybrid and battery electricity cars. This transition is important to take into account in the PED planning phase, so that the electricity supply and infrastructure are sufficient to satisfy the future power demand of EVs. Figures 9 and 10 show the EV situation in Finland, the Netherlands, Germany and Italy in 2018. The Netherlands is clearly a few steps ahead in the transition process since its share of EVs is significantly higher than in other EU countries.

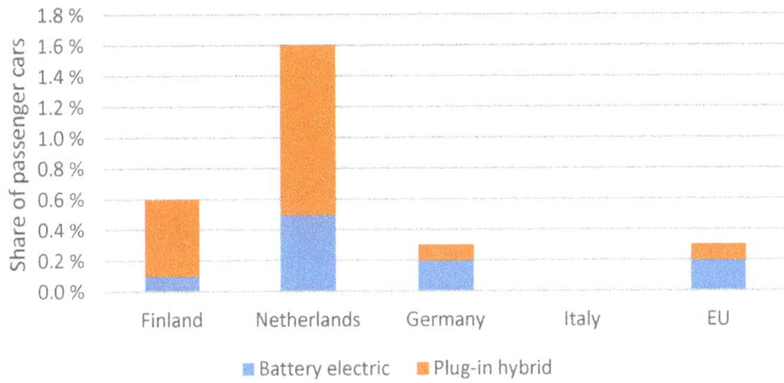

Figure 9. Battery electric and plug-in hybrid personal cars in use, 2018 [80].

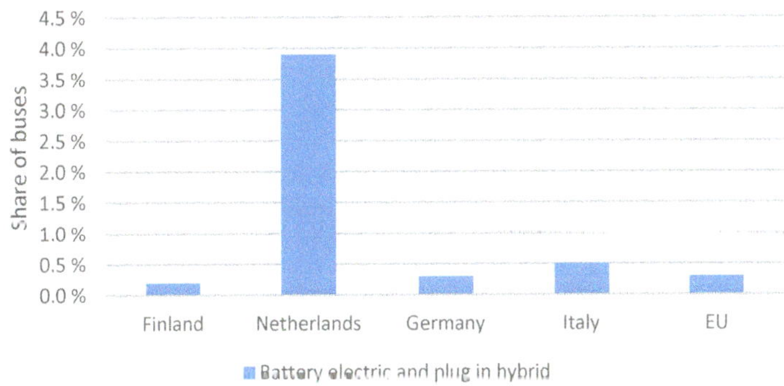

Figure 10. Battery electric and plug-in hybrid buses in use, 2018 [80].

The population density affects the area-specific energy demand, which is a highly relevant parameter in the renewable integration planning of the PED. Renewable energy integration and achieving a positive annual energy balance tend to become significantly more complicated in districts with a high area-specific energy demand.

4.4. Electricity Prices in Different EU Regions

The profitability of PEDs is highly dependent on the electricity prices of the region. The electricity spot prices are mainly influenced by two types of factors: demand side factors and supply side factors. Examples of demand side factors are industrial activity and cooling and heating demand peaks [81]. Industrial activity is highly dependent on the industry sector as well as global markets. Changes in the global market can have an impact on the electricity prices in the whole EU, but they can also only affect certain regions. Heating and cooling demand peaks, on the other hand, usually only affect the electricity prices on a regional level.

Typical electricity price supply side factors are the prices of fuels used in power plants and carbon dioxide prices. These factors are sensitive to changes in the political landscape and the economic situation in the world [81]. The impact of different supply side factors varies within the EU since different regions have different energy generation mixes. Germany is, for instance, more sensitive to variations in natural gas prices than Finland, as natural gas represents 14% of the German energy mix, while only 6% of the Finnish energy generation mix is covered by natural gas [31]. As the share of renewable energy grows throughout the EU, the renewable environment becomes an increasingly important electricity price supply side factor. Parts of the European energy market have always been highly dependent on annual precipitation, which affects the supply of hydropower [82]. Now, variations in solar radiation and wind speed are also starting to show in the electricity spot prices. In the third quarter of 2019, renewable energy reached 33%, which is the highest for a third quarter to date [83].

According to Helistö et al. (2017), an increased share of intermittent energy, such as wind power and solar PV, could lead to longer periods of low electricity prices, which would entail a lower average electricity cost [83]. The electricity price range would, however, not change since periods of low intermittent energy generation would still be covered by fuel-powered power plants and hydropower.

The electricity spot price is not the only electricity cost that the final consumers stand for. The grid tax and other levies represent a substantial part of the net electricity cost. These taxes and levies vary from country to country, and they are also different for household and non-household consumers. In the EU, Germany has the highest total cost of electricity (including the spot price, taxes and levies) for household consumers. The average total cost of electricity for household consumers in Germany was 0.3088 EUR/kWh in 2019. The corresponding cost for Latvia, the EU country with the lowest household electricity costs, was 0.1629 EUR/kWh. The average electricity prices, with and without taxes and levies, of Finland, the Netherlands, Germany and Italy are presented in Table 7.

Table 7. Energy prices for household and non-household consumers, 2019 [84,85].

Country	Electricity Price 2019 (EUR/kWh)			
	Household Consumers		Non-Household Consumers	
	Excluding Taxes and Levies	Including Taxes and Levies	Excluding Taxes and Levies	Including Taxes and Levies
Finland	0.1173	0.1734	0.0639	0.0880
Netherlands	0.1357	0.2052	0.0679	0.1138
Germany	0.1473	0.3088	0.0855	0.1958
Italy	0.1432	0.2301	0.0952	0.1913

5. Results and Discussion

5.1. Renewable Energy Generation Methods for PEDs

As noted in Section 4, the energy generation potential of renewable energy technologies varies between different regions within the EU. A renewable energy technology that excels in one region might be impossible to implement in another region. The geographical location and its properties must therefore be taken into account when planning a PED. Solar PV is a good example of an energy technology that is highly dependent on the geographical location. In northern Europe, where there are only a few hours of daylight in the winter season, solar PV generation is significantly lower than in southern Europe. Hence, the capital costs per kWh of generated solar power are significantly higher in the Nordic countries compared to the Mediterranean region. The situation is similar for wind power, which is naturally more remunerative in windy areas, such as the regions close to the northern Atlantic Ocean, the Baltic Sea and parts of the Mediterranean Sea.

Different renewable energy technologies also have different properties when it comes to flexibility, cost and service life [32]. Intermittent renewable energy generation technologies, such as solar and wind energy, are considered non-flexible energy sources, as they can only generate energy when the wind speed and solar radiation are sufficient. Run-of-river hydropower is more flexible than solar and wind energy, but not as flexible as reservoir hydropower and bioenergy.

The installation costs, costs of electricity and service lives of different renewable energy technologies are presented in Table A1 in the Appendix A. However, the installation cost is dependent on the size and location of the installation, and the cost of electricity is highly dependent on the geographic location [32]. Hence, the costs in the table are given as global weighted averages.

The diversification of intermittent renewable energy technologies is a great way to increase the demand coverage and reduce life cycle costs [86,87]. Intermittent renewable energy technologies, such as wind and solar energy, are often able to compensate each other, as windy and sunny periods are not synchronized. As the energy export price is often lower than the energy import price for small-scale energy producers [88], it might be beneficial for a PED to minimize the external grid interaction. By diversifying the intermittent renewable energy generation, it would be possible to achieve a positive annual energy balance with a lower export rate [89].

According to a study by Heide et al. (2010), wind power is, in general, more beneficial in Europe from a load-matching perspective since both the wind power generation and the energy demand are higher during the winter than during the summer [86]. Solar energy generation, on the other hand, is the highest during the summer months. Thus, from a load-matching point of view, a larger share of the PED energy generation mix should be covered by wind energy in most of Europe. This is, however, not that simple, as installing wind turbines in populated areas is complicated and solar energy is, on a global level, a more cost-effective energy generation method [32].

In most districts, especially in densely populated areas, space is also an issue. Renewable energy systems must thus be integrated in a smart way, so that energy generation does not conflict with other functions that are essential for the district. Solar power integration in urban districts is convenient since solar PVs, CSPs and solar heat collectors can be installed on rooftops and various available surfaces within districts. Solar PV panels can also be integrated into building façades. So-called building-integrated photovoltaics (BIPV) can be integrated in stable and heavy structural elements as well as in lightweight and transparent structural elements [90]. According to a study by Fath et al. (2015), building façades provide almost three times the area of roofs in a 2 km^2 urban area in Karlsruhe, Germany [91]. However, due to their angles and positions, they receive only 41% of the total solar irradiation. Hence, solar PV panels on roofs should be prioritized in PEDs, while façade solar PV panels can be considered if the solar radiation on a particular façade is sufficient. Overall, city-integrated solar PVs have a great potential and can satisfy over 60% of the electricity demand in some smaller cities in Europe [92,93].

Wind power integration in urban areas, on the other hand, does have many practicality issues and is thus less suitable for on-site energy generation in PEDs. It would be complicated to install large-scale wind turbines due to their size, aesthetics and noise as well as low and turbulent urban wind-speed and safety issues [94,95]. Small-scale wind turbines could be an option, but their cost per installed kWh is about twice as high as large-scale turbines [32,96]. Vertical axis wind turbines (VAWTs) are a popular alternative among small-scale wind turbines. These wind turbines are able to handle the higher turbulence and varied wind speeds associated with urban environments [94]. Another benefit with VAWTs is that the generator can be installed at a lower part of the so-called tower, allowing building-mounted turbines to be more easily serviced [94]. The hub height of small-scale urban wind turbines is, however, not high enough to access the same wind speeds as large-scale wind turbines [97].

Due to the many shortcomings of wind turbine installations in urban areas, wind power is best suited for virtual power plants. The distance between the district and the virtual wind power farm could, however, be relatively short and thereby ease the power transmission to the district. Wind farms could, for instance, be installed in nearby rural areas or even offshore if the district is in a coastal area.

Bioenergy and hydropower can be used to provide PEDs with flexible power when the intermittent energy generation is lower than the electricity demand [98,99]. These flexible power generation methods make the district less dependent on electricity supplied by the external grid and thereby foster a positive annual energy balance.

Bioenergy plants can be built almost anywhere in Europe, as biomass is relatively cheap to transport from biomass-producing regions, such as the Nordic countries, the Baltic countries and Austria [50], to other parts of Europe. Bioenergy generation does, however, produce emissions, which contradicts the PED's aim to provide a carbon-free energy environment and better life quality in residential areas. Even though bioenergy is carbon neutral from a life-cycle perspective (as the carbon dioxide emissions originate from carbon dioxide captured from the atmosphere by biomass), this does not change the fact that bioenergy plants pollute the air in the district where they operate.

Hydropower, on the other hand, is extremely dependent on the location of the district since hydropower can only be generated in regions that satisfy the requirements described in Section 4.1.3. Most of the potential hydropower sites in Europe are already in use or unattainable due to regulations and environmental protection [100]. Hence, hydropower is best suited for a virtual power plant for virtual PEDs, where the district boundaries are virtual instead of geographic. According to Graabak et al. (2019), a 2050 Central-West European grid with large shares of intermittent renewable energy could benefit from using Norwegian hydropower as flexible energy for grid balancing [98].

Heat pumps are expected to provide a significant share of future heating [101]. Due to the flexibility and high coefficient of performance (COP) of modern heat pumps [101], they could be a highly valued source for heating in future PEDs. Due to the relatively large operating temperature interval, heat pumps can be used to recover low temperature heat from the ground and the ambient air as well as low temperature waste heat from sewage systems, ventilation air and other waste heat flows. Heat pumps are thus able to increase the total energy efficiency of PEDs and minimize the import of externally generated thermal energy. Moreover, heat pumps provide additional flexibility to PEDs, as they can be used to transform electrical energy into heat that can be stored in TESs [88]. It is thereby possible to reach a higher utilization rate for electricity generated by on-site intermittent renewable energy technologies.

5.2. Energy Storage Methods for PEDs

Energy storage enables PEDs to store excess energy instead of exporting it. Hence, energy storage can be used to increase the on-site utilization of intermittent energy sources, such as solar and wind. This is particularly important for self-sufficient PEDs, so-called autonomous PEDs, as they are not allowed to import energy from the external grid. For dynamic PEDs, energy storage is not as crucial since they allow bidirectional interaction between the district and its surroundings, and can thereby use the external grid to balance the energy demand during periods of low on-site energy generation.

Table A2 in the Appendix A presents the installation costs, energy densities, lifetimes and round-trip efficiencies of different energy storage technologies that can be utilized in PEDs. Based on this table, the most cost-effective energy storage methods are pumped hydro and compressed air energy storage. As explained earlier in the paper, these energy storage methods are extremely dependent on the geographical characteristics of the site, and hence, they are not possible to implement anywhere [61]. Another issue with these storage methods is their low energy density, which makes it difficult to install them in densely populated districts [61].

Pumped hydro and compressed air energy storage do, however, have great potential as virtual energy storage. A virtual PED with a periodical intermittent energy surplus and shortage could, for example, interact with virtual storage located far from the geographical location of the district itself. Similar energy management strategies have, for instance, been implemented between Denmark and Norway, where excess Danish wind power is stored in pumped hydro storage in Norway [102]. This collaboration between nations is possible due to the high level of wind power generation in Denmark (>20% of the annual electricity generation) and the enormous pumped hydro storage potential in the mountains of Norway.

Batteries, on the other hand, are not so reliant on the geographical site of the PED, but they are considerably more expensive than pumped hydro storage and compressed air storage [60]. It is therefore often more cost-effective for dynamic PEDs to interact with the electricity grid than to use batteries [26]. The combination of decreasing battery prices and an increasing share of intermittent energy in the electricity grid could, however, open up more opportunities for batteries in the future.

Even if pumped hydro and compressed air energy storage would be an available option for autonomous PEDs, it could be beneficial to also install a battery for short-term energy storage. Batteries have a significantly shorter reaction time and can thereby add more flexibility to the energy system of the district and increase the utilization of on-site intermittent renewable energy [103].

Compared to electricity storage systems, TES systems are relatively cheap to install [72]. Sensible heat storage in the form of hot/warm water tanks is, by far, the most common TES method for heating and domestic hot water applications [104]. Short-term energy storage can be implemented at the building level without causing significant heat losses. The storage temperatures of these forms of storage are usually kept at 55–60 °C in order to avoid bacterial growth [104].

When heat is stored for longer periods, heat losses become an issue. As heat losses can be minimized by increasing the water volume and lowering the storage temperature, it might be beneficial to implement centralized low-temperature systems for long-term or seasonal TES [71]. The temperature of these TESs can be increased by utilizing heat pumps.

5.3. Possibility of Implementing Virtual Power Plants in PEDs

Virtual PEDs allow renewable energy systems to be installed outside the geographical boundaries of the district. Renewable energy generation systems that cannot be installed within the geographical boundaries of a PED can be implemented as so-called virtual power plants (VPPs). According to Next Kraftewerke, the operator of one of Europe's largest VPPs [105], a VPP is "a network of decentralized, medium-scale power-generating units such as wind farms, solar parks, and Combined Heat and Power (CHP) units, as well as flexible power consumers and storage systems" [106]. The power generated by the interconnected units is distributed by a centralized control system to the energy consumers. Nevertheless, the power-generating units remain independent in their operation and ownership [106].

The type and size of on-site renewable energy systems and energy storage systems are often restricted by regulations as well as limited and unsuitable conditions. By utilizing the VPP concept, a PED could own and operate renewable energy systems and energy storage outside its geographical boundaries, which would enable the PED to access a greater geographical area and more suitable conditions for renewable energy generation and energy storage. The utilization of VPPs could also be implemented through agreements with other energy market actors instead of the ownership of the renewable energy systems and energy storage [107].

VPPs could benefit PEDs in several ways. They can enable the PED to utilize a larger variety of renewable energy systems as well as long-term low-cost energy storage with low energy densities, and thereby increase the flexibility of the PED. According to a study by Vasirani et al. (2013), a combination of wind and electric vehicle energy storage in a VPP could also have a synergetic impact from an economic point of view [108].

5.4. District Heating/Cooling and Electricity Networks

Due to the surge in heat pump installations during the last decade, electricity grids and district heating and cooling networks are becoming more and more interconnected [101,109]. Thanks to heat pumps, energy systems can reach a higher degree of flexibility, as energy can be converted from electricity to heat with high COPs.

The reduction of fossil fuel CHP plants in the energy generation mix would require a more sophisticated district heating network that is better suited for decentralized heating. This field has recently received increased attention from researchers, and hence, the properties of the next generation, i.e., the fourth generation, of district heating and cooling networks have been investigated and discussed in several research papers [110–113].

The fourth generation district heating (4GDH) network will be an integrated part of smart energy systems and thus able to interact with other components, such as heat pumps, solar heat collectors and TESs [110]. Hence, the 4GDH networks rely on the optimized distribution, consumption and interaction between renewable energy sources [112]. Another key objective of the 4GDH network is to enable heat recovery from low-temperature sources and to decrease the temperature of both the supply and return district heating water [110]. The low temperature of the district heating network district also benefits heat pumps, as their efficiency is higher for lower output temperatures [101].

District cooling solutions are also a relatively new technology, and they are not as widely used as traditional district heating [101], but they can be implemented with the same operating principles as the 4GDH networks [110]. District cooling is usually supplied by natural cold resources, absorption chillers, mechanical chillers and cold storage [114]. During periods when heating and cooling demands are occurring simultaneously, synergies between the district cooling and heating networks can be utilized by using heat pumps to produce cold and warm water at the same time [114].

Both 4GDH and district cooling can be implemented as local networks (to which all energy consumers and producers are connected) in the PED with connections to the external district heating and cooling networks. This way, PEDs can balance their internal heating and cooling demands before exporting or importing energy from the external network. The same principles can also be applied to the electricity grid in the district. In order to streamline the utilization of such local energy networks, centralized control systems can be implemented. A centralized control system can optimize the energy flows between energy consumers, producers and storage in the PED so that the economic benefit of the PED is maximized.

Connections to the district heating/cooling network and electricity grid are an essential part of the PED concept, as one of the main targets of a PED is to interact with other PEDs and provide renewable energy to other parts of the metropolitan area. Hence, the energy transfer connections in and out of the district must be carefully planned and designed based on the purpose and capacity of the PED energy system.

5.5. Construction of PED Networks

Cities can be very different when it comes to size, population, population density, economic situation, public transportation, etc., and consequently, there are also significant differences in energy consumption. Cities in cold and hot climates consume a large amount of energy for heating and cooling, respectively [94]. Industrial cities also also consume more energy; however, they usually have a greater potential for district heating [94]. Even within the same city, there can be considerable variations in energy consumption between different districts [94]. According to a study by Jones and Kammen (2011), there is a clear correlation between income and household energy consumption [115]. Additionally, the energy consumption per household of big American metropolitan areas is usually higher in the suburbs than in the urban cores, due to longer driving distances and bigger homes [116]. All in all, there are numerous factors that affect the energy usage of cities and districts within cities, and therefore, it is impossible to develop specific PED construction guidelines that can be applied to every district in every city.

The high population density of urban cores complicates the installation of renewable energy systems. The population density does, however, usually decrease as the distance to the city centre grows, and therefore, it is easier to install renewable energy systems in the suburbs, where there is more space in relation to the number of residents. Hence, we propose an onion model for PED networks, where most of the PEDs are constructed in the outer-most layers, i.e., the districts furthest away from the city centre. These outer-layer PEDs produce more renewable energy than they consume and can thereby export excess renewable energy to the inner layers of the city. This way, networks of PEDs can increase the renewable energy share of the city centre and the self-sufficiency of the whole metropolitan area. A visual explanation of the onion model is depicted in Figure 11.

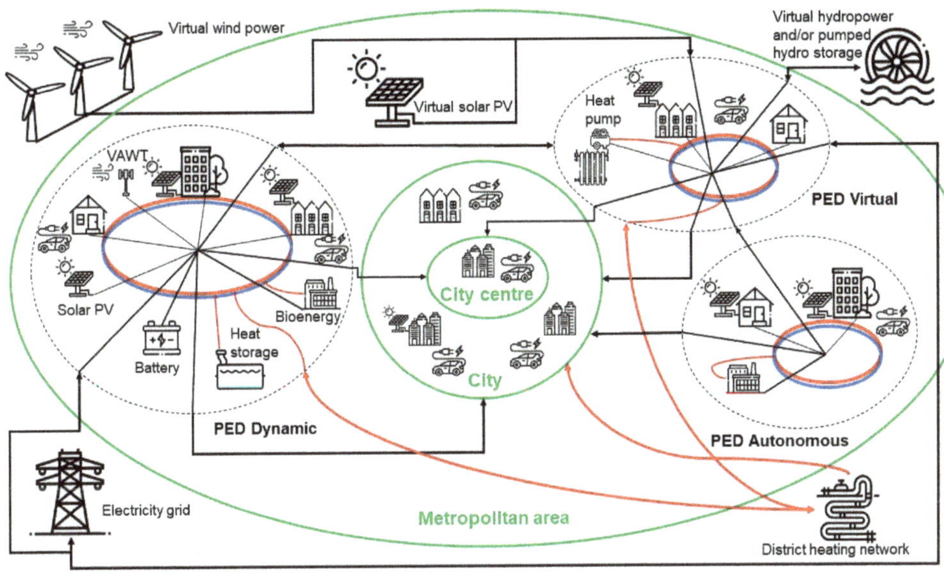

Figure 11. Different PEDs implemented in the onion model.

There is a strong correlation between the share of a country's population that lives in urban areas and CO_2 emissions [117]. Air quality might thus become an increasing problem as global urbanization continues and metropolitan areas around the world grow [117]. By ensuring that the cities are surrounded by PEDs, the amount of polluting fossil fuel power plants can be reduced in the region. This way, PEDs can improve the air quality of densely populated areas and contribute to decelerating climate change.

5.6. Regulative Aspects

The EU has, in several ways, highlighted the importance of preventing climate change and global warming. This is also noticeable from a legislative point of view. The European Green Deal, initiated by the European Commission in December 2019, aims to tackle climate- and environment-related challenges [118]. One of the main goals of this deal is for the EU to become climate neutral (no net greenhouse gas emissions) by 2050 [118]. The President of the European Commission Ursula von der Leyen has stated the importance of this deal, by calling it the EU's "new growth strategy" [118].

Since the goal of the PED concept is in line with the aim of the Green Deal, the enormous focus on the deal might benefit the development and construction of PEDs in the future. Some of the EU's Green Deal key actions, such as the "'Renovation wave' initiative for the building sector", the "Assessment of the final National Energy and Climate Plans" and the "Zero pollution action plan for water, air and soil", are directly enhancing the preconditions for the application of PEDs [119].

The Clean Energy Package proposed by the European Commission in 2016 is also a ground-breaking act for PEDs and other small-scale energy producers since it recognizes, for the first time under EU law, the rights of communities and citizens to engage directly in the energy sector [120]. As a result of this, renewable energy and energy storage could be shared within communities, using internal electricity grids [120,121]. The energy community and its shareholders cannot, however, be engaged in large-scale commercial activity in the energy sector.

The legislative features of energy communities might benefit the PEDs since they reduce the economic friction between renewable energy producers and consumers within the community. Regulations might, however, prohibit PEDs defined as energy communities from exporting energy to the external electricity grid and district heating network, as energy communities are not allowed to engage in commercial energy trading.

6. Conclusions

A general survey of the renewable energy market circumstances in different parts of Europe is provided to form a conception of the potential for implementing PEDs in the EU. The capitals of four different EU countries, representing four EU regions, were examined with extra care to highlight the variation in the renewable energy environment within the EU. Based on this survey, it can be concluded that the techno-economic potential of different renewable energy and energy storage technologies varies between different EU countries and cities. The economic viability of wind power is, for instance, greater in regions close to the Northern Atlantic than in the heart of Central Europe. Other factors that affect the renewable energy market circumstances of a region are the energy consumption behaviour and the electricity prices. High energy prices and suitable energy demand profiles might enhance the implementation of renewable energy systems and PEDs.

Three different PED definitions are presented in the paper: autonomous PED, dynamic PED and virtual PED. The difference between the definitions is their ability to interact with energy networks, consumers and producers outside the geographical boundaries of the PED. These PED definitions serve as the foundation of the PED concept in this paper.

An analysis was conducted to further investigate the available technologies and concepts that can be used for PEDs and networks of PEDs. Here, it was found that not all available renewable energy and energy storage technologies are suitable for all types of PEDs. Due to the high population density of modern European cities, some technologies are only possible to implement as VPPs for virtual PEDs. These VPPs are renewable energy generation and energy storage systems that are installed outside the geographical boundaries of the district. Examples of technologies that are best suited as VPPs are wind power and hydropower as well as large-scale energy storages, such as pumped hydro and CAES. Solar PV and batteries, on the other hand, are more suitable for an urban environment and are thus possible to install in all types of PEDs.

As a part of the analysis, the authors also proposed a unique onion model for constructing PED networks. According to this model, the majority of the PEDs are placed in the outskirts of the city, and the excess energy generated from these PEDs is exported to the more central areas in the city, where the renewable energy installations are not able to fulfil the energy demand. This way, it would be possible to increase the renewable energy share of the whole city.

In a regulation analysis, we found that there are several regulations and policies that benefit the implementation of PEDs throughout the EU. The European Green Deal and the Clean Energy Package are examples of EU initiatives that are in line with the targets of the PED. The Clean Energy Package has contributed to one of the most significant legislative advancements in favour of the PED concept, as the package recognizes the rights of communities and citizens to engage directly in the energy sector.

The PED definition is still in a conceptualization phase, and further research is therefore needed in order to initiate a discussion on a societal level. More studies on the technological and economic viability of the PED are required, as well as comparative studies with other renewable energy solutions. A comparison between centralized large-scale renewable energy systems and PED-like distributed renewable energy systems would be a particularly interesting research topic. Another topic that would need to be further investigated is the resilience of PEDs and how PEDs are able to handle various types of failures in the local energy system.

Author Contributions: Conceptualization, O.L., F.R. and H.u.R.; formal analysis, O.L.; writing—original draft preparation, O.L. and H.u.R.; writing—review and editing, O.L.; visualization, O.L.; supervision, F.R. All authors have read and agreed to the published version of the manuscript.

Funding: The research was funded by the European Union's Horizon 2020 research and innovation program LC-SC3-SCC-1-2018-2019-2020-Smart Cities and Communities under the project name SPARCS, grant number 864242. The funding body had no involvement in preparing the manuscript, methods and results etc.

Institutional Review Board Statement: Not applicable.

Informed Consent Statement: Not applicable.

Data Availability Statement: Data sharing not applicable.

Conflicts of Interest: The authors declare no conflict of interest.

Appendix A

Table A1. Techno-economic properties of different renewable energy technologies [32].

Technology	Geographical Locations with High Capacity in the EU	Weighted Average Total Installed Cost (USD/kW)	Weighted Average Cost of Electricity (USD/kWh)	Life of Investment (Years)	Average Capacity Factor
Solar photovoltaics	Southern Europe, particularly the Iberian Peninsula and the Mediterranean [122]	1210	0.085	25	18%
Concentrating solar power	Southern Europe, particularly the Iberian Peninsula and the Mediterranean [122]	5204	0.185	25	45%
Onshore wind power	Along the coast of the Atlantic Sea and the Baltic Sea as well as coastal areas in Croatia and inland areas in France, Germany and Poland [47]	1497	0.056	25	34%
Offshore wind power	The Northern Atlantic (especially the North Sea), the Baltic Sea, the Gulf of Lyon and the Aegean Sea [47]	4353	0.127	25	43%
Hydropower	EU countries with the most hydropower per capita [48]: - Sweden (6.6 kWh) Austria (4.7 kWh) - Finland (2.6 kWh) - Slovenia (2.1 kWh) - Croatia (1.7 kWh) - Latvia (1.5 kWh) - Portugal (1.2 kWh)	1491	0.047	30	47%
Geothermal energy	Italy [59]	3976	0.07	25	84%
Biomass power plants	Finland, Sweden, Norway, Estonia, Latvia, Austria, Bulgaria, Croatia, Slovenia and Slovakia [50]	2105	0.062	20	78%

Table A2. Techno-economic properties of different energy storage technologies [60].

Technology	Geographical Locations with High Capacity in the EU	Installation Cost (USD/kWh)	Energy Density (kWh/m^3)	Life of Investment (years)	Round-Trip Efficiency
Pumped hydro storage	Austria, France, Italy and Spain [61]	5–100 (avg.: 20)	0–2	30–100 (avg.: 60)	80%
Compressed air storage	Northern and central Germany, Poland, parts of the UK, Denmark, eastern and northern parts of the Netherlands, northeast Spain, eastern France, western Portugal, central Romania, eastern Bosnia and Herzegovina and western Greece [64]	0–85 (avg.: 50)	2–6	20–100 (avg.: 50)	60%
Lithium-ion batteries	-	200–800 (avg.: 350)	200–600	5–20 (avg.: 12)	95%
Lead-acid batteries	-	100–500	50–100	5–20	85%

References

1. D'Agostino, D.; Mazzarella, L. What is a Nearly zero energy building? Overview, implementation and comparison of definitions. *J. Build. Eng.* **2019**, *21*, 200–212. [CrossRef]
2. Jin, Y.; Wang, L.; Xiong, Y.; Cai, H.; Li, Y.H.; Zhang, W.J. Feasibility studies on net zero energy building for climate considering: A case of 'all Green House' for Datong, Shanxi, China. *Energy Build.* **2014**, *85*, 155–164. [CrossRef]
3. Wang, L.; Gwilliam, J.; Jones, P. Case study of zero energy house design in UK. *Energy Build.* **2009**, *41*, 1215–1222. [CrossRef]
4. Wells, L.; Rismanchi, B.; Aye, L. A review of Net Zero Energy Buildings with reflections on the Australian context. *Energy Build.* **2018**, *158*, 616–628. [CrossRef]
5. Al-Saeed, Y.; Ahmed, A. Evaluating Design Strategies for Nearly Zero Energy Buildings in the Middle East and North Africa Regions. *Designs* **2018**, *2*, 35. [CrossRef]
6. A Common Definition for Zero Energy Buildings. Available online: https://www.energy.gov/eere/buildings/downloads/common-definition-zero-energy-buildings (accessed on 6 April 2020).
7. European Parliament. Directive 2010/31/EU of the European Parliament and of the Council of 19 May 2010 on the energy performance of buildings. *Off. J. Eur. Union* **2010**, *53*, 13–35.
8. Energy Independence and Security Act of 2007. Available online: https://www.epa.gov/greeningepa/energy-independence-and-security-act-2007 (accessed on 8 May 2020).
9. Crawley, D.; Pless, S.; Torcellini, P. Getting to net zero. *ASHRAE J.* **2009**, *51*, 18–25.
10. European Parliament. Commission Delegated Regulation (EU) No 244/2012 of 16 January 2012 supplementing Directive 2010/31/EU of the European Parliament and of the Council on the energy performance of buildings by establishing a comparative methodology framework for calculating. *Off. J. Eur. Union* **2012**, *81*, 18–36.
11. Zacà, I.; D'Agostino, D.; Congedo, P.M.; Baglivo, C. Assessment of cost-optimality and technical solutions in high performance multi-residential buildings in the Mediterranean area. *Energy Build.* **2015**, *102*, 250–265. [CrossRef]
12. Congedo, P.M.; D'Agostino, D.; Baglivo, C.; Tornese, G.; Zacà, I. Efficient solutions and cost-optimal analysis for existing school buildings. *Energies* **2016**, *9*, 851. [CrossRef]
13. Atanasiu, B.; Kouloumpi, I.; Thomsen, K.E.; Aggerholm, S. *Implementing the Cost-Optimal Methodology in EU Countries*; Buildings Performance Institute Europe: Brussel, Belgium, 2013.
14. Annunziata, E.; Frey, M.; Rizzi, F. Towards nearly zero-energy buildings: The state-of-art of national regulations in Europe. *Energy* **2013**, *57*, 125–133. [CrossRef]
15. Laustsen, J. Energy Efficiency Requirements in Building Codes, Energy Efficiency Policies for New Buildings. 2008. Available online: https://www.iea.org/reports/energy-efficiency-requirements-in-building-codes-policies-for-new-buildings (accessed on 22 December 2020).
16. Panagiotidou, M.; Fuller, R.J. Progress in ZEBs-A review of definitions, policies and construction activity. *Energy Policy* **2013**, *62*, 196–206. [CrossRef]
17. Lund, H.; Marszal, A.; Heiselberg, P. Zero energy buildings and mismatch compensation factors. *Energy Build.* **2011**, *43*, 1646–1654. [CrossRef]
18. Torcellini, P.; Shanti, P.; Deru, M.; Crawley, D. *Zero Energy Buildings: A Critical Look at the Definition*; ACEEE Summer Study: Pacific Grove, CA, USA, 2006.

19. Kolokotsa, D.; Rovas, D.; Kosmatopoulos, E.; Kalaitzakis, K. A roadmap towards intelligent net zero- and positive-energy buildings. *Sol. Energy* **2011**, *85*, 3067–3084. [CrossRef]
20. Saheb, Y.; Shnapp, S.; Paci, D. *From Nearly-Zero Energy Buildings to Net-Zero Energy Districts*; JRC Technical Reports; Publications Office of the European Union: Luxembourg, 2019.
21. Ala-Juusela, M.; Crosbie, T.; Hukkalainen, M. Defining and operationalising the concept of an energy positive neighbourhood. *Energy Convers. Manag.* **2016**, *125*, 133–140. Available online: https://ec.europa.eu/jrc/en/publication/nearly-zero-energy-buildings-net-zero-energy-districts (accessed on 22 December 2020). [CrossRef]
22. Clean Energy for All Europeans Package. Available online: https://ec.europa.eu/energy/topics/energy-strategy/clean-energy-all-europeans_en#documents (accessed on 4 June 2020).
23. Laitinen, A.; Lindholm, O. Energiaomavarainen ei pärjää energiapositiiviselle. *Tekniikka Talous*. 2020. Available online: https://www.tekniikkatalous.fi/uutiset/energiaomavarainen-ei-parjaa-energiapositiiviselle/7dc87e68-ba08-4976-ac2a-d4f5f30156d4 (accessed on 22 December 2020).
24. Making City—Energy Efficient Pathway for the City Transformation. Available online: http://makingcity.eu/ (accessed on 5 January 2020).
25. Wyckmans, A.; Karatzoudi, K.; Brigg, D.; Ahlers, D. D9.5: Report on Attendance at Eventsheld by Other SCC-01 Co-Ordinators 2, +CityxChange Work Package 9 Task 9.2. 2019. Available online: https://cityxchange.eu/wp-content/uploads/2019/11/D9.5_Report-on-Attendance-at-events-held-by-other-SCC-01-co-ordinators2.pdf (accessed on 22 December 2020).
26. Greenhouse Gas Emissions from Transport in Europe. Available online: https://www.eea.europa.eu/data-and-maps/indicators/transport-emissions-of-greenhouse-gases/transport-emissions-of-greenhouse-gases-12 (accessed on 6 April 2020).
27. International Energy Agency. Global EV Outlook 2019. 2019. Available online: https://www.iea.org/reports/global-ev-outlook-2019 (accessed on 22 December 2020).
28. Kaupunkisuunnitteluviraston Esitteet | Helsingin Kaupunki. Available online: https://www.hel.fi/kaupunkiymparisto/fi/julkaisut-ja-aineistot/julkaisuarkisto/kaupunkisuunnitteluviraston-esitteet (accessed on 6 April 2020).
29. The Frankfurt School—UNEP Collaborating Centre for Climate & Sustainable Energy Finance; BloombergNEF. Global Trends in Renewable Energy Investment 2019. 2019. Available online: https://www.unenvironment.org/resources/report/global-trends-renewable-energy-investment-2019 (accessed on 22 December 2020).
30. International Renewable Energy Agency. The Power to Change: Solar and Wind Cost Reduction Potential to 2025. 2016. Available online: https://www.irena.org/publications/2016/Jun/The-Power-to-Change-Solar-and-Wind-Cost-Reduction-Potential-to-2025 (accessed on 22 December 2020).
31. Energy Statistical Datasheets for the EU Countries—Datasets. Available online: https://data.europa.eu/euodp/en/data/dataset/information-on-energy-markets-in-eu-countries-with-national-energy-profiles (accessed on 17 January 2020).
32. International Renewable Energy Agency. Renewable Power Generation Costs in 2018. 2019. Available online: https://www.irena.org/publications/2019/May/Renewable-power-generation-costs-in-2018 (accessed on 22 December 2020).
33. Bazoukis, N. *Solar Thermal Technologies: Applications and Global Markets*; BCC Research: Wellesley, MA, USA, 2016.
34. Madsen, D.N.; Hansen, J.P. Outlook of solar energy in Europe based on economic growth characteristics. *Renew. Sustain. Energy Rev.* **2019**, *114*, 109306. [CrossRef]
35. Solar Heat Europe ESTIF. Solar Heat Markets in Europe. 2019. Available online: http://solarheateurope.eu/publications/market-statistics/solar-heat-markets-in-europe/ (accessed on 22 December 2020).
36. Parida, B.; Iniyan, S.; Goic, R. A review of solar photovoltaic technologies. *Renew. Sustain. Energy Rev.* **2011**, *15*, 1625–1636. [CrossRef]
37. Müller-Steinhagen, H.; Trieb, F. Concentrating solar power: A review of the technology. *Ingenia* **2004**, *18*, 43–50.
38. Šúri, M.; Huld, T.A.; Dunlop, E.D.; Ossenbrink, H.A. Potential of solar electricity generation in the European Union member states and candidate countries. *Sol. Energy* **2007**, *81*, 1295–1305. [CrossRef]
39. JRC Photovoltaic Geographical Information System (PVGIS)—European Commission. Available online: https://re.jrc.ec.europa.eu/pvg_tools/en/tools.html#PVP (accessed on 10 January 2020).
40. Adeh, E.H.; Good, S.P.; Calaf, M.; Higgins, C.W. Solar PV Power Potential is Greatest over Croplands. *Sci. Rep.* **2019**, *9*, 11442. [CrossRef]
41. Informationen zu den SOLARWATT PV-Modulen zum Download als pdf. Available online: https://www.solarwatt.de/downloads/pv-modules (accessed on 29 January 2020).
42. Solar Panel Efficiency: What Panels Are Most Efficient? | EnergySage. Available online: https://news.energysage.com/what-are-the-most-efficient-solar-panels-on-the-market/ (accessed on 29 January 2020).
43. Cocco, D.; Migliari, L.; Serra, F. Influence of Thermal Energy Losses on the Yearly Performance of Medium Size Csp Plants. In Proceedings of the Ecos2015: 28th International Conference on Efficiency, Cost, Optimization, Simulation and Environmental Impact of Energy Systems, Pau, France, 29 June–3 July 2015.
44. Rinne, E.; Holttinen, H.; Kiviluoma, J.; Rissanen, S. Effects of turbine technology and land use on wind power resource potential. *Nat. Energy* **2018**, *3*, 494–500. [CrossRef]
45. Angelis-Dimakis, A.; Biberacher, M.; Dominguez, J.; Fiorese, G.; Gadocha, S.; Gnansounou, E.; Guarisod, G.; Kartalidis, A.; Panichelli, L.; Pinedo, I.; et al. Methods and tools to evaluate the availability of renewable energy sources. *Renew. Sustain. Energy Rev.* **2011**, *15*, 1182–1200. [CrossRef]

46. Ragheb, M.; Adam, M. Wind Turbines Theory—The Betz Equation and Optimal Rotor Tip Speed Ratio. *Fundam. Adv. Top. Wind Power* **2011**, *1*, 19–38.
47. Global Wind Atlas. Available online: https://globalwindatlas.info/ (accessed on 10 January 2020).
48. International Renewable Energy Agency. Hydropower—Technology Brief. 2015. Available online: https://www.irena.org/publications/2013/Jan/IRENA-IEA-ETSAP-Technology-Briefs (accessed on 22 December 2020).
49. International Renewable Energy Agency. Cost Analysis of Hydropower. 2012. Available online: https://www.irena.org/-/media/Files/IRENA/Agency/Publication/2012/RE_Technologies_Cost_Analysis-HYDROPOWER.pdf (accessed on 22 December 2020).
50. European Commission. Brief on Biomass for Energy in the European Union. 2019. Available online: https://ec.europa.eu/jrc/en/publication/brochures-leaflets/brief-biomass-energy-european-union (accessed on 22 December 2020).
51. Eurostat. *Energy, Transport and Environment Statistics*; Eurostat: Luxembourg, 2019; Available online: https://ec.europa.eu/eurostat/web/products-statistical-books/-/KS-DK-19-001 (accessed on 22 December 2020).
52. IEA-ETSAP and IRENA. Biomass for Heat and Power. 2015. Available online: https://www.irena.org/publications/2015/Jan/Biomass-for-Heat-and-Power (accessed on 22 December 2020).
53. Vigants, G.; Galindoms, G.; Veidenbergs, I.; Vigants, E.; Blumberga, D. Efficiency Diagram for District Heating System with Gas Condensing Unit. *Energy Procedia* **2015**, *72*, 119–126. [CrossRef]
54. Miles, T.R.; Baxter, L.L.; Bryers, R.W.; Jenkins, B.M.; Oden, L.L. Boiler deposits from firing biomass fuels. *Biomass Bioenergy* **1996**, *10*, 125–138. [CrossRef]
55. Broström, M.; Kassman, H.; Helgesson, A.; Berg, M.; Andersson, A.; Backman, R.; Nordin, A. Sulfation of corrosive alkali chlorides by ammonium sulfate in a biomass fired CFB boiler. *Fuel Process. Technol.* **2007**, *88*, 1171–1177. [CrossRef]
56. Davidsson, K.O.; Åmand, L.E.; Steenari, B.M.; Elled, A.L.; Eskilsson, D.; Leckner, B. Countermeasures against alkali-related problems during combustion of biomass in a circulating fluidized bed boiler. *Chem. Eng. Sci.* **2008**, *63*, 5314–5329. [CrossRef]
57. Searcy, E.; Flynn, P.; Ghafoori, E.; Kumar, A. The relative cost of biomass energy transport. *Appl. Biochem. Biotechnol.* **2007**, *137–140*, 639–652.
58. Heat Values of Various Fuels—World Nuclear Association. Available online: https://www.world-nuclear.org/information-library/facts-and-figures/heat-values-of-various-fuels.aspx (accessed on 19 February 2020).
59. International Renewable Energy Agency. Geothermal Power: Technology Brief. 2017. Available online: https://www.irena.org/publications/2017/Aug/Geothermal-power-Technology-brief (accessed on 22 December 2020).
60. International Renewable Energy Agency. Electricity Storage and Renewables: Costs and Markets to 2030. 2017. Available online: https://www.irena.org/publications/2017/Oct/Electricity-storage-and-renewables-costs-and-markets (accessed on 22 December 2020).
61. International Hydropower Association. The World's Water Battery: Pumped Hydropower Storage and the Clean Energy Transition. 2018. Available online: https://www.hydropower.org/publications/the-world-e2-80-99s-water-battery-pumped-hydropower-storage-and-the-clean-energy-transition (accessed on 22 December 2020).
62. Rehman, S.; Al-Hadhrami, L.M.; Alam, M.M. Pumped hydro energy storage system: A technological review. *Renew. Sustain. Energy Rev.* **2015**, *44*, 586–598. [CrossRef]
63. Gimeno-Gutiérrez, M.; Lacal-Arántegui, R. Assessment of the European potential for pumped Hydropower energy storage based on two existing reservoirs. *Renew. Energy* **2015**, *75*, 856–868. [CrossRef]
64. Caglayan, D.; Weber, N.; Heinrichs, H.U.; Linßen, J.; Robinius, M.; Kukla, P.A.; Stolten, D. Technical Potential of Salt Caverns for Hydrogen Storage in Europe. *Int. J. Hydrogen Energy* **2020**, *45*, 6793–6805. [CrossRef]
65. Maara, C. Residential Energy Storage, Blockchain and Energy Sharing Systems: Technologies and Global Market. 2019. Available online: https://www.bccresearch.com/market-research/fuel-cell-and-battery-technologies/residential-energy-storage-blockchain-energy-sharing-stems-technologies-global-market.html (accessed on 22 December 2020).
66. Energy Storage. Available online: https://www.iea.org/reports/energy-storage (accessed on 7 February 2020).
67. Chen, J. Graphite, Silicon, Metal, and Other Anode Materials for Motive Lithium Ion Batteries: Global Market and Emerging Opportunities. 2020. Available online: https://www.bccresearch.com/market-research/fuel-cell-and-battery-technologies/graphite-production-and-applications-market-report.html (accessed on 22 December 2020).
68. Molaeimanesh, G.R.; Mousavi-Khoshdel, S.M.; Nemati, A.B. Experimental analysis of commercial $LiFePO_4$ battery life span used in electric vehicle under extremely cold and hot thermal conditions. *J. Therm. Anal. Calorim.* **2020**, 1–10. [CrossRef]
69. Ghobadpour, A.; Kelouwani, S.; Amamou, A.; Agbossou, K.; Zioui, N. A brief review of plug-in hybrid electric vehicles operation in cold climates. In Proceedings of the 2019 IEEE Vehicle Power and Propulsion Conference, Hanoi, Vietnam, 14–17 October 2019.
70. Fan, R.; Zhang, C.; Wang, Y.; Ji, C.; Meng, Z.; Xu, L.; Ou, Y.; Chin, C.S. Numerical study on the effects of battery heating in cold climate. *J. Energy Storage* **2019**, *26*, 100969. [CrossRef]
71. Hesaraki, A.; Holmberg, S.; Haghighat, F. Seasonal thermal energy storage with heat pumps and low temperatures in building projects—A comparative review. *Renew. Sustain. Energy Rev.* **2015**, *43*, 1199–1213. [CrossRef]
72. International Renewable Energy Agency. Thermal Energy Storage: Technology Brief. 2013. Available online: https://www.irena.org/publications/2013/Jan/Thermal-energy-storage (accessed on 22 December 2020).
73. Mangold, D.; Schmidt, T.; Müller-Steinhagen, H. Seasonal thermal energy storage in Germany. *Struct. Eng. Int.* **2004**, *14*, 230–232. [CrossRef]

74. Xu, J.; Wang, R.Z.; Li, Y. A review of available technologies for seasonal thermal energy storage. *Sol. Energy* **2014**, *103*, 610–638. [CrossRef]
75. Gupta, D.; Zaman, M. Stability of boreholes in a geologic medium including the effects of anisotropy. *Appl. Math. Mech.* **1999**, *20*, 837–866. [CrossRef]
76. Tsemekidi, S.; Bertoldi, P.; Diluiso, F.; Castellazzi, L.; Economidou, M.; Labanca, N.; Ribeiro Serrenho, T.; Zangheri, P. Analysis of the EU Residential Energy Consumption: Trends and Determinants. *Energies* **2019**, *12*, 1065. [CrossRef]
77. EURIMA—U-Values in Europe. Available online: https://www.eurima.org/u-values-in-europe/ (accessed on 17 June 2020).
78. Final Energy Consumption in Households per Capita. Available online: https://ec.europa.eu/eurostat/databrowser/view/sdg_07_20/default/table?lang=en (accessed on 17 June 2020).
79. Share of Final Energy Consumption in the Residential Sector by Type of End-Use, 2017 (%). Available online: https://ec.europa.eu/eurostat/statistics-explained/index.php?title=File:Share_of_final_energy_consumption_in_the_residential_sector_by_type_of_end-use,_2017_(%25).png (accessed on 17 June 2020).
80. ACEA—European Automobile Manufacturers' Association. Vehicles in Use—Europe 2019. 2019. Available online: https://www.acea.be/uploads/publications/ACEA_Report_Vehicles_in_use-Europe_2019.pdf (accessed on 22 December 2020).
81. Directorate-General for Energy. Quarterly Report on European Electricity Markets. *Mark. Obs. Energy* **2019**, *13*. Available online: https://ec.europa.eu/energy/data-analysis/market-analysis_en (accessed on 22 December 2020).
82. Torró, H. Forecasting Weekly Electricity Prices at Nord Pool. *SSRN Electron. J.* **2011**. [CrossRef]
83. Helisto, N.; Kiviluoma, J.; Holttinen, H. Sensitivity of electricity prices in energy-only markets with large amounts of zero marginal cost generation. In Proceedings of the International Conference on the European Energy Market, Dresden, Germany, 6–9 June 2017.
84. Electricity Prices for Household Consumers—Bi-Annual Data (from 2007 Onwards). Available online: https://appsso.eurostat.ec.europa.eu/nui/show.do?dataset=nrg_pc_204&lang=en (accessed on 17 June 2020).
85. Electricity Prices for Non-Household Consumers—Bi-Annual Data (from 2007 Onwards). Available online: https://appsso.eurostat.ec.europa.eu/nui/show.do?dataset=nrg_pc_205&lang=en (accessed on 17 June 2020).
86. Heide, D.; von Bremen, L.; Greiner, M.; Hoffmann, C.; Speckmann, M.; Bofinger, S. Seasonal optimal mix of wind and solar power in a future, highly renewable Europe. *Renew. Energy* **2010**, *35*, 2483–2489. [CrossRef]
87. Liu, H.; Andresen, G.B.; Greiner, M. Cost-optimal design of a simplified highly renewable Chinese electricity network. *Energy* **2018**, *147*, 534–546. [CrossRef]
88. Lindholm, O. Techno-Economic Analysis of a Reversible Solid Oxide Cell and Hydrogen Gas Storage System. Master's Thesis, Åbo Akademi University, Turku, Finland, October 2019.
89. ur Rehman, H.; Reda, F.; Paiho, S.; Hasan, A. Towards positive energy communities at high latitudes. *Energy Convers. Manag.* **2019**, *196*, 175–195. [CrossRef]
90. Lai, C.M.; Hokoi, S. Solar façades: A review. *Build. Environ.* **2015**, *91*, 152–165. [CrossRef]
91. Fath, K.; Stengel, J.; Sprenger, W.; Wilson, H.R.; Schultmann, F.; Kuhn, T.E. A method for predicting the economic potential of (building-integrated) photovoltaics in urban areas based on hourly Radiance simulations. *Sol. Energy* **2015**, *116*, 357–370. [CrossRef]
92. Amado, M.; Poggi, F. Solar urban planning: A parametric approach. *Energy Procedia* **2014**, *48*, 1539–1548. [CrossRef]
93. Hofierka, J.; Kaňuk, J. Assessment of photovoltaic potential in urban areas using open-source solar radiation tools. *Renew. Energy* **2009**, *34*, 2206–2214. [CrossRef]
94. Kammen, D.M.; Sunter, D.A. City-integrated renewable energy for urban sustainability. *Science* **2016**, *352*, 922–928. [CrossRef]
95. Tummala, A.; Velamati, R.K.; Sinha, D.K.; Indraja, V.; Krishna, V.H. A review on small scale wind turbines. *Renew. Sustain. Energy Rev.* **2016**, *56*, 1351–1371. [CrossRef]
96. Small Wind Turbines: Are They Right for You? | EnergySage. Available online: https://news.energysage.com/small-wind-turbines-overview/ (accessed on 16 April 2020).
97. Wagner, R.; Antoniou, I.; Pedersen, S.M.; Courtney, M.S.; Jørgensen, H.E. The influence of the wind speed profile on wind turbine performance measurements. *Wind Energy* **2009**, *12*, 348–362. [CrossRef]
98. Graabak, I.; Korpås, M.; Jaehnert, S.; Belsnes, M. Balancing future variable wind and solar power production in Central-West Europe with Norwegian hydropower. *Energy* **2019**, *168*, 870–882. [CrossRef]
99. Haakana, J.; Tikka, V.; Lassila, J.; Partanen, J.; Rinne, S. Opportunities of bioenergy-based CHP production in balancing renewable power production. In Proceedings of the International Conference on the European Energy Market, Porto, Portugal, 6–9 June 2016.
100. International Hydropower Association. 2019 Hydropower Status Report. 2019. Available online: https://www.hydropower.org/news/iha-releases-2019-hydropower-status-report-charting-growth-in-renewable-hydro (accessed on 22 December 2020).
101. David, A.; Mathiesen, B.V.; Averfalk, H.; Werner, S.; Lund, H. Heat Roadmap Europe: Large-Scale Electric Heat Pumps in District Heating Systems. *Energies* **2017**, *10*, 578. [CrossRef]
102. International Renewable Energy Agency. Renewable Energy Cost Analysis—Hydropower. 2012. Available online: https://www.irena.org/publications/2012/Jun/Renewable-Energy-Cost-Analysis---Hydropower (accessed on 22 December 2020).
103. Haas, J.; Cebulla, F.; Nowak, W.; Rahmann, C. Palma-Behnke, R. A multi-service approach for planning the optimal mix of energy storage technologies in a fully-renewable power supply. *Energy Convers. Manag.* **2018**, *178*, 355–368. [CrossRef]

104. Parra, D.; Swierczynski, M.; Stroe, D.I.; Norman, S.; Abdon, A.; Worlitschek, J.; O'Doherty, T.; Rodrigues, L.T.; Gillott, M.C.; Zhang, X.; et al. An interdisciplinary review of energy storage for communities: Challenges and perspectives. *Renew. Sustain. Energy Rev.* **2017**, *79*, 730–749. [CrossRef]
105. Virtual Power Plant (VPP) Operator. Available online: https://www.next-kraftwerke.com/company (accessed on 11 June 2020).
106. VPP Explained: What Is a Virtual Power Plant? Available online: https://www.next-kraftwerke.com/vpp/virtual-power-plant (accessed on 11 June 2020).
107. Sector Integration Will Help to Manage the Power Balance—Fingrid-Lehti. Available online: https://www.fingridlehti.fi/en/sector-integration/ (accessed on 4 June 2020).
108. Vasirani, M.; Kota, R.; Cavalcante, R.L.G.; Ossowski, S.; Jennings, N.R. An agent-based approach to virtual power plants of wind power generators and electric vehicles. *IEEE Trans. Smart Grid* **2013**, *4*, 1314–1322. [CrossRef]
109. Heat Pumps. Available online: https://www.iea.org/reports/tracking-buildings-2020/heat-pumps#abstract (accessed on 7 September 2020).
110. Lund, H.; Werner, S.; Wiltshire, R.; Svendsen, S.; Thorsen, J.E.; Hvelplund, F.; Mathiesen, B.V. 4th Generation District Heating (4GDH). Integrating smart thermal grids into future sustainable energy systems. *Energy* **2014**, *68*, 1–11. [CrossRef]
111. Lake, A.; Rezaie, B.; Beyerlein, S. Review of district heating and cooling systems for a sustainable future. *Renew. Sustain. Energy Rev.* **2017**, *67*, 417–425. [CrossRef]
112. Lund, H.; Østergaard, P.A.; Chang, M.; Werner, S.; Svendsen, S.; Sorknæs, P.; Thorsen, J.E.; Hvelplund, F.; Mortensen, B.O.G.; Mathiesen, B.V.; et al. The status of 4th generation district heating: Research and results. *Energy* **2018**, *164*, 147–159. [CrossRef]
113. Lund, H.; Duic, N.; Østergaard, P.A.; Mathiesen, B.V. Future district heating systems and technologies: On the role of smart energy systems and 4th generation district heating. *Energy* **2018**, *165*, 614–619. [CrossRef]
114. Werner, S. District heating and cooling in Sweden. *Energy* **2017**, *126*, 419–429. [CrossRef]
115. Jones, C.M.; Kammen, D.M. Quantifying carbon footprint reduction opportunities for U.S. households and communities. *Environ. Sci. Technol.* **2011**, *45*, 4088–4095. [CrossRef]
116. Jones, C.; Kammen, D.M. Spatial distribution of U.S. household carbon footprints reveals suburbanization undermines greenhouse gas benefits of urban population density. *Environ. Sci. Technol.* **2014**, *48*, 895–902. [CrossRef] [PubMed]
117. Gapminder Tools. Available online: https://www.gapminder.org/tools/#$state$marker$axis_x$which=urban_population_percent_of_total&domainMin:null&domainMax:null&zoomedMin:null&zoomedMax:null&scaleType=linear&spaceRef:null;&axis_y$which=co2_emissions_tonnes_per_person&domainMin:null&domainMax:null&zoomedMin:null&zoomedMax:null&scaleType=genericLog&spaceRef:null;;;&chart-type=bubbles (accessed on 19 May 2020).
118. A European Green Deal. Available online: https://ec.europa.eu/info/strategy/priorities-2019-2024/european-green-deal_en (accessed on 5 August 2020).
119. European Commission. *Annex to the Communication from the Commission to the European Parliament, the European Council, the Council, the European Economic and Social Committee and the Committee of the Regions*; European Commission: Brussels, Belgium, 2019.
120. Caramizaru, A.; Uihlein, A. *Energy Communities: An Overview of Energy and Social Innovation*; Publications Office of the European Union: Luxembourg, 2020.
121. Valtioneuvoston Asetus Sähköntoimitusten Selvityksestä ja Mittauksesta Annetun Asetuksen Muuttamisesta. Available online: https://www.finlex.fi/fi/laki/alkup/2016/20160217 (accessed on 2 June 2020).
122. Global Solar Atlas. Available online: https://globalsolaratlas.info/map (accessed on 7 September 2020).

buildings

MDPI

Article

Challenges and Barriers for Net-Zero/Positive Energy Buildings and Districts—Empirical Evidence from the Smart City Project SPARCS

Daria Uspenskaia *, Karl Specht, Hendrik Kondziella and Thomas Bruckner

Institute for Infrastructure and Resources Management, Leipzig University, 04109 Leipzig, Germany; specht@wifa.uni-leipzig.de (K.S.); kondziella@wifa.uni-leipzig.de (H.K.); bruckner@wifa.uni-leipzig.de (T.B.)
* Correspondence: uspenskaia@wifa.uni-leipzig.de

Abstract: Without decarbonizing cities energy and climate objectives cannot be achieved as cities account for approximately two thirds of energy consumption and emissions. This goal of decarbonizing cities has to be facilitated by promoting net-zero/positive energy buildings and districts and replicating them, driving cities towards sustainability goals. Many projects in smart cities demonstrate novel and groundbreaking low-carbon solutions in demonstration and lighthouse projects. However, as the historical, geographic, political, social and economic context of urban areas vary greatly, it is not always easy to repeat the solution in another city or even district. It is therefore important to look for the opportunities to scale up or repeat successful pilots. The purpose of this paper is to explore common trends in technologies and replication strategies for positive energy buildings or districts in smart city projects, based on the practical experience from a case study in Leipzig—one of the lighthouse cities in the project SPARCS. One of the key findings the paper has proven is the necessity of a profound replication modelling to deepen the understanding of upscaling processes. Three models analyzed in this article are able to provide a multidimensional representation of the solution to be replicated.

Keywords: smart city; net- and nearly-zero-energy buildings; positive energy communities and districts; renewable energy integration; energy flexibility in buildings and communities; simulation and optimization methods; practical experience from demo sites

Citation: Uspenskaia, D.; Specht, K.; Kondziella, H.; Bruckner, T. Challenges and Barriers for Net-Zero/Positive Energy Buildings and Districts—Empirical Evidence from the Smart City Project SPARCS. *Buildings* **2021**, *11*, 78. https://doi.org/10.3390/buildings11020078

Academic Editors: Ala Hasan and Francesco Reda

Received: 11 December 2020
Accepted: 18 February 2021
Published: 23 February 2021

Publisher's Note: MDPI stays neutral with regard to jurisdictional claims in published maps and institutional affiliations.

1. Introduction

1.1. Background and Motivation

Cities have been considered as the major contributors to the overall greenhouse gas emissions and this tendency is only going to get stronger. By 2050, over 70% of the global population will be living in cities [1]. Increasing the share of renewables in the energy mix will not help to reduce CO_2 emissions to the limits set in the Paris agreement: [2] the consumption should also be reduced by achieving ambitious targets for energy efficiency. However, without decarbonizing cities current energy and climate objectives cannot be achieved as cities are responsible for about two thirds of energy consumption and emissions [3]. This goal of decarbonizing cities has to be facilitated by promoting net-zero/positive energy buildings and districts and replicating them, driving cities towards sustainability goals.

It is important to bear in mind that almost complete decarbonization is a key objective of the EU by 2050 [4], particularly for long-lived infrastructures such as energy infrastructures. The EU position [5] is based on the idea that a sustainable decarbonization policy cannot be built solely on reducing emissions or saving energy, but on "creating value" for cities and people, taking full advantage of the technology and the increasingly growing opportunities generated by digitalization. If such mitigation (and climate adaptation) strategy is well implemented, cities will solve many socio-economic problems by procuring creative solutions, promoting cross-sectoral cooperation, creating new business models, and partnering with the private sector and residents.

Many projects in smart cities show novel and groundbreaking approaches and are often creative in nature (for example, REMINING-LOWEX where the heat energy stored in old mine shafts was used for the heating and cooling of buildings; EnerGAware—a mobile app-based game that is linked to the actual energy consumption (smart meter data) of the game user's home and allows to transfer energy savings achieved virtually to the reality and decrease the energy consumption costs; the STORM—a project which tackles energy efficiency at district level by developing an innovative district heating & cooling network controller, based on self-learning algorithms). Detailed information about these and other EU-funded smart city projects available in the Smart Cities Information System (SCIS) database: https://smartcities-infosystem.eu/sites-projects/projects accessed on 11 December 2020. Various approaches are demonstrated in urban living laboratories, test beds, pilots, and lighthouse projects. This evidence-based approach is very useful for life-testing outcomes, procedures, strategies and insights in a particular territorial context. However, as the historical, geographic, political, social and economic characteristics of urban areas vary greatly, not only between cities, but also between districts and even communities, it is not that easy to repeat the project in another city or even district. At the same time, the findings of a demonstration or pilot should be transferable to other locations and circumstances in order to give the project a broader and deeper impact, and to promote urban transformations towards sustainability and resilience [6].

In order to avoid the situation where pilot schemes are an "on-off" exercise, there is a strong need to look for opportunities to scale up and repeat successful pilots. Sharing experiences and best practices, repeating, replicating and scaling up already implemented and life-tested projects is the key to further uptake and acceleration of low-carbon smart city solutions as such "success stories" help to build trust among stakeholders and positively influence the decision-making process [7].

1.2. Methodology

The study is designed as follows (see Figure 1). At first, different existing models and frameworks of replication in smart cities were compared, and replication models were chosen for further analysis. Using the deconstruction method of analysis, the authors prepared a matrix which can be applied to the particular case in a smart city project in order to define the project's replication type.

In parallel, in the quantitative part of the study, a model for the energy related parameters of the demo district was prepared. The input data for the modeling exercise were prepared the following way. Regarding the demand-side, the annual heat demand as settled is disaggregated to derive an hourly demand profile of the relevant buildings in the demonstration district applying the Hellwig [8] procedure. With respect to the supply-side the solar thermal plant is characterized by the following technical parameters: peak power and total annual heat generation, number of solar panels, technological specifics of the system. After that, the input data are transferred to a modeling toolbox establishing the specific energy system model of the demo district. The model output data were analyzed in order to understand the potential for the upscaling of the demand and supply side of the district's heating energy, and one of the models was applied.

Synthesizing the results of the research, the replication process in the EU smart city project SPARCS was described step-by-step in order to demonstrate the practical application of the replication models and the energy system modelling.

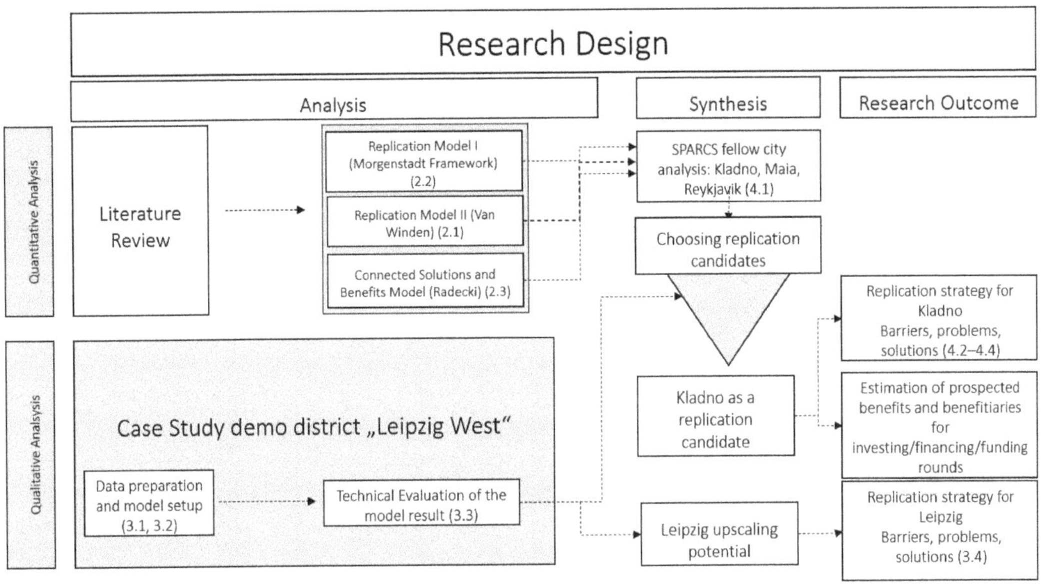

Figure 1. Research Design.

1.3. Definitions and Objectives

Since 2016, buildings have accounted for almost 40% of the EU final energy consumption according to the EU energy efficiency monitoring tool ODYSSEE-MURE [9]. Overall, the building sector accounts for 41% of the final energy consumption and 60% of the electricity consumption in the EU. Two thirds of this intake belong to residential buildings. This opens a tremendous potential for energy efficiency gains in Europe.

In 2018, the European Parliament gave its final approval on the revised Energy Performance of Buildings directive (EPBD). The EPBD is a part of the implementation of the Juncker Commission priorities to build "a resilient Energy Union with a forward-looking climate change policy". The Commission wants the EU to lead the clean energy transition. For this reason, the EU has committed to cut CO_2 emissions by at least 40% by 2030 while modernizing the EU's economy and delivering on jobs and growth for all European citizens. In doing so, the Commission is guided by three main goals: putting energy efficiency first, achieving global leadership in renewable energies and providing a fair deal for consumers [10].

In the same year, the Program on Positive Energy Districts and Neighborhoods has been established [11] by the Action 3.2 on Smart Cities and Communities of the European Strategic Energy Technology (SET) Plan, but the exact definition of a Positive Energy District (PED) and its boundaries are still under discussion. In general, PED is a "working area" in smart city projects with a focus on energy, sustainability and emission reduction or, to put it another way, the functionality of a PED is enabled by a smart city. In a SET Plan a PED is defined as following: "Positive Energy Districts are energy-efficient and energy-flexible urban areas or groups of connected buildings which produce net zero greenhouse gas emissions and actively manage an annual local or regional surplus production of renewable energy. They require integration of different systems and infrastructures and interaction between buildings, the users and the regional energy, mobility and Information and Communication Technology (ICT) systems, while securing the energy supply and a good life for all in line with social, economic and environmental sustainability." [12]

The Thematic Working Group on ICT for energy efficiency states that: "Energy-positive buildings and neighborhoods are those that generate more power than their needs.

They include the management of local energy sources (mainly renewable, e.g., solar, fuel cells, micro-turbines) and the connection to the power grid in order to sell energy if there is excess or, conversely, to buy energy when their own is not sufficient" [13].

The purpose of this paper is to explore common trends in technologies and replication strategies for positive energy buildings or districts in smart city projects. This paper discusses some of the barriers that prevent the successful replication of PEDs and suggests strategies to overcome them, taking the city of Leipzig as an example.

The paper is structured as follows. The concept of replication in smart cities is discussed and an overview of the existing models and frameworks which help to categorize different types of replication is given in Section 2. Thereafter, the case study of a particular smart district in Leipzig within the framework of SPARCS, an EU-funded smart city project, is described and preliminary results as well as suggestions for upscaling are provided (see Section 3). In Section 4 the adoption possibility of the solution demonstrated in the case study by the fellow cities of SPARCS project is examined and discusses the possible replication challenges. Finally, a conclusion is presented and results are discussed in Section 5.

2. Replication: The Collaborative Approach

What is a replication? According to the Cambridge Dictionary, replication is "the act of making or doing something again in exactly the same way, or something that is made or done in this way" [14].

The EU Smart City Information System (SCIS) definition is partly aligned with the one from the Cambridge Dictionary and claims that replication is the "possibility of transporting or 'copying' results from a pilot case to other geographical areas, albeit with potentially different boundary conditions", but it also includes "the management process that was used in the pilot scheme or the cooperation structure between critical stakeholders."

The definition prepared for the European Parliament's Committee on Industry, Research and Energy is the following: "Replication essentially means repeating successful Smart City initiatives in another locale or replicating the same type of Smart City in other cities. These replicas would be based on matching the aggregate characteristics (population, income distribution, local economic characteristics, socio-economic outcomes), and deliberately creating a similar strategic vision and portfolio of (locally relevant) initiatives." [15]

As was mentioned in the introduction, the idea of replicating smart city projects goes beyond reducing CO_2 emissions in a given district/city or refurbishing some buildings: The intention is to make the process of decarbonizing cities in Europe easier and faster. In this case, the definition given by the EU Parliament's Committee describes the replication process in the best way highlighting that this process is more about "matching the aggregate characteristics" and "creating the similar portfolio" rather than "copying results from a pilot case". Having this definition in mind, it is easy to suggest that replication requires a deep understanding of the scaling process of smart city solutions as well as the establishment of strategic cooperation between cities on European and international level.

In this section, we will have a look at two contemporary replication models which have been proposed by the literature. (The third model described in this paper (A. Radecki) is not a replication model but an economic model describing costs and benefits of a single smart city solution.) Provided by leading researchers, these models serve the above-mentioned goal: to deepen the understanding of processes of scaling up the smart city solutions and to provide a common background for establishing a collaboration between the cities.

2.1. Replication Model I (W. van Winden)

One of the most descriptive models to capture the processes of scaling up smart-city solutions is the one suggested by the Dutch urban researcher Willem van Winden [16]. He proposed a clear distinction between three types of upscaling: roll-out, expansion, and replication (see Table 1).

Table 1. Replication types of smart city solutions (W. van Winden).

Scaling Type	Description
Type 1. Roll-out Market roll-out Organizational/city roll-out	A technology or a solution that was successfully tested and developed in the pilot project needs to be brought to the consumer or to the market (Market roll-out) or applied in the entire organization/city (Organizational/city roll-out). The new technology does not fundamentally challenge the current state-of-theart technology and is easily adoptable. Does not require new partnerships, major behavioral or organizational changes.
Type 2. Expansion Quantitative expansion Functional expansion Geographic expansion	The technology/solution developed in a pilot can only be "expanded" by (a) adding partners/users (Quantitative expansion), (b) enlarging the geographical area covered by the solution (Geographic expansion), or (c) added functionality (Functional expansion). Collaborating partners create added value, or the value of the solution grows with the number of participating organizations. Also relevant for local circular economy projects.
Type 3. Replication Organizational replication Geographical replication	The most ambitious type of scaling. The solution/technology that was developed in the pilot project is replicated in another context (another organization, another part of the city, or another city) and involves the complexity of the new context (legal-, organizational- or partnership-wise). The project might be done by the same or new partners, but it never is the exact copy of the pilot. Replicate (exactly or by proxy) the solution in another context by the original partners involved in the pilot project (organizational replication), or by others (geographical replication).

Van Winden highlights that these types of replication are different but not mutually excluding: a project may scale in various directions simultaneously, having some parts of the project replicated according to one type and other parts to another (see Figure 2).

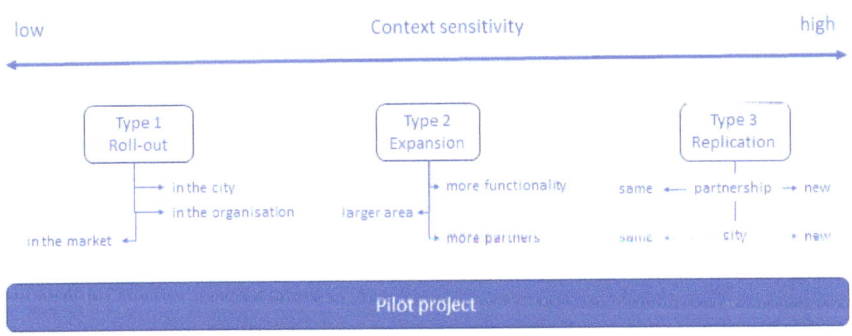

Figure 2. Three types of scaling a pilot project against the context sensitivity.

For the purpose of this paper, whenever we refer to the term "context" we consider the interrelated conditions in which a smart city solution or a project is maintained.

To apply the model to the particular case in a smart city project and define the project's replication type, the following matrix can be used (see Table 2).

Van Winden created his scheme by observing and classifying the completed projects, but his matrix can be used for mapping out the replication strategy for ongoing projects. For example, taking the Type 1 criteria as requirements, we can describe the solution which can be easily rolled-out: a successfully tested solution which does not fundamentally challenge the state-of-the art technology, behavioral patterns or organizational structure, does not require new partnership, does not anticipate regulatory challenges and does not require new budget.

Table 2. Evaluation matrix to define the replication type of the solution.

Criteria	Type		
	Type 1: Roll-out	**Type 2: Expansion**	**Type 3: Replication**
Technology: engineering potential	successfully tested and developed	added functionality	must be further developed or redesigned
Technology: commercial potential	commercialized/brought to the market (market roll-out), widely applied in an organization (organizational roll out) or rolled out in the entire city (city rollout)	the value of the solution grows with the number of participating organizations	the project is able to create value
Technology: Innovative potential	don't fundamentally challenge the current state-of-the-art and are easily adoptable	the innovation is not a single product controlled by one organization, but a co-production that depends on a close alignment of more partners	not the exact copy of the pilot, new context, the consumer reaction to the product is not known
Geographical context	within one company/city	enlarged geographical area covered by the solution (compared to the pilot)	totally new context (compared to the pilot project)
No of partners	no new partnership established	collaborating partners create added value;	can be done by the original pilot partnership but also by others, and the replication can be exact or by proxy.
Behavioral or organizational changes	no major changes	the upscaling is more complicated due to the nature of the solution that was developed and the partnership relations	the solution developed in the pilot must be re-designed by the new partners in the new context.
Changes of interests or organizational cultures	no major changes	relevant for local circular economy projects (where the waste of company x is reused as input for company y)	involves the complexity of the new context (legal, organizational or partner context)
Modifications of the product/solution required	no major modification required	a solution is a co-production that depends on a close alignment of more partners.	the solution never fully matches the original.
Process is managed by one organization	the one that initiated the pilot, based on a profitable business model; the organization has a high level of control	there cannot be a straightforward "rolled out" because there is limited control over the process and several independent organizations are involved	managed by consortium of partners, the pilot organization may or may not be present
Regulatory and legal barriers	limited	transaction and communication costs are high.	the lack of standards, open data formats and protocol poor knowledge transfer mechanisms; Communications about a project, if existing, tend to focus on the successful outcomes, rather than the design process and the difficulties that were tackled along the way
Funding scheme	subsidized/co-funded	co-funded, external investments, revenue from the project	usually no to little funding; external investments, revenue created by the project

A smart city project is a complex venture which includes different stakeholders with varying rationales, ambitions and perspectives regarding the upscaling. Moreover, the partners follow different motivations to take part in the project itself from conducting a study that might be commercialized later on to establishing close relations with the local government or improving their own sustainability policy. Therefore, more research is needed to study the dynamics in the area of upscaling where different interests meet and collide.

At the same time, a smart city project must maintain a good balance between exploration (developing new knowledge and competencies associated with Research&Development and innovation) and exploitation (implementation, scale production, refinement). Pilot projects, after all, are designed mainly for the exploration stage [17].

2.2. *Replication Model II (The Morgenstadt Framework)*

Another framework which has been used widely in smart city projects within the EU is the Morgenstadt Initiative developed by the Fraunhofer Institute for Industrial Engineering (IAO). Using innovation management methodologies and a range of tools and measures, e.g., international city surveys, "city labs", analytical tools, online assessment instruments, etc. this framework aims to accelerate development that helps to reduce energy and resource consumption while enhancing the livability and prosperity of a city.

2.2.1. Morgenstadt Model for Sustainable Urban Development

Starting from a systemic analysis of six leading cities (Singapore, Copenhagen, Freiburg, New York City, Berlin and Tokyo) Fraunhofer researchers aligned and synchronized insights from all cities in one action-oriented model—called the "Morgenstadt Model for Sustainable Urban Development".

There are three types of data available in the Model [18]:

- Urban indicators

Over 300 indicators are used for measuring the performance of the cities within the eight defined sectors and for assessing the social, economic and environmental state of the city. All indicators are put into one of the following three categories to provide a complete basis for quantitative analysis of the status quo in any city:

Pressure Indicators—indicate which pressures exist on the city system from the different sectors and from the social, economic and environmental point of view.

State Indicators—describe the current state of the environment, the society, the economy and the different technology sectors within the city.

Impact Indicators—show which impact the city system has on the environment, the society, the economy and long-term resilience.

- Key action fields

The 83 defined key action fields for sustainable development represent the Morgenstadt Model core. Assessing the state of key action fields allows to create a city profile and analyze the coherency of existing strategies and measures. Relating key action fields to indicators allows us to assess whether the response of a city is in line with pressures and state and really helps optimizing outputs for enhanced sustainability.

- Impact factors

An impact factor analysis uncovers why certain progress happens (or does not happen) in a particular way in a specific urban system. Understanding the constellation of impact factors of a city means understanding external pressures, underlying forces, dynamics, socio-cultural and historical implications that are present within a city and have an impact (often unnoticed) on decisions, structures, strategies and measures taken on the city level and on the project level. Identifying impact factors is complex and needs trans-disciplinary reflection of the researchers. It builds upon an on-site analysis, addressing specific interview questions, applying defined interview techniques, using pre-structured interaction of the researchers and working with mind maps and clustering of impact factors.

Analyzing the three main levels of urban systems (indicators, action fields, and impact factors) the Model describes the selected cities in depth and helps to understand their sustainability performance. The full list of urban indicators, key action fields and impact factors as well as detailed descriptions of their evaluation and assessment is presented in the Morgenstadt: City Insights—Final Report [18].

2.2.2. Morgenstadt City Index

A simplified version of the methodology defined within the Morgenstadt Framework is the Morgenstadt City Index (Figure 3). It provides a general, yet accurate, representation of urban performance derived from statistics and data already available and does not require costly and time-consuming surveys and analysis.

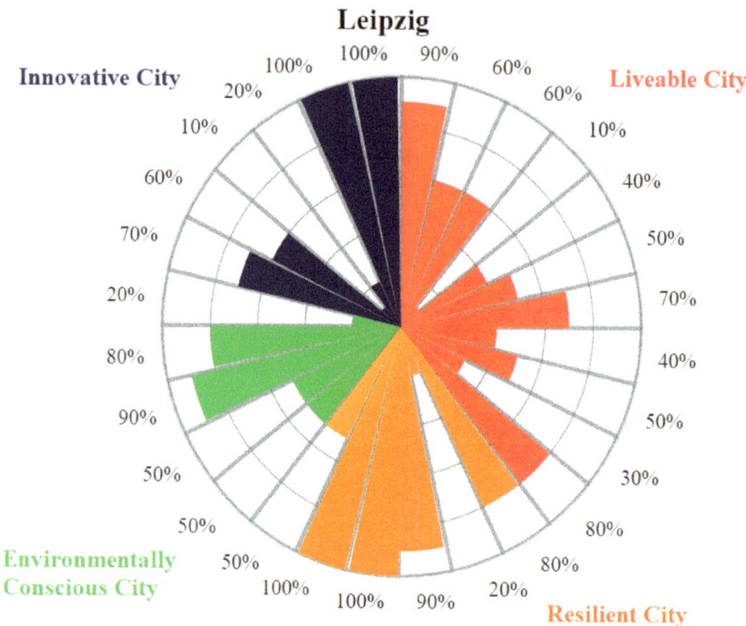

Figure 3. Leipzig City Index assessment.

To create the City Index, 28 indicators were identified and categorized into four central themes (quality of life, environmental consciousness, innovation and resilience) in order to measure the future viability of a city. (For the full list of the indicators and their values refer to the Appendix A and Morgenstadt City Index Online Documentation). Through graphic visualization of the results, the Morgenstadt City Index offers a snapshot of a city identifying its strengths and weaknesses, and presents a baseline for a city's strategy for improvement. The indicators within the City Index were classified on a scale of 0 to 10 points (or 0 to 100%). The values for each indicator and its place on the scale illustrate where the focus of the city development should lay. The only absolute benchmark is set for CO_2 emissions: The target value is CO_2 neutrality.

Below is the graphic representation of the City Index assessment [19] done for Leipzig, one of the lighthouse cities in the SPARCS project.

Interpretation of the Research Results

Liveable City

In terms of quality of life, Leipzig hits exactly the middle of the scale, with individual factors being quite extreme: On the one hand a very low poverty rate (1.1%) and quite favorable rents, but one the other hand one of the highest unemployment rates (within Germany). The bad access to medical care and polluted air contribute to low life expectancy—with 80.3 years Leipzig lies in the middle of the lower third of the scale. Despite the small number of cars, public transport only has average usage, the situation for cyclists is quite positive from the perspective of the citizens. Another plus is the high amount of green and water surfaces within the city.

Resilient City

The city has a rather low debt level, but is very strongly dependent on subsidies from outside—with 58% of self-financed expenditures. The economy is well diversified. In such areas as "risk management" and "disaster control" Leipzig does well, and the city is already thinking about a new Climate Adaptation Strategy.

Environmentally Conscious City

With a low waste volume of 291 t per head (average amount = 430.14 t per head), a rather low water consumption of 121.80 L per person per day (average amount = 120–123 L) and CO_2 emissions 6.62 t per head (average amount = 8.4 t) Leipzig scores well as an Environmentally Conscious City. The share of renewable energies is expandable (30%); the same applies also to the recycling rate—only 38% versus the average of 45%.

Innovative City

In the area of innovation, Leipzig is in the top third of the scale and offers good conditions for research and experimentation. There is an above-average number of newly founded companies, but the number of highly qualified jobs and patents can still be increased significantly. Due to the relatively high number of city inhabitants, the student share is 6.59% which is quite low.

The idea behind the Morgenstadt Framework is to create a tool-kit for the collaboration between cities, industries and research institutions. Developing the model, the Fraunhofer researchers witnessed several challenges that industry and businesses are facing while working together with cities: A single company can never meet the needs of a city, nor can a city implement innovative solutions without cooperation from business partners from different sectors. The Morgenstadt Framework is designed to address these challenges with a new collaborative approach: to initiate and accelerate the long-term transformation of cities into sustainable urban systems, and to thereby create both international and Germany based reference projects on the city-level.

2.3. Connected Solutions and Benefits Model (Alanus von Radecki)

To further deepen the understanding of the scaling process of smart city solutions, the connected solutions and benefits model was suggested by Alanus von Radecki and first used in the Triangulum project [20]. The basic principle behind the replication of smart city solutions lies within economies of scale [21] meaning that the cost advantage for a company increases with the increased output of goods or services. However, the conventional business models of scaling would not work with smart city solutions.

There are two general approaches which describe the cost-benefit relations in energy economics:

- The efficiency model, where the technological innovation itself is able to reduce external costs and to increase the socially efficient allocation through a free market allocation of money and technology at the same time; An investment is made into a key technology or solution (e.g., an efficient LED-light bulb) because the increased efficiency of the technology leads to (energy) savings over time which, in turn, leads to lower energy costs for operating the light bulb. After a time period the investment has paid itself off and money is saved.
- The policy model, where the government closes the gap for the investor with a subsidy or a regulation that makes investments into the desired technologies more profitable than investing into conventional alternatives, and therefore provides an economic rationale for incentivizing investments into clean technologies and for developing the markets of green tech. We encounter it, for example, when governments provide

subsidies for desired technologies (e.g., solar- and wind energy, electric vehicles etc.), or regulate the market through taxes and fees (e.g., for polluting cars).

With digitalization and clean and efficient technologies starting to take off, a third economic model becomes viable, and it is intrinsically linked to smart cities.

- The connected solutions and benefits model is a new economic paradigm to link the value creation of integrated socio-technical systems to a set of different beneficiaries and types of benefits. For example, an electric car-sharing solution reduces noise in cities, frees up urban space, reduces emissions and increases personal mobility for everyone. A hybrid district energy grid reduces fossil fuel consumption, maximizes clean energy use, achieves cost effective production use and storage of energy through intelligent balancing schemes and increases the livability for city dwellers that have electricity and heat at their demand at any time.

The basic concept for leveraging the additional value of connected solutions lies within identifying all the additional benefits that come on top of the conventional efficiency model, therefore all benefits the solution creates for a range of different stakeholders, need to be identified in a first step.

The identified beneficiaries then invest their money (or use corresponding investment schemes) into the solution proportionally to the benefits that they will achieve. Usually a large part of the required return of investments (ROI) will already be generated through the efficiency model. Therefore, it is estimated that the identified beneficiaries will only need to invest a smaller share of their own estimated benefits (10–30%), making the solution highly attractive to a range of beneficiaries (see Figure 4).

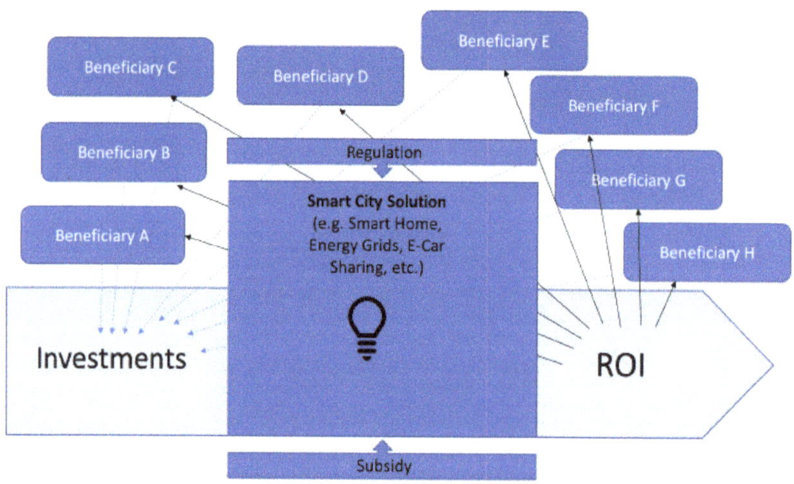

Figure 4. Connected solution and benefits model (Alanus von Radecki).

The main problem with distributed benefits and shared investments is the risk that is connected to achieving the benefits. It means that prospected benefits of smart city solutions need to be proven under reproducible circumstances in order to convince future beneficiaries to become smart city investors. This makes the connected solutions and benefits model firmly linked to smart districts—places where the positive effects of a connected solution can be proven, and where many different beneficiaries are able to create different kinds of value through the interaction of many systems and people.

3. Case Study: Demonstration District "Leipzig-West"

The following section presents preliminary technical results for a particular smart district case study in Leipzig within the framework of SPARCS, an EU-funded Smart city project [22]. An economic analysis of the case study is out of scope for the paper at hand. A short introduction is provided for the energy related parameters of the demo district and the applied operational framework (Section 3.1). Next, additional data are derived and pre-processed for the application of an optimization model (Section 3.2). In Section 3.3 the modelling results as well as deduced indicators are presented and discussed. Finally, the technical perspective of an upscaling of the demonstration district in Leipzig is outlined considering potential challenges and limitations.

3.1. Introducing the Case Study

As part of the SPARCS project, the city district "Leipzig West" is transformed into a smarter, more interactive district. To enable this process multiple deliverables are defined for the project [22]. The overarching goals comprise the set-up of a Virtual Power Plant that also embraces the integration of a community energy storage as well as community demand response measures. Additional tasks include the provision of energy services for prosumers supported by the blockchain technology, enhanced by ICT applications and user interfaces for electricity consumption monitoring.

One main element of SPARCS is to show pathways for a decarbonization of district heating. The demo district is characterized by social housing units that make up the multitude of apartment buildings. For construction of new development areas or retrofitting of existing areas multi-criteria decision making methods can be applied to achieve an optimal, energy efficient building/district design [23]. As the demo-district is made up of existing, recently retrofitted buildings, a renovation is not a viable option. The demonstration activities should enable the provision of heat based completely on renewable energy for those properties in the area that are connected to the heating grid infrastructure. For the demonstration district this condition is met by seven apartment buildings with a total of about 300 tenants, owned by the municipal housing company. According to the housing service provider, the total demand for heating and hot water supply is approx. 1300 MWh each year.

Among other measures, the installation of renewable power generation plants on the supply side of the district heating system can support the complete provision of renewable heating energy to the system. The municipal utility, Leipziger Stadtwerke (LSW), as owner and operator of the district heating grid, has decided to build and integrate a solar thermal power plant for district heating purposes. Based on the planning, the site for the solar thermal plant will take an area of about 11 ha. The solar heat supply is projected to be 25–30 GWh per year, whereas the peak power will be ca. 30 MWp. In addition, a thermal storage unit will be installed nearby with a capacity of ca. 30 MWh. To model the demand and supply behavior of the case study's energy system an Energy System Optimization Model (ESOM) is used (see Section 3.2) [24].

In [25] a review and overview of currently existing definitions of PEDs and similar concepts is given and three PED-frameworks are defined. One of them states: "PED virtual—a district that allows the implementation of virtual renewable energy systems and energy storage outside its geographical boundaries. The combined annual energy generation of the virtual renewable energy systems and the on-site renewable energy systems must, however, be greater than the annual energy demand of the district."

As the solar thermal power plant and the demonstration district are not in direct geographical vicinity, the solar thermal power plant must be viewed as a virtual renewable energy system in the above-mentioned manner. Furthermore, an annual surplus of regenerative heating energy over the district's heating demand is expected. According to this definition the considered district of the case study is a virtual PED.

In Ala-Juusela et al. [26] the concept of an energy positive neighborhood (EPN) is introduced, defined and operationalized. For this case study, the key performance indicators

(KPI) that have been developed for the operationalization of EPNs will be applied. This includes KPIs for the so-called Onsite Energy Ratio (OER), Annual Mismatch Ratio (AMR), Maximum Hourly Surplus (MHS), Maximum Hourly Deficit (MHD), and Monthly Ratio of Peak hourly demand to Lowest hourly demand (RPL). For calculating the first KPI the annual values of energy production and demand are sufficient (The KPIs are defined and measured for several forms of end-use energy, e.g., heating, cooling and electricity. We focus on heating energy in this study). However, the majority of the KPIs require input data with a higher temporal resolution, i.e., hourly values.

3.2. Input Data and Model Setup

Within this section we present a modeling approach to analyze the projected impact of the solar thermal plant on the decarbonization efforts of the demo district. In doing this, the analysis can be abstracted from the estimated date of commissioning of the plant in 2023. Moreover, the heat demand is virtually balanced against the hourly solar heat generation, enabling the calculation of district specific KPIs. In practice, the generation portfolio of the district heating system consists of several gas-fired cogeneration units, gas boilers, and thermal storage units that impede a correct allocation of single carbon-free energy sources to specific parts of demand-side.

In the first step, the input data for the modeling exercise have to be prepared. Regarding the demand-side, the annual heat demand as settled is disaggregated to derive an hourly demand profile of the relevant buildings in the demonstration district applying the Hellwig procedure [8]. This method was developed at the TU Munich in cooperation with multiple natural gas suppliers to estimate the demand of small customers and the resulting amount of gas distributed in the grid. It utilizes statistical load profiles of different customer groups to emulate the heating demand depending on ambient temperature and building parameters (age, number of tenants). Firstly, the annual heating demand is disaggregated to daily values using a normalized regression function. Secondly, the daily values are further disaggregated into hourly values by means of percentage factors contingent on ambient temperature, building parameters and the customer group, e.g., the private housing sector in this case.

With respect to the supply-side the solar thermal plant is characterized by the following technical parameters: peak power and total annual heat generation, number of solar panels, technological specifics of the system such as collector efficiency η_0 and heat loss coefficients a_1, a_2, total aperture area A as well as the temperature level of the district heating system ϑ_m—and environmental parameters—ambient temperature ϑ_a and global irradiance G at the chosen location. Given those data the hourly thermal power output $\dot{Q}$ of the plant is calculated considering the following physical relations, expressed by Equations (1) and (2):

$$\dot{Q} = A \times G \times \eta \tag{1}$$

with

$$\eta = \eta_0 - a_1 \times \frac{\vartheta_m - \vartheta_a}{G} - a_2 \times \frac{(\vartheta_m - \vartheta_a)^2}{G} \tag{2}$$

According to Ecuation (2) the optical efficiency η decreases due to thermal losses through the temperature gradient between the fluid inside the solar collector and the ambient temperature $(\vartheta_m - \vartheta_a)$ as well as the transmission losses that occur through reflection at the collector's surface. Given the hourly temperature and irradiance data from the "Deutscher Wetterdienst" for the specific location, we simulated an hourly heat output of the solar thermal power plant. [27]

In a second step, the input data are transferred to a modeling toolbox establishing the specific energy system model of the demo district. The modeling framework IRPopt (Integrated Resource Planning and optimization, version hash: 847078050cdd3535be63d564bae71b49c6b1f741; by University of Leipzig, Leipzig, Germany) is an actor-oriented techno-economical optimization program that enables the modelling of a wide variety of different energy systems. [28]. For this case study, the demand-side (property 1–7) is linked with a balancing entity (WGrid) to reflect the district heating system that is fueled by the solar thermal plant (see Figure 5). In addition, a thermal storage and a backup facility are modelled to cope with the inter-temporal variability of the solar heat supply.

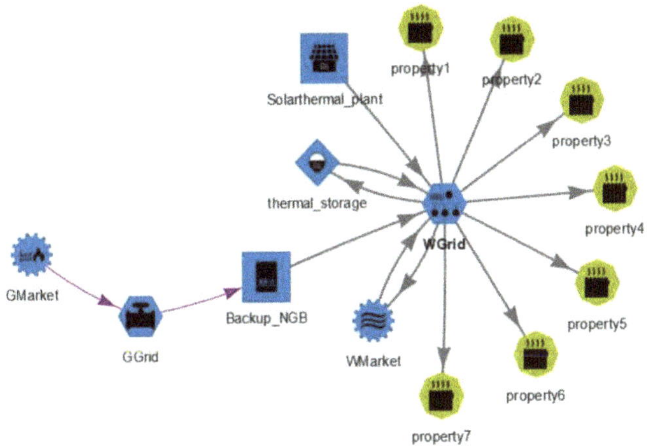

Figure 5. Model scheme of the demo district as visualized by IRPopt.

Through demand-side management (DSM), the temporal distribution of demand of the district heating system can be shaped at the municipal utility will to a certain degree. DSM utilizes the thermal storage of the buildings themselves to shift or reduce the heat demand of the customers without reducing their comfort [29]. In the demo district the technical requirements for DSM are not met. However, prospectively technical adjustments that enable DSM are possible. As a means of consideration of this possibility, the surplus thermal storage present in the system can be viewed as a proxy for DSM. Through a sensitivity analysis of the thermal storage capacity and the foresight horizon, the impact of further flexibility through storage or DSM is considered. Accordingly, the thermal storage varies from 10 MWh to 100 MWh whereas the optimization horizon which denotes the accuracy of the forecast horizon for the system operator ranges between two days (48 h) and one year (8760 h). As a reference case, this study assumes a thermal storage size of 30 MWh and an optimization horizon of 48 h.

3.3. Modelling Results and KPIs

The model output includes first and foremost the thermal energy flow between the implemented components in an hourly resolution for the reference case and the sensitivities. Based on the energy flow for one entire year the KPIs can be calculated and interpreted.

The Onsite Energy Ratio (OER) denotes the overall balance between renewable heat supply and the demand as annually aggregated values in relation to one another. For the case study the OER is 13.93, i.e., the solar thermal energy exceeds the demand almost fourteen-fold. The Maximum Hourly Surplus (MHS) and Maximum Hourly Deficit (MHD) measure by how much the hourly supply maximally exceeds the demand and vice versa. For calculating these KPIs, the hourly zero carbon heat supply and demand is balanced and normalized to the demand or supply, respectively. As a result, the MHS for the case study is 1028.33, reflecting that the supply exceeds the demand by that factor at a certain point in time. Conversely, the MHD is 20.89 indicating that the demand overshoots the supply by a factor of 20.89. The RPL denotes the maximum monthly ratio of Peak hourly demand to Lowest hourly demand. For the demo district the maximum demand ratio arises in the month of June, it is 17.76. It should be noticed that the sensitivity analysis has no influence on these four KPIs, as the total sum of supply and demand do not change by varying assumptions on the storage size or the model foresight.

The last and most elaborated KPI proposed by VTT [26] is the Annual Mismatch Ratio (AMR). For this ratio, firstly, the Hourly Mismatch Ratio is determined considering the state of charge of the storage and the demand and supply situation at every hour of the year. Secondly, the AMR is derived as an arithmetic mean of all HMR values. The AMR varies between zero—denoting a complete temporal simultaneity of energy generation and consumption—and one which hallmarks an entire mismatch of both determinants.

Given the reference case (30 MWh, 48 h), the AMR is calculated at 0.278. In Figure 6 the impact of the sensitivity analysis on the AMR is visualized depending on thermal storage size (10–100 MWh) and optimization horizon (48 h, 336 h, 1344 h, 8760 h). It shows that lower levels of AMR are achieved with a larger storage in combination with enhanced forecast accuracy (100 MWh, 1344 h). However, enlarging the storage separately does not necessarily reduce the AMR as a shorter forecast accuracy impedes the possibility to utilize the storage. As an example, fixing the optimization horizon at 48 h or 336 h, respectively, a storage capacity larger than 20 MWh or 30MWh, respectively, does not yield a further decline of the AMR.

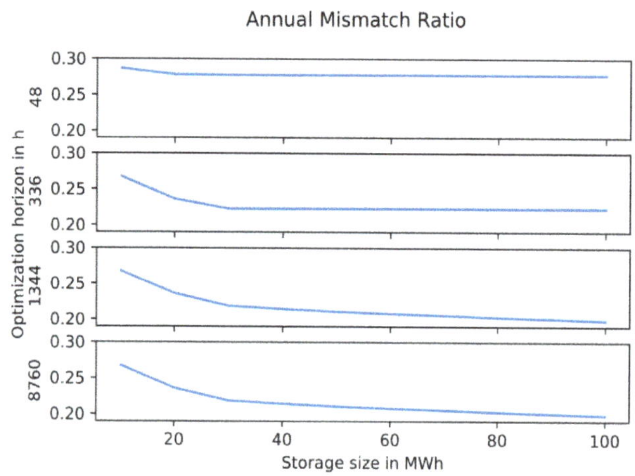

Figure 6. Annual Mismatch Ratio (AMR) for the demo district depending on the thermal storage capacity and the model foresight.

The KPIs are able to provide a first impression on the technical challenges of the specific energy system. Based on the OER, the total solar heat supply suffices to cover the total demand given. However, a higher temporal resolution reveals the daily or seasonal volatility of the heating demand of the buildings in each single month (RPL). Additionally, the MHD and MHS indicate that the solar heat overshoots the energy demand by several magnitudes and vice versa. And finally, the AMR clearly quantifies the amount of mismatched energy for the use case at hand. For the reference case there is an energy mismatch of 27.8%. By expansion of the storage volume and an improvement of the forecast horizon, this KPI can be reduced to 20%. The mismatch is triggered by the generic properties of the solar irradiation and the heat demand profile of private households in general. The installed thermal storage can be utilized as a short-term storage, to account for the time-shift between supply and demand peaks.

In the future, more housing properties in the demonstration district "Leipzig West" may be connected to the district heating system. LSW also plans to install room thermometers in the district's buildings to gain more insight into the heating behavior. The municipal utility company also develops an application to supply tenants with further information about the energy consumption and connected savings potential which may serve as a gateway to future DSM-activities. On the other hand, more or different regenerative heating solutions may be added to the district heating system, changing the structure of the demo districts energy consumption and generation and thereby the KPIs. The upscaling of the demand and supply side of the district's heating energy is partially explored in Section 3.4.

3.4. *Upscaling of the Demonstration Project in Leipzig*

As of 2017, Leipzig's district heating system relies entirely on fossil energy sources (see Figure 7). However, over the next years LSW is planning to increase the share of the regenerative heating energy production. The solar thermal power plant covered in this case study is a first step on the transition pathway. However, the mid-term strategy for 2030 requires additional investments in net zero carbon technologies for the district heating system. One option is to scale up the solution introduced in this case study's demonstration district.

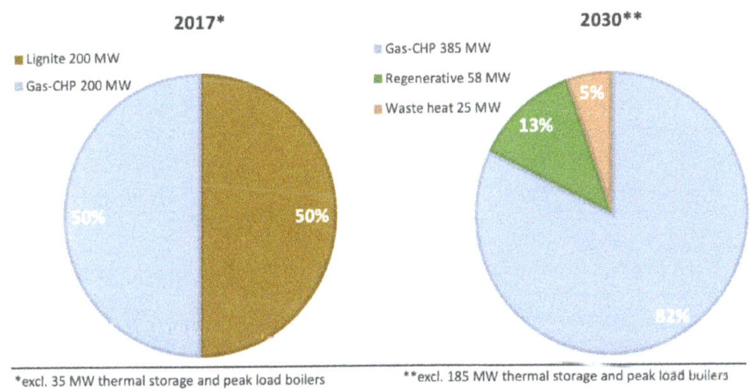

Figure 7. Projection of thermal capacity (secured power) of the district heating system in Leipzig [30]. Reproduced with permission from Leipziger Stadtwerke.

From a theoretical standpoint, the design of a pilot project is essential for the success of the scaling. Moreover, "pilots should be designed in such a way that they could be scaled up, if successful, and so that key factors which will be necessary for a scaling up decision—with what dimensions, with which approach, along which paths, etc.—are already explored during the pilot phase." [31]

3.4.1. Applying the Model of van Winden

The replication model of Willem van Winden (Section 2.1) can be applied to the upscaling process in Leipzig based on the demonstration district covered in the case study. Firstly, the upscaling Type 1—Rollout is considered. For a market rollout of solar thermal plants, the state-of-the-art technology does not have to be challenged and no new partnerships or organizational changes are necessary. In this context, it entails identifying additional districts connected to the district heating system that can potentially be supplied by the planned solar thermal power plant. The ongoing cooperation with the municipal property owner LWB ensures a smooth expansion process through repetition.

Secondly, applying an Expansion according to Type 2 of upscaling, further partners are added to the process. Regarding the demand-side, owners of existing buildings or property developers of new living quarters should be convinced to be connected to the district heating grid. Alternatively, a so-called geographical supply expansion is considered. For this approach, new properties have to be identified that fulfil the technical and geographical requirements for the installation of further solar thermal power plants, e.g., distance to the district heating grid.

The third option for upscaling the solution, a defined replication, is discussed in Section 4 of the paper at hand.

3.4.2. Challenges, Limitations and Alternative Solutions

Solar thermal plants rely on the local irradiation mainly and require only minor auxiliary fuels to run the pumps for heat generation. Thus, the operational cost in cent per kWh is relatively low whereas investment cost (€ per kW installed capacity) is quite high in comparison to natural gas boilers. A recent case study in Vienna reveals that similar large scale solar thermal plants would not be economically viable without a subsidy [32].

To increase or enable the economic feasibility of the project, it is important to identify viable business models from the beginning [33]. One possibility here is the introduction of special heating tariffs to the district heating customers by LSW. The cost of a special tariff can exceed the usual heating tariff cost while offering the customer the opportunity to contribute to a sustainable future. Research on household preferences shows that customers′ willingness to pay for district heating from renewable sources proves to be larger than for other heating options [34]. The special "green" tariffs can ensure a certain percentage of regenerative heat in the generation mix. In this case, the heat provider has to ascertain that the sales of "green" heat do not exceed the energy produced by the solar thermal plant in balance. (As district heating is a monopoly market for technical reasons, the price can basically be chosen freely. However, there is a danger of customers being dissatisfied or switching to other forms of heating, for which high market entry cost have to be paid.). As district heating is a monopoly market for technical reasons, the price can basically be chosen freely. However, there is a danger of customers being dissatisfied or switching to other forms of heating, for which high market entry cost have to be paid

Other business models on the topic of solar thermal heating—like public, financial participation or custom-made heating tariffs for commercial customers—have been developed in recent years and can further enhance the project's feasibility. [35]

The expansion of the district heating grid infrastructure plays a major role for the upscaling of the heat demand, where multiple issues can arise. The district heating system is a complex technical system, extensions have to be carefully planned both from a technical and economical perspective. Considering a yet unconnected district for linking to the system, the following aspects should be taken into account: The district should be preferably in close geographical proximity to the existing system. Furthermore, the grid operator should be aware of the city-wide distribution of heating demand [36]. Districts identified for integration featuring a high demand potential are also financially attractive. Lastly, it is also significant to be aware of socio-economic factors in the considered district. An evaluation on relevance of players and the willingness to switch to district heating, including exploration of possible partnerships or conflicts of interest should be done beforehand.

The results of the case study suggest some limitations for the heating solution. As elaborated in Section 3.3 the energy production of the solar thermal plant does not coincide with the energy consumption of the customers completely. This temporal mismatch cannot be solved by scaling up the supply and demand through roll-out or expansion. If scaled-up to the entire city of Leipzig, the AMR (Section 3.3) would still remain at values comparable to the demonstration district. To improve the simultaneity of energy production and consumption other technologies or larger thermal storages have to be implemented in the district heating system. A methodology designed to find the most fitting optimization solution is introduced in [37]. Here, an eight-step program is presented to support city planners in finding the energy performance solution to transform the considered city district into a PED.

Firstly, the baseline and structure of the considered district are analyzed, then possible energy performance measures are identified and economically evaluated. Based on this, the district's energy demand is calculated and energy system alternatives are defined. Finally, a cost benefit calculation is executed and an optimal combination of the district's buildings, energy performance and energy system measures is determined. [37] The application of this methodology supports the identification of additional regenerative heating energy solutions for the demonstration district in Leipzig West, e.g., geothermal heat, biogas power plants, heat pumps powered by green electricity or thermal power stations powered by green hydrogen. LSW plans to incorporate some of the stated technologies into the district heating system in the future. [38]

In the following section an insight into the upscaling on an international/inter-city level is presented. The replication of the case study for SPARCS fellow cities will be illuminated, continuing the consideration of challenges and introduction of possible solutions.

4. Replication Challenges and Solutions in Fellow Cities

As part of the project, SPARCS lighthouse cities need to prepare and test the solutions which later can be replicated or adapted by fellow cities. Aim of this replication strategy is to simplify the application of sustainable solutions for the follower cities.

In the previous section we demonstrated how a city district "Leipzig West" is transformed into a smarter, more interactive district, and examined the challenges which may occur during the upscaling of the district in the city of Leipzig.

In this section, we will see how the replication process in SPARCS is organized, how the demo district fits into the SPARCS replication strategy (Section 4.1), which challenges can occur during the adaptation of the demonstrated solution and what strategies can be applied in order to overcome those (Sections 4.2–4.4).

4.1. Replication Process in SPARCS

The replication process in SPARCS is based on the Morgenstadt assessment framework and divided into the following steps:

- Phase 1. Preparation: creating an individual city profile highlighting the drivers, barriers and opportunities. This phase includes the desktop research and preliminary analysis, as well as the selection of the experts and team members.
- Phase 2. Understanding: detailed analysis of specific sectors relevant to SPARCS (e.g., energy—see below). This phase consists of the analysis of strategic documents and plans of the city relevant to the energy sector and the data collection through online research and desktop analysis with the following identification of the gaps.
- Phases 3 and 4. Co-creation and Design: the goal of this phase is to formulate sustainability solutions for the city followed by the implementation plan. Ideally, the solutions are already developed and tested in the Lighthouse cities.

Since the SPARCS project is focused on energy and related impacts (e.g., building), a carefully considered selection of indicators and action fields from the original framework related to these sectors was carried out. For the detailed assessment of the energy sector the cities are presented with an indicator selection as seen in Table 3. These indicators assess municipal energy generation and distribution with respect to renewables share, networks for intersectoral resource sharing and the existence of district heating. The cities then could choose those indicators that are most relevant for them or have a consistent history of data collection.

By comparing a subset of the indicators from three other cities—Reykjavik, Kladno and Maia—and evaluating them—we can make some assumptions and suggestions for further replication (Table 4).

Reykjavik performs well in the energy sector: 100% of energy demand is covered by renewable sources, and 100% of heating demand is served by district heating systems. However, one area for improvement is decreasing energy use per household. This can be done by installation of smart meters, sensors, time switchers and other demand-management technologies as well as building refurbishment.

Table 3. Overview of indicators for data collection for SPARCS fellow cities.

Indicator	Description	Unit
Rate of building refurbishment	Annual rate of refurbishment as a percentage (%) of existing building stock	%
Total energy demand per capita	Total energy use of the city (GWh/year) divided by inhabitants	MWh/a/cap
Total electricity consumption per capita	Total energy use per capita (MWh/year) ISO 37120-7.5	kWh/a/cap
Electricity consumption per household	kWh/a/cap	kWh/a/cap
Share of power produced within the city in the grid	Share of power produced within the city in the grid	% of total electricity
Share of the renewable energy in the grid	Share of the renewable energy in the grid	% of total electricity
Electricity price	Average electricity price for private consumers	€ per kWh
Natural gas price	Average price for natural gas for private consumers	€ per kWh per square meter
Utilization of local district heating	Share of heat demand delivered by local district heating systems	%

Table 4. Sample energy indicators for SPARCS fellow cities: Reykjavik, Kladno and Maia.

Indicator	Reykjavik	Kladno	Maia
Electricity consumption per household (kWh/household/year)	3700	1006	3400
Share of energy demand covered by RES (% of end energy demand)	100	5	26.5
Share of electricity demand generated by RES (% of electricity demand)	100	4.75	45
Share of heat demand generated by district heating systems (%)	100	71	N/A

Maia also demonstrates a good degree of sustainability in its energy and electricity provision: The share of demand covered through renewable energy is 26.5%. However, to reach the target of complete decarbonization, this share could be increased. The measures that could be taken are, for instance, installing solar panels on the residential buildings, or conversion of the automobile fleet to electric mobility. Additionally, the city of Maia could concentrate on the holistic promotion of renewable energies and raising the awareness of energy saving measures among the citizens. Investors can be motivated to install or use renewable energies in new buildings or when refurbishing them with new strategic pilot projects.

In contrast to this, the performance of Kladno in the energy sector has a lot of potential for improvement. According to the energy indicators presented in Table 4, only 4.75% of electricity demand is covered by renewable sources. The city's main energy supplier providing 343 MW electrical capacity and 173 MW thermal capacity is a coal-fired power plant [39]. Even though the location of the power plant allows for disposing of most of the generated pollutants outside the city, the global environmental effect of using coal as the energy source cannot be ignored. Thus, the city should first and foremost focus on expanding the share of renewable energy sources on energy consumption supported by energy efficiency measures, e.g., building insulation, to reduce the security margin on thermal capacity and temperature levels of the district heating grid.

If we now examine whether the solution demonstrated in our case study (Section 3) can be adopted by the fellow cities it becomes evident that Reykjavik's energy demand is already completely covered by renewables and Maia does not possess a district heating system which is a crucial element of Leipzig's solution. The solar thermal plant, however, suits Kladno as it gives the city an opportunity to increase the share of renewable energy generation supplying the district heating grid.

Which possible challenges can occur during this replication?

According to van Winden's model (see Section 2.1) this type of replication falls into the Type 3—the most complex type of scaling—as the solution needs to be replicated in another context (another country and city) with possible legal-, organizational- or partnership-related obstacles.

4.2. Legislation Barriers for Replication in Kladno

In order to combat CO_2 emissions, the Czech Republic/Czechia created the 'National Energy and Climate Plan'. It aims to meet EU targets and announced to reduce its total greenhouse gas emissions by 30% by 2030 compared to 2005. This decarbonization goal also includes the shift to renewable energy sources. In the Czech Republic's/Czechia's previous "Energy Action Plan", it set renewable energy targets for 2020. It aims to meet 14% of the heating and cooling demand by renewable energy sources by 2020 and 22% by 2030. Additionally, 14% of the electricity demand shall be produced by renewable energy sources and 11% of the energy demand in the transport sector stem from renewable sources by 2020 [40].

Even though the legislation package in Kladno looks particularly well in terms of the promotion of renewable energies, the policies supporting renewables are still to be implemented. In August 2013, the Czech Parliament amended Act No. 165/2012 (Act No. 310/2013 Coll.), which de facto abolished the feed-in tariff scheme for all technologies except small hydro since the end of 2013. New PV installations and biogas plants are only being supported if put into operation before 31 December 2013 (§ 4 par. 10 Act No. 165/2012) [41].

According to the EU Smart City Information System, at present rather ordinary and sub-optimal solutions are commonly being implemented in smart city projects only because the initial (and better) solution was blocked off by legal bottlenecks [7]. This diminishes the effect of the replication, affects the development of performant business models and as a result slows down the process of driving cities towards sustainability goals.

Coming across such legislation and/or regulatory bottlenecks, policymakers need to be aware that changes in legislation will take time and involve social, cultural, political, institutional and behavioral changes that are very context sensitive.

4.3. Challenges Related to Stakeholder Engagement

In order to make the solution appealing to potential investors, the Connected Solutions and Benefits Model (see Section 2.3) should be applied, relating the economic impact to the projected social benefits and making the solution more attractive to a variety of beneficiaries. Mutual understanding is a key factor to exploit potential cooperation with investors and relevant stakeholders (e.g., developers, distributors, engineers, spatial planners). Even though Positive Energy District might seem like a logical and understandable concept, for citizens or potential partners who are not very much familiar with modern approaches in the energy sector it is necessary to explain the concept in a short and structured way.

Demonstrating the purpose and potential social benefits of the PED (a comfortable living space, a well-organized recreational area, modern urban services for citizens, opportunities for entrepreneurship, etc.) is crucial for negotiations with the relevant stakeholders and establishing a profitable business-model. Even district heating systems themselves are still only emerging in some cities and countries. People are skeptical about trying something "new" and "different" to what they are used to. If for example, the heat market is unregulated and has historically gained a bad reputation as inefficient, expensive and unreliable, it can be extra difficult to shift people's perceptions [38]. It is therefore important to engage (potential) customers as well as other key stakeholders at an early stage to facilitate the expansion of district heating and cooling systems.

4.4. Urban Planning Challenges

For PED replication, the urban area plays an important role. New development areas where no buildings exist yet are good prerequisites for spatial planning to steer PED replication, as the PED can be planned to integrate into other development interests of the area, prior to the implementation of the buildings and infrastructure. [39] When a PED project takes place in existing urban environments, there is often a vast number of stakeholders such as citizens and building-owners. In this case, the replication process strongly depends on cooperation with stakeholders. Once the implementation area is determined, financial schemes or innovative business models for the deployment take place.

5. Conclusions

This paper contributes to the existing research on replication of smart city projects by investigating the common replication models and strategies. The paper also addresses several challenges that prevent positive energy districts from replicating effectively and proposes solutions to resolve them, using a specific decarbonization measure from the city of Leipzig as an example.

One of the key premises of the paper, as explicated in Section 2, is the necessity of a profound replication planning and modelling to deepen the understanding of upscaling processes for smart city solutions, complemented by having a shared context for city-to-city collaboration. The three replication models analyzed in this article are able to provide a multidimensional representation of a potential solution for reaching PED status. Nonetheless, depending on the ambition of decision makers and the environmental conditions different types of replication are applicable.

Before implementing and commissioning the solution in the real-world, energy system modeling provides first insights into the technical potential of integrating renewable energy sources in combination with storage options. As minimum prerequisites of the analysis, we determined the hourly demand of thermal energy services of the demonstration district, as well as the heat supply pattern of the solar thermal plant. As a basis for discussing results with respect to replication compatibility we provide a set of KPIs to capture the demonstration district's provision with renewable thermal energy. It becomes evident that even though the annual total generation exceeds the consumption and supports a weak definition of PED, the temporal mismatch between supply and demand, depicted by an Annual Mismatch Ratio of 27.8%, remains despite thermal storage use. Prospectively, the utilization of actual data from metering services in the district and the solar thermal plant will enable a retrospective reality check of the model-based KPIs.

Additional findings derived by an application of van Winden's upscaling model to Leipzig based on the demo district leads to various challenges including the identification of viable business models for project feasibility (for Type 1: Roll-out) and the expansion of the district heating system in a beneficial way (for Type 2: Expansion). The introduction of a special heating tariff for providing regenerative heat supports the integration of districts on the demand side and different regenerative heating technologies on the supply side. Concluding from the KPIs of the demo district a more versatile heat generation mix is indispensable for a full transition to an environmentally sustainable heat provision and CO_2-neutral districts (PED).

In Section 4, data from the in-depth Morgenstadt assessment of the SPARCS fellow cities is used to apply van Winden's model for the replication of the solution demonstrated in the case study. However, the solution only proves possible for the city of Kladno, which utilizes a district heating system and has improvement potential in regenerative energy provision.

In addition to the challenges that occur during scaling up in Leipzig, several further potential challenges (e.g., challenges related to the stakeholder engagement, legislation barriers) were identified for Kladno as well as ways to address them in the replication process.

Concluding the article, it is important to once again to highlight the fact that replication is a very complex and context-sensitive topic. There is no one-size-fits-all solution as every single city has its own specific requirements. However, certain ways to speed up the process of repeating and scaling up successful projects, to accelerate the market for low-carbon smart city solutions, as well as certain mechanisms to allow and simplify the replication of those solutions do exist and should be more widely known/spread/used within smart city projects across the EU. Having said that, the authors would recommend planning and modelling the replication of a smart city project at the very early stage as it is important to find tailor-made solutions that fit the spatial, legislative, socio-economic conditions and historical growth of the cities.

Author Contributions: Conceptualization, K.S. and D.U.; methodology, K.S. and D.U.; validation, K.S. and D.U.; formal analysis, H.K. and T.B.; writing—original draft preparation, K.S. and D.U.; writing—review and editing, H.K.; supervision, T.B. and H.K. All authors have read and agreed to the published version of the manuscript.

Funding: This research was funded by SPARCS Project. The SPARCS Project has received funding from the European Union's Horizon 2020 research and innovation programme under grant agreement No. 864242.

Institutional Review Board Statement: Not applicable.

Informed Consent Statement: Not applicable.

Data Availability Statement: The data presented in this study are available on request from the corresponding author. The data are not publicly available due to security reasons.

Conflicts of Interest: The authors declare no conflict of interest.

Abbreviations

AMR	Annual Mismatch Ratio
DSM	Demand-Side Management
EU	European Union
EPBD	Energy Performance of Buildings directive
EPN	Energy Positive Neighborhood
ICT	Information and Communication Technology
IAO	(German: Institut für Arbeitswirtschaft und Organization) Institute for Industrial Engineering
KPI	Key Performance Indicators
LED	Light-Emitting Diode
LSW	(German: Leipziger Stadtwerke) Leipzig Municipal Utilities
MHS	Maximum Hourly Surplus
MHD	Maximum Hourly Deficit
RPL	Monthly Ratio of Peak hourly demand to Lowest hourly demand
OER	Onsite Energy Ratio
PED	Positive Energy District
R&D	Research and Development
ROI	Return of Investments
SCIS	Smart City Information System
SET	Strategic Energy Technology
SPARCS	Sustainable energy Positive & zero cARbon CommunitieS
VTT	Technical Research Centre of Finland Ltd.

Appendix A. List of urban indicators for Leipzig within the City-Index (Morgenstadt Framework)

Appendix A.1. Livable City

Poverty rate (social assistance according to SGB II, III, XII), %—1.10%
Unemployment rate, %—9.30%
Rental and service charges, % of household income—34.79%
Doctors per 100,000 inhabitants—219.11
Life expectancy of newborns—80.3
Burglary rate per 100,000 inhabitants p.a.—266
Private car per capita—0.38

Use of public transport per capita p.a.—250.24
Situation for cyclists ADFC wheel index—3.61
Air quality (according to LBI) Air pollution index—4.6
Green and water areas, hectares per 100,000 inhabitants—3063.55

Appendix A.2. Resilient City

Debt service ratio, %—1.88
Independent income (as a percentage of total income), %—58.00%
Share of the three largest employers in total employment, %—1.81%
Emergency plans for various natural disasters, green/yellow/red—yellow
Provisions in the budget for catastrophes, green/yellow/red—green
Climate adaptation strategy, green/yellow/red—yellow

Appendix A.3. Environmentally Conscious City

Greenhouse gas emissions, tons per capita—6.61
Share of renewable energies in own energy production, %—30.00%
Waste volume, kilo per capita p.a.—291,02
Water consumption, daily consumption per capita—121.8
Recycling rate for solid waste, %—38%

Appendix A.4. Innovative City

Difference between new and abandoned businesses, three-year average per 1000 inhabitants—132.42
Share of highly qualified jobs in total labor market, %—33.70%
Number of new patents per 100,000 inhabitants p.a.—15.61
Proportion of students in the total population, in % of total population—6.59%
Smart City or innovation strategy, Yes/No—yes
Research institutions for experimental and innovative technologies and applications, Yes/No—yes

References

1. United Nations. 68% of the World Population Projected to Live in Urban Areas by 2050. Available online: https://www.un.org/development/desa/en/news/population/2018-revision-of-world-urbanization-prospects.html (accessed on 7 December 2020).
2. United Nations Climate Change. The Paris Agreement. Available online: https://unfccc.int/process-and-meetings/the-paris-agreement/the-paris-agreement (accessed on 7 December 2020).
3. Garrido-Marijuan, A. The Making of a Smart City: Best Practices Across Europe 2017. Available online: https://smartcities-infosystem.eu/sites/default/files/document/the_making_of_a_smart_city_-_best_practices_across_europe.pdf (accessed on 7 December 2020).
4. IEC Annual Report 2014. Available online: https://basecamp.iec.ch/download/iec-annual-report-2014/ (accessed on 1 December 2020).
5. European Commision. Smart Cities—Smart living: EU Policy. Available online: https://ec.europa.eu/digital-single-market/en/about-smart-cities (accessed on 7 December 2020).
6. Borsboom-van Beurden, J.; Kallaos, J.; Gindroz, B.; Costa, S.; Riegler, J. Smart City Guidance Package. Available online: https://smart-cities-marketplace.ec.europa.eu/news/smart-city-guidance-package (accessed on 7 December 2020).
7. Kaiser, G.; Remmele, B. Why May Replication (Not) be Happening. Recommendations on EU R&I and Regulatory Policies. 2018. Available online: https://smartcities-infosystem.eu/sites/www.smartcities-infosystem.eu/files/document/scis_policy_paper_1_replication_web.pdf (accessed on 7 December 2020).
8. Hellwig, M. Entwicklung und Anwendung Parametrisierter Standard-Lastprofile. Ph.D. Thesis, Technical University of Munich, Munich, Germany, 2003.
9. ODYSSEE-MURE. Energy Efficiency Trends in Buildings: Policy Brief. Available online: https://www.odyssee-mure.eu/publications/policy-brief/buildings-energy-efficiency-trends.pdf (accessed on 7 December 2020).
10. European Commission. Commission Welcomes Final Vote on Energy Performance of Buildings: EU Press Release. Available online: https://ec.europa.eu/commission/presscorner/detail/en/IP_18_3374 (accessed on 7 December 2020).
11. Temporary Working Group of the European Strategic Energy Technology. Implementation Plan on Action 3.2 "Smart Cities and Communities". Available online: https://setis.ec.europa.eu/system/files/setplan_smartcities_implementationplan.pdf (accessed on 7 December 2020).

12. Hinterberger, R.; Gollner, C.; Noll, M.; Meyer, S.; Schwarz, H.-G. White Paper on Reference Framework for Positive Energy Districts and Neighbourhoods. Key Lessons from National Consultations. 2020. Available online: https://jpi-urbaneurope.eu/wp-content/uploads/2020/04/White-Paper-PED-Framework-Definition-2020323-final.pdf (accessed on 7 December 2020).
13. Davis, K. Towards an ICT Infrastructure for Energy Positive Neighbourhoods: Report from ELSA Thematic Working Group on ICT for Energy Efficiency. 2009. Available online: http://ec.europa.eu/information_society/activities/sustainable_growth/docs/elsa/elsa_1/ELSA-EnergyPositive-Report1.pdf (accessed on 7 December 2020).
14. Cambridge University Press. Definition from the Cambridge Advanced Learner's Dictionary & Thesaurus ©. Available online: https://dictionary.cambridge.org/dictionary/english/replication (accessed on 7 December 2020).
15. Manville, C.; Milliard, J.; Liebe, A.; Massink, R. Mapping Smart Cities in the EU. Policy Department A: Economic and Scientific Policy. 2014. Available online: https://www.europarl.europa.eu/RegData/etudes/etudes/join/2014/507480/IPOL-ITRE_ET(2014)507480_EN.pdf (accessed on 7 December 2020).
16. Van Winden, W.; van den Buuse, D. Smart City Pilot Projects: Exploring the Dimensions and Conditions of Scaling Up. *J. Urban Technol.* **2017**, *24*, 51–72. [CrossRef]
17. Van Winden, W. Smart City Pilot Projects, Scaling Up or Fading Out? Experiences from Amsterdam. In Proceedings of the Regional Studies Association, Annual Conference, Graz, Austria, 3–6 April 2016.
18. Radecki, A.; Tettenborn, F.; Mohr, M.; Hiessl, H.; Hedmans, T.; Knudsen, M.; Siedlarek, L.; Wahren, S.; Loleit, M.; Raiber, S.; et al. *Morgenstadt: City Insights—Final Report*; Fraunhofer Institute for Industrial Engineering IAO: Stuttgart, Germany, 2013.
19. Von Radecki, A.; Pfau-Weller, N.; Domzalski, O.; Vollmar, R. *Morgenstadt City Index*; Online Documentation; Fraunhofer Institute for Industrial Engineering IAO: Stuttgart, Germany, 2013.
20. Von Radecki, A. *Smart City Replication Framework*; Fraunhofer Institute for Industrial Engineering IAO: Stuttgart, Germany, 2015.
21. Gelles, G.M.; Mitchell, D.W. Returns to Scale and Economies of Scale: Further Observations. *J. Econ. Educ.* **1996**, *27*, 259–261. [CrossRef]
22. SPARCS: Sustainable Energy Positive & Zero cARbon CommunitieS. Available online: https://www.sparcs.info/ (accessed on 7 December 2020).
23. Mela, K.; Tiainen, T.; Heinisuo, M. Comparative study of multiple criteria decision making methods for building design. *Adv. Eng. Inform.* **2012**, *26*, 716–726. [CrossRef]
24. Scheller, F.; Bruckner, T. Energy system optimization at the municipal level: An analysis of modeling approaches and challenges. *Renew. Sustain. Energy Rev.* **2019**, *105*, 444–461. [CrossRef]
25. Lindholm, O.; Rehman, H.U.; Reda, F. Positioning Positive Energy Districts in European Cities. *Buildings* **2021**, *11*, 19. [CrossRef]
26. Ala-Juusela, M.; Crosbie, T.; Hukkalainen, M. Defining and operationalising the concept of an energy positive neighbourhood. *Energy Convers. Manag.* **2016**, *125*, 133–140. [CrossRef]
27. Corradini, R.; Sutter, M.; Leukefeld, T. (Eds.) *Solarthermie—Technik, Potenziale, Wirtschaftlichkeit und Ökobilanz für Solarthermische Systeme in Einfamilienhäuser*; Wüstenrot Stiftung: Ludwigsburg, Germany, 2014; ISBN 978-3-933249-89-0.
28. Scheller, F.; Burgenmeister, B.; Kondziella, H.; Kühne, S.; Reichelt, D.G.; Bruckner, T. Towards integrated multi-modal municipal energy systems: An actor-oriented optimization approach. *Appl. Energy* **2018**, *228*, 2009–2023. [CrossRef]
29. Kärkkäinen, S.; Sipilä, K.; Pirvola, L.; Esterinen, J.; Eriksson, E.; Soikkeli, S.; Nuitinen, M.; Aarnio, H.; Schmitt, F.; Eisgruber, C. Demand Side Management of the District Heating Systems. Research Notes No. 2247. 2004. Available online: https://www.vttresearch.com/sites/default/files/pdf/tiedotteet/2004/T2247.pdf (accessed on 4 December 2020).
30. Leipziger Stadtwerke. Umweltbericht für die Fernwärmeversorgung der Leipziger Stadtwerke. Available online: https://www.l.de/file/download/c/cdce09ec92c8b3a460dac5bd44d682.pdf (accessed on 10 December 2020).
31. Hartmann, A.; Linn, J.F. Scaling Up: A Framework and Lessons for Development Effectiveness from Literature and Practice. *SSRN J.* **2008**. [CrossRef]
32. Zinganell, C. Ökonomische Analyse von Unterschiedlichen Solarthermischen Systemen für Eine Einbindung in das Bestehende Fernwärmenetz der Fernwärme Wien. Ph.D. Thesis, University of Natural Resources and Life Sciences, Vienna, Austira, 2013.
33. Van Winden, W.; Oskam, I.; van den Buuse, D.; Schrama, W.; van Dijck, E.-J.; van den Buuse, D. Organising Smart City Projects: Lessons learned from Amsterdam. In *Role of Stakeholder Interaction in Collaborative Business Modelling for Sustainability*; Eburon: Amsterdam, The Netherlands, 2016.
34. Krikser, T.; Profeta, A.; Grimm, S.; Huther, H. Willingness-to-Pay for District Heating from Renewables of Private Households in Germany. *Sustainability* **2020**, *12*, 4129. [CrossRef]
35. Solarthermaworld. SHC Solar Award: Six Successful Business Models. Available online: https://www.solarthermalworld.org/news/shc-solar-award-six-successful-business-models (accessed on 4 December 2020).
36. Dorotić, H.; Novosel, T.; Duić, N.; Pukšec, T. Heat Demand Mapping and District Heating Grid Expansion Analysis: Case Study of Velika Gorica. In *E3s Web Conference, Proceedings of the International Conference Energy, Environment and Material Systems (EEMS 2017), Polanica-Zdroj, Poland, 13–15 September 2017*; EDP Sciences—Web of Conferences: Paris, France, 2017. [CrossRef]
37. Shnapp, S.; Paci, D.; Bertoldi, P. *Enabling Positive Energy Districts across Europe: Energy Efficiency Couples Renewable Energy*; JRC Technical Report; European Commission: Brussels, Belgium, 2020.
38. Leipziger Stadtwerke. Die Zukunft der Leipziger Fernwärme—Leipziger Stadtwerke. Available online: https://zukunft-fernwaerme.de/ (accessed on 10 December 2020).

39. Alpiq Group. Czech Power Station ECKG. Available online: https://www.alpiq.cz/en/alpiq-group/media-relations/media-releases/media-release-detail/czech-power-station-eckg (accessed on 7 December 2020).
40. Smrz, M. National Energy Action Plan of Czech Republic. Available online: https://www.eurosolar.de/en/index.php/sections-eurosolar/czech-republic/781-national-energy-action-plan-of-czech-republic-eurosolar (accessed on 10 December 2020).
41. Legal Source on Renewable Energy. Feed-in Tariff. Available online: http://www.res-legal.eu/search-by-country/czech-republic/single/s/res-e/t/promotion/aid/feed-in-tariff-act-on-the-promotion-of-the-use-of-res/lastp/119/ (accessed on 10 December 2020).

buildings

MDPI

Article

Integrating Plus Energy Buildings and Districts with the EU Energy Community Framework: Regulatory Opportunities, Barriers and Technological Solutions

Andreas Tuerk [1,*], Dorian Frieden [1], Camilla Neumann [1], Konstantinos Latanis [2], Anastasios Tsitsanis [2], Spyridon Kousouris [2], Javier Llorente [3], Ismo Heimonen [4], Francesco Reda [4], Mia Ala-Juusela [4], Koen Allaerts [5], Chris Caerts [5], Thomas Schwarzl [6], Martin Ulbrich [6], Annette Stosch [6] and Thomas Ramschak [7]

1 Joanneum Research, 8010 Graz, Austria; dorian.frieden@joanneum.at (D.F.); camilla.neumann@joanneum.at (C.N.)
2 Suite5, Limassol 3013, Cyprus; latanis@suite5.eu (K.L.); tasos@suite5.eu (A.T.); spiros@suite5.eu (S.K.)
3 CENER Renewables Energies Centre of Spain, 31621 Navarra, Spain; jllorente@cener.com
4 VTT Technical Research Centre of Finland Ltd., 02044 Oulu, Finland; Ismo.Heimonen@vtt.fi (I.H.); francesco.reda@vtt.fi (F.R.); mia.ala-juusela@vtt.fi (M.A.-J.)
5 VITO/EnergyVille, 3600 Genk, Belgium; koen.allaerts@vito.be (K.A.); chris.caerts@vito.be (C.C.)
6 Schwarzl IT, 8020 Graz, Austria; office@schwarzl-it.com (T.S.); office@raumdruck.at (M.U.); anny@schwarzl-it.com (A.S.)
7 AEE INTEC, 8200 Gleisdorf, Austria; t.ramschak@aee.at
* Correspondence: andreas.tuerk@joanneum.at; Tel.: +43-3168767650

Citation: Tuerk, A.; Frieden, D.; Neumann, C.; Latanis, K.; Tsitsanis, A.; Kousouris, S.; Llorente, J.; Heimonen, I.; Reda, F.; Ala-Juusela, M.; et al. Integrating Plus Energy Buildings and Districts with the EU Energy Community Framework: Regulatory Opportunities, Barriers and Technological Solutions. *Buildings* **2021**, *11*, 468. https://doi.org/10.3390/buildings11100468

Academic Editor: Fabrizio Ascione

Received: 1 July 2021
Accepted: 19 September 2021
Published: 12 October 2021

Publisher's Note: MDPI stays neutral with regard to jurisdictional claims in published maps and institutional affiliations.

Abstract: The aim of this paper is to assess opportunities the Clean Energy Package provides for Plus Energy Buildings (PEBs) and Plus Energy Districts (PEDs) regarding their economic optimization and market integration, possibly leading to new use cases and revenue streams. At the same time, insights into regulatory limitations at the national level in transposing the set of EU Clean Energy Package provisions are shown. The paper illustrates that the concepts of PEBs and PEDs are in principle compatible with the EU energy community concepts, as they relate to technical characteristics while energy communities provide a legal and regulatory framework for the organization and governance of a community, at the same time providing new regulatory space for specific activities and market integration. To realize new use cases, innovative ICT approaches are needed for a range of actors actively involved in creating and operating energy communities as presented in the paper. The paper discusses a range of different options to realize PEBs and PEDs as energy communities based on the H2020 EXCESS project. It concludes, however, that currently the transposition of the Clean Energy Package by the EU Member States is incomplete and limiting and as a consequence, in the short term, the full potential of PEBs and PEDs cannot be exploited.

Keywords: plus energy buildings; plus energy districts; energy communities; energy sharing; energy trading; clean energy package

1. Introduction

In 2019, the "Clean Energy for all Europeans" Package (in the following Clean Energy Package or CEP), a set of directives and regulations, was adopted, among others, to deliver on the EU's Paris Agreement commitment to reduce greenhouse gas (GHG) emissions by 40% compared to 1990 by 2030 [1]. In 2019, the EU 2030 GHG emissions target was raised to 55% as part of the EU Green Deal [2]. Given the ambitious EU climate targets for 2030 and the high relevance of the building sector for emission reductions, the concepts of Plus (or "positive") Energy Buildings and Districts (PEBs and PEDs) are gaining increasing attention. While the Nearly Zero Energy Buildings (NZEBs) concept is already part of EU legislation, there are at least indicative policy goals also for PEDs: The Strategic Energy Technology (SET) plan aims to implement about 100 PEDs in Europe by 2025 [3]. PED concepts are not new, however, there were major regulatory constraints regarding the interaction between individual buildings and with the energy system and market. Thus, except for planning and optimization from the energy system perspective, PEDs could rather be seen as the sum of their individual buildings and onsite technologies than as a highly interacting and flexible system. In addition, suitable organizational models for actors related to PEDs are not sufficiently available yet.

This may change with the Clean Energy Package that includes energy communities as new regulatory and organizational formats for, among others, collective decentralized renewable energy generation and consumption. This new concept may well serve to frame the implementation of PEBs and PEDs and support them in fulfilling their function as an active element in the energy system. Besides the mere generation of surplus energy, this may include multiple roles for using technologies and addressing the broader integration in the energy system. Thereby, the possible functions of PEBs and PEDs for decarbonisation could be optimized by going beyond just maximizing onsite self-consumption and sharing excess energy.

The concept for energy communities is enshrined in the recast of the renewable energy directive (REDII; defining "Renewable Energy Communities") and the new electricity market directive (EMD; defining "Citizen Energy Communities"). Besides providing organizational frameworks and new legal opportunities regarding specific rights for energy communities to act in the energy market, the EMD also strengthens the market access of aggregators and the provision of flexibility in the energy system in general. The EU energy community frameworks specifically include provisions on the possibility to share energy within a community, including through the public grid. Thus, energy communities provide room for new activities of PEBs and PEDs as the internal exchange of energy is no longer limited to the building level but opened to other buildings, the district and the market. Thereby, new revenue streams and optimization options could improve the business case of PEBs and PEDs and may lead to a faster market roll-out. The link between PEBs and PEDs and the new regulatory opportunities of the Clean Energy Package has already been addressed to some extent in the recent literature (see Section 3), so far, however, no comprehensive assessment has been made. While the aims of the different directives of the Clean Energy Package may be synergetic with PEB and PED aims, further analysis is needed on concrete use cases, technology solutions and organizational formats.

The aim of this paper is therefore to systematically analyse possible new opportunities and limitations for PEBs and PEDs based on the EU framework for energy communities and related national transpositions. The paper is written in the context of the H2020 project EXCESS on Plus Energy Buildings but also takes into account insights from a range of other ongoing H2020 projects including projects on energy communities in which the authors of this paper are involved.

2. Ambition and Methods

So far, assessments on the relationship between the Clean Energy Package and PEBs/PEDs were only partially made, not taking into account the full range of new regulatory options and not investigating the synergies of the new directives in the context of PEBs and PEDs. These previous assessments include Østergaard Jensen et al. who emphasize the important role of flexible buildings to provide local flexibilities to DSOs [4]. Shnapp et al. [5] relate the minimum energy requirements for districts and buildings as defined in the Energy Efficiency of Buildings Directive (EPBD) to the definition of Renewable Energy Communities. Karg [6] considers PEDs as one category of energy communities, while Tuerk [7] mentions PEDs as a potential nucleus for energy communities. Magrini et al. [8] linked PEDs to the prosumer concept of the Clean Energy Package stating that the theory of individual prosumers that cooperate for the energy needs of entire communities perfectly aligns with the concept of PEDs. Moreno et al. were more specific stating that spatial boundaries of PEDs could follow the ones that need to be defined for Renewable Energy Communities and Citizens Energy Communities [9]. The European Environmental Agency furthermore stresses the social innovation aspects of energy communities, emphasizing that communities are "a locus for innovation", providing great opportunities for learning and networks, and offering the possibility of achieving a whole-system change at local scales [10]. The JPI Urban Europe White Paper on a PED Reference Framework finally mentions the important role of the energy community concept for finding PED business models [11].

The key novelty of this paper is to systematically assess to what extent the new EU provisions on energy communities and their national transpositions could unlock the full potential of PEBs and PEDs. The research hypothesis of this paper is that these new regulatory frameworks have the potential to importantly support the implementation of PEDs and PEBs by

- fostering their internal optimization as well as system and market integration,
- unlocking new use cases, improving business cases, and, as a consequence
- improving their function for decarbonisation of the energy system.

The paper addresses both, PEBs and PEDs as in the context of energy communities no clear boundary between the concepts can be drawn. While PEBs in the past focus rather on the single building, recent definitions (see Section 3.1) also emphasize the energy exchange with other buildings. Clusters of buildings may already shift the focus from a single building to a more district-oriented (PED) approach. In addition, a PED may be formed by a group of PEBs as well as other types of buildings such as NZEBs, leading to a synergy between the different concepts. The paper however addresses NZEBs less explicitly as they address the system benefits of surplus energy generation to a lower extent. However, NZEBs are the only concept being legally defined so far and can fulfil important functions in the context of district and community-oriented initiatives. Thus, all of these concepts are interrelated and may be importantly supported by the new opportunities within energy communities.

The work was carried out in several steps. The first step was based on literature review of the NZEB/PEB/PED concepts outlining new options a community-based approach provides compared to the current building-oriented NZEB and PEB approaches. In a second step, an assessment was made to what extent the Clean Energy Package provisions on energy communities and other related provisions can enable new use cases for the optimization of PEBs and PEDs, both community-internal but also as part of the broader energy system. In a third step, work carried out under the EXCESS project was presented: this includes ICT needs that were investigated based on a range of interviews, the EXCESS ICT approach enabling PEBs and PED to become energy communities as well as an analysis and comparison on how the EXCESS demos plan to become energy communities. The approaches and limitations observed were then put into a broader context including findings from related projects in order to draw robust policy conclusions.

3. Overview of Concepts and New Regulatory Frameworks for PEBs and PEDs

3.1. From Nearly Zero Energy Buildings to Plus Energy Districts

Nearly Zero Energy Buildings (NZEBs), Plus Energy Buildings (PEBs) and Plus Energy Districts (PEDs) can very generally be distinguished in terms of (1) their energy performance/efficiency, (2) the degree to which their energy generation covers or overshoots their own demand and (3) the scale the concepts refer to (building or district). While the terms "plus energy buildings" or "plus energy districts" have not yet been defined in EU regulation, Nearly-Zero Energy Buildings have been defined in the Energy Performance of Buildings Directive (EPBD). Within the EPBD it is stated that "a nearly zero energy building means a building that has a very high energy performance" and that the remaining energy required "should be covered to a very significant extent by energy from renewable sources, including energy from renewable sources produced on-site or nearby". The directive further declares that starting 2019 all new buildings occupied and owned by public authorities are to be NZEBs. This requirement is expanded to all buildings by 2021. The exact definitions of NZEBs are, however, left to the Member States.

A range of recent literature aims to compare emerging PEB and PED definitions with the concept of NZEBs. Magrini et al. [8] for example mention that PEBs include a more complex structure of energy exchanges than NZEBs. Moreover, energy generation in PEBs and even more PEDs may not be based on a single building but rather on a system-based approach. Juusela et al. [12] emphasize the need for energy flexibility of a PEB but also the ability to integrate future technologies such as electric vehicles. Shnapp et al. [5] state that, from an energy perspective, moving beyond NZEBs to PEBs and PEDs provides opportunities to achieve better cost-effectiveness of energy efficiency and renewable energy systems. Hedman et al. [13] argue that in PEDs the renewable energy supply and demand can be unevenly distributed throughout the district, which allows a more strategical installation of renewable energy systems and energy storage. According to Hedman et al. the aim of a PED is not only to generate surplus energy, but rather to minimize the impact on the centralized grid by promoting higher self-consumption and self-sufficiency. The PED thereby should offer options to increase the onsite load matching by allowing the integration of long- and short-term storage and smart controls for improving energy flexibility [13]. Also the JPI Europe White Paper on a "PED Reference Framework" mentions the important role of PEDs as a provider of energy flexibility. Thereby, PEDs would actively contribute to the resilience and balancing of the regional energy system, to reach carbon neutrality with 100% renewable energy in the local consumption, as well as to achieve a surplus of renewable energy over the period of a year [10]. Carbon neutrality is generally gaining increased attention in PEB and PED definitions, current PED definitions often consider (embodied) greenhouse gas emissions or even require a net zero-emissions balance of the PED [3].

Table 1 shows a comparison of the three concepts based on the stated sources.

Table 1. Differences between NZEBs, PEBs and PEDs (based on [10–14]).

NZEB Model	PEB Model	PED Model
• Energy trade between one building and the grid • Power generated in individual buildings can be exported	• High energy flexibility • Complex system for trading between buildings and the grid • Optimized energy performance of the building	• Optimizing assets across the district • Minimize impact on the centralized grid • Provision of flexibility across the district and to the market

The shift of discourse from buildings to districts has important implications also for the characteristics and requirements of buildings. The buildings' ability to interact with other buildings or the grid is of high relevance for district optimization. When placing a zero or plus-energy objective on a district, the diversity of the energy interplay of the buildings' different energy performances and production capabilities determines the opportunity to share the neighbourhood's energy needs, costs and resources (see [15]). Recent literature, therefore, links new and more systemic PED concepts with possible new roles and requirements for NZEBs and PEBs referring to a needed high efficiency and flexibility [14].

3.2. *Plus Energy Buildings and Districts in the Context of the EU Energy Community Framework*

So far, there have been limited options to actively manage and optimize PEBs and PEDs. The Clean Energy Package includes two types of energy communities as defined in the recasts of the renewable energy directive and the recast of the electricity market directive (REDII and EMD). "Renewable Energy Communities" (RECs, defined in the REDII) and "Citizen Energy Communities" (CECs, defined in the recast EMD) allow citizens, public authorities and specific types of companies to collectively organize their participation in the energy system including energy generation, self-consumption, sharing, storage, and selling of energy. Both types of energy communities have to become a legal body, the legal forms allowed are defined by the Member States. Furthermore, the REDII defines "Renewables

Self Consumers" providing a basis for individual and collective self-consumption (CSC) as an activity rather than an organizational format [16].

Renewable Energy Communities address all types of renewable energy and have a local character. They could provide a suitable framework for PEBs and PEDs by defining their geographic scope, the eligibility of members and their governance, as well as the general purpose of the community. Focusing on an increased share of renewable energy in the system or collectively self-consumed within the community, RECs are supposed to receive public support. Citizen Energy Communities on the other hand can operate over a larger area, limited only by national borders or even involving international cooperation. Being defined in the recast EMD, their scope includes electricity only. Besides the inclusion of activities comparable to RECs (generation, consumption, storage, sale), Citizen Energy Communities have an emphasis on non-discriminatory access to the electricity markets, either directly or through aggregation. CECs may also engage in distribution, aggregation, and energy-related services such as energy efficiency or charging services. CECs, being entitled to freely act on the market, are not eligible for public support specific to energy communities. Both, RECs and CECs may have natural persons, SMEs and local authorities as members or shareholders that may also exercise decision-making power ("effective control") but RECs exclude large-sized companies from effective control while CECs exclude large- and medium-sized companies [16]. In the context of PEBs and PEDs, SMEs could include housing associations or building managers that may take on new roles via operating energy communities.

Two major principles of both types of energy communities that open up new possibilities for PEBs/PEDs are the access of "non-professional" actors to the energy markets as well as the right to use the public grid for community-internal sharing. The latter increases the potential for interactions and optimizations between buildings. In addition, depending on the national frameworks, RECs and CECs may be entitled to operate electricity or heating grids, which may equally support the PED internal energy exchange. PEBs and PEDs conceptually have more similarities to RECs than to CECs given their local character and a restriction to renewables in all current PEB and PED definitions. In addition, for RECs, supportive frameworks are emerging in several EU Member States, including, e.g., reduced grid tariffs for internal electricity sharing, improving the business case [17]. CECs may, however, also be an interesting frame for PEDs. They are hardly restricted in their geographical expansion and could serve as an umbrella for several PEDs and other districts but also allow to include renewable energy generation units that are not in proximity to the district but still part of a PED. CECs would not only allow for a broader exchange of electricity, but for a joint integration in the electricity market, e.g. for the collective marketing of flexibilities or surplus electricity. To this end, a CEC could itself act as an aggregator which may facilitate meeting the minimum capacity requirements that exist on many flexibility markets. However, also, the more limited energy sharing concept of the Clean Energy Package, collective self-consumption, could help optimize the use of renewables at the building level, or at the level of clusters of buildings depending on the national legislation that defines the spatial scope for such activities. Collective self-consumption may be an alternative in case the organizational and legal requirements for an energy community may be seen as too challenging.

Other provisions of the Clean Energy Package relevant to energy communities include Art 32 of the EMD that aims to provide incentives for the use of flexibility in distribution networks. Specifically, the article requires Member States to establish frameworks and incentives for grid operators (DSOs) to procure flexibility services based on, e.g., distributed (renewable) generation, demand response or energy storage, including through aggregation. PEDs may as well provide such services and could thus benefit from the establishment of suitable national frameworks.

Table 2 summarizes examples of important new regulatory features of the CEP of high relevance for PEBs and PEDs.

Table 2. Features of high relevance for PEBs/PEDs enabled through provisions of the CEP.

Key Features	REDII	EMD
Framework for citizen participation and governance	X	X
Integration of "non-professional" actors in the energy markets	X	X
Energy sharing via the public grid	X	X
Ownership and operation of local grids (if Member States allow it)	X	X
Provision by energy communities of a range of services to markets, such as for energy efficiency		X
Provision of local/small scale flexibilities to markets		X

There are several other directives or provisions in the Clean Energy Package that can be seen as enablers for PEBs/PED becoming energy communities. This includes Art 15 of the REDII requiring a minimum level of renewables in new and renovated buildings. The Energy Performance of Buildings Directive at the same time requires Member States to develop comprehensive long-term renovation strategies, including initiatives to "promote smart technologies and well-connected buildings and communities" [18]. As the optimisation potential of energy communities strongly depends on the ability of different building units to operate in a flexible way the EPBD's provisions to create smart readiness indicators (SRI) are highly relevant. The SRIs consider the flexibility of a building's overall electricity demand, including its ability to enable participation in active and passive as well as implicit and explicit demand response in relation to the situation of the public grid.

4. Implementing PEBs and PEDs as Energy Communities: Approaches of the EXCESS Project

This section first provides an overview of suitable technologies for PEBs and PEDs and then presents the EXCESS ICT framework including key innovations that are needed to realize energy communities. Furthermore, the section presents approaches of EXCESS demonstrations to become energy communities outlining technological innovations combined with ICT solutions applied as well as regulatory barriers the demos faced.

4.1. Technology Outlook for PEBs and PEDs in the Energy Community Context

While energy efficiency should be a key consideration to reach a positive energy balance, also renewable and supporting technologies in plus energy buildings and districts have a major role in increased self-consumption and flexibility provision. A higher share of renewable energy sources for heating, cooling and electricity, compared to current business-as-usual practices can be achieved by integration of renewable energy sources into buildings and their immediate surroundings, such as PV on facades and stand-alone RES production facilities [14]. An advantage of PEDs is the opportunity to install more centralized and locally shared technologies and infrastructure such as centralized storage systems, improving the business case. Unlocking flexibilities from individual, small-scale devices, at the same time, can be very complex due to social and technical barriers and therefore costly. The energy community concept can help to optimize the use of infrastructure via the exchange of energy and flexibilities, pooling flexibilities offers to markets and enabling multi-use cases of technologies.

There is a range of literature on suitable technologies and technology combinations for PEBs and PEDs [11,19,20]. PEBs/PEDs implemented as an energy community should align the technology choice with new use cases that are enabled by the CEP. In case PEBs/PEDs have a high focus on seasonal self-sufficiency, this could lead to issues for the existing energy market system and actors, such as in the case a high share of solar energy is fed into the grid in summer, while a large amount of energy is taken from the public grid in winter, thus using the grid as virtual storage. Seasonal storage solutions therefore may become valuable for a good interaction with the surrounding energy system [21]. In this context, geothermal energy is not only a renewable energy resource delivering heating and cooling anytime and anywhere but soil and bedrock offer a suitable solution for the integration of solar energy and residual heat in all the EU heating-dominated climates that need seasonal storage for solving the seasonal energy mismatch. Geothermal heat pumps connected to a deep borehole recharging the underground as tested in EXCESS have a high COP and could represent a viable solution for heating-dominated climates, such as the Nordic countries [22]. Solar energy is a viable solution towards Nearly Zero Energy Buildings even in locations with low solar irradiance [23]. However, in central and northern Europe, PEBs/PEDs may not reach enough onsite energy generation and storage to achieve a net-zero yearly balance (yearly onsite produced renewable energy equals building energy demand), if conventional solar technologies are used (either PV or solar thermal) [24]. Thus, while conventional solar technologies can be used for achieving plus energy buildings in continental and mediterranean climates, solar hybrid solutions, such as combined photovoltaic and solar-thermal panels (PVT), are the products best positioned to respond to oceanic and nordic climates as they maximize the energy use of the solarized surface.

Different technologies can be used as energy flexibility sources at the building and district level. Energy storages are the main conventional source of flexibility. Others are aggregated controllable Heating, Ventilation and Air Conditioning (HVAC) loads, realizing fast balancing or frequency response, or power-to-heat demand response solutions (such as heat pumps combined with storages, boreholes and/or water tanks and/or floor heating as thermal mass solutions) [19,22,25].

Assuming that a future official PED definition will include a net-zero emission balance, an allowable limit of GHG emissions per unit of generated/consumed thermal/electrical energy will also gain importance in technology choice. Pucker et al. [26], for example, showed that storage solutions to increase the self-sufficiency in decentralized renewable energy systems not always reduce GHG emissions due to embodied emissions, but also due to direct emissions caused by the energy needed to produce and operate the devices as well as the related losses. Thus, renewable energy sources and technology portfolios have to be selected based on a number of factors: local climate, market and regulatory conditions (e.g., on energy transactions), the ability for multi-use applications, the onsite building characteristics in order to match building load profiles and renewable energy sources and, finally, direct and indirect GHG emissions.

4.2. ICT Framework for Energy Communities based on PEBs/PEDs

The Clean Energy Package empowers a range of new actors in the energy sector but also provides opportunities for new roles of traditional actors. These include aggregators, who will get better market access, service providers, such as ESCOs or IT companies, building managers, municipalities that may build the grid infrastructure and, finally, building occupants. Some of them can take on several roles. All of them have information and data needs but also require new IT tools and approaches to enable the activities shown in Table 2. This section therefore first describes ICT needs of different actors, and then outlines the EXCESS ICT framework.

Within EXCESS, actor's ICT needs in the context of PEB use cases were assessed based on 43 interviews in the demo countries (among them building occupants, building managers and aggregators [27]).

Aggregators highlighted the need to be able to manage and optimize the assets in the PEB/PEDs and enable trading including to markets. In order to optimize services they could offer on markets, analysing occupants' flexibilities on a district/community level are highly relevant towards increasing monetary benefits for both sides through the provision of ancillary services to network operators [27].

Building/district managers have shown within the interviews a high interest in moving beyond their traditional activities enabling trading the building's non self consumed energy in local flexibility and energy markets towards monetary gains via cooperating with aggregators. Preconditions, however, are careful consideration of comfort standards and as stated by consumers, clear, transparent and consumer-protecting regulations that ensure customer rights as well as savings in the energy bills of residents [27].

Service providers may control the district network, manage sharing the buildings/districts energy and flexibility for optimizing the performance of the neighbourhood.

Building occupants showed interest to monitor their energy consumption and flexibility via mobile apps. This may stimulate prosumer engagement towards increasing self-consumption and energy performance of building blocks and districts [27].

In order to equip the above-mentioned actors with the needed tools and realize new use cases under energy communities, the ICT framework plays a crucial role. Key innovations of the EXCESS ICT framework are:

- Advanced data analytics for actors controlling the system and providing services to markets;
- Advanced software solutions for sharing within the community;
- Different data visualization approaches either for aggregators and building managers or for occupants.

The EXCESS ICT framework, therefore, includes the Data Management Platform, the Data Analytics Framework, the Model Predictive Control component, a Data Visualizations Framework and blockchain infrastructure and applications.

4.2.1. EXCESS Data Management Platform

The EXCESS Data Management Platform is responsible for the collection and management of all types of data coming from various sensors, submeters and energy components of the distributed information systems in the EXCESS demo sites. It aims, among others, to enable interoperability and data protection, key issues in setting up energy communities [28]. The Data Management Platform consists of different components that perform the various necessary data management processes including ingestion, mapping, cleaning, anonymization and storage, so that the collected data can then be used for analytical purposes by the Data Analytics Framework, for control strategies optimization by the Model Predictive Control component and for the operations of the visualization and blockchain applications [29]. An EXCESS user management service facilitates the authentication and authorization of users in the EXCESS Data Management Platform. It performs the necessary authentication mechanisms for entering the platform, while setting the different types of users and user groups. It also defines the access rights control in the EXCESS Data Management Platform, specifying which datasets can be accessed by certain users or user roles and ensuring that no unauthorized data access is performed [27]. In the context of the Data Management Platform, the EXCESS Common Information Model is developed in order to semantically model all the necessary information regarding the development, installation, deployment and operation of the PED system. The EXCESS Common Information Model constitutes a common language for all the different datasets that will reside in the EXCESS Data Management Platform enhancing their alignment and interoperability.

4.2.2. EXCESS Data Analytics and Control Framework

The EXCESS Data Analytics Framework is responsible for the performance of analytical activities using the data residing in the EXCESS Data Management Platform providing important information to actors controlling the system and providing services to market such as aggregators. The framework exploits a series of mostly pre-trained algorithms in order to include meaningful analytical results that will be used by the Model Predictive Control component where the constraints and various variables are controlled to achieve the PEB and PED concepts, as well as the various visualization and blockchain applications. A key innovation of EXCESS is the comfort profiling component of building occupants, which enables unlocking households' flexibilities [27]. The comfort preferences of the building occupants will be extracted based on the sensor measurements and other metrics of the buildings. Through the performance of profiling analytics, their comfort profiles will be derived subsequently. The combination of the comfort profiles with the energy and demand forecasts for the building provides context-aware flexibility profiles of the devices and loads in the building through the respective profiling mechanisms of the context-aware flexibility profiling and analytics component. These are the flexibilities of the devices and loads, such as a small change in the temperature set point of an HVAC system, which is created due to a dynamic adaptation of the energy demand according to a corresponding slight change within the comfort bounds preferences of building occupants without actually affecting their comfort. These context-aware flexibility profiles will also be utilized by a dynamic Virtual Power Plant (VPP) configuration component that will run multiple alternative scenarios in order to find the optimal flexibility-based surplus energy scheme that can be communicated to an aggregator for possible trading on the energy market.

4.2.3. EXCESS Data Visualizations Frameworks

Data visualization frameworks were developed for aggregators, building managers and for occupants aiming to change their behavior. Visualization applications provided to aggregators and building managers, use data coming either directly from the EXCESS Data Management Platform or from the analytical results of the EXCESS Data Analytics Framework. The *flexibility analytics visualizations* enable aggregators to view demand flexibility and energy generation forecasts from buildings through corresponding analysis of data, allowing them to understand which buildings can provide flexibilities and select the optimal VPP configuration for provision to the energy grid. In that way, the aggregators address the grid requirements set by DSOs through the constant monitoring of the performance of their established clusters/VPPs and identification of potential flexibility overrides that are tackled by the appropriate modification of VPP schemes. The *energy consumptions visualizations* give building managers the opportunity to monitor the energy profiles and patterns in the building through dashboards and enriched energy analytics. The visualizations are produced by data coming from sensors and submeters regarding the buildings' energy consumption and demand based on the building occupants' energy behaviors. In that way, the building managers will understand what energy behaviors and patterns can lead to the achievement of energy savings in the building. For occupants, a mobile app was developed that comprises energy performance indicators that stimulate ecological and economic energy consumption including production and consumption values, energy score, energy savings, and corresponding monetary values, where applicable.

4.2.4. EXCESS Blockchain Infrastructure and Applications

The emerging blockchain mechanisms have proven successful for the transparent, secure, reliable and timely management of energy flexibility transactions by adapting energy demand profiles of consumers of distributed energy resources to all the stakeholders involved in the flexibility markets [30]. The EXCESS blockchain infrastructure comprises the basis for the design and implementation of standardized energy exchange rules, mechanisms, agreements and demand response smart contracts towards enabling automated settlement and verification when underlying events occur. Two distinct applications are developed for operation in the EXCESS demo sites:

- An application for facilitating the establishment of energy communities and benefit-sharing among prosumers out of the deployment of energy optimization strategies (Austrian demo);
- An application for the execution of explicit demand response programs through flexibility trading with energy market actors (Belgian demo).

The *Objective Benefit Sharing Application* (OBS) utilizes blockchain technology to enable automated governance of community installations and trading shares of assets and the resulting generation. As shareholders, producers can get involved in the decision-making process as well as in the definition of contract rules, and they can trade their shares. By using different indicators, shareholders can stimulate different energy optimization strategies of their choice is assisted by visualisation approaches of the energy app as described above. Energy optimization strategies are built upon thresholds and parameters governed within smart contracts. For every billing cycle, these thresholds and parameters are combined with metering data. Subsequently, rules defined by the shareholders are applied. This process enables automatic invoicing, incentivizing beneficial consumer behaviour, and dividend pay-out. The *Explicit Demand Response Application* enables flexibility sharing and pooling for the realization of demand response services resulting in maximization of benefits for prosumers through their participation in energy markets. Using the analyses and forecasts coming from the Data Analytics Framework as input, aggregators are able to publish their demand response offers (flexibility requests) and associated strategies to attract prosumers and engage them in demand response services through the signing of blockchain-powered smart contracts.

4.3. Approaches of EXCESS Demonstrations to Become Energy Communities

This section provides an overview of the EXCESS demos presenting approaches to implement use cases enabled by the Clean Energy Package with a focus on energy communities. The four EXCESS demos are in different climate zones, have different technological concepts and are also significantly different in their scope to establish energy communities as Table 3 shows. All demos aim at making use of the national energy community or self-consumption frameworks.

Table 3. Overview of the spatial scope of the EXCESS demos.

Country	Scope
Finnish demo	Cluster of two multistory buildings
Spanish demo	Cluster of several multistory buildings
Belgian demo	Cluster several multistory buildings
Austrian demo	Commercial area with 19 buildings

4.3.1. Austrian Demo

A former commercial zone is transformed into an area with mixed use including offices, recreation zones, and a student hostel. In total, the 19 buildings in the area are being refurbished towards passive house standards while increasing the share of locally produced renewable energy (solar energy, small hydropower). Through the integration of innovative elements for load shifting, storage, user integration, interaction with the local electricity grid as well as smart predictive control, maximum energy flexibility will be achieved and stress on the central grid will be reduced. Several energy efficiency measures will be integrated, including a multifunctional façade (electricity generation, heating and cooling) that can be mounted to the exterior of an existing building to improve its energy performance. Flexibility is also maximized through a cascadic heat pump system for heating, water heating and cooling. Depending on the load situation or the availability of renewable electricity generated onsite, a wide range of heat and cooling capacity can be retrieved flexibly and temporarily stored in connection with the heat-side flexibilization elements (activated building mass, decentralized buffer storage, screed mass of the underfloor heating). The three heat pumps are reversible and can therefore also be used for cooling buildings.

Within EXCESS, the focus is on a plus energy building, but a nationally-funded project extends the scope to an energy community. The blockchain-based "Objective Benefit Sharing" application (see Section 4) will be a key innovation to facilitate the creation of an energy community. The application allows constant monitoring and verification of energy savings at the prosumer and the community levels and facilitates the transparent distribution of benefits arising from energy optimization among prosumers based on energy measurements handled through blockchain. By automated voting, shareholders can specify their level of involvement in the decision-making process. The system, therefore, is not only an innovative way to share benefits but also allows for a virtual decision-making mechanism as an innovative way to operationalize governance structures in energy communities.

4.3.2. Finnish Demo

The EXCESS demo building at Kalasatama, Helsinki, consists of eight floors, located next to another similar demo that receives national research funding. The energy system in place for the Kalasatama PEB is a hybrid geothermal energy system. It combines semi-deep geothermal energy wells with coaxial collectors in ~800 m deep boreholes, heat pumps, PV and PVT panels that will produce electricity and heat for the building. To increase temperature levels to a suitable level for space heating and domestic hot water, the hybrid energy system utilises heat from the PVT panels, ventilation, and ground source heat from heat pumps. The building structures, heating, ventilation and air conditioning have been designed as energy-efficient as possible. To optimise the overall energy system performance, an integrated smart control system enables demand response and bi-directional electricity trade. The energy community in Kalasatama will consist of two housing cooperatives representing the two neighbouring condominiums. The technical operation of the energy community will be made by a specialized service provider. The locally produced energy is used onsite as much as possible. The local PV production is primarily used for electric appliances inside the cooperatives (HVAC: fans, pumps, heat pumps, lighting, etc., electrical vehicle (EV) charging on-site). Excess electricity will be shared among the building residents. The buildings have flexibility options by intermittent use of heat pumps and storage capacities of geothermal wells and domestic hot water systems support. The integration into the electricity market, however, will be a long-term goal and needs an energy broker between the housing cooperatives and the electricity market. Finland has not yet presented a regulatory framework for local flexibility procurement. At the beginning of 2021, a new regulation on local energy communities came into force. It enables the self-consumption and sharing of locally produced electricity. It has already been criticized for its inflexibility regarding the sharing of electricity between the participants, as it entails pre-defined shares for the participants [31].

4.3.3. Belgian Demo

In the demo site in Hasselt, several PEBs are part of a larger residential area with 68 apartments intended for social housing. The demo site includes a central heating system to allow better integration of renewable sources than traditional individual heating systems. Geothermal heat pumps will provide heat for heating applications and domestic hot water production. Thermal energy is distributed to the different apartments by a small district heating network. A smart substation installed in each apartment extracts the necessary thermal energy from the heating grid. PVT panels and a small wind turbine on the roof of the demo building provide renewable electricity. The energy system is controlled, monitored and optimized by a central Building Energy Management System. A data-centric-model multi-agent system architecture has been developed that allows the optimal coordination and control of a dynamic cluster of buildings by implementing a "Flex Trading" concept that goes beyond the traditional demand response concept. Each of the buildings determines its available flexibility and communicates this information to a community-level platform. Thereby, a 'virtual' energy community will be created, where each member can generate flexibility through its boiler (power-to-heat, thermal storage). The central (virtually shared) generation is used for shared infrastructure, the remainder (excess energy) is optimally used by the community members. The system is managed by a centralized manager ('middleman'). The platform aggregates all this information and determines the optimal community-level consumption plan. This is then disaggregated to an optimal consumption plan per community member. Such a bottom-up aggregation of information provided by the buildings themselves addresses the consumption forecasting challenges associated with small clusters of buildings (e.g., limitation of statistical approaches) and buildings that actively control their consumption (e.g., acting on dynamic prices, performing PV self-consumption, etc.). At the community or district manager level, the bottom-up aggregated baseline and flexibility information is used to decide on a given collective objective (e.g., community-level self-consumption) and/or the aggregated flexibility can be offered to external stakeholders, e.g., for congestion management for the local DSO. By proper interactions with the local DSO, it can be assured that only local grid secure flexibility activations are offered to external stakeholders. While Belgium (both Flanders and Wallonia) has a basic regulatory framework for energy communities in place, regulatory details for active prosumers and energy communities are still unclear and there are insufficient incentives for flexibility control. Proper incentives for flexibility activations such as dynamic tariffs (implicit demand response) or a better regulatory framework for explicit demand response would strongly assist the creation of a business case.

4.3.4. Spanish Demo

The Spanish demo case is located in the metropolitan area of Granada. The planned four-story apartment complex consists of 30 dwellings with the ground floor to be used for commercial activities. The complex was originally planned to meet NZEB standards, but will be upgraded to become a PEB as part of the EXCESS project. For this purpose, the apartment complex will be equipped with a number of technical and architectural design features to minimize the energy demand while maximizing the self-consumption of onsite generated renewable energy, making it one of the very first plus energy buildings in Spain. The demo case building is part of the larger urban development project NIVALIS, Spain's first planned zero energy district.

The energy concept of the building prioritizes direct self-consumption of renewable energy generated by PV panels. Additional key features of the PEB include a geothermal system that has been designed for optimal seasonal performance as well as an integrated state-of-the-art control system at different levels (apartment, building and renewable technologies) to reduce the energy consumption to the minimum, while maintaining indoor comfort conditions. Surplus electricity is stored in a battery and can also be fed into the grid in case it is not consumed on the demo site. Electric vehicle charging stations for cars of residents are another feature to encourage low-carbon living and high self-consumption. The building manager will be responsible for the control system operation and the supervision of the proper renewables integration.

While Spain has no regulatory framework for energy communities, the Spanish government, in 2019, approved the Royal Decree 244/2019 that regulates the administrative, technical and economic conditions of self-consumption. The maximum distance between the production and consumption meters is 500 m and should be connected on the low voltage grid (LV) level, limiting the spatial scope of the PED. Excess electricity fed into the public grid from production facilities not exceeding 100 kW can benefit from the compensation of costs on a monthly basis. These facilities (below 100 kW with cost compensation) will be exempt from the obligation to register as an electricity supplier and will be subject only to technical regulations [16]. In the case of the EXCESS demo, the PV power installed will be 150 kW. A split of the installation into two with a peak power below 100 kW could be defined, in case certain technical preconditions are fulfilled. Next to technical issues, first experiences made in the EXCESS demo showed that novel self-consumption facilities require a high effort also related to non-technical issues, such as designing the contract agreement between the building owners regarding sharing of energy.

5. Comparison of the Different Approaches and Discussion

This section compares and discusses the planned use cases, sharing approaches, organizational set-ups and the possible roles of actors in the four EXCESS demos considering also findings from related research and demonstration projects. Figure 1 displays the current and additionally planned use cases of the demos (time horizon 3–5 years), most of them improved or enabled by the Clean Energy Package. The first category includes maximizing self-consumption and self-sufficiency, which will be improved by a minimum required share of renewables in buildings or energy efficiency requirements. Also, the access to exchange energy and flexibility on markets will be improved by the Clean Energy Package. Sharing and trading of energy using the grid as well as provision of local services to the DSO would not be possible without the Clean Energy Package.

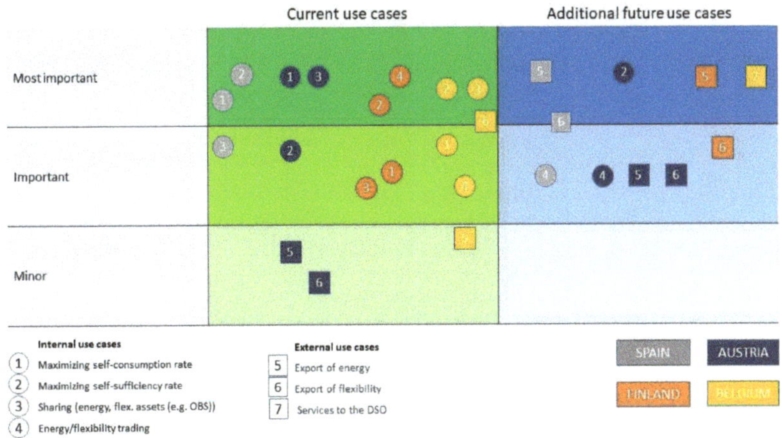

Figure 1. EXCESS PEB/PED use cases facilitated or enabled by the Clean Energy Package.

The figure above shows that there is a high focus on self-consumption/self-sufficiency in the short term, which is a key element of current PEB and PED definitions. The provision of services to market actors is currently not the main aim in most of the demos, except for the Belgian demo. Market integration, however, is a longer-term goal of all the EXCESS demos but is currently hardly possible due to a lack of supporting national regulations and incentives. This was also observed in the +CityxChange project that concludes that due to the absence of incentives even cheap flexibilities of PEDs may be left idle [32]. The strong focus on self-consumption in the context of energy communities was also reflected in recent interviews with European regulators that fear a high amount of fluctuating renewable energy being injected into the grid, while there was a lack of understanding of the need for local flexibility procurement at the level of energy regulators and DSOs [33]. Collective self-consumption is promoted by policymakers in some Member States implementing reduced grid tariffs for locally shared electricity in RECs (Austria, Portugal) or operational support for collectively self-consumed electricity (Italy) [34].

Sharing and internal energy trading is a key element in all EXCESS demos already in the short term. While in the Spanish and Finnish demo electricity is planned to be traded among buildings, in the Belgian and Austrian demos, a virtual energy community will be created. While in the Belgian demo flexibilities are shared, the Austrian demo plans to share benefits arising from energy optimization among prosumers, awarding top-performing prosumers. While electricity sharing and trading is a key concept of the Clean Energy package it still faces significant barriers in many EU Member States that have energy community frameworks in place. These include a lack of access of energy communities to real-time data of smart meters or, as mentioned in the Finnish EXCESS demo and by the H2020 project syn.ikia for Spain, yearly fixed distributions rules preventing flexible energy sharing [18].

Regarding the organizational set-up in the Spanish and Finnish demos, a cooperative structure will provide a suitable governance model. Actually, cooperatives are quite a common model for many emerging energy communities and the EU framework for energy communities is partly based on cooperative principles [17]. In Belgium, a housing complex is seen as a suitable framework for an energy community having existing organizational structures in place that can be used. Housing associations are in general seen as an important enabler for energy communities, assisting the implementation of shared infrastructure [17]. Several ongoing PED projects such as Positive4North mention as barriers that multiple and diverse stakeholders are in the districts [35], the energy community framework thus could provide a basic organizational and governance structure for PEDs, assigning clear roles to different actors.

The EXCESS demos and other related projects, such as syn.ikia, illustrate the strong limitations for PEBs and PEDs in some member states that have already a REC framework in place such as a maximum generation power (200 kW in Italy for RECs or 150 kW in Spain in case of collective self-consumption). Other countries with REC frameworks in place are imposing system-related limitations to the low voltage (LV) level such as in Slovenia or Croatia that may limit the scope of activities and the participation of districts in energy communities [16]. In such national frameworks, energy communities would rather consist of a few buildings than larger districts. In most EU Member States, however, the geographical scope for RECs is broader, often being based on the medium voltage (MV) level or municipal boundaries [17]. Regulatory frameworks for CECs hardly exist in Member States [17]. Collective self-consumption is limited in most member states to the multi-apartment level limiting the use of this concept for PEBs and PEDs [17]. There is a range of other barriers mainly at the national level that will limit the deployment of energy communities in the short term, such as complexities in setting up the legal structure or designing sharing agreements [17], as EXCESS and related projects illustrate.

6. Conclusions

This article outlined opportunities the Clean Energy Package and, in particular, energy communities provide for PEBs and PEDs regarding their economic optimization and market integration, possibly leading to new use cases and revenue streams. At the same time, insights into regulatory limitations at national level in transposing the set of Clean Energy Package provisions are presented. The paper showed that the concepts of PEBs and PEDs are in principle synergetic with the energy community concepts as PEBs/PEDs relate to technical characteristics and optimizations while energy communities provide a legal and regulatory framework for the organization and governance of a community, at the same time providing new regulatory space for specific activities and market integration. While the hypotheses of the paper assumed that the Clean Energy Package provisions could improve PEBs and PEDs in their function for the decarbonisation of the energy system, including both, a high renewable self-sufficiency rate as well as flexibility provisions to the grid, it was observed that there is a strong policy aim to use energy communities to increase self-sufficiency, as also illustrated in the case of the EXCESS demos. Collective self-consumption within Renewable Energy Communities often receives financial benefits from Member States, while providing services to balancing markets by pooling small-scale flexibilities via aggregators or providing local flexibilities to DSOs is hardly yet possible in most EU countries. As a consequence, in the short term, some of the proposed new use cases will not be available and the full potential of PEBs and PED for decarbonisation cannot be fully exploited. The focus on self-consumption may result in an overinvestment in local infrastructure and may counterbalance least-cost decarbonisation for the society.

As this paper illustrated, there is a range of additional barriers mostly at the national level that will hamper the roll-out of energy communities in general and PEBs and PEDs as energy communities in particular, such as a lack of access to real-time data or generation capacity and spatial limitations. Overall, the heterogeneity with which Member States are implementing the EU energy community (and self-consumption) provisions may prevent a clear blueprint for an organizational and business model for PEBs and PEDs across the EU. Even though Member States are likely to adjust their national regulatory frameworks after having gained first experiences, new revenue streams for PEBs and PEDs will not fully be available in the next years, and PEBs and PEDs may strongly remain to rely on public support. Importantly, EU Member States need to get a better picture of the future role of energy communities in general and PEBs/PEDs in particular for decarbonizing the energy system. The corresponding design of both, appropriate market rules and support schemes will be of high importance.

Author Contributions: Writing—original draft: A.T. (Andreas Tuerk), D.F., K.L., A.T. (Anastasios Tsitsanis), S.K., F.R.; formal analysis: C.N., J.L., C.C., M.A.-J., A.S., T.S., M.U., K.A.; writing—review and editing: I.H., T.R. All authors have read and agreed to the published version of the manuscript.

Funding: The research was supported by the European Union's Horizon 2020 research and innovation program LC-EEB-03-2019—New developments in plus energy houses (IA) under the project name EXCESS "FleXible user-CEntric Energy poSitive houseS" (grant number 870157). The funding bodies had no involvement in preparing the manuscript, methods and results, etc.

Institutional Review Board Statement: Not applicable.

Informed Consent Statement: The study quotes interviews made under the EXCESS project. Following the GDPR, the answers to the questionnaires/interviews were anonymous (no personal data were obtained).

Data Availability Statement: Data sharing not applicable. No new data were created or analyzed in this study.

Conflicts of Interest: The authors declare no conflict of interest.

References

1. European Commission. *Clean Energy for All Europeans;* European Commission: Brussels, Belgium, 2019.
2. European Commission. *The European Green Deal;* European Commission: Brussels, Belgium, 2019.
3. JPI Urban Europe. *SET Plan Action 3.2 White Paper on PED Reference Framework for Positive Energy Districts and Neighbourhoods;* JPI Urban Europe: Vienna, Austria, 2020.
4. Jensen, S.Ø.; Madsen, H.; Lopes, R.; Junke, R.G.; Aelenei, D.; Rongling, L.; Metzger, S.; Lindberg, K.B.; Marszal, A.; Kummert, M.; et al. *Annex 67: Position Paper of the IEA Energy in Buildings and Communities Program (EBC) Annex 67 "Energy Flexible Building";* IEA: Paris, France, 2017.
5. Shnapp, S.; Paci, D.; Bertoldi, P. *Enabling Positive Energy Districts across Europe: Energy Efficiency Couples Renewable Energy;* European Commission: Brussels, Belgium, 2020.
6. Karg, L. Taskforce on Energy Communities. In *Mission Innovation Austria Online Event: Energy Communities—Findings from Innovation Programs and Pilots;* Austrian Federal Ministry for Climate Action, Environment, Energy, Mobility, Innovation and Technology: Vienna, Austria, 2020.
7. Tuerk, A. Energy Positive Buildings/Districts as Nucleus for Energy Communities. In *Mission Innovation Austria Online Event: Energy Communities—Findings from Innovation Programs and Pilots;* Austrian Federal Ministry for Climate Action, Environment, Energy, Mobility, Innovation and Technology: Vienna, Austria, 2020.
8. Moreno, A.G.; Vélez, F.; Alpagut, B.; Hernández, P.; Montalvillo, C.S. How to Achieve Positive Energy Districts for Sustainable Cities: A Proposed Calculation Methodology. *Sustainability* **2021**, *13*, 710. [CrossRef]
9. European Environmental Agency. *Sustainability Transitions: Policy and Practice;* Publications Office of the European Union: Luxembourg, 2019.
10. Schwarz, H.G. *SET Plan Action 3.2/PED-Programme Implemented by JPI Urban Europe Definition and Further Operationalisation of PEDs;* JPI Urban Europe: Vienna, Austria, 2021.
11. Magrini, A.; Lentini, G.; Cuman, S.; Bodrato, A.; Marenco, L. From nearly zero energy buildings (NZEB) to positive energy buildings (PEB): The next challenge—The most recent European trends with some notes on the energy analysis of a forerunner PEB example. *Dev. Built Environ.* **2020**, *3*, 100019. [CrossRef]
12. Ala-Juusela, M.; ur Rehman, H.; Hukkalainen, M.; Tuerk, A.; Trumbic, T.; Llorente, J.; Claes, S.; Tsitsanis, S.; Tsitsanis, T.; Latanis, K.; et al. *Deliverable 1.1: PEB as Enabler for Consumer Centred Clean Energy Transition: Shared Definition and Concept;* EXCESS: Graz, Austria, 2020.
13. Hedman, Å.; Rehman, H.; Gabaldón, A.; Bisello, A.; Albert-Seifried, V.; Zhang, X.; Guarino, F.; Grynning, S.; Eicker, U.; Neumann, H.-M.; et al. IEA EBC Annex83 Positive Energy Districts. *Buildings* **2021**, *11*, 130. [CrossRef]
14. Vandevyvere, H.; Ahlers, D.; Alpagut, B.; Cerna, V.; Cimini, V.; Haxhija, S.; Hukkalainen, M.; Kuzmic, M.; Livik, K.; Padilla, M.; et al. *Positive Energy Districts Solution Booklet;* Smart Cities Information System: Brussels, Belgium, 2020.
15. Amaral, A.R.; Rodrigues, E.; Gaspar, A.R.; Gomes, Á. Review on performance aspects of nearly zero-energy districts. *Sustain. Cities Soc.* **2018**, *43*, 406–420. [CrossRef]
16. Frieden, D.; Tuerk, A.; Neumann, C.; d'Herbemont, S.; Roberts, J. Collective Self-Consumption and Energy Communities: Trends and Challenges in the Transposition of the EU Framework. 2020. Available online: https://www.rescoop.eu/uploads/rescoop/downloads/Collective-self-consumption-and-energy-communities.-Trends-and-challenges-in-the-transposition-of-the-EU-framework.pdf (accessed on 22 August 2021).
17. Tuerk, A.; Neumann, C.; Rakocevic, L. Energy Community Monitor 2021. DECIDE: Leuven, Belgium. Available online: https://decide4energy.eu/fileadmin/user_upload/Resources/DECIDE_D3.1_final.pdf (accessed on 2 August 2021).

18. Boll, J.R.; Dorizan, V.; Broer, R.; Toth, Z. *Policy Mapping and Analysis of Plus-Energy Buildings and Neighbourhoods: Barriers and Opportunities of Plus Energy Neighborhoods in the National and Local Regulatory Framework*; syn.ikia: Trondheim, Norway, 2021.
19. Laitinen, A.; Lindholm, O.; Hasan, A.; Reda, F.; Hedman, Å. A techno-economic analysis of an optimal self-sufficient district. *Energy Convers. Manag.* **2021**, *236*, 114041. [CrossRef]
20. Kohlhepp, P.; Harb, H.; Wolisz, H.; Waczowicz, S.; Müller, D.; Hagenmeyer, V. Large-scale grid integration of residential thermal energy storages as demand-side flexibility resource: A review of international field studies. *Renew. Sustain. Energy Rev.* **2018**, *101*, 527–547. [CrossRef]
21. Weitemeyer, S.; Kleinhans, D.; Vogt, T.; Agert, C. Integration of Renewable Energy Sources in future power systems: The role of storage. *Renew. Energy* **2015**, *75*, 14–20. [CrossRef]
22. Kukkonen, I. Geothermal Energy in Finland. In Proceedings of the World Geothermal Congress, Tohoku, Japan, 28 May–10 June 2000.
23. Heiskanen, E.; Nissilä, H.; Lovio, R. Demonstration buildings as protected spaces for clean energy solutions—The case of solar building integration in Finland. *J. Clean. Prod.* **2015**, *109*, 347–356. [CrossRef]
24. Reda, F.; Fatima, Z. Northern European nearly zero energy building concepts for apartment buildings using integrated solar technologies and dynamic occupancy profile: Focus on Finland and other Northern European countries. *Appl. Energy* **2019**, *237*, 598–617. [CrossRef]
25. Gjorgievski, V.Z.; Markovska, N.; Abazi, A.; Duić, N. The potential of power-to-heat demand response to improve the flexibility of the energy system: An empirical review. *Renew. Sustain. Energy Rev.* **2020**, *138*, 110489. [CrossRef]
26. Pucker-Singer, J.; Aichberger, C.; Zupančič, J.; Neumann, C.; Bird, D.; Jungmeier, G.; Gubina, A.; Tuerk, A. Greenhouse Gas Emissions of Stationary Battery Installations in Two Renewable Energy Projects. *Sustainability* **2021**, *13*, 6330. [CrossRef]
27. Latanis, K. *Deliverable 3.1: EXCESS ICT Architecture Blueprint*; EXCESS: Graz, Austria, 2020.
28. Frieden, D.; Vallant, H.; Neumann, C.; Bozic, J.; Tuerk, A.; Nacht, T.; Pratter, R.; Ganglbauer, J. *Förderung Sektorengekoppelter Energiegemeinschaften Durch Digitalisierung und Automatisierung, Forthcoming 2021*; SEED: Graz, Austria, 2021.
29. Ivask, N. *Introduction to the COMETS Project Study*; COMETS: Torino, Italy, 2021.
30. Hajizadeh, A.; Hakimi, S. Chapter 8—Blockchain in decentralized demand-side control of microgrids. In *Blockchain-Based Smart Grids*; Elsevier: Amsterdam, The Netherlands, 2020; pp. 145–167.
31. Finnish Government Government Decree Amending the Government Decree on the Settlement and Metering of Electricity Supplies 1133/2020. Finish Government: Helsinki, Finland, 2020. Available online: https://www.finlex.fi/fi/laki/alkup/2020/20201133 (accessed on 22 September 2021).
32. Hackett, S.; Kvaal, B.; Runnerstrøm, M.F.; Grøttum, H.H.; Økstad, N.; Livik, K.; Danielsen, S.; Wright, S.; van Vuuren, L.M.; Stewart, D.; et al. *D2.3: Report on the Flexibility Market*; +CityxChange: Trondheim, Norway, 2019.
33. Peeters, L. *Economies of Energy Communities—Review of Electricity Tariffs and Business Models*; European Commission: Brussels, Belgium, 2020.
34. Frieden, D.; Türk, A.; Roberts, J.; Herbemont, S.; Gubina, A. Collective Self-Consumption and Energy Communities: Overview of Emerging Regulatory Approaches in Europe. 2019. Available online: https://www.compile-project.eu/wp-content/uploads/COMPILE_Collective_self-consumption_EU_review_june_2019_FINAL-1.pdf (accessed on 7 July 2021).
35. Gollner, C.; Hinterberger, R.; Bossi, S.; Theierling, S.; Noll, M.; Meyer, S.; Schwarz, H.G. Europe towards Positive Energy Districts: A Compilation of Projects towards Sustainable Urbanization and the Energy Transition. 2020. Available online: https://jpi-urbaneurope.eu/wp-content/uploads/2020/06/PED-Booklet-Update-Feb-2020_2.pdf (accessed on 15 July 2021).

buildings

MDPI

Article

Obstacles to Developing Net-Zero Energy (NZE) Homes in Greater Toronto Area

Ghazal Makvandia [1] and Md. Safiuddin [1,2,3,*]

[1] Angelo DelZotto School of Construction Management, George Brown College, 146 Kendal Avenue, Toronto, ON M5T 2T9, Canada; gmakvandia@gmail.com
[2] Department of Civil Engineering, Faculty of Engineering and Architectural Science, Ryerson University, 350 Victoria Street, Toronto, ON M5B 2K3, Canada
[3] Department of Civil and Environmental Engineering, Faculty of Engineering, University of Windsor, 401 Sunset Avenue, Windsor, ON N9B 3P4, Canada
* Correspondence: msafiuddin@georgebrown.ca

Abstract: Efforts have been put in place to minimize the effects of construction activities and occupancy, but the problem of greenhouse gas (GHG) emissions continues to have detrimental effects on the environment. As an effort to reduce GHG emissions, particularly carbon emissions, countable commercial, industrial, institutional, and residential net-zero energy (NZE) buildings were built around the globe during the past few years, and they are still operating. But there exist many challenges and barriers for the construction of NZE buildings. This study identifies the obstacles to developing NZE buildings, with a focus on single-family homes, in the Greater Toronto Area (GTA). The study sought to identify the technical, organizational, and social challenges of constructing NZE buildings, realize the importance of the public awareness in making NZE homes, and provide recommendations on how to raise public knowledge. A qualitative approach was employed to collect the primary data through survey and interviews. The secondary data obtained from the literature review were also used to realize the benefits, challenges, and current situation of NZE buildings. Research results indicate that the construction of NZE buildings is faced with a myriad of challenges, including technical issues, the lack of governmental and institutional supports, and the lack of standardized measures. The public awareness of NZE homes has been found to be very low, thus limiting the uptake and adoption of the new technologies used in this type of homes. The present study also recommends that the government and the academic institutions should strive to support the NZE building technology through curriculum changes, technological uptake, and financial incentives to buyers and developers. The implementation of these recommendations may enhance the success and popularity of NZE homes in the GTA.

Keywords: buildings; construction; efficient homes; energy efficiency; GHG emissions; homes; net-zero energy (NZE); NZE building technology

Citation: Makvandia, G.; Safiuddin, M. Obstacles to Developing Net-Zero Energy (NZE) Homes in Greater Toronto Area. *Buildings* **2021**, *11*, 95. https://doi.org/10.3390/buildings11030095

Academic Editors: Ala Hasan, Francesco Reda and Umberto Berardi

Received: 5 January 2021
Accepted: 25 February 2021
Published: 4 March 2021

Publisher's Note: MDPI stays neutral with regard to jurisdictional claims in published maps and institutional affiliations.

1. Introduction

Every year, numerous commercial, industrial, institutional, and residential buildings are constructed around the world. Iyer-Raniga [1] reported that almost 235×10^9 m^2 of building floor area was constructed by 2016, and an additional 230×10^9 m^2 will be constructed by 2056. A large amount of energy is required to operate the buildings, which may significantly add to greenhouse gas (GHG) emissions if non-renewable energy sources are used. Ontario consumes a significant portion of primary energy in its building sector, and as a developing region, this share is estimated to continue to grow in the future due to today's fast urbanization. Buildings account for around 33.3% of the final energy demand and 30% of GHG emissions in the globe [2]. As of 2018, buildings were the largest source of GHG emissions in Toronto, accounting for about 55% of the total community-wide emissions [3]. This is concerning with respect to the effects of GHG emissions on the environment. Therefore, more efficient buildings are required for energy savings and reduced environmental pollution. The goal of net-zero energy (NZE) buildings is to produce enough renewable energy to meet buildings' annual energy consumption requirements by zeroing the use of non-renewable energy [4–6], thus decreasing GHG emissions [7]. The annual onsite renewable energy for NZE buildings should be equal to or greater than the actual annual energy demand [8,9]. Achieving this energy condition is not very straightforward. This paper analyzes the different obstacles to constructing commercial, industrial, institutional, and residential NZE buildings across the Greater Toronto Area (GTA) with a focus on single-family homes.

NZE buildings require both energy-saving and energy-producing approaches. Passive solar design, effective building envelope, energy-efficient appliances, and renewable energy sources are essential for these buildings. The energy requirement of buildings can be reduced by orienting the structure properly so that it can gain more solar heat during the winter but less during the summer, and getting daylight is also important as it can reduce the energy spent for indoor lighting [10]. Moreover, substantial energy can be conserved using high-performance insulation in wall, floor, and roof systems [11]. Energy consumption in buildings can also be significantly decreased using energy-efficient appliances, HVAC (heating, ventilation, and air conditioning) equipment, and light fixtures [12]. On the other hand, sunlight and solar heat are commonly used as a source of renewable energy for buildings [13,14]. Besides, wind, biomass, and geothermal energy can be used for various building services [15]. The other aspects such as advanced framing, triple-pane windows, and open-web trusses are also considered for constructing NZE buildings [16].

Research on NZE buildings recognizes the importance of design and acknowledges the role of construction in the development of these buildings to reduce GHG emissions and reverse the adverse effects of global warming. Certain studies focused the design and technical requirements of NZE buildings, and how they impact the environment [17,18]. Many published works show that NZE buildings have been built in different parts of the globe, including Canada, and they are still operating [14,19,20]. However, in Ontario, the past research on NZE buildings mainly focused on energy consumption rather than evidence-based approaches to NZE housing in different areas, including the GTA. Necessary steps were not taken to effectively minimize foremost challenges such as technological barriers, and to increase the governmental and institutional supports for NZE homes. Also, the importance of the public awareness was somehow not prioritized. To build more NZE homes in communities, people who live in those communities and are paying to buy properties must be educated about efficient homes. More awareness will lead to an increase in demand for NZE homes. This study has examined the degree of the public awareness of NZE homes in the GTA.

Natural Resources Canada (NRC) took initiatives in 2013 to enhance the development of NZE buildings in Canada, starting with a pilot project called "The R-2000 Net-Zero Energy Pilot" [21]. The purpose was to encourage and support the Canadian construction companies to build NZE homes, and rate these homes using NRC's EnerGuide Rating System and mark them under the R-2000 standard. Thereafter, by 2016, six construction companies built twenty-three NZE homes in three provinces, including Ontario, and the feasibility of these homes on a community level in Canada was also demonstrated to increase their acceptance in the housing market [21]. With all these efforts, NRC has developed the methodology to evaluate the NZE aspects of buildings that includes modified home energy rating scale, whole house energy analysis, assessment of emerging technologies, feasibility studies, mechanisms for enabling construction industry to adopt renewable energy sources, establishment of energy model and guidelines, and field appraisal [22].

Despite the vast resources and technologies in today's construction industry, there are many challenges along the way to develop energy-efficient buildings [23]. NZE homes are built based on certain new technologies, and thus are not yet popular. It is, therefore, important to identify the constraints which may hinder the development of NZE homes. The public awareness is also important because it creates opportunities for the acceptance of new technology. This study first identifies and analyzes the technical, organizational, and social challenges of building NZE homes. Secondly, the study is intended to establish the importance of public knowledge in promoting NZE homes. Finally, the study aimed at providing recommendations on how to raise public knowledge on NZE homes. These recommendations, when applied, could help increase the public awareness of NZE homes and consequently eliminate some of the obstacles during implementation.

2. Literature Review

The intensification of commercial and residential construction within the City of Toronto and close to major transit nodes is contributing to high energy consumption and environmental degradation. Efforts have been put in place to minimize the effects of construction activities and occupancy, but the problem of carbon emissions continues to have harmful effects on the environment. To alleviate this problem, Toronto's leading developers have developed mechanisms that reduce emissions and improve the energy performance of buildings [24]. One such mechanism is the adoption of NZE technology for making homes. Various definitions of NZE homes exist but the overall concept is having a building which generates as much energy as it consumes with zero carbon emissions. Iyer-Raniga [1] defines the NZE home as a building that has annual net-zero energy consumption from the power grid and net-zero carbon emissions. NRC agrees that an NZE home focuses on producing at least as much energy as it requires on an annual basis [22]. Numerable new buildings have already been built in the GTA and other cities or towns of Ontario using NZE technology [25–27]. However, more pace is required to help the government achieve the targets of its climate action plan through reduced carbon emissions.

There are barriers for the development and construction of NZE buildings in the GTA. The lack of knowledge on how to integrate conventional electrical supply to the green energy system of the home by the electric utility company personnel may undermine the performance of energy components and systems. Pettit et al. [16] observed that NZE design requirements such as advanced framing, open-web floor trusses, high-performance wall and roof enclosure, and solar power integration need advanced skills and expertise. Piderit et al. [28] agreed that NZE home encompasses a complex system that needs skilled professionals with knowledge across different engineering, scientific, and technical disciplines. The construction industry in the GTA is still adjusting to the multi-disciplinary design approach for NZE buildings. The principal design requirement is to integrate many energy-saving and energy-producing techniques that are suitable and economically feasible for a green construction project. For this, an efficient energy use analysis mechanism helps in the projection of the total annual energy consumption of a building.

Improving energy efficiency is initially costly due to the complex technical requirements of NZE buildings. Also, the energy use in the household may vary from one place to another depending on climate conditions, thus affecting the design and development of NZE homes. Hamilton [29] reported that surveys on homes with similar energy-saving systems showed variations in energy use across different climates and regions in Canada; for example, the average annual home energy cost in Ontario is $2358 and that in Nova Scotia is $2903. Moreover, the initial construction cost becomes higher for NZE homes than traditional homes. The case study of NRC [22] showed that their NZE home is not yet market-feasible because of the large first cost ($100,000 to $150,000) to achieve it. Initial construction costs increase significantly due to the lack of technological knowledge and industry experts who possess the technical skills and experience to maximize the benefits of passive design and install energy-producing components in homes. Hence, certified NZE home builders should be available in plenty to enhance the growth of NZE homes. But green energy developers lack an effective certification and assessment system solely for NZE housing in the GTA. Although Toronto's leading developers have developed mechanisms to decrease GHG emissions from buildings and enhance their energy performance, the study by Pettit et al. [16] revealed that Chile and Canada have challenges in establishing the best practices and standards for NZE homes in the construction industry. Although developers are incorporating LEED Canada (LEED means Leadership in Energy and Environmental Design) certification for homes [30], it is not specially formulated for NZE construction. Without a standardized certification system exclusively for NZE homes, the construction of these homes may fail.

Government policy plays a major role in the development and popularity of NZE homes. The City of Toronto has green standards on energy efficiency and renewable energy for low-rise residential and mid- to high-rise residential and non-residential buildings [31,32]. The City also has guidelines regarding the district energy system and operational systems for mid- to high-rise residential and non-residential development [32]—these standards and guidelines could be useful to support the development of NZE buildings in the GTA. However, McVey et al. [33] stated that regional regulations limit the ability of local distribution companies to establish district energy networks and serve as aggregators of distributed energy generation and storage resources. Their analysis of the policies indicated that the existing rules on the participation of combined heat and power projects in electricity markets create further barriers for coordinating renewable energy sources into the power system. The lack of policies supporting renewable energy development at the district level undermines the development of NZE homes. Currently, Toronto lacks NZE infrastructures that provide capital and technical support to contractors and specialists. The withdrawal of federal and provincial funding in the Port Lands project debilitated the sustainable development of NZE community buildings. The lack of capital and funding destabilizes the process of innovation in the construction industry through research and development.

Recent trends in building construction create limitations for the development of NZE housing in the GTA. The high costs of labor and land are contributing to unfavorable building design strategies [24]. For instance, developers are using low-cost building materials and minimizing their residential suite size and floor-to-floor height to address the high land and labor costs. However, these strategies have only served to minimize the costs by compromising the quality of buildings, and they are not conducive to promote NZE homes. Eliminating the barriers to NZE housing by improving micro-power grid access, easing strict permits, and enhancing the process of certification for construction companies will significantly reduce the cost of construction. Zhang et al. [23] argued that the introduction of utility energy pricing policy will offer developers realistic economic returns for energy efficiency and contribute to reducing GHG emissions. Besides, Wills et al. [34] identified NZE as a performance target, which has not been implemented by the government. The revision of Canada's energy policies and new targets for the building sector to facilitate the development of net-zero ready building codes will improve the standardization processes. In addition, facilitating the technological development through research grants will improve the quality of solar, wind, and heating technologies for use in NZE households in the GTA and other areas of Ontario. Piderit et al. [28] supported the claim that government funding provides incentives for construction experts to test new designs, products, and mechanisms in a risk-free environment.

Along with technical barriers and government policy issues, the lack of awareness among residents on the benefits of NZE homes has limited their market penetration in the GTA. People are reluctant to invest in projects whose benefits they do not clearly understand, especially in such a case where the alternative is cheaper. The results of personal interviews reported by Hamilton [29] demonstrated the lack of public education on the benefits of NZE homes in promoting energy sustainability in Ontario. Whereas more citizens are becoming conscious of their role in household energy emissions, they lack the basic information on how NZE housing can reduce GHG emissions and improve sustainability.

3. Methodology

This study mostly uses a qualitative research approach to examine the barriers to the development of NZE buildings, particularly single-family NZE homes, in the GTA. Qualitative research is exploratory in nature, and hence it was better suited for this study whose main aim was to identify the obstacles for NZE homes by gathering information from the public and building experts. Surveys and interviews were used to collect information on the degree of the public awareness of NZE homes. Major barriers or challenges were identified largely from the interview results and considerably from the survey data. The data collected were mainly subjected to descriptive analyses to provide an overall picture of the research findings. The significance of the survey data was also judged using statistical means. The overall methodology is schematically shown in Figure 1.

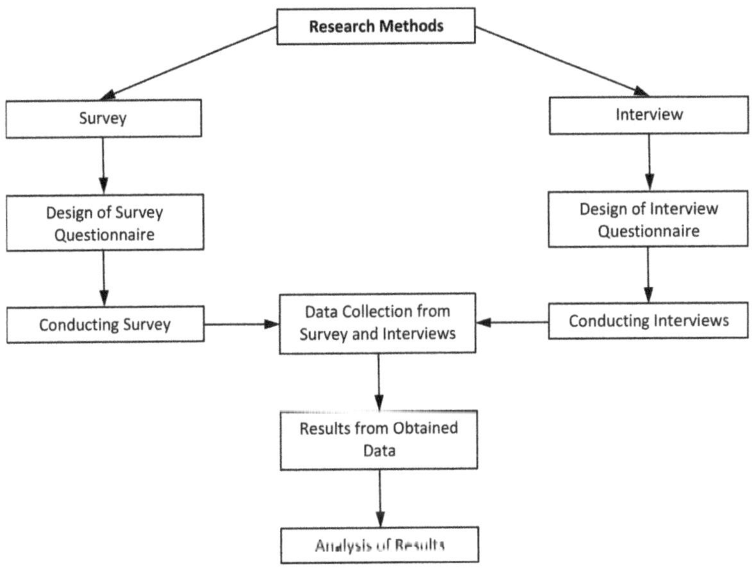

Figure 1. Overall research methodology.

3.1. Survey

This study used survey as a primary method of data collection to achieve the research objectives. Survey allows to gather more data for nonexperimental research—researchers describe the findings by observing the collected data and analyzing how variables are related to each other [35]. It is also widely used in nonexperimental building research [12,29,36]. For the present study, the survey questionnaire comprising nine questions (without any abbreviations), as shown below, was distributed through SurveyMonkey. The link for the survey questionnaire was also shared on LinkedIn. More emphasis was given on NZE homes while designing the questionnaire for survey.

(1) What is your age? Please indicate.
(2) What is your occupation? Please select the best answer.
(3) How would you describe your household annual income?
(4) How much do you know about net-zero energy homes?
(5) Would you be interested to purchase a net-zero energy home?
(6) If you are interested to purchase a net-zero energy home, do you think the Ontario government would support that financially?
(7) Would you be willing to pay extra for a home that helps keep our environment greener?
(8) Do you think net-zero energy homes will help our environment?

(9) If you already have a house, would you be interested to install solar panels on your roof?

For the sake of clarity and the ease of analysis, close-ended questions were used to obtain information from the survey participants. The participants included varying categories of people in terms of age, occupation, and financial status. The survey collected information on age, occupation, income level, knowledge on NZE homes, the views on government involvement, and the level of interest in NZE homes among participants. The respondents fell within the range of 18 to over 65 years of age with variable household incomes.

According to Statistics Canada's 2016 Census, the total population of the GTA in 2016 was 5,928,040; among them, the population in the age range of 25–64 years was 3,305,380 (55.76%) and the working population with various occupations was 3,286,350 (55.44%) [37]. The employed population might have the financial capability to buy a home, as most of them are probably earning annually $50,000 and above. Indeed, they are the main driver for the growth of NZE homes in the GTA, because they are most capable of buying these homes due to their economic solvency. Therefore, this population was considered in the present study to determine the sample size. The necessary sample size was calculated based on the following Equation (1) [38,39]:

$$n = \frac{Nz^2p(1-p)}{e^2(N-1)+z^2p(1-p)} \tag{1}$$

where,

N = Population size (employed population)
n = Required sample size
p = Expected population proportion (prevalence)
z = Statistical parameter (z-score) for a confidence level (CL)
e = Margin of error (level of precision)

The value of p needs to be estimated before conducting any new survey. To get a larger sample size, researchers should use 0.5 as an estimate of p [39–41]. The z-score depends on CL: it is 1.65 for 90% CL, 1.96 for 95% CL, and 2.58 for 99% CL [42,43]. In survey research, 95% CL is commonly used; however, 90% and 99% CLs are also used depending on the type of survey work [44,45]. For a survey that deals with a smaller sample size, a 90% CL is acceptable [42,44]. Moreover, a margin of error (e) needs to be assumed to calculate the sample size from Equation (1). The commonly used range for the margin of error is 1–5% [45], but it can be in the range of 1–10% [42,44,46]. In some cases, the margin of error can be even more than 10%. Naing et al. [41] mentioned that researchers may use a larger margin of error (e.g., >10%) in the case of a preliminary study and where there is a resource limitation. In the present study, a 10% margin of error at 90% CL was specified to calculate the sample size. For this margin of error, the necessary sample size found from Equation (1) was 68. However, it should be mentioned that this sample size was found using $p = 0.5$, which maximizes the sample size. A smaller sample size will be obtained when $0.5 < p < 0.5$ [41], which practically can occur in survey work. Also, the sample size will be reduced if a margin of error >10% is allowed [47].

3.2. Interview

This study used interview as another method of data collection to attain the research objectives. Interview is an important method of data collection for research, where researchers typically gather data through verbal communication with the participants [48]. This technique is also used in building research [49,50]. For the present study, three interviews were conducted that targeted professionals in the construction industry who have strong backgrounds and experience in the design and construction of NZE buildings. Candidates were selected based on the number of years of experience, exposure to the fundamental elements of design, and application of those elements to the construction of buildings. Andrew Bowerbank and Tony Cupido are two of the interviewees who played critical roles in the Mohawk College project, "The Joyce Centre for Partnership and Innovation", one of the few recently known projects with the focus of committing to high performance targets like NZE and zero emissions [26]. They were interviewed over the phone. Andrew Bowerbank is a former Executive Strategist at EllisDon, Mississauga, Ontario, for sustainability, energy, and innovation. Tony Cupido, Ph.D., P.Eng., is a Research Chair, Sustainability at Mohawk College, Hamilton, Ontario, and led the planning, design, construction, and early operation of the Joyce Centre for Partnership and Innovation. The other interviewee was Adrian Wang, who worked as the former Director of Innovation and Sustainability at Tridel, which is a leading developer in the GTA. The interview with Wang was conducted in person during the tour of one of the first residential NZE suits built in Toronto.

The interviews were guided by the nine key questions, as listed below, but more elaborate information was gathered through probing questions. The questionnaire for interviews was designed giving more focus on public knowledge and challenges regarding NZE buildings (e.g., single-family homes):

(1) How would you describe NZE buildings?
(2) Name key benefits of NZE buildings.
(3) From your experience, do you believe the public have a good knowledge of this subject?
(4) How do you describe the role of the government and academic institutions to advertise sustainability?
(5) How do you think we can improve public knowledge?
(6) Tell me about the challenges of designing NZE buildings?
(7) Tell me about the challenges of constructing NZE buildings?
(8) Can you tell me about the financial impact on the investors? What is the average payback of the additional costs?
(9) Where do you see the future of NZE buildings?

4. Results

The results obtained from the survey are presented in Figures 2–10. These are further explained in the text referring to the corresponding figure. Additionally, the results obtained from the interviews are presented as text paragraphs with each question in the interview covered separately. Section 4.1 hereafter presents the survey results while Section 4.2 presents the results of the interviews.

4.1. Survey Results

There were 58 survey respondents who answered all the survey questions, though the expected number was more than 68, which is required for a 10% margin of error at 90% CL. This is 85.3% of the necessary sample size (e.g., 68). Moser and Kalton [51] stated that the results of a survey can be biased and of minimal value if the response rate is <30–40%. Hence, the above respondent number was considered acceptable for the analysis of the survey results. The actual margin of error was calculated for the obtained survey data using the following Equation (2):

$$e = z\sqrt{\frac{p(1-p)}{n}} \quad (2)$$

where,

n = Sample size, it is 58 in this study
p = Expected population proportion (prevalence), a value of 0.5 has been assumed
z = Statistical parameter (z-score) for a confidence level (CL), it is 1.65 in this study
e = Margin of error (level of precision)

The margin of error obtained from Equation (2) is 10.8% at the 90% confidence level. In terms of standard error (margin of error/z-score), it is 6.6%. These error values are obtained for $p = 0.5$. The level of error decreases with the larger sample size and higher response rate [52]. If the number of the respondents were 68, which is the calculated sample size, the margin of error would have been 10%, as specified. Moreover, the positiveness of the people and their greater participation in a survey lessen the level of error with increased response rate. For instance, considering the response rate of 85.3% based on the necessary sample size, if $p = 0.853$ instead of $p = 0.5$ is used in Equation (2) for $n = 58$, it gives 7.6% margin of error, which is equivalent to 4.7% standard error. Nevertheless, to reduce the level of error, the authors conducted several interviews in addition to the survey. It was also done to ensure the reliability of the survey data. Indeed, the interview and survey results were consistent for related questions, particularly in the viewpoint of the public awareness, and no serious discrepancies were observed.

4.1.1. Age Distribution of Participants

The respondents were divided into seven groups based on their age, as shown in Figure 2. The percentages of the respondents for different age groups were as follows: 0% (under 18 years), 13.79% (18–24 years), 39.66% (25–34 years), 25.86% (35–44 years), 10.34% (45–54 years), 5.17% (55–64 years), and 5.17% (65+ years).

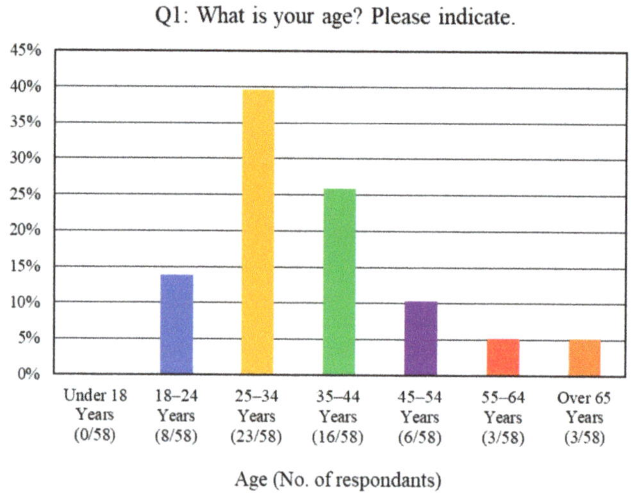

Figure 2. Participants' age distribution.

The few respondents (5.17%) who were aged above 65 years may not be really interested in new home ownership since most of them are already settled in their retirement homes and moving homes may be quite inconvenient for them. Consequently, they may not be interested in NZE homes.

4.1.2. Occupation of Participants and Potential Home Buyers

Potential homebuyers constituted the largest percentage (32.76%) of the sample, as can be seen from Figure 3. This number represents the respondents from the various age brackets planning to buy homes in the future. Developers and professors constituted the smallest percentages of the samples with 1.72% and 3.45%, respectively. Of the respondents, 25.86% were students, 10.34% were consultants, and 25.86% were construction professionals. The respondents may fall into multiple categories. However, the question regarding occupation only allowed each of them to choose one category, which best describes them.

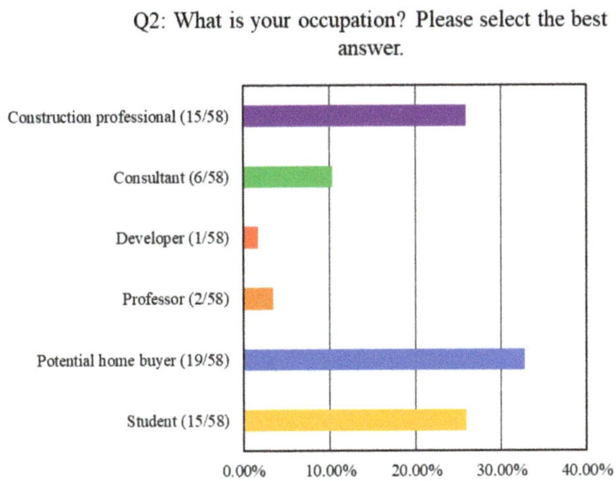

Figure 3. Participants' occupation and potential home buyers.

4.1.3. Annual Income of Participants

The respondents came from varying financial categories. Most of the respondents (53.44%) were in the upper categories ($75,000 and above), as shown in Figure 4. The respondents with earning below $15,000 annually constituted the smallest fraction (5.17%) of the sample. A section of the respondents (13.79%) earns above $150,000 per year. The middle-income group earning between $50,000 and $74,999 made up 17.24% of the sample. Of the respondents, 24.13% earn between $15,000 and $49,999 per year.

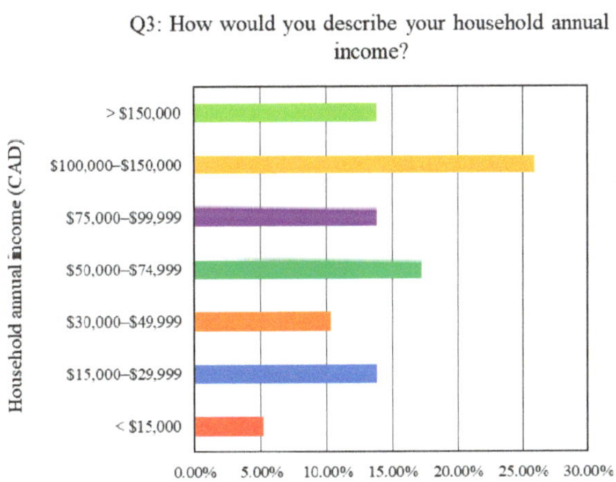

Figure 4. Participants' annual income (amounts are in Canadian dollars).

4.1.4. Knowledge of Participants on NZE Homes

The respondents had varying knowledge levels on NZE homes. Only 17.24% of the survey sample was sufficiently familiar with the concept of NZE homes, with only 6.90% stating that they were extremely familiar with these efficient homes, as can be seen from Figure 5. On the other hand, 22.41% of the sample had no idea about NZE homes.

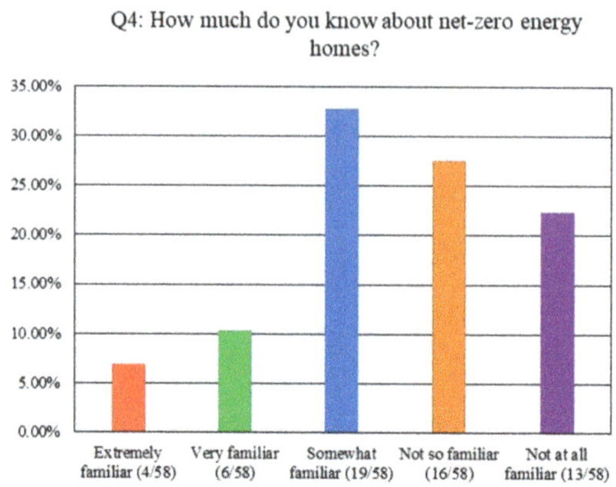

Figure 5. Participants' knowledge on net-zero energy homes (degree of familiarity).

4.1.5. Level of Interest of Participants in NZE Homes

The respondents' level of interest in buying NZE homes is shown in Figure 6. Among the respondents, 10.34% and 29.31% were extremely interested and very interested, respectively. Half of the respondents were somewhat interested in NZE homes, whereas 5.17% were not so interested and another 5.17% were not at all interested. In total, 39.65% of the participants were greatly interested in NZE homes.

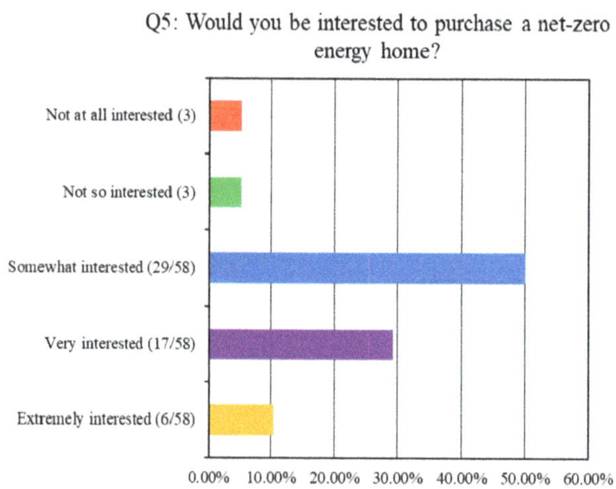

Figure 6. Participants' interest in NZE homes (degree of purchase interest).

4.1.6. Thought of Participants on Governmental Support for Purchase of NZE Homes

In the survey questionnaire, the respondents were asked whether they think the government would support individuals interested in buying NZE homes or not. The survey results are shown in Figure 7. Of the survey respondents, 39.66% said 'yes', 22.41% said 'no', while 37.93% said 'I do not know'. The overall data confirm that many people believe they would get governmental support to purchase NZE homes, while a significant portion thinks they will not get such support.

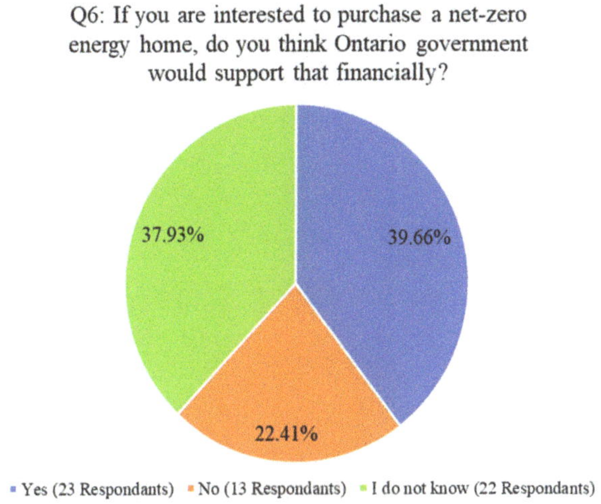

Figure 7. Participants' thoughts on governmental support towards purchase of NZE homes.

4.1.7. Willingness of Participants to Pay More for NZE Homes

NZE homes require extra installations that would lead to additional costs on the home. The respondents were asked whether they would be willing to pay more for these homes or not, and the majority (74.14%) said 'yes', while the rest (25.86%) said 'no', as can be seen from Figure 8. Although many respondents were not interested initially, they became willing to pay more for NZE homes after knowing the benefits of NZE technology.

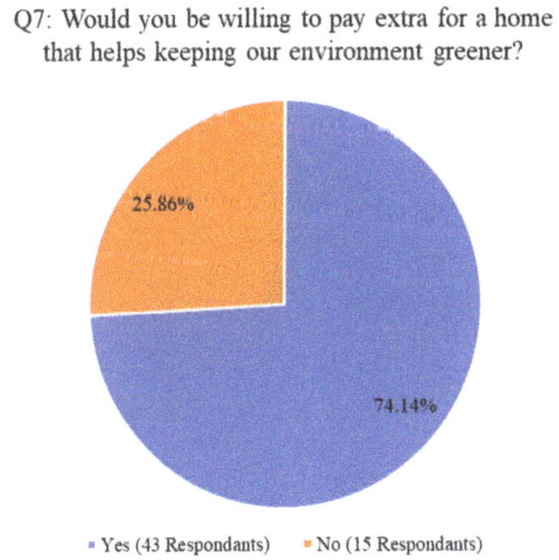

Figure 8. Participants' willingness to pay more for NZE homes.

4.1.8. Thought of Participants on Environment-Friendliness of NZE Homes

After explaining the concept of NZE homes, the respondents were asked whether they agree or not that these homes would help protect the environment. Among the respondents, 22.41% were neutral—they neither agreed nor disagreed—while the rest fully agreed that these homes would help save the environment, and 29.31% agreed strongly while 48.28% agreed, as can be seen from Figure 9.

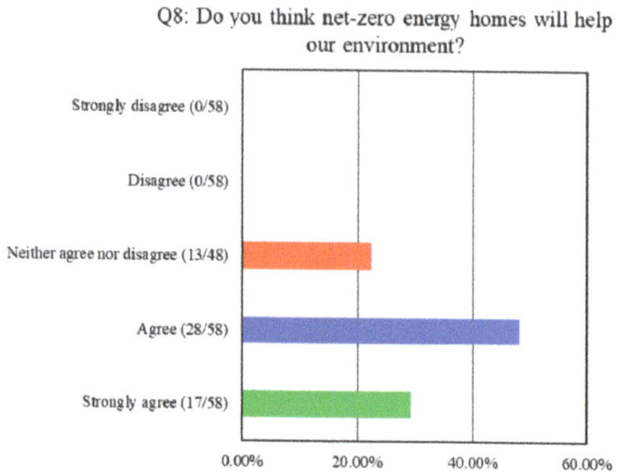

Figure 9. Participants' agreeance or disagreeance on environment-friendliness of NZE homes.

4.1.9. Interest of Participants in Installation of Solar Panels on Existing Homes

After explaining the concept of solar panels, the respondents who own homes were asked whether they would consider installing solar panels on their homes or not. The majority were clearly interested, as can be seen from Figure 10. Of the respondents, 24.14% and 20.69% were extremely interested and very interested respectively, with 32.76% responding that they were somewhat interested. However, 6.9% were not so interested, and 1.72% were not interested at all. Also, 13.79% answered that it is not applicable for them. The respondents may have varying energy requirements and they might also understand green energy differently. These variations may bring about the differences in their interests in solar energy.

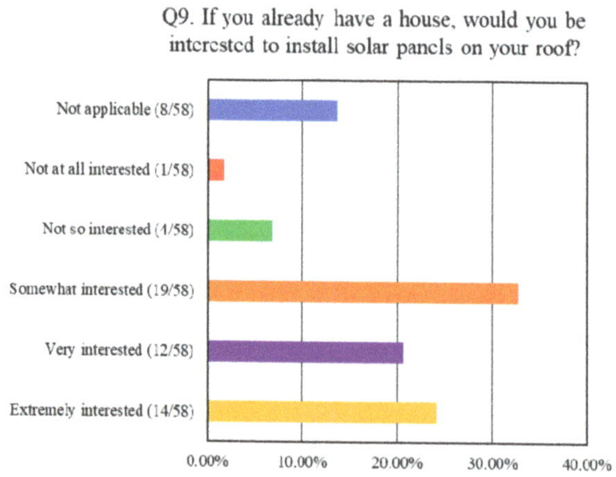

Figure 10. Participants' interest in installation of solar panels on their homes.

4.2. Interview Results

The three interviewees revealed a range of information from a professional perspective regarding various aspects of NZE buildings. The information gathered provides some valuable insight into this new building technology. The interview results are summarized in the following sections.

4.2.1. Definition of NZE Buildings

When the interviewees were asked what is meant by NZE buildings, the responses culminated in one definition: a building that produces as much energy as it consumes annually, which is consistent with the definitions discussed in Liu et al. [6] and the one given by Iyer-Raniga [1]. In defining "net-zero", the "net" is measured annually since there are peaks and troughs in both production and consumption of energy in different seasons of the year; thus, they balance out when an annual analysis is done.

4.2.2. Key Benefits of NZE Buildings

The most important benefit mentioned by all interviewees is that NZE buildings do not consume any energy from the local power grid. This means that a power utility bill is eliminated when occupying an NZE home. Besides, they stated that an NZE home is a net-zero carbon building as well. This is because the house relies entirely on the green energy generated on-site and does not require any additional energy from the power grid, thereby eliminating any possibility of using non-renewable energy whose manufacture and consumption lead to carbon emissions [53]. NZE homes contribute to the long-awaited distribution grid systems. Power is generated and distributed from different points, thus decentralizing the power grid. NZE requires buildings to be constructed in a high quality and efficient manner, thus leading to homes that guarantee better occupant health, better building performance, and increased durability and longevity of the building.

4.2.3. Public Knowledge on NZE Homes

The interviewees agreed that the public lacks knowledge on NZE homes, which was also obvious from the survey results (refer to Section 4.1.4 and Figure 5). They stated various probable reasons, including the lack of interest, the high technicality of these buildings, poor sales and marketing strategies, slow adoption of the technology by the industry, and the newness of this technology. Additionally, it was hinted that construction companies are limited by capital incentives as to how far they can spread their wings. Therefore, most construction companies are localized, and consequently, new technologies may not expand to other areas.

4.2.4. Role of Government and Academic Institutions to Advertise Sustainability

The interviewees feel that the government is not doing enough to incentivize the public towards newly developed NZE homes, although in general, many people believe they would get governmental support to purchase this type of home (refer to Section 4.1.6 and Figure 7). The use of green energy has been widely identified as an effective way to protect the environment. However, efficient homes had not been upheld by the government, particularly the provincial or local municipal government, as essential efforts towards environmental protection despite their clear benefits. Additionally, academic institutions have been slow in the uptake of NZE technology despite their clear advantage over other institutions to raise the public awareness. Training institutions have the capacity to build NZE buildings to be used as demonstration centers while incorporating different modules of NZE homes in their curriculums to train the future generations of developers and construction experts on this technology.

4.2.5. Measures for Improving Public Awareness

On ways to improve the public awareness of NZE homes, the building experts stated that it might not be very easy. They stated that creating mass acceptance may be quite challenging. The public awareness can only be created gradually and patiently. Climate change advocacy groups can incorporate green homes as part of their suggested solutions to climate change issues. At the same time, educationists can teach NZE building technology in colleges, universities, and other training institutions to cultivate the acceptance in learners so that they can apply this when their time comes.

4.2.6. Challenges of Designing and Constructing NZE Buildings

The interviewees stated that there were not any significant challenges reported regarding the design of NZE buildings. However, the lack of skills and professional capacity was mentioned as one of the major problems in the construction process. This is consistent with the challenges and barriers reported by McNabb [54]. One of the building experts stated that, for a successful project, every player must understand the importance of quality, specificity, and the variation of NZE homes from conventional buildings to play his or her part as per expectation. This understanding is currently lacking in the industry.

4.2.7. Financial Impact on the Investors

There are additional costs in the construction of NZE homes. It translates to a higher market price when compared to a similar conventional home, and this is consistent with what Hamilton reported in the past [29]. His report revealed that new homebuyers would need to spend at least $35,000 more for a single-family NZE home. However, the extra cost will be catered for by the savings on energy, which will be realized by the occupant over time. One of the building experts pointed out that the payback on the extra cost will depend on how the occupant uses the energy-efficient features of a house and live on NZE, thus eliminating the cost of home energy. Also, another building expert indicated that the payback period on the extra cost will significantly decrease if more units of NZE homes are built.

4.2.8. Future of NZE Buildings

On the future of NZE buildings, the building experts projected that the technology would spread and be widely adopted. One of the interviewees mentioned that many of his colleagues in colleges and universities are designing net-zero energy or zero-carbon buildings. More people will focus on energy conservation, NZE, and net-zero carbon, thus increasing the popularity of NZE homes. Consequently, it will bring forth enormous energy savings and mitigate GHG emissions. Such positive impacts were also discussed by Feng et al. [55] in the context of China. They predicted that the overall primary energy savings in the range of 270–320 million tons of coal equivalent (Mtce) annually in 2030 and 740–1060 Mtce annually in 2050 will be achieved in China, mostly through energy efficiency measures.

5. Analyses of Survey and Interview Results

5.1. Participants' Age Distribution

The participants were in the age range of 18 to over 65 years. The respondents aged between 18 and 24 years are naturally expected to be students or in their entry careers. They may have varying degrees of knowledge about NZE homes, depending on their field of education, expertise, and exposure. Individuals aged between 25 and 44 years are expected to be the first-time or potential homebuyers. These are the people who may have a plethora of knowledge on homeownership, including prices and new developments in the industry. Hence, the people in the age groups of 25–34 and 35–44 years shall be more interested in purchasing NZE homes. Participants aged between 45 and 54 years are most likely homeowners and already settled down. But they may be termed as potential homebuyers if they plan to buy other residential properties. They have gained quite an experience in homeownership and the real estate business. Some from this age group may decide to purchase an NZE home. Individuals aged over 54 years of age may not have vast experience and knowledge in NZE homes because they are relatively new in the housing market. However, some of them, especially those in the construction and real estate industries, may be well-versed with the idea, and go for purchasing an NZE home.

5.2. Participants' Occupation and Potential Home Buyers

Participants fell into different occupational categories. However, participants may have fallen into multiple categories, but the study limited each participant to one category.

The participants were to choose what describes them best. Potential homebuyers (32.76%) represent the respondents from the various age brackets and economic statuses planning to buy homes in the future. These are mostly people in the age groups of 25–34 and 35–44 years who are already working, probably in their early career years or those who have been working but have not yet managed to buy homes. These individuals may be professionals in different fields. The respondents who identified themselves as consultants and construction professionals may also be potential homebuyers.

5.3. Participants' Annual Income

The survey participants were from different financial categories. The different financial abilities indicate the disparity in the purchasing power of individuals. An individual's financial ability dictates the kind of housing s/he would consider and the features that would make up such an individual's ideal housing. Most participants (53.44%) have an annual income above $75,000. These individuals could be more interested to purchase NZE homes due to their higher financial ability.

5.4. Participants' Knowledge on NZE Home

Most participants had not enough knowledge on NZE homes, as can be seen from Figure 5. Among the respondents, 32.76% were somewhat familiar and 27.59% were not so familiar with NZE buildings. Only 17.24% knew most about NZE homes, and many of these individuals are probably consultants and construction professionals. Most of them are in the age groups of 25–34 and 35–44 years. Furthermore, 22.41% of the respondents were totally unaware of NZE homes. Many of this group may not be interested in these energy-efficient homes unless they rightly know their benefits.

5.5. Participants' Level of Interest in NZE Homes

The participants had varying degrees of interest in buying NZE homes, as obvious from Figure 6. The degree of interest may vary according to the degree of awareness and understanding of the concept of NZE homes. Among the respondents, 39.65% were highly interested in NZE homes. These people perhaps know and understand NZE homes and the benefits of having such a home, possibly because of their building-related occupation (e.g., consultants, construction professionals). They are likely to be in the age groups of 25–34 and 35–44 years. Moreover, many participants were slightly interested in NZE homes and they comprise the largest segment (50%), as can be seen from Figure 6. They may have an idea of what NZE homes are all about but have not understood the concept clearly. Some of them may shift their interest once they fully understand the concept of NZE homes. Those who were not so interested and disinterested collectively constituted 10.34% of the individuals participated in the survey. They are the smallest in number, as evident from Figure 6. The lack of awareness is a major contributing factor to the lack of interest.

5.6. Participants' Willingness to Pay More for NZE Homes

The initial cost of purchasing NZE homes is higher compared with conventional homes [29]. But most of the individuals who participated in the survey became willing to pay more for NZE homes when they were informed of the benefits of having these homes. It is evident from Figure 8 that 74.14% of the respondents became willing to spend more money for purchasing an NZE home. They are mostly from the age groups of 25–34, 35–44, and 45–54 years, with an annual income of $50,000 and above.

5.7. Participants' Thoughts on Environmental Friendliness of NZE Homes

The idea of NZE homes was brought up as a way of protecting the environment from harmful GHG emissions and exploitation of natural resources [24]. NZE homes produce as much energy as they consume [4]. Also, NZE homes produce energy from renewable sources, such as sunlight and solar heat, and they consume less energy [11–13]. After knowing these facts, 77.59% of the participants agreed that NZE homes are environmentally friendly, as obvious from Figure 9. This number would rise with the appropriate dissemination of knowledge and enhanced public awareness of NZE buildings.

5.8. Participants' Willingness to Install Solar Panels

One way of achieving NZE homes is through using solar energy [13,14]. Solar panels convert solar energy (e.g., sunlight) into electrical energy. This is a comparatively new technology. Earlier on, the use of solar energy was not commonly applied in residential construction. Therefore, the owners of old homes can only take the advantage of solar energy through the installation of solar panels. After knowing this fact, most of the respondents (77.59%) became interested in installing solar panels on existing homes, as evident from Figure 10. These individuals are mostly different professionals who are earning annually $50,000 or more and prepared to install solar panels. Within the remaining 22.41%, 8.62% were not interested and 13.79% said that it is not applicable to them. Some of these individuals might not need solar panels depending on where they are living, while the others may not have the financial capability to install these panels.

5.9. Definition of NZE Homes

Various definitions exist to describe the concept of NZE home [4,6]. The three experts interviewed defined NZE homes as buildings that produce as much energy as they consume and agreed that these homes are either zero- or low-carbon as well. However, Andrew Bowerbank noted that net-zero is an ideal measure, but the idea is to come as close to this as possible. The primary benefit stated for these buildings is that they use minimal to no energy from the power grid, and thus significantly reduce the energy cost. On the other hand, Tony Cupido, a building expert, stated that sometimes an NZE building may add some energy to the power grid if it produces more renewable energy than it consumes.

5.10. Challenges Faced in Design and Construction of NZE Homes

Various challenges exist against the development and construction of NZE homes in the GTA. They are discussed below.

5.10.1. Technological Barriers

The interviewed experts pointed out the lack of technological know-how as a major challenge faced in the construction of NZE homes. This corroborates Piderit et al. [28], who observed that NZE home involves an intricate system that requires competent professionals with experience across various engineering, scientific, and technical fields. The common challenges in design and construction include finding the right interdisciplinary team as well as getting suppliers to deliver the specific materials or products as required for NZE homes. This is because the idea is quite new and qualified professionals educated with NZE building technology are not yet abundantly available in the construction industry. The deficit of knowledge on how to integrate the regular electrical supply with the green energy system of homes by the electric utility company staffs may emasculate the performance of energy components and systems. The principal design requirement is to integrate many energy-saving and energy-producing techniques that are suitable and economically feasible for a particular construction project. Pettit et al. [16] asserted that net-zero design requirements such as advanced framing, high-performance wall and roof assemblies, open-web floor trusses, and solar power integration require advanced competences and expertise. For this, architects and engineers should be well-versed with NZE building technology before applying it to make homes. Also, the trade people need to be trained beforehand for the installation of various elements and systems of NZE homes.

Along with a construction team comprising various professionals with different expertise, advanced materials and products with good quality are required for constructing NZE homes. For example, solar units (e.g., solar panel, solar tile or shingle, photovoltaic glass) are intended to be used for generating green electricity [13,14,56]. The manufacturers should have adequate knowledge for producing these advanced products. The contractors should also know enough about NZE building technology. Moreover, for the economic use of such products, the total energy requirement for a home needs to be known. An effective energy use analysis mechanism helps in the prediction of the total yearly energy consumption of a house. By executing Building Information Modeling (BIM)-based energy performance analysis, it is possible to optimize the energy consumption of a house [57]. Trained professionals and special software (e.g., Autodesk Revit, Green Building Studio, EnergyPlus) are required to fulfill this purpose.

5.10.2. Lack of Governmental and Institutional Supports

The interviewees stated that both the government and academic institutions have a role to play in improving NZE homes. They think that the government should provide good support to individuals interested in buying NZE homes. This may be attributed to the fact that NZE homes are not popular amongst the public and therefore, the government intervention is required. According to McVey et al. [33], the existing regulations on the participation of combined heat and power projects in electricity markets create barriers for coordinating renewable sources into the energy system. Federal and municipal policy frameworks ought to eliminate the capital and financial barriers to the construction of NZE homes. According to an interviewee, Tony Cupido, the provincial and local municipality governments have been less supportive of energy-efficient buildings than the federal government. In the GTA, the City of Toronto has established green standards on energy efficiency and renewable energy with limits of GHG emissions in the cases of low-rise and mid- to high-rise residential as well as non-residential buildings [31,32]. Low-carbon district energy system for mid- and high-rise buildings is also addressed in one of these green standards [32]. However, regional regulations constrain local distribution companies from establishing district energy networks and serving as collectors of distributed energy generation and storage resources [33]. The lack of policies for enhancing renewable energy at the district level undercuts the development of NZE buildings. Furthermore, single-family homes are not yet focused by the green standards of Toronto. Hence, the government policies supporting the growth of NZE homes in the GTA are necessary. According to an interviewee, Andrew Bowerbank, the government should listen more to the industry to develop the frameworks and regulations for the advancement of NZE homes.

Academic institutions have recognized the need for change in curriculums, but it seems to be a lengthy and complex process and is taking a long time just to change some of the curriculums. According to Tony Cupido, this is unacceptable, and the colleges and universities should move quicker to adopt to the industry needs and what the marketplace demands. Academic institutions need to do more and should be more progressive. On the other hand, Andrew Bowerbank, an interviewee, thinks the colleges and universities focus on sustainability as much as they can. Academic institutions can build the buildings to demonstrate NZE technology, and they can also access the funds required to accomplish these projects, which are happening now, as he said. According to him, academic institutions "are the conduit to train the next generation of professionals, the millennial generation, or the one coming after". Colleges and universities should understand the impacts of climate change and the importance of green economy as soon as possible, and act accordingly to educate the people.

The survey results revealed that many participants (37.93%) do not know whether they will get the governmental support or not for purchasing NZE homes. This may be due to the lack of public knowledge and inadequate distribution of information regarding the government's programs on NZE technology. However, a significant portion (22.41%) of the participants said that the government will not support them (refer to Section 4.1.6 and Figure 7) to buy an NZE home. Two interviewees also pointed out that the government is not adequately supportive of NZE homes: Adrian Wang said, "For the consumers there are really no funds available to purchase a more energy-efficient unit." However, Canada Mortgage and Housing Corporation (CMHC) gives a rebate of up to 25% on the CMHC mortgage loan insurance premium when an individual buys or builds an energy efficient home (new home) or buys an existing home and makes energy-saving renovations [24,58]. At present, some rebates and incentives are also available for Toronto homeowners through a multi-partner program "BetterHomesTO" launched by the City of Toronto on 5 November 2019 [59] to help the residents undertake home energy retrofits.

There are no incentives and rebates currently from the City of Toronto to encourage people to purchase new NZE homes. This is perhaps due to the reason that the City first wants to transform the existing homes (old homes) to the net-zero emissions level as they are collectively consuming more energy and emitting more GHG. The City Council of Toronto professed a climate emergency on 2 October 2019 and urged that the City should comply with net-zero emissions by 2050 or earlier. The City wants all new homes to be built to net-zero emissions by 2030 and all existing homes to be retrofitted to net-zero emissions by 2050 [60]. To achieve the target of net-zero emissions from buildings, attractive incentives and rebates should also be available for Toronto homebuyers to go for new NZE homes.

5.10.3. Lack of Public Awareness

The survey results revealed that NZE homes were not well-known to many people, as can be seen from Figure 5. This is probably because the concept of NZE homes is relatively new. It has started as an effort towards climate change adaptation. The interviewed building experts agreed that the public lacks enough awareness of NZE homes. They stated that the public is not familiar with energy-efficient homes except for a few people and the fact that it is difficult to create significant awareness among the public. It was observed from the survey results that 82.76% of the participants were not well-familiar with NZE homes (refer to Section 4.1.4 and Figure 5). Also, 60.34% of the survey respondents were not fairly interested in NZE homes (refer to Section 4.1.5 and Figure 6). In these perspectives, it should be mentioned that 25.86% (from Section 4.1.2 and Figure 3) of the survey participants were construction professionals. In general, these data conclude that the public is not quite familiar with NZE homes. Such lack of the public awareness of NZE homes has limited their market penetration. The report of Hamilton [29] also revealed that the lack of public knowledge on NZE homes is acting as a barrier for promoting energy sustainability in Ontario. The conception of NZE homes was introduced as a means of minimizing the exploitation of natural resources and protecting the environment from detrimental GHG emissions. Most of the survey respondents agreed that these homes would help save the environment. Nevertheless, although more citizens are becoming conscious of their role in household energy emission, they lack the basic information on how net-zero housing can reduce GHG emissions and improve sustainability. As shown in Figure 9, 22.41% of the survey participants were not sure whether NZE homes are one of the mechanisms contributing to the green environment.

Significantly more than 50% of the survey participants were not quite interested in purchasing NZE homes. This is linked with their shortfall of knowledge and understanding of the concept of NZE homes. A good understanding of the benefits of NZE homes may contribute positively to a growing interest in such homes. On the other hand, NZE homes have extra installations that would lead to additional costs on the home. The issue of extra cost may be clarified further while enhancing the public awareness of NZE homes. After knowing about NZE technology, most of the participants said they would be willing to pay more for NZE homes. Additionally, most of the participants indicated that they would be willing to install solar panels on their homes. This may be an indicator that more people are willing to purchase greener homes for a greener environment despite their high initial cost. However, the building experts observed that while the initial cost may be higher, the return on investment should be enough to offset this cost. Advocacy and public campaigns may enhance the public's understanding of NZE homes, especially the associated benefits. The research findings point to the fact that increased awareness of the benefits of NZE homes may increase the public's interest in these homes. Consequently, this will open more market as demand increases and help save the environment from harmful GHG emissions. The government, academic institutions, and the industry should work harmonically together to increase the public awareness of NZE homes.

6. Conclusions and Recommendations

The study sought to find the obstacles associated with NZE homes, realize the public awareness of these homes, and recommend ways in which their popularity can be achieved.

6.1. Conclusions

Research findings revealed that there are various challenges associated with the construction of NZE homes, including insufficient technological experts, the deficient governmental and institutional supports, and the lack of public knowledge, stated broadly. The shortage of enough professionals who understand the NZE technology exists in the GTA construction industry. The construction of an energy-efficient building calls for interdisciplinary collaboration. Finding experts from the different disciplines is an uphill task. The lack of public awareness is also hindering the growth of NZE homes in the GTA. In these scenarios, the building experts (interviewees) have various opinions on what should be done and how it should be done to overcome the visible obstacles to the development of NZE homes.

The interviewed building experts mentioned that technological barriers are the most critical obstacles to developing NZE homes in the GTA. The lack of technological know-how is hindering the promotion of NZE homes. This is because the idea of NZE homes is relatively new and the availability of skilled professionals educated with NZE building technology is currently limited in the construction industry. To promote NZE homes, technological barriers must be removed urgently. Architects and engineers should be well-educated about NZE building technology before designing and constructing NZE homes. The manufacturers should also be trained to make energy-saving building materials, products, and appliances as well as to create energy-producing elements. In addition, the trade people need to be trained in advance for the installation of different energy-producing components and systems on NZE homes. There should also be standardized measures to evaluate the performance of NZE homes.

Next to technological barriers, the lack of governmental and institutional supports plays a critical role against the promotion of NZE homes. Many survey participants residing in the GTA think that they will not get governmental support to purchase an NZE home. In this perspective, the interviewed building experts indicated that the government should be adequately supportive to individuals who are interested in purchasing NZE homes. At present, there are no incentive and rebate programs in the GTA for buying new NZE homes, although some financial supports are available for energy retrofitting of old homes. The federal and municipal policy frameworks should also be conducive to the construction of NZE homes. The lack of good policies for developing renewable energy may undermine the progress of NZE houses.

Lastly, the lack of the public awareness is also acting against the advancement of NZE homes. At present, the public awareness of NZE homes is seriously lacking in the GTA. Both interview and survey results disclosed that the GTA public lacks enough awareness of NZE homes, although construction professionals and consultants know about these homes. Many survey participants were not quite familiar with NZE homes and they were not interested in purchasing these energy-efficient homes. This is because they were not aware of the benefits of living in an NZE home. Such lack of knowledge is limiting the sale of NZE homes. The popularity of NZE homes depends to a greater extent on the knowledge that the public have towards them. The lack of the public awareness may lead to a poor level of interest in NZE homes despite their perceived benefits. The government, academic institutions, and the industry all have an important role to increase the public awareness of NZE homes.

Many residents in the GTA are financially capable of purchasing a new NZE home, having an annual income above $75,000. They also have the capability to install solar panels on old homes for energy efficiency. Most of the survey participants residing in the GTA became willing to pay more for NZE homes and install solar panels on old homes after knowing their benefits. They agreed that NZE homes are environmentally friendly. These findings suggest that more public awareness of NZE homes is required to enhance the growth of such homes in the GTA.

6.2. Recommendations

The government, the academic institutions and centers, and the industry should work together to remove technological barriers, which are the most critical obstacles on the way of NZE homes for their growth in the marketplace. Skilled professionals should be produced through necessary training to remove the technological barriers for constructing NZE homes. For this, academic institutions and training centers should take the leading role and the government should support them. In addition, the manufactures ought to understand the essence of NZE building technology and ensure the availability of the materials and products required for NZE homes. Necessary tools, such as equipment and software, required for analysis, design, and construction should also be obtainable for the creation of NZE homes.

The education sector should do more to incorporate newer NZE building technology into the curriculum to prepare students for the real world that they will be going to face after graduation. This sector needs to uptake any technological change as rapidly as it occurs to prepare the students to be adaptable in the ever-changing world, especially in the matters related to construction technology. Universities and colleges should produce technologically knowledgeable graduates who can work competently in the construction industry to build NZE homes.

More efforts need to be made for the public awareness of NZE homes. Different approaches to the creation and enhancement of the public awareness on matters related to NZE homes and green energy, in general, should be sought and implemented. The government should engage the green energy construction experts in policymaking and curriculum development to enhance the applicability and effectiveness of NZE homes. In addition, the government should offer incentives or rebates to encourage the public to purchase NZE homes. An incentive program should also be created for the developers to construct more and more NZE homes. Moreover, the federal and municipal policy frameworks should eradicate the capital and financial obstacles to the construction of NZE homes.

Additional studies should be carried out, although the survey and interview findings of the present study are coherent. More people should be surveyed, and more building experts should be interviewed to minimize the level of errors. Especially, further investigation should be conducted to find effective ways (e.g., technical means, training, knowledge dissemination, policies) of resolving various obstacles or challenges for the advancement of NZE homes.

Author Contributions: Conceptualization, G.M.; methodology, G.M. and M.S.; validation, G.M. and M.S.; formal analysis, G.M. and M.S.; investigation, G.M.; resources, M.S.; data curation, G.M.; writing—original draft preparation, G.M.; writing—review and editing, M.S.; visualization, G.M. and M.S.; supervision, M.S.; project administration, M.S. All authors have read and agreed to the published version of the manuscript.

Funding: This study was not funded by any external research institute. G.M. performed the study as a capstone industry research project to fulfill the requirements for her Honours Bachelor of Technology (Construction Management) degree.

Institutional Review Board Statement: Not applicable.

Informed Consent Statement: Not applicable.

Data Availability Statement: The data presented in this study are not publicly available. The data were gathered through survey and interviews.

Acknowledgments: The authors are grateful to Angelo DelZotto School of Construction Management for giving the opportunity to conduct the research. They are also thankful to Tom Orman for his valuable instructions regarding technical writing.

Conflicts of Interest: The authors declare no conflict of interest.

References

1. Iyer-Raniga, U. Zero energy in the built environment: A holistic understanding. *Appl. Sci.* **2019**, *9*, 3375. [CrossRef]
2. Frappé-Sénéclauze, T.-P.; Kniewasser, M. *The Path to "Net-Zero Energy" Buildings in BC: The Case for Action and the Role of Public Policy*; Pacific Institute for Climate Solutions, University of Victoria: Victoria, BC, Canada, 2015; p. 24.
3. Toronto's 2018 Greenhouse Gas Emissions Inventory. Available online: https://www.toronto.ca/wp-content/uploads/2020/12/9525-2018-GHG-Inventory-Report-Final-Published.pdf (accessed on 9 February 2021).
4. Torcellini, P.; Pless, S.; Deru, M.; Crawley, D. Zero energy buildings: A critical look at the definition. In Proceedings of the 2006 ACEEE Summer Study on Energy Efficiency in Buildings, Pacific Grove, CA, USA, 14–18 August 2006; p. 12.
5. Pless, S.; Torcellini, P. *Net-Zero Energy Buildings: A Classification System Based on Renewable Energy Supply Options*; Technical Report: NREL/TP-550-44586; National Renewable Energy Laboratory (NREL): Golden, CO, USA, June 2010; p. 21.
6. Liu, Z.; Zhou, Q.; Tian, Z.; He, B.; Jin, G. A comprehensive analysis on definitions, development, and policies of nearly zero energy buildings in China. *Renew. Sustain. Energy Rev.* **2019**, *114*, 109314. [CrossRef]
7. Baden, S.; Fairey, P.; Waide, P.; de T'serclaes, P.; Laustsen, J. Hurdling financial barriers to low energy buildings: Experiences from the USA and Europe on financial incentives and monetizing building energy savings in private investment decisions. In Proceedings of the 2006 ACEEE Summer Study on Energy Efficiency in Buildings, Pacific Grove, CA, USA, 14–18 August 2006; p. 15.
8. Peterson, K.; Torcellini, P.; Grant, R. *A Common Definition for Zero Energy Buildings*; U.S. Department of Energy: Washington, DC, USA, September 2015; p. 22.
9. Hoque, S.; Iqbal, N. Building to net zero in the developing world. *Buildings* **2015**, *5*, 56–68. [CrossRef]
10. Albayyaa, H.; Hagare, D.; Saha, S. Energy conservation in residential buildings by incorporating passive solar and energy efficiency design strategies and higher thermal mass. *Energy Build.* **2019**, *182*, 205–213. [CrossRef]
11. Baetens, R. High performance thermal insulation materials for buildings. In *Nanotechnology in Eco-Efficient Construction: Materials, Processes and Applications*; Woodhead Publishing Limited: Cambridge, UK, 2013; pp. 188–206. [CrossRef]
12. Ohler, A.M.; Loomis, D.G.; Ilves, K. A study of electricity savings from energy star appliances using household survey data. *Energy Policy* **2020**, *144*, 111607. [CrossRef]
13. Shayan, M.E. Solar energy and its purpose in net-zero energy building. In *Zero-Energy Buildings—New Approaches and Technologies*; IntechOpen Limited: London, UK, 2020; p. 14.
14. Musall, E.; Cory, S.; Donn, M.; Garde, F.; Lenoir, A.; Voss, K.; Weiss, T. Net zero energy solar buildings: An overview and analysis on worldwide building projects. In Proceedings of the EuroSun 2020: 13th International Conference on Solar Energy for Buildings and Industry, Athens, Greece, 1–3 September 2020, p. 9.
15. Chang, Y.; Wei, Y. The utilization of renewable energy for low-carbon buildings. In *Renewable-Energy-Driven Future: Technologies, Modelling, Applications, Sustainability, and Policies*; Academic Press: Cambridge, MA, USA, 2021; pp. 289–309. [CrossRef]
16. Pettit, B.; Gates, C.; Fanney, A.H.; Healy, W.M. *Design Challenges of the NIST Net-Zero Energy Residential Test Facility*; NIST Technical Note 1847; National Institute of Standards and Technology (NIST), U.S. Department of Commerce: Gaithersburg, MD, USA, March 2015; p. 71.
17. Todd, J.; Chen, J.; Clogston, F. *Creating the Clean Energy Economy: Analysis of the Net-Zero Energy Home Industry*; International Economic Development Council (IEDC): Washington, DC, USA, 2013; p. 140.
18. Abdellah, R.H.; Masrom, M.A.N.; Chen, G.K.; Mohamed, S.; Omar, R. The potential of net zero energy buildings (NZEBs) concept at design stage for healthcare buildings towards sustainable development. *IOP Conf. Ser. Mater. Sci. Eng.* **2017**, *271*, 012021. [CrossRef]
19. Quinn, E. Green acres: Developer to build 'zero-energy' homes in New Paltz. *New Paltz Times*, 7 August 2008.
20. Torcellini, P.; Pless, S.; Lobato, C. Main street net-zero energy buildings: The zero-energy method in concept and practice. In Proceedings of the ASME 2010 4th International Conference on Energy Sustainability, Phoenix, AZ, USA, 17–22 May 2010; p. 12.
21. NetZero: Future Building Standards. 2020. Available online: https://www.nrcan.gc.ca/energy-efficiency/energy-efficiency-homes/buying-energy-efficient-new-home/netzero-future-building-standards/20581 (accessed on 2 February 2021).
22. Net Zero Energy (NZE) Housing. 2015. Available online: https://www.nrcan.gc.ca/energy/efficiency/data-research-and-insights-energy-efficiency/housing-innovation/net-zero-energy-nze-housing/5131 (accessed on 5 January 2021).
23. Zhang, Y.; Wang, H.; Gao, W.; Wang, F.; Zhou, N.; Kammen, D.M.; Ying, X. A survey of the status and challenges of green building development in various countries. *Sustainability* **2019**, *11*, 5385. [CrossRef]
24. City of Toronto. *Zero Emissions Buildings Framework*; The Planning Division, the City of Toronto: Toronto, ON, Canada, March 2017.

25. Netzero Home: The Ultimate in Comfort and Efficiency. 2021. Available online: https://www.chba.ca/CHBA/BuyingNew/Net-Zero-Homes.aspx/#:~{}:text=Net%20Zero%20Homes%20produce%20as,the%20remaining%20energy%20they%20need.&text=The%20result%3A%20exceptional%20energy%20performance,at%20the%20forefront%20of%20sustainability (accessed on 6 February 2021).
26. Bowerbank, A.; Agustin, K.; Arsenijevich, N. *The Journey to Netzero Energy*; A Lessons Learned Report; EllisDon Corporation: Mississauga, ON, Canada, 2018; p. 48.
27. The Beaches Net-Zero Home. 2021. Available online: https://craigrace.com/portfolio/the-beaches-net-zero-home/ (accessed on 1 February 2021).
28. Piderit, M.; Vivanci, F.; van Moeseke, G.; Attia, S. Net-zero buildings—A framework for an integrated policy in Chile. *Sustainability* **2019**, *11*, 1494. [CrossRef]
29. Are Net-Zero Homes the Future of Ontario Living? 2016. Available online: https://www.tvo.org/article/are-net-zero-homes-the-future-of-ontario-living (accessed on 31 January 2021).
30. LEED®Canada for Homes. 2021. Available online: https://www.cagbc.org/CAGBC/Programs/LEED/LEED_Canada_Rating_System/Homes/Leed_Canada_for_Homes.aspx (accessed on 12 February 2021).
31. Energy/GHG & Resilience for Low-Rise Residential Development. 2021. Available online: https://www.toronto.ca/city-government/planning-development/official-plan-guidelines/toronto-green-standard/toronto-green-standard-version-3/low-rise-residential-version-3/energy-ghg-resilience-for-low-rise-residential-development/ (accessed on 9 February 2021).
32. Energy/GHG & Resilience for Mid to High-Rise Residential & Non-Residential Development. 2021. Available online: https://www.toronto.ca/city-government/planning-development/official-plan-guidelines/toronto-green-standard/toronto-green-standard-version-3/mid-to-high-rise-residential-all-non-residential-version-3/energy-ghg-resilience-for-mid-to-high-rise-residential-all-non-residential-development/ (accessed on 10 February 2021).
33. McVey, I.; Farbridge, K.; Calvert, K. *On the Path to Net-Zero Communities: Integrating Land Use and Energy Planning in Ontario Municipalities*; Community Energy Knowledge—Action Partnership (CEKAP): Toronto, ON, Canada, May 2017; p. 86.
34. Wills, A.; Beausoleil-Morrison, I.; Ugursal, I. Evaluation of source net-zero performance for residential community-scale retrofit photovoltaic systems in Canadian locations. In Proceedings of the eSim 2016 Building Performance Simulation Conference, Hamilton, ON, Canada, 3–6 May 2016; p. 10.
35. Bonds-Raacke, J.M.; Raacke, J.D. *Research Methods: Are You Equipped?* 2nd ed.; Kendall Hunt Publishing: Dubuque, IA, USA, 2014.
36. Martinez-Molina, A.; Boarin, P.; Tort-Ausina, I.; Vivancos, J.-L. Post-occupancy evaluation of a historic primary school in Spain: Comparing PMV, TSV and PD for teachers' and pupils' thermal comfort. *Build. Environ.* **2017**, *117*, 248–259. [CrossRef]
37. Census Profile, 2016 Census. Toronto (Census Metropolitan Area), Ontario and Ontario (Province). 2017. Available online: https://www12.statcan.gc.ca/census-recensement/2016/dp-pd/prof/details/page.cfm?Lang=E&Geo1=CMACA&Code1=535&Geo2=PR&Code2=35&Data=Count&SearchText=Caledon%20East&SearchType=Begins&SearchPR=01&B1=All (accessed on 1 February 2021).
38. Cochran, W.G. *Sampling Techniques*, 3rd ed.; John Wiley & Sons: New York, NY, USA, 1977; p. 448.
39. Krejcie, R.V.; Morgan, D.W. Determining sample size for research activities. *Educ. Psychol. Meas.* **1970**, *30*, 607–610. [CrossRef]
40. Bartlett, J.E.; Kotrlik, J.W.; Higgins, C.C. Organizational research: Determining appropriate sample size in survey research. *Learn. Perform. J.* **2001**, *19*, 43–50.
41. Naing, L.; Winn, T.; Rusli, B.N. Practical issues in calculating the sample size for prevalence studies. *Arch. Orofac. Sci.* **2006**, *1*, 9–14.
42. Diez, D.; etinkaya-Rundel, M.; Barr, C.D. *OpenIntro Statistics*, 4th ed.; OpenIntro: San Francisco, CA, USA, 2019; p. 422.
43. Fellows, R.; Liu, A. *Research Methods for Construction*, 3rd ed.; Wiley-Blackwell Publishing: Hoboken, NJ, USA, 2008; p. 322.
44. Mustafa, F.A. Performance assessment of buildings via post-occupancy evaluation: A case study of the building of the architecture and software engineering departments in Salahaddin University-Erbil, Iraq. *Front. Archit. Res.* **2017**, *6*, 412–429. [CrossRef]
45. Taherdoost, H. Determining sample size: How to calculate survey sample size. *Int. J. Econ. Manag. Syst.* **2017**, *2*, 237–239.
46. Petersen, J.S.; Allman, P.H. The margin of error on damages calculations based on sample survey data in class action wage and hour cases. *J. Leg. Econ.* **2019**, *25*, 139–155.
47. Liang, Z.; Shen, H. Determining sample size for building energy consumption surveys using statistical theory. *Energy Build.* **2012**, *47*, 533–539. [CrossRef]
48. Fox, N. *Using Interviews in a Research Project*; The NIHR RDS for the East Midlands/Yorkshire & the Humber: Nottingham, UK; Sheffield, UK, 2009; p. 39.
49. Galvin, R. How many interviews are enough? Do qualitative interviews in building energy consumption research produce reliable knowledge? *J. Build. Eng.* **2015**, *1*, 2–12. [CrossRef]
50. Bavaresco, M.V.; D'Oca, S.; Ghisi, E.; Lamberts, R. Methods used in social sciences that suit energy research: A literature review on qualitative methods to assess the human dimension of energy use in buildings. *Energy Build.* **2020**, *209*, 109702. [CrossRef]
51. Moser, C.A.; Kalton, G. *Survey Methods in Social Investigations*, eBook ed.; Taylor and Francis: London, UK, 2017; p. 576.
52. Biemer, P.P. Total survey error: Design, implementation, and evaluation. *Public Opin. Q.* **2010**, *74*, 817–848. [CrossRef]
53. CaGBC. *Zero Carbon Buildings Framework: For Commercial, Institutional, and Multi-Family Buildings in Canada*; Canada Green Building Council (CaGBC): Ottawa, ON, Canada, November 2016; p. 48.

54. McNabb, N. *Strategies to Achieve Net-Zero Energy Buildings: A Framwork for Future Guidelines*; Workshop Summary Report, NIST Special Publication 1140; National Institute of Standards and Technology: Gaithersburg, MD, USA, 2013; p. 48.
55. Feng, W.; Khanna, N.Z.; Fridley, D.; Ke, J.; Ke, H.; Zhou, N. Impact analysis of developing net zero energy buildings in China. In Proceedings of the 2016 ACEEE Summer Study on Energy Efficiency in Buildings, Pacific Grove, CA, USA, 21–26 August 2016; p. 14.
56. Bahaj, A.S. Photovoltaic roofing: Issues of design and integration into buildings. *Renew. Energy* **2003**, *28*, 2195–2204. [CrossRef]
57. Jung, D.K.; Lee, D.H.; Shin, J.H.; Song, B.H.; Park, S.H. Optimization of energy consumption using BIM-based building energy performance analysis. *Appl. Mech. Mater.* **2013**, *281*, 649–652. [CrossRef]
58. Energy-Efficient Housing Made More Affordable with Mortgage Loan Insurance. 2021. Available online: https://www.cmhc-schl.gc.ca/en/finance-and-investing/mortgage-loan-insurance/the-resource/energy-efficient-housing-made-more-affordable-with-mortgage-loan-insurance (accessed on 11 February 2021).
59. City Council Declares Climate Emergency and Commits to Accelerating Action to Address Climate Change. 2021. Available online: https://wx.toronto.ca/inter/it/newsrel.nsf/11476e3d3711f56e85256616006b891f/49a94e0367747ad58525848700578157?OpenDocument (accessed on 2 February 2021).
60. City of Toronto Launches BetterHomesTO to Help Residents Shift to 'Net Zero' Homes. 2021. Available online: http://wx.toronto.ca/inter/it/newsrel.nsf/7017df2f20edbe2885256619004e428e/a2c6f3fa7e7b3047852584a9005c0bb1?OpenDocument (accessed on 1 February 2021).

buildings

MDPI

Review

Characterizing Positive Energy District (PED) through a Preliminary Review of 60 Existing Projects in Europe

Xingxing Zhang [1,*], Santhan Reddy Penaka [2], Samhita Giriraj [2], Maria Nuria Sánchez [3], Paolo Civiero [4] and Han Vandevyvere [5,6]

1 School of Technology and Business Studies, Dalarna University, SE-79131 Falun, Sweden
2 Department of Energy and Community Building, Dalarna University, SE-79188 Falun, Sweden; santhanreddypenaka@gmail.com (S.R.P.); samhita4597@gmail.com (S.G.)
3 Center for Energy, Environmental and Technological Research (CIEMAT), Department of Energy, 28040 Madrid, Spain; nuria.sanchez@ciemat.es
4 IREC—Catalonia Institute for Energy Research, 08930 Sant Adrià de Besòs, Barcelona, Spain; pciviero@irec.cat
5 Faculty of Architecture and Design, Norwegian University of Science and Technology, NO-7491 Trondheim, Norway; han.vandevyvere@vito.be
6 VITO—Flemish Institute for Technological Research, BE-2400 Mol, Belgium
* Correspondence: xza@du.se; Tel.: +46-(0)-23-77-87-89

Abstract: Positive Energy District (PED) is recently proposed to be an integral part of a district/urban energy system with a corresponding positive influence. Thus, the PED concept could become the key solution to energy system transition towards carbon neutrality. This paper intends to report and visualize the initial analytical results of 60 existing PED projects in Europe about their main characteristics, including geographical information, spatial-temporal scale, energy concepts, building archetypes, finance source, keywords, finance model and challenges/barriers. As a result, a dedicated date base is developed and it could be further expanded/interoperated through an interactive dashboard. It is found that Norway and Italy have the most PED projects so far. Many PED projects state a 'yearly' time scale while nearly 1/3 projects have less than 0.2 km^2 area in terms of spatial scale. The private investment together with regional/national grants is commonly observed. A mixture of residential, commercial and office/social buildings are found. The most common renewable energy systems include solar energy, district heating/cooling, wind and geothermal energy. Challenges and barriers for PED related projects vary from the planning stage to the implementation stage. Furthermore, the text mining approach is applied to examine the keywords or concentrations of PED-related projects at different stages. These preliminary results are expected to give useful guidance for future PED definitions and proposals of 'reference PED'.

Keywords: PED; characterization; review; text mining

Citation: Zhang, X.; Penaka, S.R.; Giriraj, S.; Sánchez, M.N.; Civiero, P.; Vandevyvere, H. Characterizing Positive Energy District (PED) through a Preliminary Review of 60 Existing Projects in Europe. *Buildings* **2021**, *11*, 318. https://doi.org/10.3390/buildings11080318

Academic Editor: Francesco Nocera

Received: 24 May 2021
Accepted: 20 July 2021
Published: 24 July 2021

1. Introduction

Recently, the Positive Energy District (PED) concept has been discussed substantially as it could become the key solution to energy systems in transition towards carbon neutrality. According to European Strategic Energy Technology (SET) Plan Action 3.2 [1], PED could be defined as an energy-efficient and energy-flexible urban area with surplus renewable energy production and net-zero greenhouse gas emission in a certain time frame. Some PED initiatives aim to create a knowledge base and a roadmap to achieve the energy transition of cities according to established time horizons [2].

Most of the studies and practical experiences about PEDs are based on newly built districts or planning of future districts. Monti et al. [3] described the process of adaption and the challenges/barriers faced by the PED decision makers. They also proposed how simulation, optimization, ICT approaches and business models are combined in a holistic and pragmatic way. Lindholm et al. [4] defined three types of PEDs (i.e., PED autonomous, PED dynamic, and PED virtual), depending on the system boundary and energy import/export conditions. They also pointed out that PED is highly dependent on local context with many impacting factors, such as the available renewable energy sources, energy storage potential, population, energy consumption behavior, costs and regulations, which affect the design and operation of PEDs in different regions. A series of technical solutions, such as the integration of batteries, electric vehicles (EV), and grid-responsive control, were discussed to promote the development of PEDs [5]. Samadzadegan et al. [6] developed a framework to design energy systems for PED or zero-carbon districts, by focusing on estimating heating and cooling demand and sizing related renewable energy systems, e.g., solar photovoltaic (PV) and heat pumps. Shnapp et al. [7] proposed handling the energy performance targets by transferring to the district level the minimum energy requirements imposed by the energy performance of buildings directives to individual buildings. Gabaldón Moreno et al. [8] proposed a methodology for calculating the energy balance at the district level and energy performance of those districts with the potentials to become PEDs. A "double density" simulation scenario was studied further by Bambara et al. [9] to test residential densification potential for PED, where each existing detached house in a community is replaced with two energy-efficient houses of equal living area on the same land lot. From economical and technical points of view, Laitinen et al. [10] concluded that it is more feasible to achieve PED or net-zero energy district, rather than full energy self-sufficiency after they studied a series of technologies (e.g., local centralized wind power, solar PV, battery, heat storage and heat pump), using Helsinki as a case study. Moreover, Soutullo et al. [11] suggested that urban living labs could be a driver to achieve PED. Fatima et al. [12] studied PED's implementation potential from a citizen engagement aspect. Uspenskaia et al. [13] recommended planning and modeling the replication of PED at the very early stage because it is important to find tailor-made solutions to fit spatial, legislative, socio-economic conditions and historical growth of the cities.

Apart from the newly built districts, an explanatory study was carried out as the first step to support the complex planning urban refurbishment, in order to achieve PED [14]. In their study, the key information on the different district types (e.g., energy consumption) was simulated to identify the districts with the highest potential for energy refurbishment. Civiero et al. [15] provide a view of a district simulation model able to analyze a reliable prediction of potential business scenarios on large scale retrofitting actions and to evaluate a set of parameters and co-benefits resulting from the renovation process of a cluster of buildings. Gouveia et al. [16] also argued that the transformation of the existing districts is essential, including historic districts, which present common challenges across EU cities, such as degraded dwellings, low-income families, and gentrification processes due to massive tourism flows. In their report, they discussed how the PED model can be an opportunity for historic districts to reduce their emissions and mitigate energy poverty. Moreover, a methodology for the evaluation of positive energy buildings and neighbourhoods is proposed in the report [17], where a set of Key Performance Indicators (KPIs) are defined with details on the calculation procedure for categories of Energy and Environmental, Economic, Indoor Environmental Quality (IEQ), Social, Smartness and Energy flexibility.

A research gap is thus observed that there are many studies starting to address technical, economic, social aspects of PED, but very limited studies are found in characterizing PED. The Joint Programme Initiative Urban Europe (JPI UE) [18] plays an important role in coordinating PED projects across Europe, it actively engages the interests of different stakeholders, particularly, cities in PEDs. To accomplish its objectives, only Bossi et al. [19] summarized part of PED's characteristics in aspects of geographic distribution, implementation status, building structure, land use, energy typology, success factors/challenges, and barriers. While Brozovsky et al. [20] identified different terminologies of PED, and related focused aspects (i.e., energy, social, climate). JPI UE needs more comprehensive scientific advice on the knowledge and methods for guiding the design, monitoring the operation and evaluating the performance of PED projects. Therefore, many other PED characteristics need to be abstracted and categorized for further development of PED, such as district size, finance source, energy concepts, building archetypes, spatial/temporal scale and keywords. Moreover, as PED projects are expanding all the time, it is necessary to use a common tool/database to increase the semantic interoperability among different stakeholders, for an updated summary of PED's main characteristics.

In the framework of both International Energy Agency—Energy in Buildings and Communities (IEA EBC) Programme Annex 83 [21] and EU Cost action CA19126 [22], the working groups are now collecting data of PEDs and characterizing them for potential proposal of reference and replication of PEDs in different contexts. This paper, therefore, reviews the existing 60 projects within the European area from the JPI Urban Europe PED booklet, establishes the database, and further analyze/visualizes them for the main characteristics. The paper aims to illustrate the basic characteristics of existing PED projects in the EU, and then deliver the information to the targeted stakeholders, such as municipality, urban planner, real estate developer, utility company, policy/regulation maker, renewable energy provider, energy engineer etc., for them to further define, design, promote and implement potential PED projects. As the PED concept is new to most of the stakeholders, this paper intends to transfer the knowledge to the targeted groups through the review/analysis and the development of a database. The result will be also used for the iterative definition of PED in the two initiatives of IEA and EU Cost action.

2. Data Source and Research Methods

2.1. Data Source

The data of PED related projects is collected from the PED booklet [23] by JPI UE updated latest on 2019. JPI Urban Europe is conducting a programme on 'Positive Energy Districts and Neighbourhoods [24] for Sustainable Urban Development' with an implementation plan, SET (Strategic Energy Technology) Plan Action 3.2 [1], participated by about 20 European member states, in the context of Europe commitment towards clean energy transition and carbon neutrality. The total databank consists of 60 projects' data that have similar goals to PED projects in Europe. These projects have been identified and updated by the participated cities of workshops conducted by JPI Urban Europe. The database is divided into several key parameters shown in Table 1.

Table 1. Table parameters for data collection.

Key Parameters	Type of Data
Project characteristics	Location, initiated year, development stage, project area, finance model, etc.
Type of buildings involved	Residential, commercial, social, industry, etc.
Common energy technologies	Solar Thermal, geothermal, PV, heat pumps, etc.
Key energy concepts	Energy combinations and strategies to meet the goals
Keywords	Positive energy district, smart city, etc.
EV/E-mobility	Included/Excluded in energy strategies
Temporal scale	Hourly/monthly/yearly, etc.
Driving stakeholders	Municipality, citizens, real estate developers, etc.
Others	Supporting regulations, barriers, key success factors, etc.

However, it has been challenging to understand the energy typology and detailed strategies due to unclear/insufficient information for many projects from the JPI Urban Europe booklet. The data for the temporal scale of the projects are only available for very few projects. Due to this insufficient information, external sources, such as the website/publication of the specific project, have been studied and reviewed in order to collect more detailed information [25–42].

2.2. Research Methods

2.2.1. Development of Database

A comprehensive critical review was conducted based on the JPI Urban Europe booklet and the related academic literature. The essential data of literature was broken down into thematic categories as shown in Table 1. The important characteristics for PED were either discussed by experts in IEA EBC Annex 83 and EU Cost action CA19126 or extracted from the literature. All the information was observed, recorded and summarized in the excel sheet, which forms up the basic database for this review.

The key thematic parameters for the database are described in detail as below:

- Project characteristics include the location of the project, initiation year, the status of the project in 2019, which is further divided into stages 'in planning', 'in implementation', 'implemented/in operation'. Such categorization refers to the projects where construction of the energy systems is completed and yet to be commissioned or integrate into the existing energy networks. The amount of area is being consumed by the cumulative of all energy systems installed with this project implementation. The appropriate financing source of each project is also checked.
- The type of buildings involved in the PEDs consist of residential, commercial and industrial, etc. In most cases, renewable energy systems are installed on building components (e.g., roofs, envelopes) to reduce local energy demands and further supply excess energy generation to the neighbourhoods.
- The common energy technologies used in PED are reviewed, including energy supply and storage.
- Key energy concepts are examined with strategies and detailed planning to reach the project goals. The selection of energy system combinations with different technologies is crucial, which needs intensive investigation and planning.
- The keywords used in the projects are identified and the most common keywords are abstracted. These keywords vary between the projects with different names, comparing to PED, such as smart city, positive energy blocks, zero energy building, smart grid, zero energy district, urban energy transition, etc.
- Inclusive strategies of EV/e-mobility are identified and included in the data collection. The strategies aim to encourage clean transport solutions within PED scope and integrate with energy systems to provide energy flexibility.
- The temporal scale of the project refers to achieving the project goals, relative to the time period in a day/month/year scale. Since most of the projects are still under planning and implementation stages and due to insufficient information from the sources, the data for temporal scale is only available for less than 50% of the identified projects.
- Stakeholders in each project are summarized, such as a regional municipality, citizens, real-estate developers etc. They are involved in a different stage of project development. The key drivers vary between every project and have analyzed the common driving stakeholders to understand the trends.
- The key success factors with supporting regulations along with challenges are collected. Every project would come across challenges/barriers or have key success factors while planning and implementing the project.

2.2.2. Text Extraction and Mining Method for Keywords Abstraction

The data used for extracting word clouds and sentiments are collected from the JPI Urban Europe booklet available in .pdf (portable document format) format. The projects are grouped according to the PED ambition and the development phase they are in, as shown in Table 2.

Table 2. Project groups according to PED ambitions and their development phase.

Project Phase	Description
PED Implemented	Indicate PED ambition and are implemented
PED in Implementation	Indicate PED ambition and are amidst implementation
PED Planning	Indicate PED ambition and are still being planned
Towards-PED Implemented	Did not declare a PED ambition but present interesting features for the PED Program and are implemented
Towards-PED in Implementation	Did not declare a PED ambition but presents interesting features for the PED Program and are amidst implementations
Towards-PED Planning	Did not declare a PED ambition but presents interesting features for the PED Program and are still being planned

Step 1: Text extraction and mining methods were firstly applied in Python with the aid of Pandas library (version 1.2.4, GitHub, Inc., San Francisco, CA, USA) [43] to transform this data from an unstructured mix of tables and text into clean and structured data frames. These cleaning methods involved extracting the data from '.pdf' format into '.txt' (Text) format (since it is more friendly for running analysis), setting up of the text as structured data frames, removal of extra spaces, special characters, line breaks, website protocols, formatting the cases, stemming [44] and removal of stop words. Hence, the resultant is a data frame consisting of 6 cleaned records (belonging to the 6 groups of projects mentioned in Table 2), each record containing consolidated transcripts of all the project descriptions belonging to the respective groups.

Step 2: Natural Language Processing (NLP) method using text mining in Python with the aid of the Natural Language Toolkit (NLTK) libraries (Version 3.5, O'Reilly Media Inc., Sebastopol, CA, USA) [45] was subsequently used to extract the most used words from the 60 projects. Each word from each of the 6 records of the cleaned data frame is tokenized into its own variable, and the number of times the word repeats itself is the count value of that token. A new data frame is created to capture the tokenized word and its count value. This is repeated for each of the 6 groups and the top 50 words from each group are extracted along with their count value and plotted on a word cloud. A word cloud is a method of visualizing the most used words in transcripts of text data by using the count value of the tokenized words for the sorting. The words in a word cloud are displayed in a specific spatial format: the font size of the words indicate relevance to the magnitude of their use and colours vary for aesthetic reasons.

Step 3: TextBlob library (Version 0.16.0, Steven Loria, New York, NY, USA) [46] was then used to carry out a sentiment analysis study [45] on the dataset in order to determine the polarity and subjectivity of the groups of projects. The polarity value is used to indicate the positive or negative sentiments of a sentence, for example, "happy", "nice", "sad", "bad" and such. Each word has a certain polarity value (positive or negative) and aggregated results of the values of words in an entire transcript are used as the key indicator of the opinion of that transcript [47]. Subjectivity and objectivity are the next measures determined wherein subjectivity is the expression of opinion in a text, and objectivity is the expression of facts.

2.2.3. Data Visualization

Given that the dataset contains several projects across different cities in Europe, a spatial visualization of the location of these projects was deemed vital. QGIS software (Version 3.10, Open Source Geospatial Foundation, Beaverton, OR, USA) [48] is a Geographic Information System (GIS) based open-source software used here to display the cities on a map. Each project is appended with the latitude and longitude of the city it lies in, and these latitudes and longitudes are wrapped over a European base map.

Another visualization technique used to plot the dataset in this project is an interactive dashboard (for non-spatial variables only) developed using the open-source Konstanz Information Miner Analytics Platform (Knime) (Version 4.3.2, KNIME AG, Zurich, Switzerland) [49]. Variables across the dataset are plotted against each other using interactive graphs and charts, for example, for visualizing the type of financing against the year of initiation of the project, and other such co-relations. Interactive means that a user can click on a project in one plot to highlight characteristics about that specific project in other plots across the dashboard as well.

3. Results

3.1. Characteristics of Existing PED Projects

3.1.1. Initiation Year

The section shows the year of initiation of the first phase of all the 60 collected PED related projects in Europe. From Figure 1, the first project was initiated in 1970 and the second project in 1995, both in France. There have been very few projects, less than one project each year until before 2014, where 5 projects took place in that year. The momentum has increased from then with 8 projects in 2016, 9 projects in 2017, 11 projects in 2018, 6 projects in 2019, 4 projects in 2020, and no data for 5 projects.

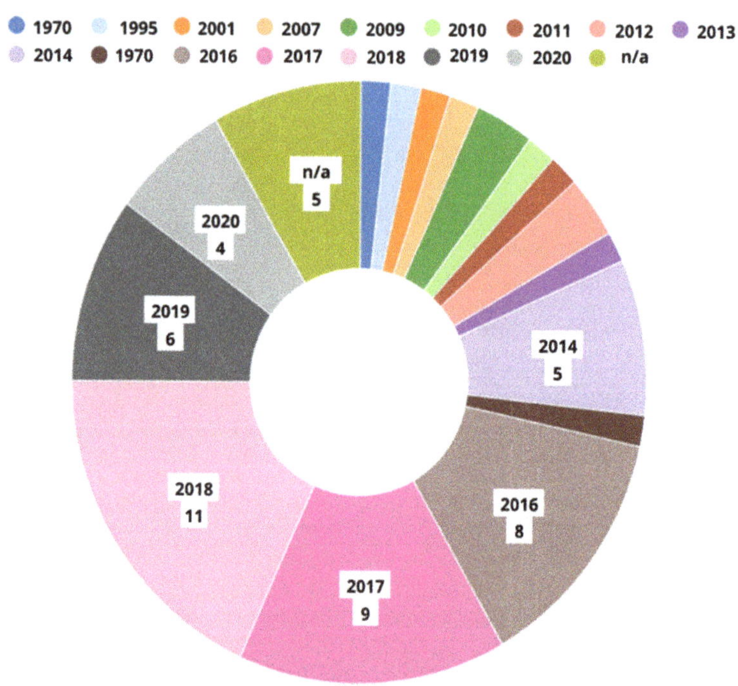

Figure 1. Initiated year of PED related projects.

3.1.2. Location of Identified 60 PED Related Projects

This location of the identified 60 PED related projects is displayed in Figure 2. The most amount of projects are located in Norway, i.e., 9 projects, followed by 8 projects, 7 projects, 6 projects, 5 projects in Italy, Finland, Sweden and The Netherlands, respectively. There are 4 projects in Spain, Germany and Austria, 2 projects in both France and Denmark. There is one project in each of the remaining countries, Portugal, Turkey, Ireland, Belgium, Hungary, Switzerland, Greece, Estonia and Romania.

Figure 2. Locations of 60 PED related projects.

3.1.3. Status of the Identified Projects

This section reports the current development stage of 60 PED projects divided into categories mentioned in the development of the database. From Figure 3, the results clearly indicate that majority of the projects are under the implementation stage i.e., 26 projects. There are 11 projects under the planning stage, and 6 projects under both the planning and implementation stages. In total, 16 PED related projects are already implemented or in operation, among which 5 projects have completed implementation but have yet to integrate the energy systems into the existing local energy networks of the specific projects, while 11 projects are finally in operation stage. Information is not available for one project.

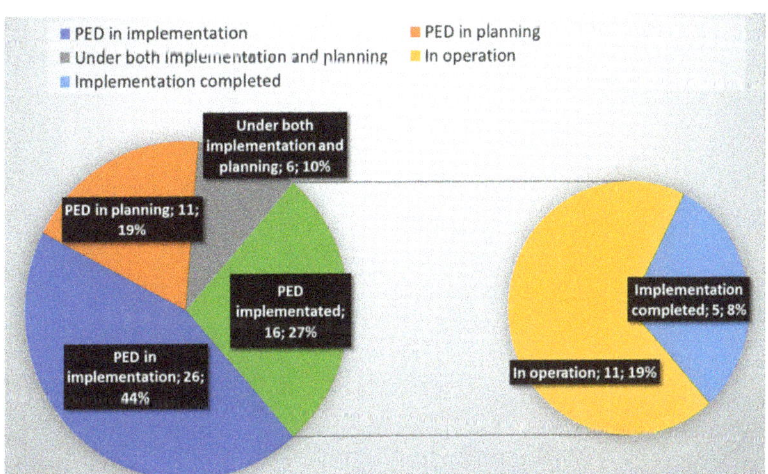

Figure 3. Development stage of collected 60 PED related projects.

3.1.4. Project Area (Spatial Scale)

The amount of project area (spatial scale) is counted by considering the installation of the planned energy systems in their locality. These energy systems might be installed on the residential, commercial or industrial roofs, or flat ground-mounted in open fields, or even through the virtual presence of an energy system. From Figure 4, most of the projects, i.e., 19 projects are claimed to be using less than 0.2 km^2 area, 7 projects between 0.21 and 0.4 km^2 area, 8 projects consuming area between 0.81 and 3.0 km^2, and there is one project claim to be consuming more than 25 km^2 area.

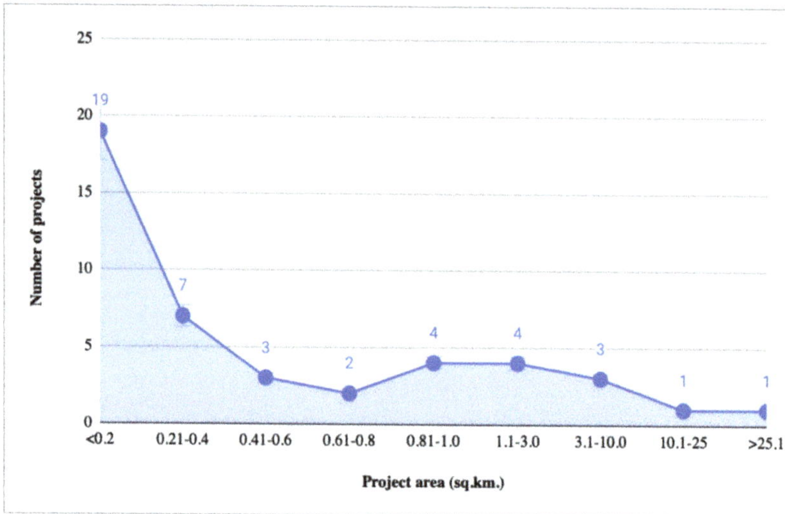

Figure 4. Project area of the 60 PED related projects.

3.1.5. Finance Models Used in PED Projects

In order to meet the project goals and bring clean energy transition, the finance model plays a vital role. Whereas this section demonstrates the common trends being deployed in 60 PED related projects shown in Figure 5. The combination of public, private and others, such as national or regional grants, has been the most common strategy in 20 projects. Only public financing in terms of EU grants or municipality funding is observed in 14 projects out of 60 projects in Europe, 5 projects which solely depend on private financing strategy, and there are 8 projects forwarding with private and public finance combination. However, there are more than 6 projects which do not have proper information about the financial model in the PED booklet by JPI Urban Europe.

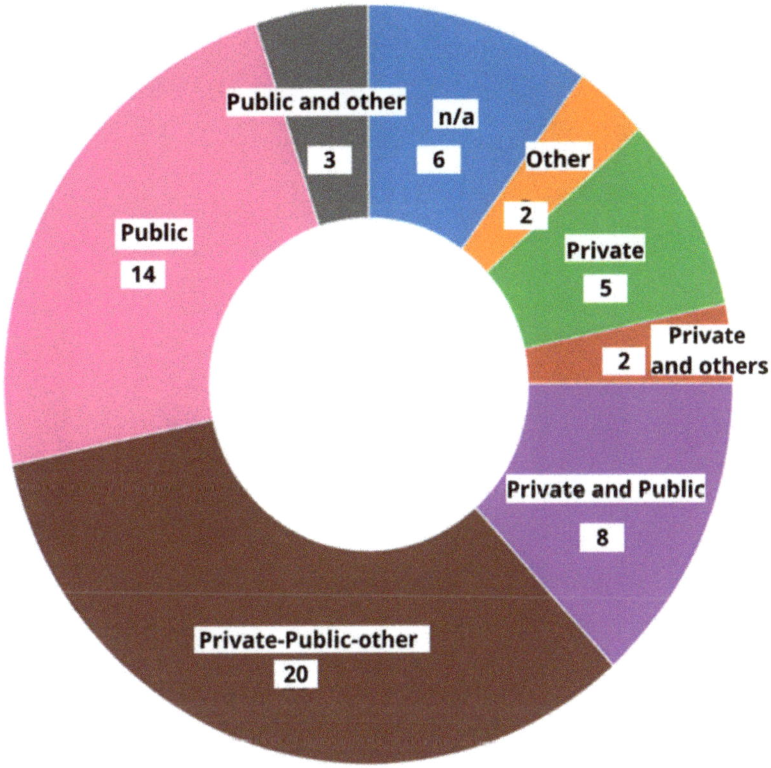

Figure 5. Commonly used type of finance models.

3.1.6. Type of Buildings Involved

This section presents the commonly involved building types for installation of energy systems to supply local energy demand and also to generate excess energy to increase energy flexibility according to the specific project goals. Figure 6 illustrates that the residential sector appears to be predominantly used in the majority of the projects to install energy systems on available roof areas as it is being the primary focus for 39 projects. Office and social buildings are identified to be the main focus in around 24 projects and also followed by commercial buildings spaces for more than 20 projects. Other types of buildings such as institutional, cultural etc., are utilized as secondary spaces for implementing the energy systems.

It is also observed that almost all the projects have considered a mixture of different building types, depending on the major type of buildings existing in the locality. However, the overall trend focuses on involving the citizens as key drivers with the right motivating strategies which eventually address the spatial challenges to install energy systems required for local energy demand.

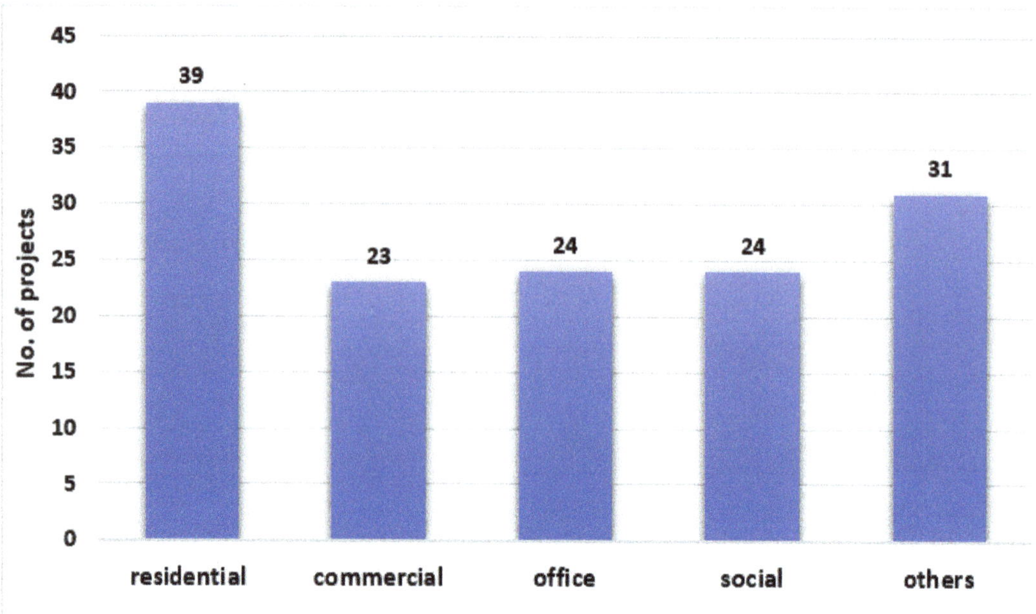

Figure 6. Type of buildings involved for space utilized by energy systems.

3.1.7. Major Energy Technologies

The commonly used energy technologies in these PED projects are examined and referred to as the three pillars of Energy Generation Energy Flexibility Energy Efficiency.

These energy technologies are divided into categories as solar, district heating/cooling, heat pumps, geothermal energy, combined heat and power (CHP), energy storage, wind, e-mobility and others present in the inner circle of the pie chart shown in Figure 7. Solar energy technology is identified to be the primary source of energy supply in almost all projects, specifically photovoltaics (PV) and thermal are the main contributors for producing electricity and heating applications respectively. There are five situations where projects claimed to use solar technology but have not been specific about the type of solar energy. Other new/innovative forms of solar such as hybrid photovoltaic/thermal (PVT), building integrated photovoltaics (BIPV), floating solar and solar roads technologies also have been considered in few projects.

District heating/cooling has been founded in 45 projects, in which heating is used in 43 projects and cooling in 2 projects. Heat pumps, geothermal energy and CHP plant used in 37 projects, 27 projects and 21 projects respectively. Electro-chemical energy battery technology storage for electricity application and seasonal thermal energy storage technology for heating/cooling application are explored as under the energy storage category. Wind energy and E-mobility technologies are identified using in 6 projects and 8 projects respectively. Other technologies, such as bioenergy, green hydrogen, hydropower and natural/mechanical ventilation etc., have also been integrated partly in few PED related projects in Europe.

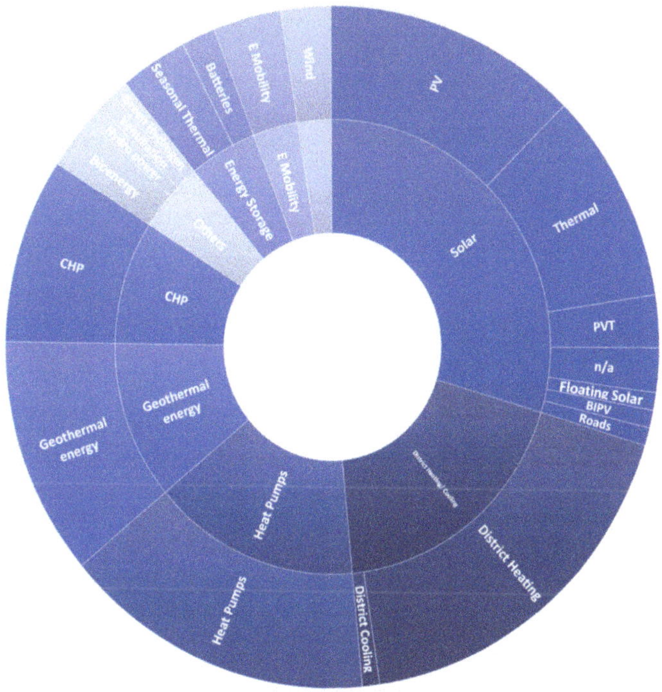

Figure 7. Commonly used energy technologies.

Figure 8 represent the diversity of energy technologies in each country. Solar energy, district heating/cooling and heat pumps technologies are commonly considered in almost of the countries, geothermal energy and CHP plant are being used in nearly half of the countries as represented in Figure 8. Wind energy is integrated in a smaller number of countries such as Denmark, Finland, Germany, The Netherlands and Turkey, and energy storage is only seen in few countries such as Austria, Finland, Germany, Italy, Norway and Turkey.

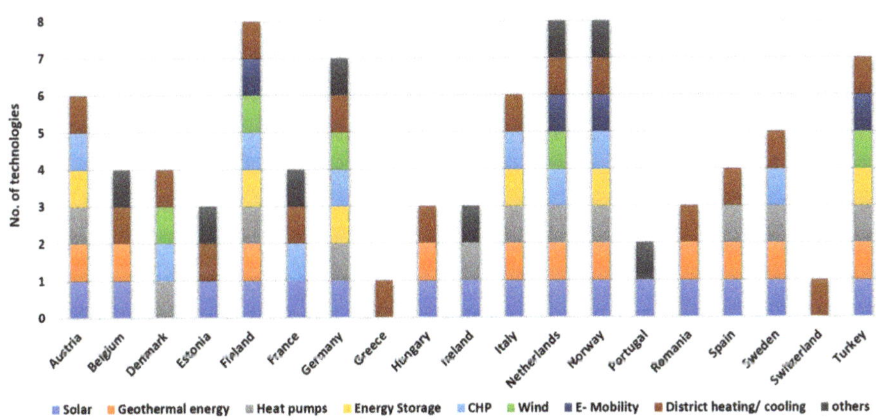

Figure 8. Country-wise approach of energy typology.

Furthermore, the results indicate that Finland, The Netherlands and Norway have high diversity of using more types of energy technologies, followed by Germany, Austria, Italy and Turkey.

3.1.8. Challenges under Different Implementation Stage

The data collection focuses on challenges/barriers that are categorized into 'under planning', 'under implementation stage' and 'implemented/in operation' stages shown in Table 3. The gathered information on challenges/barriers reveals the following main topics: Administrative and policy (A&P), Legal and Regulatory (L&R), Technical. Environmental, Social and Cultural, Information and Awareness, Economical and Financial, and Stakeholders interest perspective [50].

Table 3. Challenges and barriers in different stages of PED projects according to the main topics.

Topic	PED in Planning	PED in Implementation	PED Implemented/in Operation
Administrative & Policy	Conflicts between different authorities involved in the project	Political management	Approvals and permits from municipality and other entities might lead to project timeline extension
Legal & Regulatory		Regulatory framework which governs involved actors throughout Europe	Regulatory barriers for piloting/testing
Technical	System boundary conditions defined Coping with rapid growth of new technologies	Identification and deployment of local feasible clean energy systems	Analysis required for hybrid energy system operations Analysis required for underground seasonal energy storage Energy generation system is far away from the consumers Thermal mining challenges in the urban areas to reduce the distance from energy generation system far away The electricity supply examined properly above 90 degrees
Environmental			Disallowing inefficient and high polluting energy generation systems
Social & Cultural			Cultural differences between different cities involved in the partnership
Information & Awareness		Local citizen acceptance towards new things in rural areas	
Economical & Financial	Economic feasibility Finance availing according to the project timeline Overlapping implementation with local ongoing constructions	Finance dependence on private investors Local finance	
Stakeholders interest	Encouragement of project drivers like real estate developers Uncertainty in stakeholder's commitment	Stakeholders and involved actor's commitment towards project goals Creating interest in project drivers like building owners and landlords	Conflicts due to lack of common interest between different landowners Strong collaborations needed between energy companies and real estate developers for fast implementation
Others	Active consideration of local knowledge Lack of supporting studies/knowledge for planning	Lack of supporting studies/knowledge for implementation	

Challenges associated with stakeholders' involvement, administrative, and technical issues had great relevance in all PED stages. The economic and financial feasibility was crucial in both planning and implementation stages as well as supporting studies or knowledge. However, legal and regulatory barriers were important in the implementation and operation stages. Finally, only in the operation stage environmental and social and cultural aspects were considered possible barriers.

3.2. Most Commonly Used Words and Sentiment Analysis

Figure 9 shows the most commonly used words in the project description transcripts according to their classification from Table 2. As seen from the figure, projects that are already implemented (both PED and towards PED) show high use of words like 'consumption', 'passive', 'heating', and 'industry'. On the other hand, projects that are yet planning (both PED and towards PED) use words such as 'urban', 'solutions', 'quarter', 'research', and 'residential'. Projects that are in implementation (both PED and towards PED) mostly repeat words like 'citizen', 'planning', 'urban', heating', 'supply', and 'cost'. Finally, both implemented and in implementation towards PED projects use heating, cost and supply words.

Figure 9. Most commonly used words for PED.

Figure 10 displays the sentiments portrayed by the 6 groups of projects in the context of polarity (positivity and negativity) and subjectivity-objectivity (opinions-facts). In general, PED implemented projects have very positive feedback, reflecting by the text. We see both PED and towards PED implemented projects have higher subjectivity than objectivity, compared to their planning phase counterparts. This could be interpreted as the implemented projects are mostly influenced by diverse factors, such as dynamic data, citizens and other stakeholders, while those projects in planning stages emphasize more on objective learning experience from literature, simulation data and the related estimations.

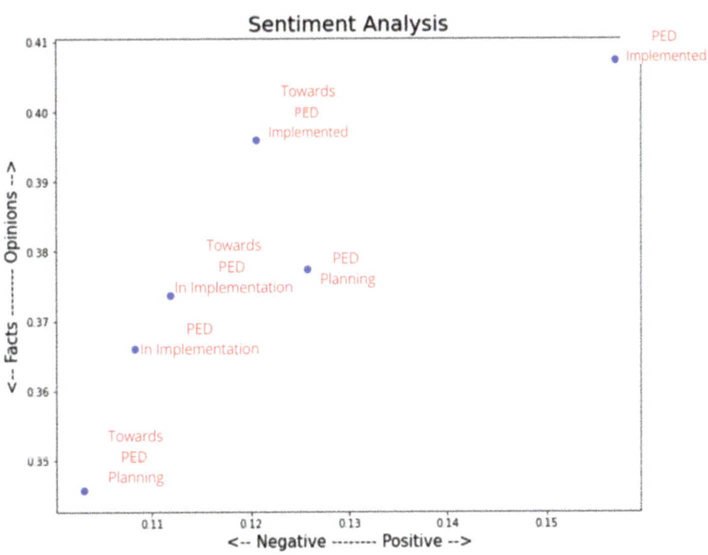

Figure 10. Sentiment Analysis.

3.3. Interactive Dashboard

The interactive dashboard consists of five visualization charts in total (as shown in Figure 11). The display begins with a pie chart that visualizes the proportions of projects initiated across the years. The respective colour scheme index displays the corresponding year in which the project was initiated. The displayed values across the pie chart can be toggled between the number of projects and proportions in the form of a percentage. Below the yearly distribution chart, on the left is a horizontal bar chart that shows the proportions of the projects based on their grouping from Table 2 (i.e., PED ambition and phase of implementation). On the right, a second pie chart visualizes the types of investments received by the projects and their respective proportions. Finally, two scatter plot charts are displayed at the bottom of the dashboard. The left chart shows the co-relation between the initiation year of the projects and the phase it is in today, and the right chart displays the co-relation between the initiation year of the projects and the financial model it observes. Multiple colours for the data points across the y-axis on these two charts are for ease of visualization for the viewer. Selecting any segment or data point from any of the plots highlights all the characteristics covered by those selected projects in the remaining 4 plots.

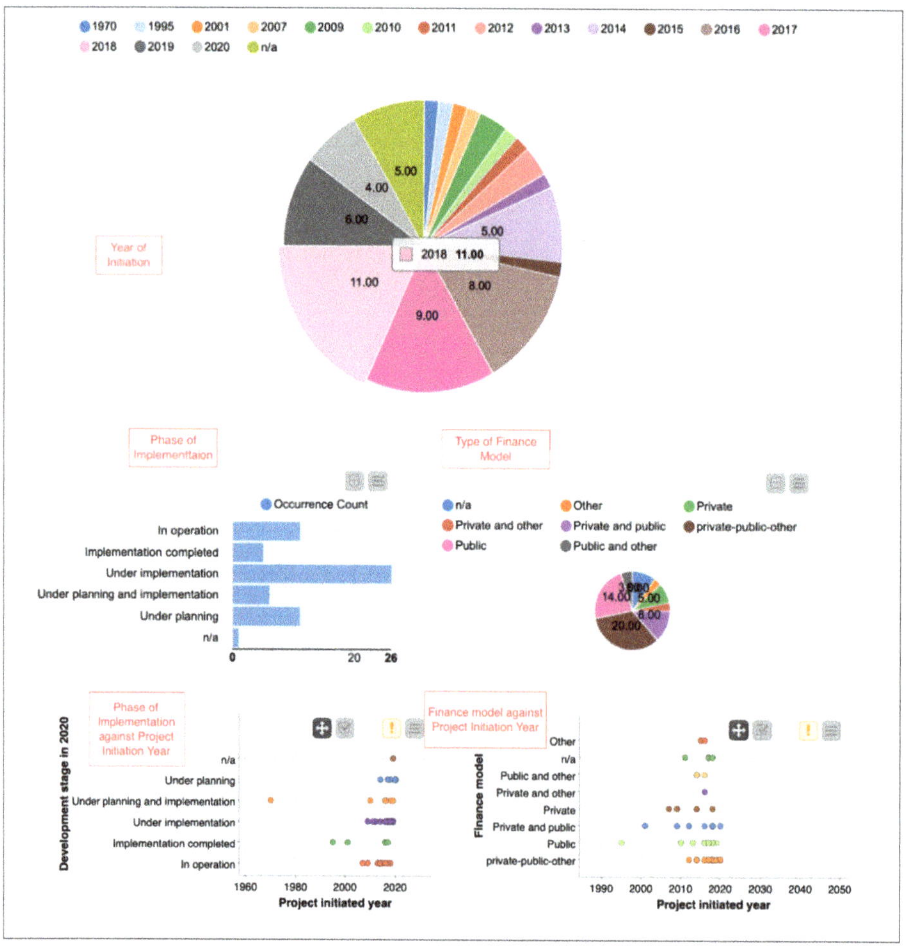

Figure 11. Interactive Knime Dashboard.

Such a dashboard is built upon the database developed in Section 3.1 and can be further extended and updated automatically once there is new project information in the database. It is also possible to upload the dashboard online, to increase the ease of sharing the knowledge, data and experience in PED related projects, as well as to enable interoperable interaction with different stakeholders when they plan or implement PED projects.

4. Discussion

In this study, the projects have been taken from the PED book by JPI Urban Europe, which invited voluntary input data over the project experience and knowledge. It should also be noted that this is not an overview of the PEDs in Europe, as countries have contributed unequally to the development of the book. Since most of the projects are still under planning and implementation stages, it has been challenging to understand the updated information/data of many projects. In addition, due to the insufficient information, there are little data, such as energy technologies for PED, which is unclear during data collection. These bring certain uncertainty to the analysis result.

However, it is interesting to examine the main characteristics of the collected 60 PED related projects, and the results shall have certain guidelines for the final PED definition and the proposal of 'reference PED'. The non-existence of a standard and consolidated definition of the PED concept is in fact one of the main limitations to its development and deployment in European cities, so as to boost the energy transition within a common reference framework [51] for sustainable urban development. So, different approaches and aspects related to the realization of PEDs will be aligned taking into account European cities diversity.

According to results, the identified 60 projects are constituted in Europe with a large number of projects in Norway (9 projects) and Italy (8 projects) respectively. Although the first project took place in 1970, the momentum for such climate neutral goals has started in 2014.

According to the database, most PED related projects choose 'yearly' as the time scale. However, it is not possible to identify the temporal scale for many projects since they are still under the planning stage. Regarding the project area (spatial scale), the general trend is to include residential, commercial and industrial buildings for installation of renewable energy systems in a city or district, which is to avoid the deployment of large energy systems in open fields. This might need supporting policies that support direct consumers to involve in adapting implementation on their premises. However, this strategy would need to consider providing economic feasibility or encouraging policies that attract private investments. The analysis observes that public, private with regional/national grants is a commonly used financial model which reflects active involvement from the private sector. In addition, there are some projects that do not have many local renewable energy sources, but they purchase energy from outside of the district boundary (so-called 'virtual PED').

Based on the results, residential, commercial and office/social buildings are highly involved in the installation of energy systems, which depends on citizens commitment towards project goals (but the goals might deviate from the designed timeframe of the project). Meanwhile, the stakeholders, such as the municipality, would need to address overcoming the policy restrictions to further ease the process of adapting the energy system, and also need to conduct necessary activities to bring awareness in consumers and motivate for participation.

The energy mix for project goals includes solar energy, district heating/cooling, wind and geothermal energy are primary technologies, where solar technologies show dominance because of its potential. However, due to the unavailability of solar energy during most half of the day and during winter seasons, exploration towards other forms of renewable energy sources, such as geothermal energy, wind, etc., yet may not be totally reliably options during peak demands. In this context, energy storage might be the alternative way. Apparently, energy storage has not been part of the major energy strategies, which might be due to the unavailability of enough planning, economic feasibility, high maintenance etc. This also might be part of the reason for PED related projects choosing a yearly temporal scale rather than daily/monthly or seasonally.

In terms of the most used words in the project descriptions, it is observed that projects that are already 'implemented' (both PED and towards PED) tend to concentrate highly on 'consumption', 'production', 'heating'- characteristics that are generally repeatedly showed interest in when the project is implemented and running. On the other hand, projects that are yet 'planning' (both PED and towards PED) tend to concentrate on 'solutions', 'research'- characteristics that are generally repeatedly discussed when a project is being planned. Projects that are in the middle, i.e., 'In Implementation' (both PED and towards PED) mostly repeat words like 'planning' and 'solution', like the 'planning' stage projects, but given they are closer to 'implementation' they also display interest in 'heating' and 'supply'. In the sentiment analysis plot, we deduce that while the X-axis does not reflect a particular pattern, it is observed that projects that are still in the planning phase are more akin to depend on established facts for their documentation, whereas the implemented projects lean towards expressing more opinions (that hint their documentation is developed through experience) and do not have to depend solely on facts. The lessons learned from the preliminary analysis of these PED projects provide a starting point for achieving the objective of reducing the existing research gap in the characterization of PEDs. A key aspect is facing the complexity of the urban system and the resulting interrelationships between social inclusion, energy systems, infrastructure, circular economy and mobility for sustainable urbanization. This calls up building or PED-related simulation tools or platforms to tackle such challenges [52,53].

Moreover, a short summary of a few PED projects with a good level of detailed data has been further analyzed in terms of their energy balance/flows. Table 4 provides the main energy concept/flows and some of them in the implementation/operation stage have clear energy flows, such as Åland Island in Finland, Stor-Elvdal and Drammen in Norway. The annual energy flows in the year 2030 for two scenarios (2030—100% sustainable mobility: (1) 2030 SM Syn scenario—Domestic production of sustainable fuels 2030, (2) 2030 SM EI scenario—High Electrification 2030) at Åland Island are illustrated in Figure 12 [54,55].

Table 4. Summary of major energy concepts and flows of a few PED projects.

City/District	Country	Development Stage in 2020	Temporal Scale	Major Energy Flows
Åland Island	Finland	Under implementation	Yearly	• Target: 100% self-sufficient and 100% fossil-free. • Solar PV now: 1.7% to 0.7% of power demand. • Wind now: about 20% of total power demand. • Other sources, such as waste heat and CHP, bioenergy, wave power are still under implementation
Stor-Elvdal Municipality	Norway	In operation	n/a	• The demand for heat on the campus is covered by on-site heat production through the CHP plant. • One-third of the electricity demand is covered. • The rest is supplied by solar PV with batteries.

Table 4. *Cont.*

City/District	Country	Development Stage in 2020	Temporal Scale	Major Energy Flows
Drammen	Norway	In operation	Yearly	• 85% of the heating needs are met by the large-scale fjord source heat pump (13 MW). The rest of the 15% heating needs are met by gas fired boiler. • The average annual energy supply is 67 GWh. • The heat pump is significantly cheaper than a gas heating system, saving the city around €2.7 m a year. • 1.5 million tonnes of CO_2 have already been saved by switching from gas to the ammonia heat pump.
Oulu	Finland	Under implementation	Yearly	• District heating system supplemented with solar PV and geothermal energy technologies. • PV installations on the roof and geothermal heat pump and thermal borehole storage underneath the shopping mall. • Surplus heat shall be used for refrigeration and seasonal energy storage tanks increasing self-reliance during peak loads.
Turku	Finland	Under planning	n/a	• Aim to become carbon neutral by 2029 • 515 solar PV panels installed on new residential buildings will supply energy more than consumption in summer. • Utilizing the ground source heat with waste heat recovery extracted from 30 other buildings nearby. • 1 MW solar park is installed in the district by energy company, where the company rents out solar panels and reduces consumer electricity bills • Solar thermal collectors are used to produce heat and store underground to use for winter needs. • Further two-way heat trading facility is provided.
Tampere	Finland	Under implementation	n/a	• Solar PV farm installed outside the city will be used for energy needs inside the city along with geothermal local district heating and heat pumps.
Bodø	Norway	Under planning	Yearly	• Although this municipality has excess power production capacity, distribution networks is the main drawback in several places. Therefore, smart city goals are focused on achieving energy efficiency, creation of stable and sustainable energy systems, and reducing of peak demands. • This energy system uses local renewable energy productions, supply and optimization with regional, national, Nordic and EU electricity networks

Table 4. *Cont.*

City/District	Country	Development Stage in 2020	Temporal Scale	Major Energy Flows
Elverum	Norway	Under planning and implementation	n/a	• Firstly, reducing the energy demand in buildings and depending energy production on local renewable energy sources. • Energy storage in the form of batteries or thermal energy storage
Trondheim	Norway	Under implementation	n/a	• Conventional electricity is being provided by largely hydropower with 21 g CO_{2eq}/kWh, and district heating through burning local waste. • Installation of solar PV arrays, heat pumps integration. • Large 1500 kWh battery storage would attribute to reaching the energy peak demands and surplus energy supply.
Bergen	Norway	Under planning	n/a	• Primarily improving energy efficiency to reduce energy demand. • Individual energy systems based on renewable energy sources such as PV, thermal technologies are developed. • Further surplus power will be supplied to EV mobility solutions
Odense	Denmark	Under implementation	Yearly	• To eliminate fossil fuels by 2025 and reach to top 3 cheapest district heating prices in Denmark. • District heating supply with waste heat, energy power production from renewables such as wind power. • Further strategically investing in smaller energy units which include 10–20 MW heat pumps, 30–50 MW biomass boilers and +50 MW electric boilers etc.
Osterby	Denmark	Under implementation	Yearly	• The project aims to reduce the heating costs from district heating with other networks. • Connecting and sharing energy with the large district heating facilities with neighbourhoods reflecting energy flexibility. • 2.07 MWp PV roof mounted installation that will operate the cooling machines in the mall.
Lund	Sweden	Under implementation	n/a	• Producing heat through local waste is enough to provide heating for the whole area. • Large scale district heating is installed to provide low temperature applications with renewable energy systems integration.
Lund (Brunnshög)	Sweden	Under implementation	Yearly	• Existing district heating used by biomass will be replaced by large scale biofuel CHP plant along with geothermal energy unit, waste heat combustion and district cooling heat pumps etc.

Table 4. *Cont.*

City/District	Country	Development Stage in 2020	Temporal Scale	Major Energy Flows
Lund (Medicon Village)	Sweden	Implementation completed	Yearly	• Primarily trying to reduce the energy needs yearly by improving energy efficiency. • Installing solar power on rooftop of buildings for more sustainability.

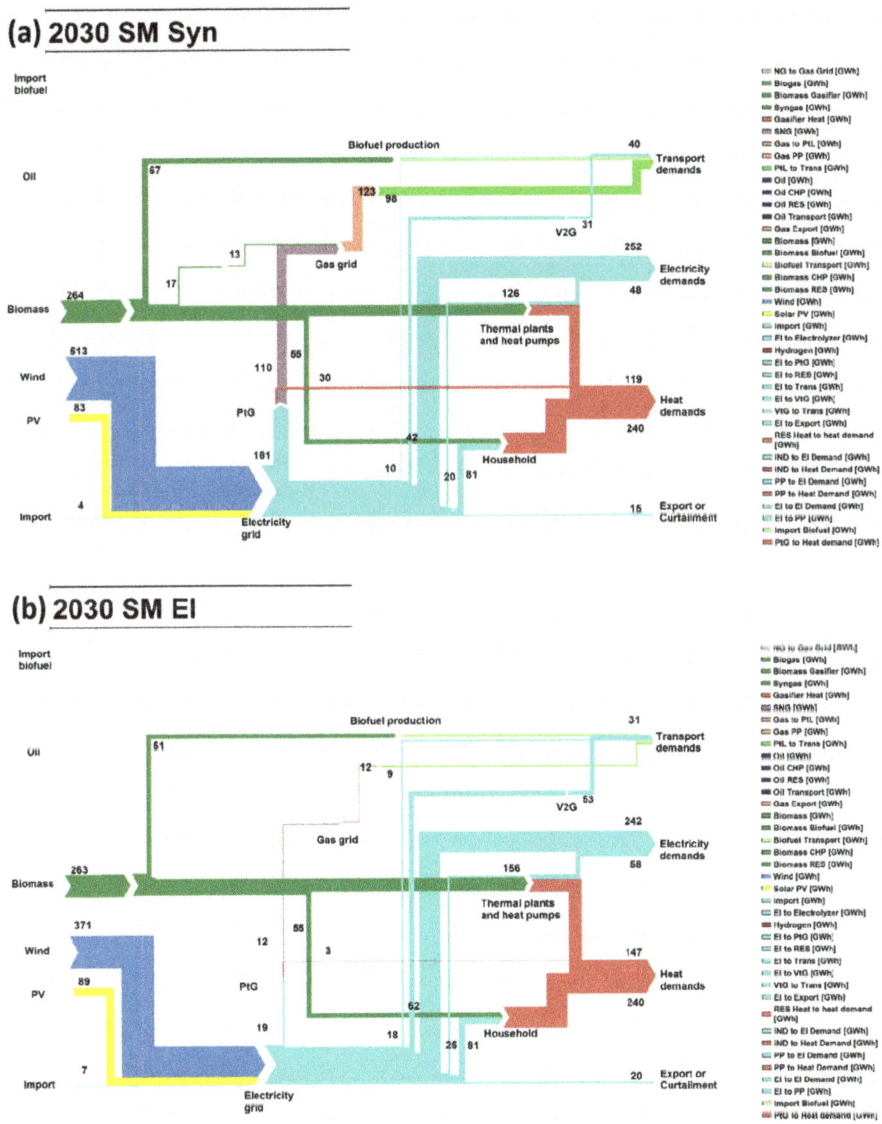

Figure 12. The annual energy flows in the year 2030 at Åland Island towards 2030 target—100% sustainable mobility for two scenarios: (**a**) SM Syn scenario—domestic production of sustainable fuels and (**b**) SM El scenario—high electrification. Reprinted from [55].

It is observed the major energy contributions varies from district to district. For instance, Åland Island replies on biomass and wind power a lot, while Stor-Elvdal municipality prefers CHP plant, and in Drammen municipalities, a heat pump is used mostly. However, these districts are not fully self-sufficient, and they have to import energy to cover peaks. For instance, as shown in Figure 12, Åland Island has to import 4 or 7 GWh of electricity in 2030. It is not easy to judge whether they are PEDs or not at this stage since there is no standard and KPIs available now. According to the mentioned work from EERA JPSC and JPI UE, four categories of PEDs have been established based on two main aspects: the boundaries and limits of the PED in order to reach a net positive yearly energy balance and the energy exchanges (import/export) in order to compensate for energy surpluses and shortages between the buildings or the external grid. All the four described categories of PEDs (PED autonomous, PED dynamic, and PED virtual, Candidate-PED) are based on the accomplishment of a yearly positive energy balance, measured in greenhouse gas emissions, with use of renewables within the defined boundaries, and considering both building energy use and non-building energy use in a neighbourhood. Auto-and Dynamic-PEDs are the only categories where a net positive energy balance is achieved and Candidate-PED should compensate the energy difference with imported certified energy from outside the boundary. According to the boundaries descriptions aligned to the draft definition of PEDs from EERA JPSC working group and JPI Urban Europe, the net positive yearly energy balance is assessed within the functional or virtual boundaries. Thus, PEDs will achieve a net positive energy balance and dynamic exchanges within the functional/virtual boundaries, but in addition, it may provide a connection between buildings within the virtual boundaries of the neighbourhood.

It is necessary to pay specific attention to the differences between cities across different regions when promoting the development of PEDs. This is because cities differ from each other at the local, national and international levels from the perspectives of geography, resources, social, economy, culture, infrastructure, and progress for the carbon-neutral target. This would bring a difference in planning, technology selection/implementation, investment portfolio, stakeholders involvement, regulations, keywords etc., during the PED development. However, it is important to have a commonly recognized definition of PED, and its related KPI framework for evaluation. By learning the main characteristics from those existing PED projects in the EU, it is helpful to define PED or propose 'reference PED' in other cultural and geographical contexts, which will bring significant common values in terms of replicability and potential generalization of PED across the globe.

5. Future Work

This paper focuses on preliminary analysis of identified PED projects, including projects with insufficient information. In order to understand the detailed analysis, the number of projects might be filtered based on projects with sufficient information to conduct the detailed analysis. Given that only 11 of the evaluated projects are at an advanced (operational) stage, a continuous evaluation of the progress of the PEDs currently in the planning and implementation phase is foreseen in order to update the initial database in subsequent stages. Collecting this additional information will extend and improve the PED characterization especially in aspects such as energy technologies and boundaries definition. Besides, more PED related projects have to be identified with sufficient data to support more comprehensive analysis. Such a task is ongoing in both IEA EBC Annex 83 and EU Cost action CA19126. This preliminary study of PED characteristics based on key parameters will be deepening and widening with a particular focus on key energy concepts, EV mobility, driving stakeholders and temporal scale. Furthermore, it is necessary to identify the potential projects with daily or monthly temporal scales, in order to discover the energy combinations to achieve a net positive energy balance and dynamic exchanges within the functional/virtual boundaries. In addition, a PED may provide a connection between buildings within the virtual boundaries of the neighbourhood.

In the context of text mining, the current analysis is developed using the cleaned dataset for the transcripts. However, when it comes to data cleaning, there are several more layers of refining and cleaning that can be carried out on the current transcripts to gain results that are even more accurate and finely assessed. To narrow down the uncertainty of the overall word cloud results, a deeper and multi-layered approach to designing the most used word cloud along with other clouds, such frequency and unique words used, can provide deeper insights. It is also planned to expand the scale of text mining, from the current PED booklet to comprehensive literature, project websites/reports, and so on. Furthermore, the Knime dashboard can include multi-variate plots across more than two variables (as is currently), allowing more significant insights on patterns of co-relation between the variables. An online version of such a dashboard will further enhance the interoperable interaction with different stakeholders when they plan or implement PED projects.

Additionally, within the same framework of developing a PED, different areas across the globe must not only take into account specificities at the local level but also have a common definition of PED for standardized assessment. Ongoing works in the EU Cost action CA19126 also consider the integration of PED-Labs characteristics in mapping PEDs projects and initiatives framework. The PED mapping activities are also related to providing a very practical tool needed to guide PEDs implementation as well as to exchange knowledge and information. Potential integration of such a GIS data driven platform with the Knime dashboard could greatly support the involvement of cities stakeholders, and show the feasibility and impact of certain strategies that can pave the way to PED and climate-neutral cities. The alignment of these pilot initiatives could enhance the knowledge not only in the planning and deployment of PEDs in all aspects such as social, technical, financial, regulatory, etc., but also in the PED characterization/definition/KPIs, as well as showing ground for new methodologies, technical solutions and services to be developed in the future implementation of PEDs. These databases thus constitute an integrated approach to deploy an optimal integration in the technical, evaluation and management infrastructures of the city in different contexts.

6. Conclusions

This paper conducts a preliminary analysis of the main characteristics for 60 identified PED projects in Europe. A dedicated database is developed by considering a series of key parameters. It is found that a large number of PED projects locates in Norway and Italy. Although the first PED project took place in 1970, the momentum for such climate-neutral goals started in 2014. Most PED related projects choose 'yearly' as the time scale. Nearly 1/3 of projects have less than 0.2 km^2 area as their spatial scale. In this case, the definition of the project area and the information regarding its boundaries calculation are both very relevant to evaluate the PEDs features of the projects and the business model adopted. Different financing mechanisms and innovative procurement solutions are required to support different large scale actions. The private investment together with regional/national grants is a commonly used financial model which reflects active involvement from the private sector. Residential, commercial and office/social buildings are mostly involved in the installation of renewable energy systems, which includes solar energy, district heating/cooling, wind and geothermal energy are primary technologies, where solar technologies show dominance. Substantial challenges and barriers for PED related projects vary from planning stage to implementation stage.

The non-technological PED solutions (e.g., solution for Governance, Economic, Social, Environmental, Spatial, Legal/Regulatory) are not clearly considered in the Booklet analysis. This is why the next interactive PEDs mapping tools will take into account those aspects that could help to share information and boost the PEDs replication within the main target groups, and according to a local broader perspective.

In addition to the development of the database, the text mining approach is applied to further examine the keywords of PED-related projects. It is observed that projects that are already 'implemented' (both PED and towards PED) concentrate highly on 'consumption', 'production', 'heating'. While the projects that are yet 'planning' (both PED and towards PED) focus on 'solutions', 'research'. Projects that are 'In Implementation' (both PED and towards PED), mostly repeat words of 'planning' and 'solution', but given they are closer to 'implementation' they also display interest in 'heating' and 'supply'. We also deduce that the projects that are still in the planning phase are more akin to depend on established facts for their documentation, whereas the implemented projects lean towards expressing more opinions by high involvement of stakeholders.

Although there is uncertainty due to limited data at the initial stage, the results are expected to give useful guidance for the final PED definition and proposal of 'reference PED'. It is confident that the alignment among ongoing initiatives will represent the best way and very practical solution to step forward and facilitate the PEDs implementation in the next years, with more useful guidance and tools.

Author Contributions: Conceptualization, X.Z.; methodology, S.R.P. and S.G.; formal analysis, S.R.P. and S.G.; investigation, X.Z., S.R.P., S.G., M.N.S., P.C. and H.V.; writing—original draft preparation, X.Z., S.R.P., and S.G.; writing—review and editing, X.Z., M.N.S., P.C. and H.V. All authors have read and agreed to the published version of the manuscript.

Funding: This research was funded by Swedish Energy Agency with grant number 8569501 and by the Spanish Ministry of Science and Innovation and co-financed by FEDER funds, under TOGETHER (RTC-2017-5926-3) project. Paolo Civiero acknowledges the funding received from the European Union's Horizon 2020 re-search and innovation programme under the Marie Skłodowska-Curie grant agreement No 712949 (TECNIOspring PLUS) and the Agency for Business Competitiveness of the Government of Catalonia.

Acknowledgments: The authors would like to acknowledge discussions and comments given by experts from subtask A in IEA EBC Annex 83, and working group 1 in EU Cost action CA19126.

Conflicts of Interest: The authors declare no conflict of interest.

References

1. SET-Plan ACTION n°3.2 Implementation Plan. Europe to Become a Global Role Model in Integrated, Innovative Solutions for the Planning, Deployment, and Replication of Positive Energy Districts. Available online: https://setis.ec.europa.eu/system/files/2021-04/setplan_smartcities_implementationplan.pdf (accessed on 26 April 2021).
2. Civiero, P.; Sanmartí, M.; García, R.; Gabaldón, A.; Chicote, M.A.; Ferrer, J.A.; Ricart, J.E.; Franca, P.; Escobar, G.J.; PTE-ee ITP—Iniciativa Tecnológica Prioritaria. Positive Energy Districts (PEDs). 2019. Available online: https://static.pte-ee.org/media/files/documentacion/itp-01-2019-distritos-de-energia-positiva-peds-tZU.pdf (accessed on 26 April 2021).
3. Energy Positive Neighborhoods and Smart Energy Districts. Methods, Tools, and Experiences from the Field. ScienceDirect. Available online: https://www.sciencedirect.com/book/9780128099513/energy-positive-neighborhoods-and-smart-energy-districts#book-description (accessed on 26 April 2021).
4. Lindholm, O.; ur Rehman, H.; Reda, F. Positioning Positive Energy Districts in European Cities. *Buildings* **2021**, *11*, 19. [CrossRef]
5. Zhou, Y.; Cao, S.; Hensen, J.L.M. An energy paradigm transition framework from negative towards positive district energy sharing networks—Battery cycling aging, advanced battery management strategies, flexible vehicles-to-buildings interactions, uncertainty and sensitivity analysis. *Appl. Energy* **2021**, *288*, 116606. [CrossRef]
6. Samadzadegan, B.; Abolhassani, S.S.; Dabirian, S.; Ranjbar, S.; Rasoulian, H.; Sanei, A.; Eicker, U. Novel energy system design workflow for zero-carbon energy district development. *Front. Sustain. Cities* **2021**, *3*, 23. [CrossRef]
7. Shnapp, S.; Paci, D.; Bertoldi, P. Enabling Positive Energy Districts across Europe: Energy Efficiency Couples Renewable Energy. Available online: https://publications.jrc.ec.europa.eu/repository/handle/JRC121405 (accessed on 12 June 2021).
8. Gabaldón Moreno, A.; Vélez, F.; Alpagut, B.; Hernández, P.; Sanz Montalvillo, C. How to Achieve Positive Energy Districts for Sustainable Cities: A Proposed Calculation Methodology. *Sustainability* **2021**, *13*, 710. [CrossRef]
9. Bambara, J.; Athienitis, A.K.; Eicker, U. Residential Densification for Positive Energy Districts. *Front. Sustain. Cities* **2021**, *3*, 3. [CrossRef]
10. Laitinen, A.; Lindholm, O.; Hasan, A.; Reda, F.; Hedman, Å. A techno-economic analysis of an optimal self-sufficient district. *Energy Convers. Manag.* **2021**, *236*, 114041. [CrossRef]

11. Soutullo, S.; Aelenei, L.; Nielsen, P.S.; Ferrer, J.A.; Gonçalves, H. Testing Platforms as Drivers for Positive-Energy Living Laboratories. *Energies* **2020**, *13*, 5621. [CrossRef]
12. Fatima, Z.; Pollmer, U.; Santala, S.S.; Kontu, K.; Ticklen, M. Citizens and Positive Energy Districts: Are Espoo and Leipzig Ready for PEDs? *Buildings* **2021**, *11*, 102. [CrossRef]
13. Uspenskaia, D.; Specht, K.; Kondziella, H.; Bruckner, T. Challenges and Barriers for Net-Zero/Positive Energy Buildings and Districts—Empirical Evidence from the Smart City Project SPARCS. *Buildings* **2021**, *11*, 78. [CrossRef]
14. Nzengue, Y.; du Boishamon, A.; Laffont-Eloire, K.; Partenay, V.; Abdeloua, Y. Planning city refurbishment: An exploratory study at district scale how to move towards positive energy districts—approach of the SINFONIA project. In Proceedings of the 2017 International Conference on Engineering, Technology and Innovation (ICE/ITMC), Madeira Island, Portugal, 27–29 June 2017; pp. 1394–1400. [CrossRef]
15. Civiero, P.; Pascual, J.; Arcas Abella, J.; Bilbao Figuero, A.; Salom, J. PEDRERA. Positive Energy District Renovation Model for Large Scale Actions. *Energies* **2021**, *14*, 2883. [CrossRef]
16. Gouveia, J.P.; Seixas, J.; Palma, P.; Duarte, H.; Luz, H.; Cavadini, G.B. Positive Energy District: A Model for Historic Districts to Address Energy Poverty. *Front. Sustain. Cities* **2021**, *3*, 16. [CrossRef]
17. Salom, J.; Tamm, M.; Andresen, I.; Cali, D.; Magyari, Á.; Bukovszki, V.; Balázs, R.; Dorizas, P.V.; Toth, Z.; Mafé, C.; et al. An Evaluation Framework for Sustainable Plus Energy Neighbourhoods: Moving Beyond the Traditional Building Energy Assessment. *Energies* **2021**, *14*, 4314. [CrossRef]
18. JPI Urban Europe—ERA-LEARN. Available online: https://www.era-learn.eu/network-information/networks/urban-europe (accessed on 20 May 2021).
19. Bossi, S.; Gollner, C.; Theierling, S. Towards 100 Positive Energy Districts in Europe: Preliminary Data Analysis of 61 European Cases. *Energies* **2020**, *13*, 6083. [CrossRef]
20. Brozovsky, J.; Gustavsen, A.; Gaitani, N. Zero Emission Neighbourhoods and Positive Energy Districts—A State-of-the-Art Review. *Sustain. Cities Soc.* **2021**, *72*, 103013. [CrossRef]
21. Hedman, Å.; Rehman, H.U.; Gabaldón, A.; Bisello, A.; Albert-Seifried, V.; Zhang, X.; Guarino, F.; Grynning, S.; Eicker, U.; Neumann, H.-M.; et al. IEA EBC Annex83 Positive Energy Districts. *Buildings* **2021**, *11*, 130. [CrossRef]
22. COST Action CA19126 Positive Energy European Network. Available online: https://www.cost.eu/actions/CA19126/#tabs\T1\textbar{}Name:overview (accessed on 26 April 2021).
23. Gollner, C.; Hinterberger, R.; Noll, M.; Meyer, S.; Schwarz, H.G. *Booklet of Positive Energy Districts in Europe*; JPI Urban Europe and Austrian Research Promotion Agency FFG, Sensengasse 1, 1090 Vienna. Available online: https://nws.eurocities.eu/MediaShell/media/Booklet_of_PEDs_JPI_UE_v6_NO_ADD.pdf (accessed on 26 April 2021).
24. Positive Energy Districts (PED). JPI Urban Europe. Available online: https://jpi-urbaneurope.eu/ped/ (accessed on 26 April 2021).
25. Évora, POCITYF. Available online: https://pocityf.eu/city/evora/ (accessed on 5 May 2021).
26. La Fleuriaye, Nature D'avance—Carquefou (44). Available online: http://www.quartierlafleuriaye.fr/ (accessed on 5 May 2021).
27. Medicon Village, E. ON Ectogrid. Available online: https://ectogrid.com/use-cases/medicon-village/ (accessed on 5 May 2021).
28. Municipal District Heating Company of Amindeo (D.H.C.A.)—ΔΕΤΕΠΑ. Available online: http://detepa.gr/dhca/ (accessed on 5 May 2021).
29. "Om Smart Energy Åland", Smart Energy Åland. Available online: https://smartenergy.ax/om-smart-energy-aland/ (accessed on 5 May 2021).
30. "Oulu Northern Lights", Oulu Northern Lights. Available online: https://www.northernlights.fi/ (accessed on 5 May 2021).
31. Red de Calor Urbana (Mieres-Asturias). Aula Hunosa de la Geoetermia y la Biomasa. Available online: http://www.aulahunosa.es/red-de-calor-mieres/ (accessed on 5 May 2021).
32. RHC—2050 Vision for 100% Renewable Heating and Cooling in Europe. Available online: https://www.rhc-platform:content/uploads/2019/10/RHC-VISION-2050-WEB.pdf (accessed on 5 May 2021).
33. Samen Werken Aan Minder CO_2: Sustainable Energy and Environment. Available online: https://www.han.nl/over-de-han/onze-focus/sustainable-energy-and-environment/ (accessed on 5 May 2021).
34. Sinfonia Project. Available online: http://www.sinfonia-smartcities.eu/ (accessed on 5 May 2021).
35. Stadtteil Dietenbach. Available online: https://www.freiburg.de/pb/495838.html (accessed on 5 May 2021).
36. Stadtwerke Hennigsdorf—Forschungsprojekte. Available online: https://www.swh-online.de/aktuell/forschungsprojekte (accessed on 5 May 2021).
37. Stardust. Available online: https://stardustproject.eu/ (accessed on 5 May 2021).
38. Giourka, P.; Apostolopoulos, V.; Angelakoglou, K.; Kourtzanidis, K.; Nikolopoulos, N.; Sougkakis, V.; Fuligni, F.; Barberis, S.; Verbeek, K.; Costa, J.M.; et al. An Empirical Evidence of Two European Union (EU) Smart Cities, Evora and Alkmaar. *Smart Cities* **2020**, *3*, 32. [CrossRef]
39. Olivadese, R.; Alpagut, B.; Revilla, B.P.; Brouwer, J.; Georgiadou, V.; Woestenburg, A.; van Wees, M. Towards Energy Citizenship for a Just and Inclusive Transition: Lessons Learned on Collaborative Approach of Positive Energy Districts from the EU Horizon2020 Smart Cities and Communities Projects. *Proceedings* **2020**, *65*, 20. [CrossRef]
40. Urban Innovation Lab—Competence and Network Regarding Sustainable Societal Development. Available online: https://urbaninnovationlab.se/ (accessed on 5 May 2021).

41. Werksviertel—Werksviertel München. Available online: https://werksviertel.de/?page_id=410&lang=en (accessed on 5 May 2021).
42. ZEN-Report-no-7. Zero Emission Neighbourhoods in SmartCities. Available online: https://fmezen.no/wp-content/uploads/2018/11/ZEN-Report-no-7-Bilingual.pdf (accessed on 5 May 2021).
43. Reback, J.; Jbrockmendel; McKinney, W.; Van Den Bossche, J.; Augspurger, T.; Cloud, P.; Hawkins, S.; Gfyoung; Sinhrks; Roeschke, M.; et al. pandas-dev/pandas: Pandas 1.2.4. *Zenodo* **2021**. [CrossRef]
44. Jivani, A.G. A Comparative Study of Stemming Algorithms. *Int. J. Comp. Tech. Appl.* **2011**, *2*, 1930–1938.
45. Loper, E.; Bird, S. NLTK: The Natural Language Toolkit. May 2002. Available online: http://arxiv.org/abs/cs/0205028 (accessed on 5 May 2021).
46. Loria, S. *Textblob Documentation*; p. 73. Available online: https://buildmedia.readthedocs.org/media/pdf/textblob/latest/textblob.pdf (accessed on 5 May 2021).
47. Agarwal, B.; Mittal, N.; Bansal, P.; Garg, S. Sentiment Analysis Using Common-Sense and Context Information. *Comput. Intell. Neurosci.* **2015**, *2015*, 715730. [CrossRef] [PubMed]
48. References—Citing QGIS in Formal Publications? Geographic Information Systems Stack Exchange. Available online: https://gis.stackexchange.com/questions/23622/citing-qgis-in-formal-publications (accessed on 5 May 2021).
49. FAQ KNIME. Available online: https://www.knime.com/faq#q1_1 (accessed on 5 May 2021).
50. Civiero, P.; Sanmartí, M.; García, R.; Gabaldón, A.; Chicote, M.A.; Ferrer, J.A.; Ricart, J.E.; Franca, P.; Escobar, G.J. Distritos de Energía Positiva (PEDs) en España, Una propuesta de Iniciativa Tecnológica Prioritaria de la PTE-EE. In Proceedings of the VII Congreso Edificios Energía Casi Nula, Madrid, Spain, 5 November 2020; pp. 109–114, ISBN 9798554378225.
51. Reference Framework for Positive Energy Districts and Neighbourhoods Key Lessons from National Consultations. Available online: https://jpi-urbaneurope.eu/wp-content/uploads/2020/04/White-Paper-PED-Framework-Definition-2020323-final.pdf (accessed on 5 May 2021).
52. Harkouss, F.; Fardoun, F.; Biwole, P.H. Optimization approaches and climates investigations in NZEB—A review. *Build. Simul.* **2018**, *11*, 923–952. [CrossRef]
53. Hong, T.; Langevin, J.; Sun, K. Building simulation: Ten challenges. *Build. Simul.* **2018**, *11*, 871–898. [CrossRef]
54. Child, M.; Nordling, A.; Breyer, C. Scenarios for a sustainable energy system in the Åland Islands in 2030. *Energy Convers. Manag.* **2017**, *137*, 49–60. [CrossRef]
55. Child, M.; Nordling, A.; Breyer, C. Scenarios for a Sustainable Energy System in the Åland Islands in 2030, First Results for Åland Smart Energy Platform Stakeholders, 5 September 2016. Available online: https://flexens.com/wp-content/uploads/2021/04/Aland_Scenarios.pdf (accessed on 14 July 2021).

buildings

MDPI

Article

IEA EBC Annex83 Positive Energy Districts

Åsa Hedman [1,*], Hassam Ur Rehman [1], Andrea Gabaldón [2], Adriano Bisello [3], Vicky Albert-Seifried [4], Xingxing Zhang [5], Francesco Guarino [6], Steinar Grynning [7], Ursula Eicker [8], Hans-Martin Neumann [9], Pekka Tuominen [1] and Francesco Reda [1]

1 VTT Technical Research Centre of Finland, Tekniikantie 21, 02150 Espoo, Finland; hassam.rehman@vtt.fi (H.U.R.); pekka.tuominen@vtt.fi (P.T.); francesco.reda@vtt.fi (F.R.)
2 Fundación CARTIF Centro Technologico, Energy Division, Parque Tecnológico de Boecillo, No: 205, 47151 Boecillo, Spain; andgab@cartif.es
3 Eurac Research, Institute for Renewable Energy, Viale Druso 1, 39100 Bolzano, Italy; adriano.bisello@eurac.edu
4 Fraunhofer Institute for Solar Energy Systems, Heidenhofstr. 2, 79110 Freiburg, Germany; vicky.bo.ki.albert-seifried@ise.fraunhofer.de
5 Energy and Community Building, Dalarna University, 79188 Falun, Sweden; xza@du.se
6 Department of Engineering, University of Palermo, Viale delle Scienze, Building 9, 90128 Palermo, Italy; francesco.guarino@unipa.it
7 SINTEF, Strindvegen 4, 7034 Trondheim, Norway; steinar.grynning@usintef.no
8 Concordia Unviersity, 1455 De Maisonneuve Blvd. W. Montreal, QC H3G 1M8, Canada; ursula.eicker@concordia.ca
9 Austrian Institute of Technology—AIT, Giefinggasse 4, 1210 Vienna, Austria; hans-martin.neumann@ait.at
* Correspondence: asa.hedman@vtt.fi; Tel.: +358-40-570-3798

Citation: Hedman, Å.; Rehman, H.U.; Gabaldón, A.; Bisello, A.; Albert-Seifried, V.; Zhang, X.; Guarino, F.; Grynning, S.; Eicker, U.; Neumann, H.-M.; et al. IEA EBC Annex83 Positive Energy Districts. *Buildings* **2021**, *11*, 130. https://doi.org/10.3390/buildings11030130

Academic Editors: Matheos Santamouris, Ambrose Dodoo, Ravi Srinivasan and Paulo Santos

Received: 31 December 2020
Accepted: 16 March 2021
Published: 20 March 2021

Publisher's Note: MDPI stays neutral with regard to jurisdictional claims in published maps and institutional affiliations.

Abstract: At a global level, the need for energy efficiency and an increased share of renewable energy sources is evident, as is the crucial role of cities due to the rapid urbanization rate. As a consequence of this, the research work related to Positive Energy Districts (PED) has accelerated in recent years. A common shared definition, as well as technological approaches or methodological issues related to PEDs are still unclear in this development and a global scientific discussion is needed. The International Energy Agency's Energy in Buildings and Communities Programme (IEA EBC) Annex 83 is the main platform for this international scientific debate and research. This paper describes the challenges of PEDs and the issues that are open for discussions and how the Annex 83 is planned and organized to facilitate this and to actively steer the development of PEDs major leaps forward. The main topics of discussion in the PED context are the role and importance of definitions of PEDs, virtual and geographical boundaries in PEDs, the role of different stakeholders, evaluation approaches, and the learnings of realized PED projects.

Keywords: PED; stakeholder engagement; urban energy transition; bi-directional grid

1. Introduction

The increase in the energy consumption, the intensification of global warming and policies to reduce the need of fossil fuels have created interest in renewable energy sources (RES). The 2015 Paris Agreement has put more emphasis on international efforts to reduce carbon dioxide (CO_2) emissions [1]. According to the International energy agency (IEA), the use of RES has increased significantly in recent decades. For instance, photovoltaic (PV) energy generation increased from 91 GWh in 1990 to 554,382 GWh in 2018 and wind energy has increased from 3880 GWh in 1990 to 1,273,409 GWh in 2018 [2]. According to IEA, around 26% of the global energy was provided through RES in 2018 [2]. With plans to increase the share of the RES by 32% in the European Union (EU) by 2030 [3] and to further reduce CO_2 emissions by 80% by 2050 [4], it is expected that the share of RES will increase on a yearly basis. However, the challenge still remains on the variability of the RES generation, which could result in putting pressure on the grid and ultimately, in compromising the stability of the grid. Thus, power grids and energy systems have to be designed in a way that can regard such issues and challenges.

1.1. Impact of Buildings and Districts on Greenhouse Gas Emissions

An important sector that contributes significantly towards climate change and global warming is the building sector. Buildings account for 30–40% of global final energy consumption [5] and nearly 40% of the global CO_2 emissions. In the last decade, policies such as the Directive on Energy Performance of Buildings (EPBD) have been introduced to address the issue, aiming to decarbonize the building stock by 2050 and to reach nearly zero energy buildings (NZEBs) [6]. In 2009 ambitious energy and climate targets were set for 2020 (20% greenhouse gas emission reduction, 20% increase in efficiency and 20% increase renewable energy). After ten years, the EU in general is on track to achieve these targets, showing that GDP can be increased while reducing carbon emissions. In fact, by 2017 the EU's greenhouse gas (GHG) emissions decreased by 21.7% compared to the 1990 GHG emission levels [7]. In Canada, the residential sector is responsible for 16.6% of the energy consumption and 12.9% of GHG emissions [8]. Between 1990 and 2016, the residential sectors emissions have been reduced by 30.2 Mt CO_2 (27% of total) [8] through enhancing building codes, applying minimum energy performance standards for appliances, improving energy monitoring systems and home retrofits. Under the Paris Agreement, Canada committed to reducing its GHG emissions up to 30% below the 2005 level by 2030 [8]. Moreover, Canada announced a plan to set Canada on a net-zero emissions pathway by 2050. Canada's 2030 GHG emissions target is 511 Mt CO_2 eq, given a 2015 level of 815 Mt CO_2 eq [8]. Between nine principal sectors, buildings are committed to a 47 Mt CO_2 eq reduction [8]. The key priorities are increasing clean electricity, developing and implementing greener buildings and communities, and developing and implementing nature-based climate solutions.

The 2015 Paris Agreement has put more emphasis on international efforts to reduce CO_2 emissions, where urban areas with a 70% share of global emissions have a key role. Accordingly, the United Nations (UN) Sustainable Development Goals include as goal 11 "sustainable cities and communities" with the aim of supporting the transition towards low-carbon cities, in a general framework which also points towards, e.g., climate action, affordability, and clean energy. In 2015, when the Paris agreement was signed, the EU planned to move further ahead and reduce greenhouse gas emissions by 40% by 2030. In order to tackle this challenge and to lead the global energy transition, the EU Commission proposed in 2016 a set of new and ambitious rules known as the Clean energy package for all Europeans [5]. Therefore, to reach the emission reduction goals it is important to focus both at the energy systems level and at the buildings or district level.

1.2. Near and Net Zero Energy Building/District Concepts

Different NZEB, net zero energy or even zero energy building (ZEB) concepts have been developed and implemented in the building sector all over the world. According to the ZEB definition, "the building can be considered as ZEB after showing through actual measurements that the energy delivered to the building is less than or equal to the onsite renewable exported energy" [9]. Similarly, according to Article 2 of the Energy Performance of Building Directive (EPBD), the Nearly Zero Energy Building (NZEB) concept states that "'nearly zero-energy building' means a building that has a very high energy performance, as determined in accordance with Annex I. The nearly zero or very low amount of energy required should be covered to a very significant extent by energy from renewable sources, including energy from renewable sources produced on-site or nearby'" [10]. A similar concept exists in the United States of America (USA) that is called net ZEB [11] it states that annually a building uses less or equal energy generated from the renewable energy on a primary energy basis. Similarly, concepts exists in other places [12] such as net zero energy (NZE) housing in Canada [13], zero emission building in Australia [14] and in Korea [15], etc. However, ZEBs or Nearly Zero Energy Building (NZEBs) mostly relate to individual building scales and do not consider the interaction with other energy consumers and producers. Nevertheless, ZEBs and NZEBs have recently received high academic and political interest around the world, as shown in Figure 1 [10,16–20]. Figure 1, x-axis shows the publication year, and the y-axis shows the cumulative number of documents published each year at the global level, for the keyword. The search is carried out using the keywords such as "nearly zero energy building", "zero energy building", "zero energy districts", "positive energy buildings" and "positive energy districts". The search is carried out from 1990 onwards until 2020 on Scopus [21]. It can be observed that interest and research are increasing each year for the ZEBs and NZEBs at the global level. Furthermore, since 2018, the positive energy district (PED) concept has come into the scene globally.

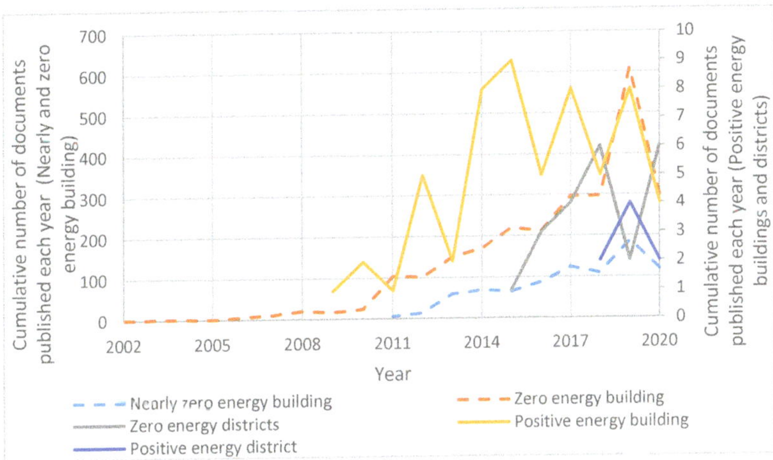

Figure 1. Cumulative number of documents published each year on the "Nearly Zero Energy Building", "Zero Energy Building", "Zero Energy Districts", "Positive Energy Buildings" and "Positive Energy Districts" according to Scopus at the global level [21].

1.3. *Reasons for Positive Energy District (PED) Solutions Instead of Building Level and ZEB Soultions*

ZEB or NZEB buildings do not only consume, but also produce energy onsite. The energy grids have to be designed in a way that allows consuming from the grid and injecting energy from RES to the grid, which can be applied to all types of grids: district heating and cooling networks (DHCN), natural gas grids, and/or power grids. Regarding power grids, bi-directional grids are needed to solve the issue of flexibility. If such issues are not considered, curtailment of the excess energy produced by buildings will be needed to avoid frequency and grid issues. For instance, in Germany [22] and in Belgium [23], the excess Photovoltaic (PV) generation has a power restriction on the export to the grid. When it comes to heat, buildings can be heated by DHCN, but it does not always allow export and production of heat by buildings (only if substations allow prosumers). In cases where this is technically possible, the financial compensation for exported heat is low. In some countries, buildings are just heated by means of an on-site generation system that consumes from the natural gas grid. However, in the future it is expected that an RES transition would occur in DHCN with the introduction of electrical heating systems (e.g., heat pumps) and prosumers, as well as the injection of hydrogen in natural gas grids. Utilizing waste heat streams from buildings and selling the waste heat to the DHCN is expected to grow [24,25]. Another aspect raised in the building sector is the inclusion of electro mobility within buildings and districts, such as a charging station for an electric-vehicle (EV). However, although the transport and mobility sector contributes 27% of the emissions in Europe, the NZEB/ZEB concepts [26] and EPB certificates (such as the ISO52000) usually omit the

EV load in the calculation process. Nevertheless, the transport and mobility sector are becoming increasingly important factors in the energy supply of populated districts, since the share of EVs is increasing rapidly at the global level. In fact, it is expected that by 2030, EV usage will increase to 44 million cars globally [27]. Many cities around the world are thus already including electrification of mobility in their city plans [28]. These EVs would increase the demand and load on the grids and they can also be used as peak savings with batteries. Other aspects, such as building mass and energy storage, have to be included in future energy systems [29]. Energy storage can provide needed flexibility and resilience to buildings [30]. All the above-mentioned issues and challenges call for large changes and renovation of the grids and energy systems so that all the issues can be addressed in an integrated and holistic way. These changes are needed at the district level rather than at the building level. Moreover, the aspects on better coordination between sectors (energy, building, mobility, etc.) and better integration of technologies (e.g., RES, EVs and other NZEB technologies) are other reasons to move from the building to the district level [31].

Various studies on buildings, smart grids and intelligent buildings have been carried out [32–34]. The flexibility and use of new technologies such as RES and storage can be increased by focusing at the district level, rather than at the building level. The research and testing of solutions are already moving from the building level to the district level. This would not only provide technically feasible solutions but also economically viable ones [35]. For example, the district energy refurbishment approach, already tested in some EU projects, leans on a set of innovative system integration activities at the district level and is geared to make the targeted district model robustly scalable and replicable and to maximize the multiple benefits creation [36,37]. Moreover, this would solve the grid and building-related emissions issues at a larger level. However, although the building level research on such topics has become well-structured in the past few years, the district level or, in particular, the positive energy district (PED) field is quite new, and it is developing on academic, scientific and business levels with time [38] as also shown in Figure 1.

1.4. Positive Energy District (PED) Concepts, Aims and Connection with Zero Energy Concepts and International Energy Agency Energy in Building and Communiy (IEA EBC) Annexs

A PED can be generally described as a district within a city that generates more energy than it consumes on an annual basis [39]. The aim of PED is not only to generate surplus energy, but rather to minimize the impact on the centralized grid by promoting higher self-consumption and self-sufficiency. The PED should offer options to increase the onsite load matching by allowing the integration of long and short-term storage and smart controls for improving the energy flexibility. This district level concept and its impact on flexibility, RES and storage integration are still in early stages globally, as shown in Figure 1. Therefore, a holistic approach is needed to define, develop, model and validate the PED concept in order to consolidate the PEDs. Moreover, as shown in Figure 2, the past research focus has been mainly on the ZEBs, intelligent buildings, energy efficiency, NZEB, RES, etc., at the global level. Therefore, Annex 83 will provide the needed platform to discuss and create a framework of PEDs considering the different urban contexts of the globe. According to the solar district heating database [40], there are approximately 195 pilot cases of different capacities of solar-based district heating systems operating in Europe. However, there is currently no insight into how the PEDs and their use in the future districts and cities would be able to provide the consumer-centric, bi-directional grids and districts that are emission-free and flexible. The PEDs can utilize the benefits of the building thermal mass, different typologies of energy storages, RES, electric mobility, demand side management, and flexibility options [30,41–43]. The district can also provide the advantage of shifting the demand, based on the functionality of the various buildings present in the district and this may assist in improving the energy flexibility at the building [44] and grid levels [45]. International Energy Agency Energy in Buildings and Communities (IEA EBC) Annex 83-Positive Energy Districts was developed to provide research contributions towards these fields, based on the outcomes of other IEA Annexes such as Annex 51 [46], Annex 60 [47], Annex 64 [48], Annex 67 [34], and Annex 73 [49].

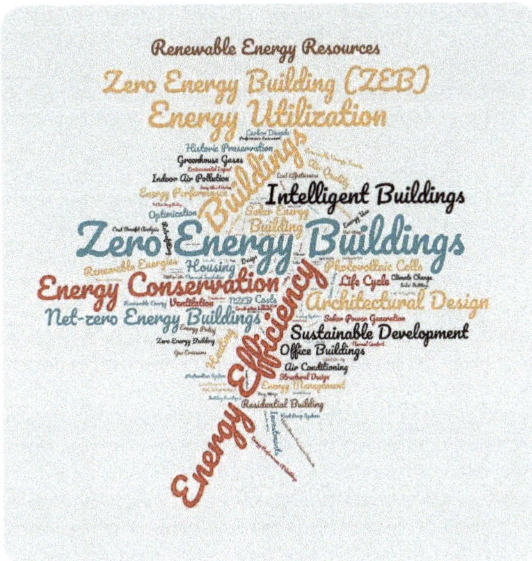

Figure 2. The keywords used globally in the literature according to Scopus [21].

The PED concept introduces an opportunity to develop a framework that introduces energy positivity on a district level, with clear guidelines for grid interaction, energy storage and renewable integration for both buildings and Electric Vehicles (EVs). The main principle of a PED is to create a district within the city that is capable of producing higher energy than it consumes, it is flexible to respond to the energy market situation and in addition to this, it contributes by improving the quality of life and wellbeing of the residents.

The PED conceptual framework will be in line with the nearly or net zero energy building/district concept. The detailed conceptual framework will be planned and designed in this Annex 83. The framework has to be designed in such a way that it can accommodate and consider the local challenges, urban context and regulations, etc. This will provide a basis to analyze various PEDs in different geographical locations. Annex 83 will focus and support the research and development of the PED concept, principles, and frameworks, while keeping the global perspective.

1.5. Challenges, Opportunites, and Global Perspectives towards PEDs

The PED includes all types of buildings present in the district environment and are connected with the energy grid. PED is a growing concept within the research community at the global level in order to create carbon neutral cities for the future. This Annex is one of the first initiatives that aims to coordinate such research on PEDs at the global level. The PED concept may have its limitations depending on the location, local regulations, technology and urban contexts. For example, Denmark's regulations force buildings to be connected to the green district heating network (which will be imports for the PED), leading to higher investments to achieve a positive balance. At the same time, Denmark allows the creation of district heating cooperatives, allowing them to lower the prices of the user [50]. The prosumer regulations are also different within the European Union (EU). While Spain's self-consumption regulation makes the balance on a monthly basis, Latvia does so on a yearly basis and both of them do not reimburse users if the exported energy exceeds their electricity consumption. The Netherlands make the balance yearly, reimbursing prosumers for their exports at the end of the year [51]. Spain and Latvia limit the installed capacity per each user (except if they form a cooperative) while the

Netherlands enable the concepts of aggregators, virtual power plants or peer-to-peer energy exchange. Looking to global contexts, China has very large regions, and not all of them are connected by means of an electric grid [52], which can mean that it is harder in some contexts to apply PEDs. A potential solution to this may be micro-grids with self-sufficient districts (also known as energy islands in some of the literature). The focus of Annex 83 is on the development of the PED concept and its application at the global level, therefore this Annex involves researchers and experts from around the world (i.e., from Europe, USA, Canada, China, Australia, Japan, United Kingdom, South Korea, Turkey etc.) to include the global perspective and challenges as discussed above.

In order to reach the PED, the district first requires higher energy-efficient buildings, secondly the use of carbon free energy renewable energy sources to meet the remaining demand and thirdly cascading local energy flows by making use of any surpluses. Better and smarter controls are needed to match the demand and supply locally and also to minimize the liability on the grid and maximize the effectiveness of PED on the grid. Moreover, since the objective is also to go beyond the fulfillment of a mere mathematical positive energy balance, a wide spectrum of initiatives and parallel objectives are included in the definition of PEDs including social concerns, inclusiveness, solutions to energy poverty, spatial and civic planning of the person building in addition to district wide considerations on the transportation networks and design optimization.

The intrinsic multi-dimensional nature of the design of PEDs requires the contemporary involvement on different levels: mathematical and energy modeling, social, environmental, economic performance assessment, interaction with stakeholders, diffusion of know-how in the territory investigated and creation of PEDs are able to spark the diffusion of these concepts on large scales.

The transition towards carbon neutral districts require multisector and multidimensional solutions. It embraces a synchronized and parallel development of instrumental technologies, public perceptions of building energy technologies, new economic paradigms, assessment approaches, and tailored business models. In this case cities can provide and act as a living lab to facilitate and incubate new technologies and solutions. This is needed in order to co-design all-inclusive packages of citizen's centric carbon-free energy solutions. A common platform is needed to facilitate such collaborations and Annex 83 will focus on providing it with the ultimate aim to generate opportunities for creating such interdisciplinary solutions. Table 1 shows some of the practical application of PEDs or zero energy concepts available.

Table 1. Positive and Zero Energy Concept application across the world including ZEBs and NZEBs.

City	Project	Level of Application	Technologies [2]	Status [1]	Website, (Accessed Date)
Aland	Flexens	island	PV, W, CHP, BE, GB, WP, ES	O	flexens.com/the-demo/, (11 November 2020)
Carquefou	Quartier la fleuriaye	district	PH, EM, CE, PV	O	www.quartierlafleuriaye.fr/, (11 November 2020)
Groningen	MAKING-CITY	district	PV, BIPV, PVT, DHN, BE, WH, GB + HP	I	makingcity.eu/, (11 November 2020)
Oulu	MAKING-CITY	district	GB + HP, PV, DHN, STES	I	makingcity.eu/, (11 November 2020)
Limerick	+CITYXCHANGE	district	CE, HP, ST, EM	I	cityxchange.eu/, (11 November 2020)
	+CITYXCHANGE	district		I	cityxchange.eu/, (11 November 2020)
Amsterdam	ATELIER	district	PV, MG, GB + HP, DHN, BE, EM	I	smartcity-atelier.eu/, (11 November 2020)
Bilbao	ATELIER	district	GDN, PV, HP, EM	I	smartcity-atelier.eu/, (11 November 2020)
Alkmaar	PoCiTYF	district	ST, GE, HP, DHN, PV, EM, CE	I	pocityf.eu/, (11 November 2020)

Table 1. *Cont.*

City	Project	Level of Application	Technologies [2]	Status [1]	Website, (Accessed Date)
Évora	PoCiTYF	district	PV, BIPV, EM	I	pocityf.eu/, (11 November 2020)
Eespo	SPARCS	district	PV. GB, DHN, EM, VPP, STES	I	www.sparcs.info, (11 November 2020)
Leipzig	SPARCS	district	PV, VPP, DHN, ES, STES	I	www.sparcs.info, (11 November 2020)
Santa Coloma Gramenet	SYN.IKIA	district	DHN, PV	I	synikia.eu, (11 November 2020)
Loopkantsestraat (Area wonen)	SYN.IKIA	district	SH, GB + HP, PV	I	synikia.eu, (11 November 2020)
Gneis	SYN.IKIA	district	No info yet	I	synikia.eu, (11 November 2020)
Ammerud	SYN.IKIA	district	DHN, PH, High efficient PV, EM	I	synikia.eu, (11 November 2020)
Aarhus	RESPOND	district	PV, DHN, SH,	I	project-respond.eu/, (11 November 2020)
Aaran	RESPOND	district	PV, HP, ST,	I	project-respond.eu/, (11 November 2020)
Turku Graz	RESPONSE	city	GE, DHN, HP, WH, PV	I	
Brunnshög (Lund)	-	district	ST, HP, DHN, WH, PV	I	futurebylund.se/, (11 November 2020)
Aarhus	READY	city	Low DHN, WH, CE, PV, BIPV, VPP, Sea-HP, EM, BE	P	www.smartcity-ready.eu/about-aarhus/, (11 November 2020)
Okotoks, Alberta	DLSC	district	ST, PH, STES	O	https://www.dlsc.ca/how.htm, (10 January 2021)
Arvada, Colorado	Zero Energy Districts Accelerator	district	PH, REB, PV, GB, HP	I	https://zeroenergy.org/project-profiles/districts-communities/, (10 January 2021)

[1] O = In Operation, P = In planning stage, I = In implementation stage, [2] ST = Solar thermal, PV = photovoltaic panels, PVT = photovoltaic-thermal hybrid panels, W = wind turbine, MW = Micro-wind turbine, DHN = District heating network, GDN = Geothermal district network, SH = Social Housing, PH = Passive House, NZEB = Nearly zero energy building, REB = Retrofitted efficient buildings, BE = Bioenergy, WP = Wave power, GB = Geothermal boreholes, ES = Electric storage, EM = e-mobility (cars/bykes), WH = Waste Heat, MG = Micro grid, HP = Heat pumps, CHP = Cogeneration heat-power unit, CE = Circular economy perspective, STES = Seasonal thermal energy storage.

1.6. Aim and Scope of This Article

The aim of this article is to show the benefits and importance of scientific global level cooperation on the topic of PEDs. This article lays down and presents all the activities planned under each task and the objective of each tasks/subtasks under the proposed Annex 83 platform at the global level. This article provides an introduction to readers about the activities that are planned and in progress in IEA EBC Annex 83. It presents and provides for the Annex 83 project plan. Moreover, it introduces the current global interest of researchers in PEDs. This Annex will be conducted at the global level, in order to include a global perspective to the PEDs, as the topic is novel and requires a global collaboration.

The activities and tasks have recently started in Annex 83 from November 2020 and will continue up until the end of 2024. Under the planned Annex, all the outcomes, findings, new tools, and results will be presented and disseminated on various platforms and in scientific journals, reports, and books. As Annex 83 will progress for the next four years, all the challenges, such as climactic, geographical, regulatory framework, boundary conditions, stakeholders, technological approaches, as well as findings, will be disseminated. A flexible working definition of PED, case studies, development of methodologies, and tools will also be provided and discussed under the Annex.

2. IEA ECB Annex83 Positive Energy Districts: Objectives of the Annex

The International Energy Agency (IEA) has established an Implementing Agreement on Energy in Buildings and Communities (EBC). The function of the EBC program is to undertake research and provide an international focus for buildings and districts energy efficiency. Tasks are undertaken through a series of "Annexes", so called because they are legally established as annexes to the EBC Implementing Agreement. (https://www.iea-ebc.org/ebc/about, accessed date: 20 November 2020).

The largest benefits arising from participation in EBC are those gained by national programmes, such as leverage of R&D resources, technology transfer, training and capacity-building. Countries lacking knowledge can benefit from the experiences of those with more expertise, thereby avoiding duplicated research efforts (https://www.iea-ebc.org/ebc/about, accessed date: 20 November 2020).

The IEA EBC Research strategy states as an objective the following: "the creation of holistic solution sets for district level systems taking into account energy grids, overall performance, business models, engagement of stakeholders, and transport energy system implications". Annex 83, Positive Energy District answers directly this objective (https://www.iea-ebc.org/strategy, accessed date: 20 November 2020).

The international cooperation between research institutions from different parts of the world brings many opportunities. Sharing of experiences between different climate regions, cultures and economic systems enables researchers to develop globally sustainable solutions that are implementable around the world. The overall knowhow and understanding, not only on PEDs, but on society and city development, is increased by an active inclusion of a wide range of stakeholders.

2.1. Objectives

The aim of Annex 83 is to develop an in-depth framework for the devising of PEDs including analyzing the technologies, planning tools and the planning and decision-making processes related to positive energy districts. Experience and data to be used in the Annex will be gained from demonstration cases.

Annex 83 aims to enhance the cooperation of PED development at an international level through collaboration within the initiatives of the IEA. The main objectives of Annex 83 are:

- Objective 1: Map the relevant city, industry, research, and governmental (local, regional, national) stakeholders and their needs and roles to inform the work for Objectives 2, 3, 4, and 5. The main purpose of this is to ensure the involvement of the main stakeholders in the development of relevant definitions and recommendations.
- Objective 2: Create a shared in-depth framework of the definition of PED by means of a multi-stakeholder governance model. So far, international activities have developed generalized definitions that leave many questions open.
- Objective 3: Develop the needed information and guidance for implementing the necessary technical solutions (on the building, district, and infrastructure levels) that can be replicated and gradually scaled up to the city level, giving an emphasis to the interaction of flexible assets at the district level and also economic and social issues such as acceptability.
- Objective 4: Explore novel technical and service opportunities related to monitoring solutions, big data, data management, smart control, and digitalization technologies as enablers of PEDs.
- Objective 5: Develop the needed information and guidance for the planning and implementation of PEDs, including both technical planning and urban planning. This includes economic, social, and environmental impact assessments for various alternative development paths.

2.2. Organisation and Methodology of Annex 83

From various projects (as shown in Table 1) it is found that firstly, for each project, there is a certain definition of the concept, framework, and key performance indicators which are defined and laid down based on the local conditions and regulations, etc. Secondly, the technical framework, technology such as buildings, renewable energy sources, storage technology and simulation models are defined. Thirdly, the socio-economic and social impact assessment criteria are defined. Fourthly, the real physical PED demo is planned, implemented, operated, and measured. Lastly, the outcomes and learnings are communicated for future learnings. A similar method and approach is used to design the Annex 83 project plan and is communicated in this paper.

The Annex 83 is divided into four subtasks: Subtask A: definitions and context; Subtask B: methods, tools and technologies for realizing positive energy districts; Subtask C: organizing principles and impact assessment; and Subtask D: demos, implementation, and dissemination.

Each subtask has a subtask leader (or co-leaders) and a vice-subtask leader. The subtask leaders meet regularly to ensure a good coordination and communication between the tasks. This is imperative since many of the activities and results are dependent on the other tasks.

3. Activities, Subtasks, and Expected Results in the Annex 83

The Annex 83 execution phase started in December 2020. This chapter describes the expected results in the four-year project. The expected results are divided into the four subtasks but highlight the need for interaction between the subtasks. Annex 83 integrates work done in other projects and initiatives (see Table 1) through workshops, questionnaires, discussions, and joint publications among the experts who are working on the different projects. On the European level, the main PED-related activities are within the SCC1 projects under the framework Horizon2020. EERA Smart Cities is an active European platform for the collaboration among researchers within the topic of PEDs. The Annex work helps to share lessons and sets a stage for scientific discussion, bringing forward the lessons from projects and initiatives on a global level.

The annex aims to achieve a shared and internationally viable PED definition through a synthesis effort between the previous experiences, to develop new and integrated modeling approaches of PEDs through different techniques and resolutions, develop methodological advances to the sustainability assessment of PEDs, and test in real district environments the knowledge developed.

The interdependencies of the four subtasks are shown in Figure 3.

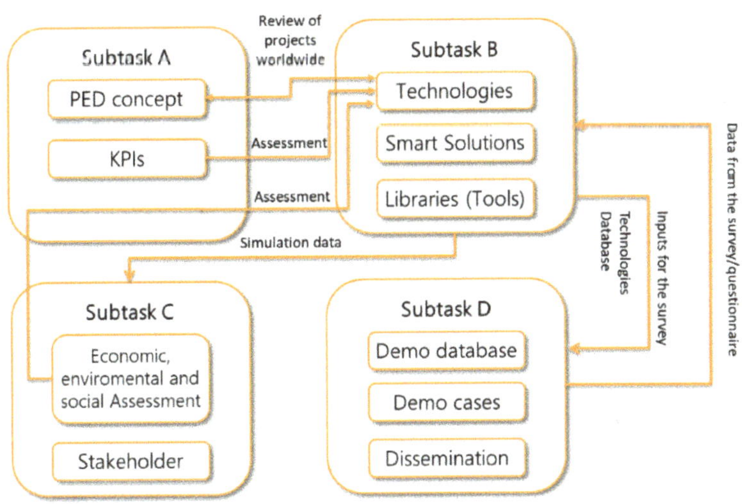

Figure 3. Subtask dependencies.

3.1. Subtask A: Activities and Planned Methodology

This activity will start by identifying the main aspects in the definition context, including system boundaries, different localities, timeframes, energy carriers, etc. The PED scope will be then defined to narrow down the focus in this Annex, so that it is practical while including the crucial elements. Furthermore, a literature review will be conducted to map the existing studies, projects, and initiatives at the international level. This task will be conducted in collaboration with other subtasks to align the scopes and foci within Annex 83.

Subtask A will examine the evolvement of the PED concepts and outline the crucial topics in the development of PEDs such as the spatial and temporal scales to be considered, essential technologies and system components, and regulations and implementation barriers. Visualization of PEDs in the form of infographics will be developed to enhance understanding of the PED concept. In addition, the stakeholders usually involved in the development of PEDs will be identified and categorized. These stakeholders can be urban planners, decision makers, energy system operators and planners, investors, construction companies, housing cooperatives, inhabitants, NGO's, etc. The stakeholders are a fragmented group with varying interests and levels of knowledge about energy and sustainability topics. A framework will be defined for different objectives of PED from different aspects of energy, economics, environment, and social context. Both the stakeholders and the objectives will be input into the Subtask C activities. Meanwhile, the KPI framework for PED will be finally developed in order to compare the performance of the different PED archetypes and PED solutions (developed by Subtask B) and define assessment models (Subtask C).

Based on case studies, common characteristics will be identified and summarized in a "Reference" PED or PED archetype. The "Reference" PED will be used for simulation and demonstration in other subtasks throughout the Annex. Moreover, this activity also plans to establish a common process flow for PED development, using the "Reference PED" as a case study to guide the development of PED projects.

3.2. *Subtask B: Activities and Planned Methodology*

The objective of Subtask B is to review which methods, tools, and technologies are necessary for realizing PEDs. The work is divided into three sub activities: B1, aimed at mapping the technical solutions (energy systems, infrastructures, etc.); B2, investigating how flexibility can help to balance the energy flows; and B3, identifying data and tools for modelling PEDs. The latter will model from two to three case studies to demonstrate different control strategies at the district level.

From the case studies identified in subtask D, an inventory of the current PED technologies applied in PEDs will also be analyzed in sub-activity B1. Through this exercise, the needed data for modelling and the best experiences of the different technologies will be identified. From the revision and analysis of the different technologies applied, the technologies can be classified into different topics/areas (heating, cooling, electricity, storage) and scopes (building, district, city). In each topic/area the technologies can be compared and evaluated (using KPIs from subtask A and assessment evaluation from subtask C) in terms of costs (LCoE, etc.) and regulatory, environmental, energy efficiency, and social acceptance indicators, among others.

The focus of sub-activity B2 is to investigate how flexibility management can help to balance energy flows within and beyond the PED boundaries. To do so, different decision-making processes (algorithms) and control strategies will be reviewed. The results from the previous task (B1) and demo cases (Subtask D) will highlight and reveal the practical challenges regarding the implementation of smart solutions at different levels, and also future research and development needs in PEDs.

This activity will conduct a literature review on decision-making process (solutions for decision makers, architects, citizens, energy experts, etc.) such as algorithms for planning a PED.

A literature review on control strategies and algorithms will be conducted from the information obtained in Subtask B1 and Subtask D. Research on data analysis techniques and control strategy techniques (more advanced control systems, forecasting, load shifting, peak saving, demand management, virtual power plants) is needed and will be conducted. Other issues, such as demand response, flexibility, and data management (block chain) which are useful for managing a PED will be considered. The result will be a comprehensive inventory of the different control solutions (depending on the technology) that can be applied at the building, district, and city levels. The different control strategies will be assessed to identify the barriers/enablers of the different smart solutions.

The focus of sub-activity B3 is to investigate and identify the data and tools for modelling a PED (from demand to the energy balance calculation) that can be used for designing and operating a PED. Activity B3 will mainly focus on data libraries and how these libraries can be used to model a PED. The idea is to generate a framework on how to standardize libraries for urban/district data models (such as City GML) and how to structure it.

To validate these urban scale models and to use data from subtask D case studies, the libraries from B3 will be used for modelling district scale case studies. To do so, existing tools and city platforms such as INSEL or City Energy Analyst, will be used. This will help to analyze how to extract attributes from data libraries, to parametrize urban scale models, and to apply different control strategies and assess them.

The result of sub-activity B1 will be a guideline of the best technologies applied in PEDs in different urban scenarios. As an output, sub-activity B2 provides ideas for the PED planning phase by city planners, citizens, etc. Furthermore, a prototype implementation of interface algorithms for decision-making solutions for PED will be developed. Finally, a report on urban scale modelling of PED districts (control-focused) and how flexibility management can help to balance energy flows within and beyond the PED boundaries will be carried out in sub-activity B3. Moreover, as an output, open-source libraries will be created.

3.3. Subtask C: Activities and Planned Methodology

The objective of Subtask C is to investigate potential sustainable pathways towards PED implementation. It aims at investigating both the impact assessment perspective as well as the organizational aspects within PEDs: the idea is to investigate through a harmonized and parallel approach the three different dimensions of sustainability (economic, environmental, social) of PEDs while ensuring that all three directions are developed through cultural contaminations and connections among them in a holistic and integrated way.

The activities are organized within a common framework that develops on three different levels, following the approach towards the sustainability of PEDs. The structure is vertically integrated.

The three major sub-activities are respectively.

- Economic Assessment (Activity C1);
- Environmental Assessment (Activity C2);
- Humanities and Social Impact Assessment (Activity C3).

The sub-activity C1 will investigate the potential of economic impact assessment methods for PED development and investigation. Key Performance Indicators (KPIs) will be used and tested for PEDs and market strategies and initiative potential will be assessed. Particular interest will be paid to renewable energy self-consumption models that are based on sharing and trading (and financing) approaches.

The development of the activity will also encompass the listing of the most relevant stakeholders to mobilize through the use of organizational models as well as the main barriers and drivers to the implementation of PEDs from an economic perspective.

The sub-activity C2 focuses on the environmental impacts of PED, taking into consideration the different stages of the life cycle of PEDs (e.g., construction, operation, end-of-life). To do so, various factors will be considered, both related to abiotic elements, as well as biotic. The sub-activity will face the challenging task of framing impacts within adequate time limits and scales by identifying relevant boundaries, such as environmental impacts, which widely vary, from climate change to local air quality or biodiversity. More likely,

PEDs are going to deliver a combination of environmental benefits hardly able to be isolated. On the other hand, to realize a PED, regardless of whether they are done by new buildings or the rehabilitation of an existing brownfield, new technologies are going to be installed and natural resources consumed. Therefore, unwanted impacts on the ecosystem may arise and resources consumed. Adequate KPIs and assessment tools are going to be selected and applied to cope with this, and also the life cycle environmental perspective has to be taken in consideration in order to identify potential trade-offs and avoid burden shifts across impact categories or life-cycle stages.

The sub-activity C2 will investigate positive and negative impacts arising from the implementation and diffusion of PEDs, their social acceptance and social inclusiveness. It will also address organizational models and stakeholder engagement in PED development. Once again, social impacts are expected to be found at different levels, from the single householder (e.g., enhanced well-being due to improved indoor comfort) to the local community (social cohesion, social capital) or on a larger population. The peculiarity of the PED's energy system, calling for advanced and innovative solutions, energy sharing and synergies among prosumers, but also implying some behavioral changes due to new technologies, is an interesting and so far unexplored research field for social scientists. Social impacts may be much more relevant as in previous smart energy transition projects and needs specific KPIs. This includes both positive as well as unwanted negative impacts, as, for example, gentrification because of an enhanced attractiveness of the district.

The main outcome of this subtask is to perform the synthesis of the lessons learned and methodological developments by integrating the outcomes of previous ones into innovative and interdisciplinary KPIs—connected to the three spheres of sustainability—and develop sustainability inspired early design tools. Such tools may be based on life cycle sustainability assessments or consider multiple benefits to provide evidence of the contribution of PEDs toward the achievement of selected sustainable development goals.

A PED early design tool for sustainability assessment will be offered to support the decision-making process of policy makers and stakeholders, and also try to leverage investments (e.g., by exploiting the impact investing approach). Substantial collaboration will be carried out with other subtasks and case studies.

3.4. Subtask D: Activities and Planned Methodology

Subtask D spans all the objectives by testing and demonstrating their operationalization in demonstration cases, reaching objective 5 to develop the needed information and guidance for the planning and implementation of PEDs, including both technical planning and urban planning. This includes economic, social, and environmental impact assessment for various alternative development paths.

Firstly, the subtask will start the work with the scoping phase, with the aim to create a framework for data collection from demo cases. References to other initiatives (e.g., SCIS, JPI UE Booklet, and other references from outside EU) will be considered in order to take inspiration for creating the data collection framework collaborative process and to fix the main aspects to be built upon.

The data collection framework will be further elaborated into a template, which will be structured to collect relevant information from demo cases. This activity has a twofold purpose: identifying relevant demo cases and creating a knowledge mass for the whole Annex.

A demo case call will be launched periodically (every 6 months) for the Annex partners and supporters to identify demonstration activities at building blocks, districts, and city levels relevant to the Annex. These can be related also to non-PED demonstrations as long as they show a concrete value for the Annex activities. This is needed to gather detailed information on the best practices, KPIs, stakeholder assessment data, technological data, and key learnings from practical sites.

Secondly, the main outcomes of subtasks A, B, and C will be elaborated into a collection of cross-domains best practices accessible for professionals, city planners, and municipal stakeholders. They will be consulted in the early stages of this activity to identify their burning needs and where they would need support for planning PEDs. This will give an input to create the PED value chain from design and construction to operation, verification, maintenance, renovation and end of life, etc. The guidelines will support the PED planning in different dimensions: urban, suburban, and rural. The integration of PED in the existing urban environment and its role in the city energy transition will also be addressed.

Lastly, a communication and dissemination plan is created with the purpose of outlining the communication, networking, and dissemination strategy, identifying relevant initiatives (associations of cities, professionals, research organizations, initiatives organized by institutions, etc.) for the Annex, describing how the Annex intend to keep up the communication and networking activities. In this regard, the Annex Subtask leaders and Operating Agents will nominate a set of ambassadors to be Annex representatives to the selected initiatives. They will be responsible for setting up collaborative interactions and cooperation events.

As discussed in Section 1, the Annex will seek continuous collaboration with other networks, projects and IEA tasks/Annexes. It is planned to periodically (every 12 months) launch initiatives and conference scouting calls for the Annex partners and supporters to map the relevant PED communication and dissemination opportunities. Under subtask D, the responsibility for all the latest information, updates, relevant content, and outcomes from all the subtasks will be communicated through the Annex website.

4. Discussion and Conclusions

PEDs are seen as an ensemble of buildings of different typologies and functions (residential, commercial, industrial, public-owned, etc.) that are interconnected and produce more energy than what is needed to cover the buildings' demand on an annual basis. Whether the energy demand of the infrastructure (water and waste management, transportation, street lighting etc.) should be included in the PED demands calculations, and the mapping of the boundaries is a topic for discussion. So far, only the building's energy demands have been considered.

Becoming a PED is seldom the overall goal of a district being planned. Elements to be considered (e.g., the type of Renewable Energy Source (RES), number of buildings, etc.) and characteristics to be investigated (how it is organized, what is the governance model, etc.) should be selected and adjusted according to the main objectives and aims identified for creating a PED (improving the circular economy, ensuring high quality of life, etc.) beyond the technical goal of optimizing the energy balance.

Depending on the selection and the definition, the calculated annual energy balance will change. However, currently, all Energy Performance of Building Directive (EPBD) and building standards such as ISO52000 are applied at the building level, not at the district level, making calculation of the annual energy balance more complex and subject to interpretations.

The elements considered within the boundaries will determine how the Positive Energy District (PED) is defined and which loads should be considered for the calculation. The majority of PEDs in Europe apply the dynamic-PED concept, with geographical boundaries (such as PEDs in the projects ATELIER and MAKING-CITY as mentioned in Table 1), which means that buildings are close to each other and dynamically exchange energy (consuming and producing) with the energy grids. However, it is true that, when no space is available within the district boundaries, it could be useful to apply detached geographical patches or virtual boundaries. The main concern when applying the latter is the ownership of the energy solutions and the business models of trading energy to the PED over the virtual boundaries and how to guarantee the energy origin.

An effort to solve these challenges was done in the Sustainable Energy Positive & Zero Carbon Communities (SPARCS) project (as mentioned in Table 1). To upgrade the interaction between energy producing, storing and consuming entities, a virtual positive energy community is created. It is understood as a "variety of energy related actions virtually connecting the multiple buildings across the district on various locations within and across the city". The entities can exchange energy based on "advanced control functionalities and dedicated communication channels (Information and Communications Technology (ICT) model, block chain infrastructure and prediction of the demand)".

Some European projects are treating the PED concept in a different way. For example, the MAKING-CITY project (as mentioned in Table 1) characterized their PEDs by local renewable energy systems (RES) that interact dynamically with the grids (thermal and electrical) and are located within the district boundaries, and aim to achieve an annual positive energy balance incorporating building-related consumption. To do so, retrofit measures to improve the energy efficiency of the buildings as well as including mature technologies such as photovoltaics (PV), photovoltaic-thermal hybrid collectors (PVT), building-integrated PV (BIPV), PV on water, waste digestion, geothermal heat pumps, district heating, and thermal energy storage (such as boreholes, seasonal storage tanks, etc.) are implemented. The concept will be tested in the two lighthouse cities (LH), Groningen and Oulu, and replicated then in six follower cities (FC), taking into account the city needs and priorities, on-site resource availability, MAKING-CITY PED (as mentioned in Table 1) solutions and their business models through a decision-making journey emphasizing citizen engagement. The ATELIER project (as mentioned in Table 1), on the other hand, has two LHs, one district in Amsterdam and another one in Bilbao, with a number of very ambitious building groups (retrofitted and new) of different typologies (tertiary, residential, etc.) that are connected by means of grids (thermal and/or electric ones). Amsterdam will participate and interact with the existing energy communities, as well as with the grid, and will use the local waste for the production of biogas. Bilbao will retrofit an industrial old district and connect its buildings with a geo-exchange loop. In a similar way as in MAKING-CITY (as mentioned in Table 1), it will include Renewable Energy Source (RES), retrofit building measures, electro-mobility and digitalization. Both follow the Smart Energy Transition (SET) plan short definition, but their approach to replicate the concept in FCs is made in a softer way, allowing each city to adapt the PED definition to their own urban context.

Furthermore, several European networks are actively working on the topic of positive energy districts. These include JPI Urban Europe, a network of European funding agencies actively promoting and funding projects on Positive Energy districts, the Urban Europe Research Alliance, a network of Research Organizations and Universities closely working with and informing JPI Urban Europe, the Joint Program Smart Cities of the European Energy Research Alliance (EERA JPSC), the group of the European Smart City Lighthouse Cities, and the COST (European Cooperation in Science and Technology) action on Positive Energy Districts. All of the European members of IEA Annex 83 are also involved in at least one other initiative. This creates huge potential for collaboration, and many synergies will be created by organizing joint meetings and conferences and by writing joint publications and policy guidelines.

As discussed above and in Table 1, most of the districts and projects are under construction that can represent PEDs. However, few of the districts are partially operational, as shown in Table 1. For instance, the drake landing solar community (DLSC) in Canada is able to meet around 96% of the space heating demand of the district via renewable (solar) energy and seasonal storage. The Flexens project in Åland is aggressively targeting to meet 100% of the demand using renewables and to become a fossil-free island. Under the Quartier la fleuriaye project in Carquefou, currently 6000 m^2 of the roof area of the buildings (almost 300 houses) in the district are populated with the solar photovoltaic panels, which cover almost 80% of the energy demand of the district. It is planned to increase the total covered area of 15,000 m^2 (almost 600 h) to provide excess energy to the district so that it can become a PED. Similarly, under the Zero Energy District Accelerator project in Arvada, Colorado, the buildings in the district are designed to be zero energy and as a result the building's life cycle costs are lower than the traditional buildings. Therefore, the building owners are saving not only in terms of reduced emissions, but also in terms of electricity price inflation and tax incentives. The owners are open to invest in new technologies such as passive building design, solar panels and ground source heat pumps etc., to become better.

The challenges raised above in the introduction section indicates that the PEDs are complex and multi-disciplinary in nature. Moreover, it has various challenges depending on the climate, location, regulations, technologies, key performance indicators, and urban context etc., and this requires a scientific global discussion. Many learnings can be done by exchanging experiences from different projects, knowledge, and cases around the world.

In the future, climate adaptation will become more important, which will bring new challenges to the planning of the urban environment and PEDs. Energy poverty might also become a bigger challenge than today due to increased immigration levels caused, among other things, by climate change. The detailed definition, key performance indicators and framework of PEDs will be discussed, developed and published in the future work as the Annex 83 progresses up until 2024. Moreover, all the issues, challenges, methodologies, technological solutions, and roles of the stakeholders will be discussed and developed under the subtasks (mentioned above) which will carried out in Annex 83.

Annex 83 is the main platform for this scientific discussion in the coming years. Different urban contexts will be covered in the different subtasks within Annex 83 to create a global framework of the concept, as well as to identify the barriers and enablers of PEDs. This will be possible thanks to cooperation between the partners involved in the Annex, with expertise from different fields and from all over the world. Canada, with the involvement of Concordia University, will give a perspective on urban scale modelling. Japan, thanks to Tokyo University, will give an overview of the different decision-making methodologies and on flexibility management. Different expertise from Europe and around the world, with the involvement of key actors and coordinators of current PED projects, will contribute to translating theory into practice and to test on the ground the latest findings.

Author Contributions: Conceptualization, Å.H., H.U.R.; methodology, Å.H., H.U.R.; investigation, H.U.R, Å.H., A.G., A.B., V.A.-S., X.Z.; resources, Å.H., A.B., V.A.-S., X.Z., A.G.; writing—original draft preparation, Å.H., H.U.R.; writing—review and editing, H.U.R., Å.H., A.G., A.B., V.A.-S., X.Z., F.G., U.E., S.G., H.-M.N., F.R., P.T.; visualization, H.U.R.; supervision, Å.H., H.U.R.; project administration, Å.H. All authors have read and agreed to the published version of the manuscript.

Funding: This research received no external funding. The authors would like to thank the VTT Technical Research Center of Finland (Finland), Eurac Research (Italy), the technology research center of CARTIF, Fraunhofer Institute for Solar Energy Systems (Germany), Swedish Energy Agency (Sweden), University of Palermo (Italy) for providing the resources, technical, and administrative support.

Institutional Review Board Statement: Not applicable.

Informed Consent Statement: Not applicable.

Data Availability Statement: Not applicable.

Conflicts of Interest: The authors declare no conflict of interest.

References

1. United Nations Framework Convention on Climate Change The Paris Agreement | UNFCCC. Available online: https://unfccc.int/process-and-meetings/the-paris-agreement/the-paris-agreement (accessed on 19 May 2020).
2. International Energy Agency (IEA) Renewables-Fuels & Technologies. Available online: https://www.iea.org/fuels-and-technologies/renewables (accessed on 6 October 2020).
3. European Commisssion 2030 Climate & Energy Framework | Climate Action. Available online: https://ec.europa.eu/clima/policies/strategies/2030_en (accessed on 4 November 2019).
4. European Commisssion 2050 Long-Term Strategy | Climate Action. Available online: https://ec.europa.eu/clima/policies/strategies/2050_en (accessed on 4 November 2019).
5. Lutsch, W. European Commission Clean energy for all Europeans. *Euroheat Power Engl. Ed.* **2019**, *14*, 3. [CrossRef]
6. The European Parliament and the Council of the European Union Directive (EU) 2018/844 of the European Parliament and of the Council of 30 May 2018 Amending Directive 2010/31/EU on the energy performance of buildings and Directive 2012/27/EU on energy efficiency. *Off. J. Eur. Union* **2018**, *L 156/75*, 75–91.

7. European Union Europe 2020 Indicators-Climate Change and Energy. Available online: https://ec.europa.eu/eurostat/statistics-explained/index.php/Europe_2020_indicators_-_climate_change_and_energy (accessed on 23 July 2020).
8. Environment and Natural Resources Progress Towards Canada's Greenhouse Gas Emissions Reduction Target. Available online: https://www.canada.ca/en/environment-climate-change/services/environmental-indicators/progress-towards-canada-greenhouse-gas-emissions-reduction-target.html#DSM (accessed on 25 January 2021).
9. D'Agostino, D.; Mazzarella, L. What is a Nearly zero energy building? Overview, implementation and comparison of definitions. *J. Build. Eng.* **2019**, *21*, 200–212. [CrossRef]
10. European Commission Directive 2010/31/EU of the European Parliament and of the Council of 19 May 2010 on the energy Performance of Buildings. Available online: https://eur-lex.europa.eu/legal-content/EN/TXT/PDF/?uri=CELEX:32010L0031&from=EN (accessed on 30 May 2020).
11. Torcellini, P.; Grant, R.; Taylor, C.; Punjabi, S.; Diamond, R.; Colker, R.; Moy Aecom, G.; Kennett, E. *A Common Definition for Zero Energy Buildings*; U.S. Department of Energy: Washington, WA, USA, 2015.
12. Liu, Z.; Zhou, Q.; Tian, Z.; He, B.J.; Jin, G. A comprehensive analysis on definitions, development, and policies of nearly zero energy buildings in China. *Renew. Sustain. Energy Rev.* **2019**, *114*, 109314. [CrossRef]
13. Natural Resources Canada Net Zero Energy (NZE) Housing. Available online: https://www.nrcan.gc.ca/energy/efficiency/data-research-and-insights-energy-efficiency/housing-innovation/net-zero-energy-nze-housing/5131 (accessed on 25 January 2021).
14. Chris, R.; Aleta, L.; Nicky, I. *Defining Zero Emission Buildings-Review and Recommendations: Final Report*; Institute for Sustainable Futures, UTS: Sydney, Australia, 2011.
15. Liu, Y.; Zhang, S.; Xu, W.D. Cho Study of zero energy building development in Korea. *Build. Sci.* **2016**, *32*, 171–177.
16. Jin, Y.; Wang, L.; Xiong, Y.; Cai, H.; Li, Y.H.; Zhang, W.J. Feasibility studies on net zero energy building for climate considering: A case of "all Green House" for Datong, Shanxi, China. *Energy Build.* **2014**, *85*, 155–164. [CrossRef]
17. Wang, L.; Gwilliam, J.; Jones, P. Case study of zero energy house design in UK. *Energy Build.* **2009**, *41*, 1215–1222. [CrossRef]
18. Wells, L.; Rismanchi, B.; Aye, L. A review of Net Zero Energy Buildings with reflections on the Australian context. *Energy Build.* **2018**, *158*, 616–628. [CrossRef]
19. Al-Saeed, Y.; Ahmed, A. Evaluating Design Strategies for Nearly Zero Energy Buildings in the Middle East and North Africa Regions. *Designs* **2018**, *2*, 35. [CrossRef]
20. U.S. Department of Energy (DOE) A Common Definition for Zero Energy Buildings | Build Up. Available online: http://www.buildup.eu/en/practices/publications/common-definition-zero-energy-buildings (accessed on 1 April 2020).
21. Elsevier B.V. Scopus. Available online: https://www.scopus.com/ (accessed on 12 October 2020).
22. Wirth, H. Recent Facts about Photovoltaics in Germany. Available online: https://www.ise.fraunhofer.de/en/publications/studies/recent-facts-about-pv-in- (accessed on 12 October 2020).
23. Consortium, G.B. Study on "Residential Prosumers in the European Energy Union" JUST/2015/CONS/FW/C006/0127 Framework Contract EAHC/2013/CP/04. Available online: https://ec.europa.eu/commission/sites/beta-political/files/study-residential-prosumers-energy-union_en.pdf (accessed on 11 May 2020).
24. Buffa, S.; Cozzini, M.; D'Antoni, M.; Baratieri, M.; Fedrizzi, R. 5th generation district heating and cooling systems: A review of existing cases in Europe. *Renew. Sustain. Energy Rev.* **2019**, *104*, 504–522. [CrossRef]
25. Abugabbara, M.; Javed, S.; Bagge, H.; Johansson, D. Bibliographic analysis of the recent advancements in modeling and co-simulating the fifth-generation district heating and cooling systems. *Energy Build.* **2020**, *224*, 110260. [CrossRef]
26. European Environment Agency Greenhouse Gas Emissions from Transport in Europe. Available online: https://www.eea.europa.eu/data-and-maps/indicators/transport-emissions-of-greenhouse-gases/transport-emissions-of-greenhouse-gases-12 (accessed on 6 April 2020).
27. International Energy Agency (IEA) Global EV Outlook 2019. Available online: https://www.iea.org/reports/global-ev-outlook-2019 (accessed on 7 February 2020).
28. City of Helsinki Kaupunkisuunnitteluviraston Esitteet | Helsingin Kaupunki. Available online: https://www.hel.fi/kaupunkiymparisto/fi/julkaisut-ja-aineistot/julkaisuarkisto/kaupunkisuunnitteluviraston-esitteet (accessed on 30 December 2020).
29. Marino, C.; Nucara, A.; Panzera, M.F.; Pietrafesa, M. Towards the nearly zero and the plus energy building: Primary energy balances and economic evaluations. *Therm. Sci. Eng. Prog.* **2019**, *13*, 100400. [CrossRef]
30. Zhou, Y.; Cao, S.; Hensen, J.L.M.; Lund, P.D. Energy integration and interaction between buildings and vehicles: A state-of-the-art review. *Renew. Sustain. Energy Rev.* **2019**, *114*, 109337. [CrossRef]
31. Becchio, C.; Bottero, M.C.; Corgnati, S.P.; Dell'Anna, F. Decision making for sustainable urban energy planning: An integrated evaluation framework of alternative solutions for a NZED (Net Zero-Energy District) in Turin. *Land Use Policy* **2018**, *78*, 803–817. [CrossRef]
32. Hamdy, M.; Hasan, A.; Siren, K. Applying a multi-objective optimization approach for Design of low-emission cost-effective dwellings. *Build. Environ.* **2011**, *46*, 109–123. [CrossRef]
33. Cao, S.; Hasan, A.; Sirén, K. On-site energy matching indices for buildings with energy conversion, storage and hybrid grid connections. *Energy Build.* **2013**, *64*, 423–438. [CrossRef]
34. Jensen, S.Ø.; Marszal-Pomianowska, A.; Lollini, R.; Pasut, W.; Knotzer, A.; Engelmann, P.; Stafford, A.; Reynders, G. IEA EBC Annex 67 Energy Flexible Buildings. *Energy Build.* **2017**, *155*, 25–34. [CrossRef]

35. Rehman, H. Ur Techno-Economic Performance of Community Sized Solar Heating Systems in Nordic Conditions. Available online: https://aaltodoc.aalto.fi/handle/123456789/34808 (accessed on 4 November 2019).
36. Bisello, A.; Vettorato, D. Multiple Benefits of Smart Urban Energy Transition. In *Urban Energy Transition*; Elsevier: Amsterdam, The Netherlands, 2018; pp. 467–490.
37. Bisello, A. Assessing Multiple Benefits of Housing Regeneration and Smart City Development: The European Project SINFONIA. *Sustainability* **2020**, *12*, 8038. [CrossRef]
38. Shnapp, S.; Paci, D.; Bertoldi, P. *Enabling Positive Energy Districts Across Europe: Energy Efficiency Couples Renewable Energy*; Publication Office of the European Union: Luxembourg, 2020.
39. Lindholm, O.; Ur Rehman, F. Positioning Positive Energy Districts in European Cities. *Buildings* **2021**, *11*, 19. [CrossRef]
40. Solar District heating Plant Database–Solar District Heating. Available online: https://www.solar-district-heating.eu/en/plant-database/ (accessed on 12 October 2020).
41. Ur Rehman, H.; Reda, F.; Paiho, S.; Hasan, A. Towards positive energy communities at high latitudes. *Energy Convers. Manag.* **2019**, *196*, 175–195. [CrossRef]
42. Chen, Y.; Xu, P.; Chen, Z.; Wang, H.; Sha, H.; Ji, Y.; Zhang, Y.; Dou, Q.; Wang, S. Experimental investigation of demand response potential of buildings: Combined passive thermal mass and active storage. *Appl. Energy* **2020**, *280*, 115956. [CrossRef]
43. Majdalani, N.; Aelenei, D.; Lopes, R.A.; Silva, C.A.S. The potential of energy flexibility of space heating and cooling in Portugal. *Util. Policy* **2020**, *66*, 101086. [CrossRef]
44. Hu, M.; Xiao, F. Quantifying uncertainty in the aggregate energy flexibility of high-rise residential building clusters considering stochastic occupancy and occupant behavior. *Energy* **2020**, *194*, 116838. [CrossRef]
45. Amaral Lopes, R.; Grønborg Junker, R.; Martins, J.; Murta-Pina, J.; Reynders, G.; Madsen, H. Characterisation and use of energy flexibility in water pumping and storage systems. *Appl. Energy* **2020**, *277*, 115587. [CrossRef]
46. Jank, R. IEA EBC ANNEX 51 Energy Efficient Communities. Available online: https://iea-ebc.org/projects/project?AnnexID=51 (accessed on 23 October 2020).
47. Michael Wetter; Christoph van Treeck IEA EBC Annex 60 New Generation Computational Tools for Building & Community Energy Systems. Available online: http://www.iea-annex60.org/ (accessed on 23 October 2020).
48. Dietrich Schmidt IEA EBC ANNEX 64 LowEx Communities-Optimised Performance of Energy Supply Systems with Exergy Principles. Available online: https://iea-ebc.org/projects/project?AnnexID=64 (accessed on 23 October 2020).
49. Rüdiger Lohse; Alexander Zhivov IEA EBC | | Annex 73 | | Towards Net Zero Energy Public Communities | | IEA EBC | | Annex 73. Available online: https://annex73.iea-ebc.org/ (accessed on 23 October 2020).
50. Danish District Heating Association The Danish District Heating Model. Available online: https://www.danskfjernvarme.dk/sitetools/english/the-danish-model (accessed on 4 December 2020).
51. Inês, C.; Guilherme, P.L.; Esther, M.G.; Swantje, G.; Stephen, H.; Lars, H. Regulatory challenges and opportunities for collective renewable energy prosumers in the EU. *Energy Policy* **2020**, *138*, 111212. [CrossRef]
52. Peter Fairley China's Ambitious Plan to Build the World's Biggest Supergrid-IEEE Spectrum. Available online: https://spectrum.ieee.org/energy/the-smarter-grid/chinas-ambitious-plan-to-build-the-worlds-biggest-supergrid (accessed on 4 December 2020).

buildings

MDPI

Article

A MILP Optimization Method for Building Seasonal Energy Storage: A Case Study for a Reversible Solid Oxide Cell and Hydrogen Storage System

Oscar Lindholm [1,*], Robert Weiss [1], Ala Hasan [1], Frank Pettersson [2] and Jari Shemeikka [1]

1 VTT Technical Research Centre of Finland, FI-02044 VTT Espoo, Finland; robert.weiss@vtt.fi (R.W.); ala.hasan@vtt.fi (A.H.); jari.shemeikka@vtt.fi (J.S.)

2 Laboratory of Process and Systems Technology, Faculty of Science and Engineering, Åbo Akademi University, Fi-20500 Turku, Finland; frank.pettersson@abo.fi

* Correspondence: oscar.lindholm@vtt.fi

Received: 10 June 2020; Accepted: 3 July 2020; Published: 9 July 2020

Abstract: A new method for the optimization of seasonal energy storage is presented and applied in a case study. The optimization method uses an interval halving approach to solve computationally demanding mixed integer linear programming (MILP) problems with both integer and non-integer operation variables (variables that vary from time step to time step in during energy storage system operation). The seasonal energy storage in the case study uses a reversible solid oxide cell (RSOC) to convert electricity generated by solar photovoltaic (PV) panels into hydrogen gas and to convert hydrogen gas back to electricity while also generating some heat. Both the case study results and the optimization method accuracy are examined and discussed in the paper. In the case study, the operation of the RSOC and hydrogen storage system is compared with the operation of a reference system without energy storage. The results of the study show that installing an RSOC and hydrogen storage system could increase the utilization of onsite renewable energy generation significantly. Overall, the optimization method presents a relatively accurate solution to the case study optimization problem and a sensibility analysis shows a clear and logical pattern.

Keywords: energy storage; hydrogen; power-to-gas; reversible solid oxide cell; optimization; mixed integer linear programming; MILP

1. Introduction

During the last decade, people have become increasingly more aware of the environmental impact of their daily routines and consumption behavior. Minimizing the carbon dioxide footprint by replacing fossil fuels with renewable energy sources is therefore a change process driven by the end consumers of products and services. Meanwhile, in the building sector, there has been a trend towards "greener" solutions and more energy efficient constructions. Concepts like net zero energy buildings (NZEB) and nearly zero energy buildings (nZEB) have been studied in several research projects [1,2]. These concepts are striving towards covering a higher share of the building energy demand with on-site generated renewable energy.

Seasonal energy storage can be used to increase the on-site utilization of solar energy installations, like photovoltaic (PV) panels and solar thermal collectors. Storing energy during periods of high on-site energy generation and utilizing the stored energy during periods of low on-site energy generation will increase the utilization rate of the solar energy installations for individual buildings. Hence, seasonal energy storage is an important contributing factor for buildings that are targeting a net zero annual energy balance.

Investing in energy storage also has an economic driving factor. Energy import prices are namely often significantly higher than export prices for household energy consumers [3–5]. By storing energy, household consumers can avoid selling cheap excess energy in the summer and buying it back with a significantly higher price in the winter. The economic benefits of seasonal energy storages are additionally boosted by the falling installation costs of solar PV technologies. Between 2010 and 2018, the total installation cost of solar PV panels dropped by 74% [6].

Optimization-based design methods can be used to maximize the utilization and minimize the cost and environmental impacts of seasonal storage systems. These problems are usually formulated as time series optimization problems, where the design variables set constraints on the system operation in every time step of the annual operation. Different kinds of stochastic optimization method have been used for this purpose. Durão et al. (2014) studied the advantages of using genetic algorithms to optimize a seasonal energy storage of solar thermal energy [7]. Other approaches, such as particle swarm optimization [8] and simulated annealing (SA) [9], have also been used for optimizing similar energy storage systems. Zhang et al. (2018) used harmony search and chaotic search methods based on a SA approach to optimize renewable energy systems including different energy storages [10]. The objective of the optimization by Zhang et al. was to minimize the life-cycle cost of the renewable energy system.

Mixed integer linear programming (MILP) is a demanding optimization problem category. MILP problems include both integer and real variables and tend to become computationally expensive as the number of integer variables increases. Kotzur et al. (2018) proposed a clustering method to solve MILP problems related to the optimization-based design of energy storage systems [11]. The integer variables in the method by Kotzur et al. are of a binary nature and define whether certain components in an energy system exist or not. Steen et al. (2014) solved similar MILP energy storage problems to minimize both the cost and the emissions of a thermal energy storage system by optimizing the system operation and setup [12]. Moreover, Wang et al. (2015) proposed a MILP-based control method that uses day-ahead pricing, weather forecasts and customer preferences to minimize the energy expenditures of a building energy system comprised of a battery and building-integrated solar PV [13]. Pinzon at al. (2017) propose a similar MILP-based control method for smart buildings with integrated solar PV and batteries [14]. Both of these control methods showed favorable results when testing them against simulation software and measured data, respectively.

This paper aims to present and evaluate a novel seasonal energy storage optimization method that uses a time interval halving approach to solve computationally expensive MILP problems. The method is unique since it accepts integer variables in every time step of the annual operation. This implies that the method can handle multi-mode devices. The use of integer control variables for multi-mode devices is essential when developing accurate operational models that are able to separate between different operational modes. The authors have not found any other optimization methods in the literature that can solve seasonal storage optimization problems that include this type of integer control variable.

A case study is presented in this paper in order to examine and evaluate the presented method. The case study examines the optimal operation and design of a seasonal energy storage system for an office building with an over-production of solar energy during the summer season. The seasonal energy storage system uses reversible solid oxide cell (RSOC) technology to convert electrical energy generated by PV to hydrogen gas and to convert hydrogen gas back to electricity, while also generating some heat. The case study is based on a preceding study presented in [15], where the energy storage was only used for short-term grid balancing and did not yet cover the seasonal aspects.

Earlier studies on hydrogen storage system optimization have been done by Castañeda et al. (2013), Luta and Raji (2018), as well as Carapellucci and Giordano (2011) [16–18]. The hydrogen storage systems in these studies use separate electrolyzers and fuel cells and are thus not comparable with the RSOC and hydrogen storage system used in the case study in this paper.

2. Methods

2.1. Optimization Method

The novelty of the optimization method is that it can solve computationally expensive seasonal storage operation optimization problems with both integer and non-integer operation variables (variables that vary from time step to time step during operation). This means that the method can optimize energy storage systems containing multi-mode devices (e.g., on-off mode devices and reversible fuel cells) where integer variables are used to control operation modes. This type of optimization problem tends to become too computationally expensive for conventional optimization as the number of integer variables increases.

The basic idea of the method is to first perform a rough optimization of the annual operation, and then gradually fine-tune the solution by re-optimizing smaller time intervals. The first step of the method is, hence, to optimize the annual operation using a relatively small number of time steps, N, ($N = 26$ time steps are used in the case study). Thereafter, the energy storage load at the beginning, the middle and the end of the one-year period is fixed. The one-year period is then bisected, so that the first half of the year forms one new time period and the second half of the year forms another new time period.

In the second optimization step, the new time intervals are re-optimized separately, using the same number of time steps (N) as in the first optimization step, and the storage loads at their center points (not the starting points and the end points) are fixed. Then, the time intervals are bisected, and new intervals are formed. The second optimization step is repeated until the time steps reach the desired size (seven hours in the case study). The time step will reduce in size for every new interval bisection since the number of time steps remains the same, while the time interval size shrinks.

A graphical explanation of the first and second optimization steps are presented in Figure 1. In the figure, $m_{i,j}$ is the energy storage load at the central point of time interval j in optimization step i. Step 1 also shows the storage load at the beginning, a, and the end, b, of the year. To guarantee a net zero annual storage balance, the storage loads at point a and b must be equal to one another. Otherwise, the energy storage cannot operate in a sustainable way for several years.

After each optimization step, the new center points of the new time intervals will be added to the final solution of the annual system operation. This means that the storage load for all the points that are fixed will remain constant throughout the optimization process. Hence, the storage load for $2^{(n-1)}$ (where n is the number of the optimization step) new points will be added to the final solution from each optimization step, except for the first optimization step, where three points (center point, starting point and end point) are added to the final solution.

Each time interval optimization is solved with a Matlab R2019b solver called intlinprog, which uses a branch-and-bound approach to tackle MILP problems [19]. The solver is also able to identify and eliminate futile subproblem candidates by adjusting the constraints of the optimization problem.

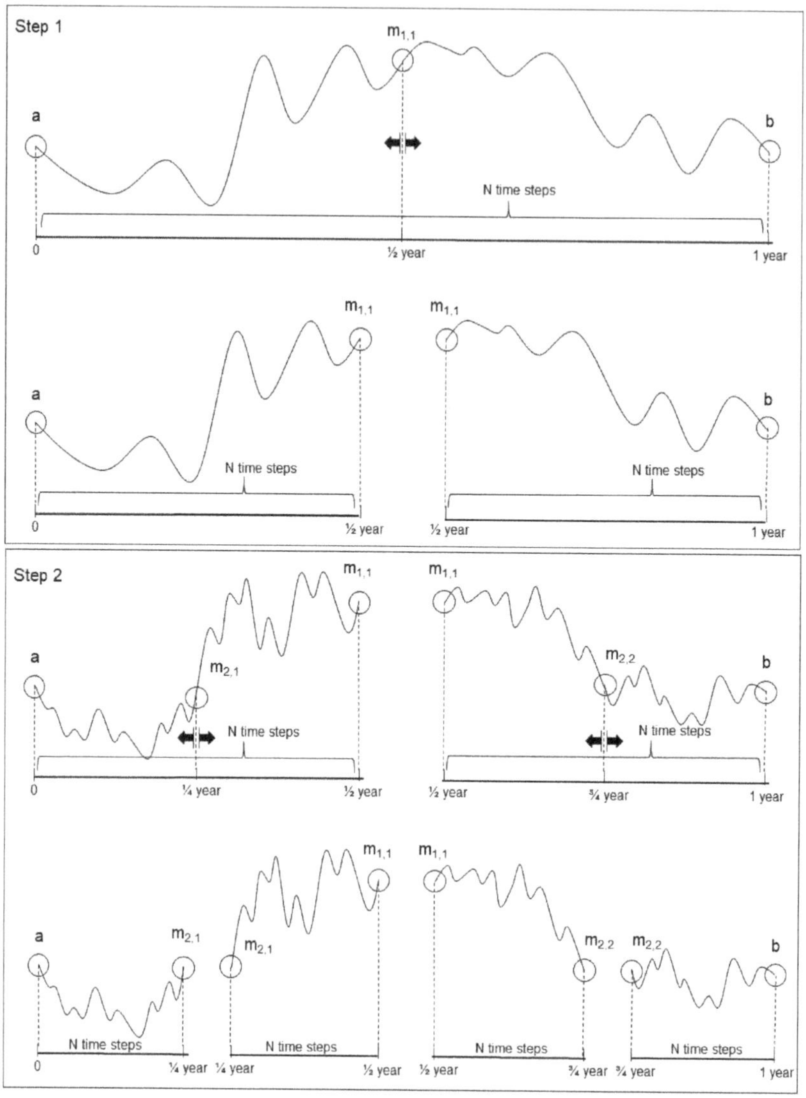

Figure 1. Graphical description of the first and second optimization step in the interval halving optimization method.

2.2. *Case Study: Reversible Solid Oxide Cell and Hydrogen Storage System for Seasonal Storage of Solar Energy*

A reversible solid oxide cell (RSOC) is an electrochemical device that can operate either as a solid oxide electrolysis cell (SOEC) or as a solid oxide fuel cell (SOFC). In SOEC mode, the RSOC consumes electricity and water in order to produce hydrogen and oxygen through a redox reaction. In SOFC mode, the redox process is run in reverse, so that the RSOC consumes hydrogen and produces electricity, heat and water [20,21]. An RSOC together with a hydrogen compressor and a compressed hydrogen storage can be used as an energy storage system which is able to store energy in the form of hydrogen gas. Such a system is hereinafter referred to as an RSOC and hydrogen storage (RSOCHS) system. The power input capacity of the RSOC in SOEC mode is significantly higher than the power output capacity in SOFC mode [22]. This means that discharging the RSOCHS storage is more time consuming than charging the storage.

In the context of the case study, the RSOCHS system is used in an energy system to balance out large seasonal variations in solar PV generation for a 7874 m^2 floor-area office building, VTT FutureHub, (Figure 2), located in Espoo, Finland. The building characteristics are presented in Table 1. The location of the building is 60°11′11.4″ N 24°48′49.0″ E and the Köppen climate classification of the site climate is Dfb [23].

Figure 2. The office building, VTT FutureHub, in Espoo, Finland.

Table 1. VTT FutureHub building characteristics.

Floor Area	7874 m^2
Volume	29,483 m^3
Roof area	1163 m^2
Construction year	2020

The assumed energy system of the office building (depicted in Figure 3) includes solar PV panels, an RSOCHS system, a water tank, a boiler with a hydrogen combustor for heat generation as well as connections to the electricity grid and the district heating network (for both import and export). The annual heating and electricity demand profiles of the office building are produced by IDA ICE simulations [24]. These profiles are presented in Figure 4.

The optimization method presented in this paper is used to optimize the system operation, where the objective function is to minimize the annual operating expense (OPEX). Due to the differences in the two RSOC operation modes, they must be modelled as separate functions in the optimization problem. A binary decision variable, which determines the operation mode, must also be introduced at each time step in the optimization problem. The other variables in the optimization problem consist of energy transfer rates, which can be described as non-integer variables. The optimization problem is considered as a computationally expensive MILP problem due to its high number of integer and non-integer variables. The problem characteristics are, hence, ideal for the interval halving optimization method presented in this paper. Table 2 presents the dimensions of the case study optimization problem.

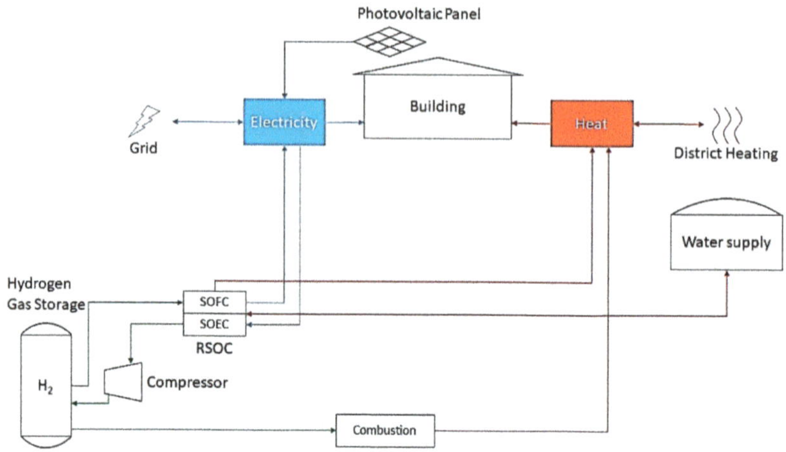

Figure 3. The reversible solid oxide cell and hydrogen storage system (RSOCHS).

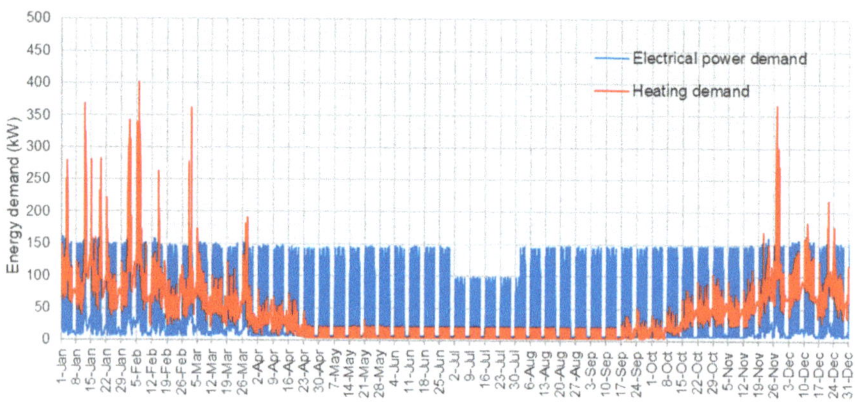

Figure 4. Electricity and heating demand profiles of the office building.

Table 2. Statistics of the optimization problem as well as the central processing unit (CPU) time required to solve the problem.

Number of Binary Variables	1248
Number of non-integer variables	9985
Number of constraints	6240
CPU time	15 s

2.2.1. Calculated Cases

The optimal operation of three RSOCHS systems with different RSOC sizes is generated for different solar PV areas and hydrogen storage capacities. The RSOC sizes used for the RSOCHS systems are 20/80 kW (20 kW maximum power output in SOFC mode and 80 kW maximum power input in SOEC mode), 50/200 kW and 100/400 kW. All tested case parameters are presented in Table 3.

Table 3. Tested case parameters.

Case Parameter	Tested Values
RSOC sizes (kW)	20/80, 50/200, 100/400
Solar PV areas (m^2)	500, 1000, 2000, 3000, 4000, 5000, 6000, 7000, 8000, 9000, 10.000
Hydrogen storage (kWh)	10, 20, ... , 600

The RSOCHS systems are compared with a reference system without any energy storage in order to identify the strengths and weaknesses of the RSOCHS system. By comparing the energy systems, it is also possible to examine whether the RSOCHS system is a suitable solution for seasonal storage or not.

2.2.2. Objective Function

The objective of the optimization problem is to minimize the annual OPEX of the RSOCHS system by using the operating mode and the different energy transfer rates as variables. The OPEX of the system in this context is defined as the sum of the electricity and district heating costs minus the income generated by energy export to the grid and the district heating network. The objective function is, thus, expressed as follows:

$$\min \sum_{i=1}^{N} (\dot{Q}_{im,i} C_{Qim,i} + P_{im,i} C_{Pim,i} - \dot{Q}_{ex,i} C_{Qex,i} - P_{ex,i} C_{Pex,i}) t, \tag{1}$$

where $C_{Qim,i}$ is the district heating import price at time step i, and $C_{Pim,i}$ is the electricity import price at time step i, which is sum of the electricity spot price, the electricity distribution tariff and the grid tax. $C_{Qex,i}$ and $C_{Pex,i}$ are the export prices for heat and electricity at time step i. The variables $\dot{Q}_{im,i}$ and $\dot{Q}_{ex,i}$ are the imported and exported heat rate at time step i, while the variables $P_{im,i}$ and $P_{ex,i}$ are the imported and exported electrical power at step i, respectively. The time step size, t, and the number of time steps, N, are selected so that the sum of all time steps is equal to one year.

2.2.3. RSOC Functions

In SOEC mode, the RSOC is modelled with one single function that describes the hydrogen output as a linear function of the electrical power input, $P_{E,in,i}$, and the binary decision variable, δ_i, for each time step i. The SOEC function is expressed as:

$$f_E(P_{E,in,i}, \delta_i) = a_E \delta_i + k_E P_{E,in,i}, \tag{2}$$

where the coefficients a_E and k_E are dependent on the size and operating range of the RSOC device.

In SOFC mode, two different linear functions are used; one for the hydrogen input, f_F, and one function for the thermal energy output, f_G. Both of these functions are dependent on the electrical power output of the RSOC. These functions are expressed as follows:

$$f_F(P_{F,out,i}, \delta_i) = a_F \delta_i + k_F P_{F,out,i}, \tag{3}$$

$$f_G(P_{F,out,i}, \delta_i) = a_G \delta_i + k_G P_{F,out,i}. \tag{4}$$

The coefficients a_F, a_G, k_F and k_G are here also dependent on the size and operating range of the RSOC device.

2.2.4. Constraints

There are five constraints for each time step in the optimization problem. These constraints consist of two energy balances, one for electrical energy and one for thermal energy, two constraints for the capacity of the RSOC and one constraint for the hydrogen storage.

The electrical energy balance is the sum of all electrical energy transfer rates and is expressed as:

$$P_{F,out,i} - P_{E,in,i} + P_{im,i} - P_{ex,i} - P_{d,i} + P_{PV,i} - \frac{h_{c,out} - h_{c,in}}{\eta_c \text{LHV}} f_E(\delta_i, P_{E,in}) = 0 \; \forall i \tag{5}$$

where the parameters $P_{d,i}$ and $P_{PV,i}$ are the electrical power demand of the building and the power generated by the solar PV panels at time step i. The last part of the equation describes the power used by the hydrogen compressor at time step i, where η_c is the total efficiency of the hydrogen compressor, LHV is the lower heating value of hydrogen gas. $h_{c,in}$ and $h_{c,out}$ are the specific enthalpies of the hydrogen gas at the inlet and outlet of the hydrogen compressor.

The thermal energy balance is expressed as follows:

$$f_G(\delta_i, P_{F,out,i}) + \dot{Q}_{b,i} + \dot{Q}_{im,i} - \dot{Q}_{ex,i} - \dot{Q}_{d,i} = 0 \; \forall i \tag{6}$$

where $\dot{Q}_{b,i}$ is the heat generated through combustion of hydrogen, and $\dot{Q}_{d,i}$ is the heating demand of the building at time step i.

The input, $P_{E,in,i}$, and output, $P_{F,out,i}$, power ranges of the RSOC device are dependent on the size of the RSOC device as well as the operating mode. The RSOC power input and output constraints are thus controlled by the decision variables for each time step. These constraints are expressed as:

$$P_{E,in,max}\delta_i \geq P_{E,in,i} \geq P_{E,in,min}\delta_i \; \forall i \tag{7}$$

$$P_{F,out,max}(1 - \delta_i) \geq P_{F,out,i} \geq P_{F,out,min}(1 - \delta_i) \; \forall i \tag{8}$$

The hydrogen constraint is dependent on the storage load at the beginning of each time step, which is the cumulative sum of the produced hydrogen plus the initial stored hydrogen, $E_{H,0}$, minus the cumulative sum of consumed hydrogen. The hydrogen storage constraint is thus dependent on the system operation of earlier time steps, which is the reason why each time step cannot be optimized individually. Hydrogen storage constraints are expressed as:

$$E_{H,cap} \geq E_{H,0} + t\sum_{s=1}^{i}\left(f_E(P_{E,in,s}, \delta_i) - f_F(P_{F,out,s}, \delta_i) - Q_{b,s}\right) \geq 0 \; \forall \; i \in \{1 \ldots (N-1)\} \tag{9}$$

$$E_{H,cap} \geq E_{H,0} + t\sum_{s=1}^{N}\left(f_E(P_{E,in,s}, \delta_i) - f_F(P_{F,out,s}, \delta_i) - Q_{b,s}\right) \geq E_{H,0} \tag{10}$$

where the parameter $E_{H,cap}$ is the maximum capacity of the hydrogen storage. The constraint for the last time step in Equation (10) is different from the constraint for the other time steps in Equation (9) since it only accepts a final storage load above the initial stored energy.

2.2.5. Assumptions

The parameters in the case study are comprised of the annual energy demand, the solar PV generation and the energy price profiles as well as some technical data regarding the RSOC, the hydrogen compressor and the solar PV panels. The RSOC performance data are based on confidential data provided by the RSOC development team at the Technical Research Centre of Finland and are therefore not presented in this paper.

According to the US Department of Energy, the isentropic efficiency, η_{is}, of hydrogen compressors used for small hydrogen gas terminals is about 65% [25]. The electrical motor efficiency, η_m, of the compressor is assumed to be 95%. The total efficiency of the hydrogen compressor is thus assumed to be:

$$\eta_c = \eta_{is}\eta_m \approx 62\% \tag{11}$$

The solar PV production capacity is assumed to be 0.17 kWp/m^2 and the PV generation profile is based on simulated data for a system where 50% of the solar panels are facing east and 50% are facing west [26].

As a power and heat price scenario base, we use current prices and tariffs for Espoo, Finland. As mentioned earlier, the electricity import price is the sum of the electricity spot price, the distribution tariff and the grid tax. The electricity spot price used in this study is based on the Nord Pool spot price data from 2017 (Figure 5) [27] and the distribution tariff is selected according to the pricing of Caruna Oy, which is the electricity distributor in Espoo [28]. The grid tax in Finland is 2.253 c/kWh for non-industrial consumers [29]. The electricity export price is based on the price the Finnish energy company Fortum Oy pays for excess electricity generated by households [3]. This price is the hourly spot price minus a 0.24 c/kWh transfer fee. The average export and import electricity prices are summarized in Table 4.

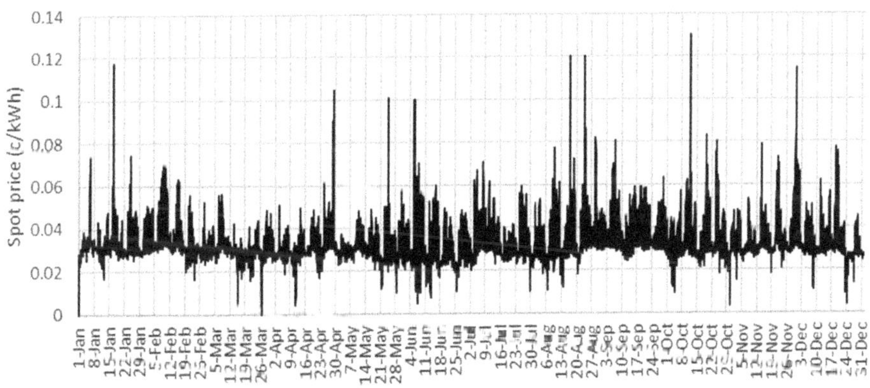

Figure 5. Nord Pool electricity spot prices 2017 [27].

Table 4. Electricity price and tariffs for the case building in Espoo, Finland.

	Export (c/kWh)	Import (c/kWh)
Electric energy	Nord Pool spot price Average: 3.30	Nord Pool spot price Average: 3.30
Daytime distribution tariff, winter [1,2]	-	2.42
Other time distribution tariff [1]	-	1.15
Transfer fee	−0.24	-
Grid tax	-	2.25
Total average price	**3.06**	**9.12**

[1] Electricity distribution tariff for a 400 V connection; [2] Daytime distribution tariff, winter: Mon–Sat 7am–10pm, Nov–Mar.

The district heating import and export prices presented in Figure 6 are also based on pricing by Fortum Oy, who operates the district heating network in Espoo [4]. These prices vary from month to month, and are usually much higher during the winter months, when the heating demand is higher. The export prices are dependent on the temperature of the provided heat, but since temperatures are not considered in the optimization model, the export prices in Figure 6 are based on the average export price for each month.

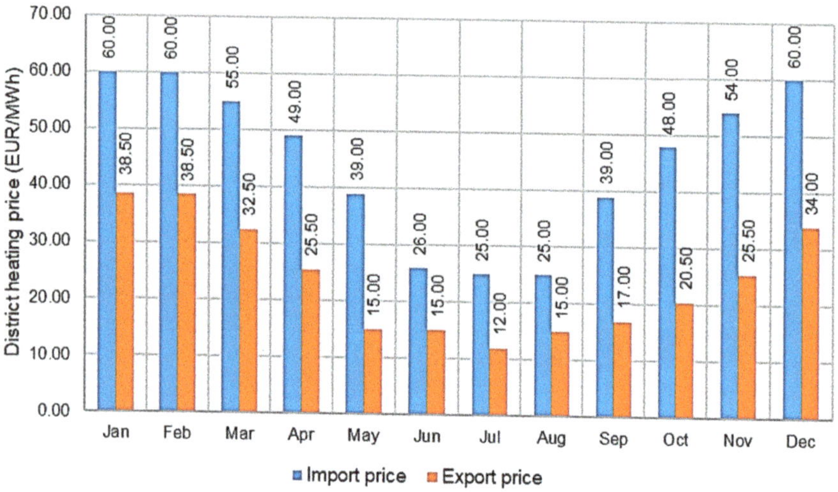

Figure 6. District heating import and export prices in Espoo, Finland [4,5].

3. Results

3.1. Results of the Case Study

Three RSOCHS systems as well as one reference system were optimized for different solar PV generation capacities using the interval halving optimization method presented in the paper. The reference system has no onsite energy generation or storage. It is only comprised of connections to the electricity grid, the district heating network and the solar PV panels. The idea of the optimization was to optimize the operation of the system by minimizing the OPEX. The results of the optimization are presented and analyzed in this chapter in order to form a conception of the behavior and benefits of a RSOCHS system as a seasonal energy storage for buildings.

The optimal operation of the different RSOCHS systems was analyzed for three different solar PV installation areas (1000, 2000, and 3000 m^2), with results presented in Figures A1–A9 in the Appendix A. All of the systems show the same trend; storing energy during summer and consuming the stored energy during winter. One surprising remark, which also holds true for all storage systems, is that the system uses large amounts of electricity from the grid to fill up the hydrogen storage, even during the summer, when there is a surplus of energy generated by the solar PV panels. This indicates that it is cost-effective to convert electricity from the grid to hydrogen gas and then use it for heating during the winter.

By comparing the optimal operation of different system setups, it is shown that seasonal energy storage has some benefits for energy systems with large solar PV installations. This remark is supported by the results presented in Figures 7–10. It shall, however, be pointed out that the optimization model is formulated so that it does not allow the RSOC device to be turned off, since it might be complicated to ramp the RSOC up and down due to the high operation temperature. This is the reason why the RSOCHS systems perform worse than the reference system for smaller PV installations.

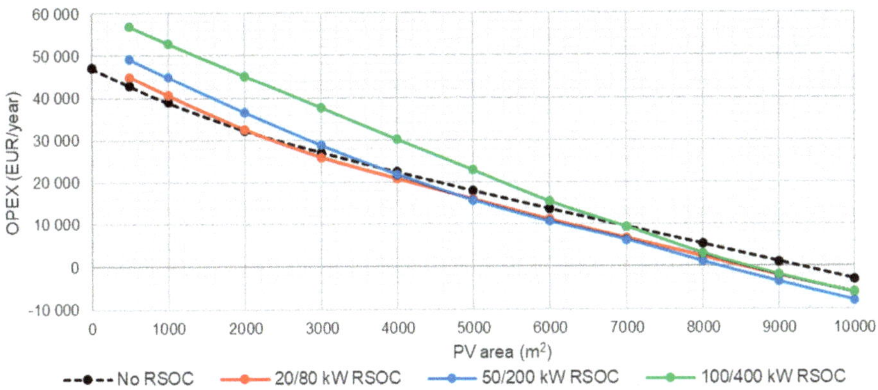

Figure 7. Optimal annual operating expense (OPEX) (for the optimal hydrogen storage size) as a function of installed solar photovoltaic (PV) panel area for the three RSOCHS systems and the reference system.

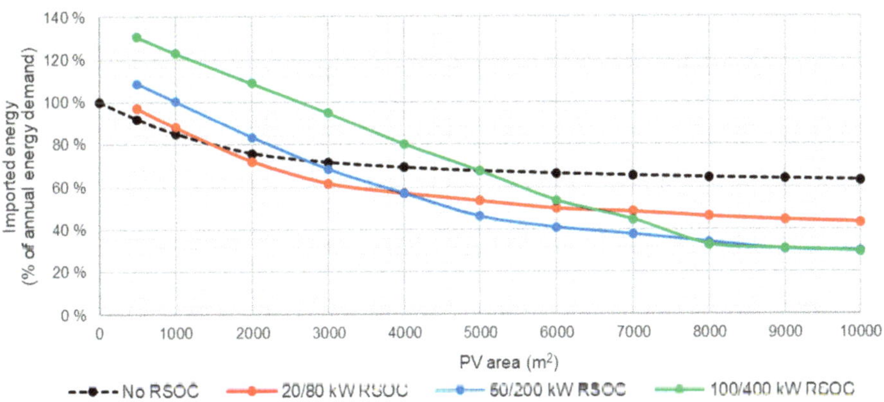

Figure 8. Annual imported energy (electricity and heat) (for the optimal hydrogen storage size) as a function of installed solar PV panel area for the three RSOCHS systems and the reference system.

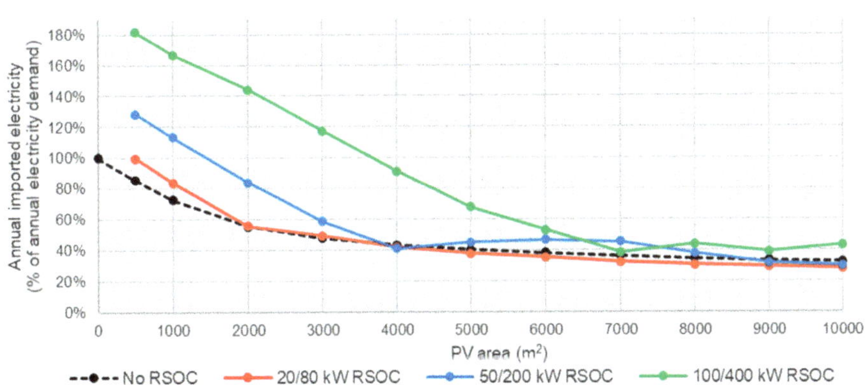

Figure 9. Annual imported electricity (for the optimal hydrogen storage size) as a function of installed solar PV panel area for the three RSOCHS systems and the reference system.

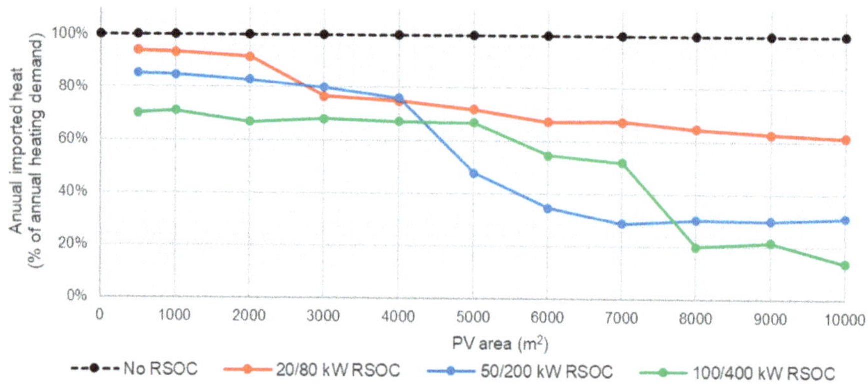

Figure 10. Annual imported heat (for the optimal hydrogen storage size) as a function of installed solar PV panel area for the three RSOCHS systems and the reference system.

Figure 7 shows the optimal OPEX (for the optimal hydrogen storage size) as a function of installed PV area for the RSOCHS storages systems and the reference energy system (no RSOC). By examining the graph in Figure 7, it can be observed that increasing the solar PV area has a higher cost saving impact on the RSOCHS system than on the reference system. More installed PV panels entail a lower OPEX and an increased advantage of the RSOCHS system. It can also be observed that the optimal size of the RSOC is dependent on the installed PV area, as small PV areas appear to be more suitable for small RSOC devices, while the power generation of larger PV areas can be utilized to a greater extent by a big RSOC device.

The same trend can be observed when the amount of imported energy is examined for the different energy systems. Figure 8 shows that the amount of imported energy can be reduced by investing in an RSOC and hydrogen storage system. It can also be observed that the energy savings increase when more solar PV panels are added to the system. This proves that the RSOCHS system works as intended, halving the need for annually imported energy for energy systems with large solar PV installations.

Figures 9 and 10 show that reduced district heating consumption accounts for most of the energy savings generated by the RSOC and hydrogen storage system. The figures describing the optimal RSOC operation in the Appendix A show that most of the time, heat is generated by the RSOC, but the peaks in the heat generation are caused by hydrogen combustion. Some of these peaks may be a consequence of the interval halving approach in the optimization method.

Without hydrogen storage, the operation of the RSOCHS systems and the reference system would be the same. Hence, the difference in OPEX between the systems in Figure 7 is only induced by the hydrogen storage size. It can be noted that the impact of the hydrogen storage size on the OPEX is marginal for PV installations below 5000 m^2. For big PV installations, however, the significance of the hydrogen storage size is more crucial, especially when the OPEX of the systen drops below zero.

Figure 11 shows the optimal hydrogen storage capacity for different RSOCHS systems and solar PV areas. The figure indicates that the optimal size of the hydrogen storage increases with the size of both the solar PV installation and the RSOC device. The energy content of the hydrogen stroage is calculated using the LHV of hydrogen gas.

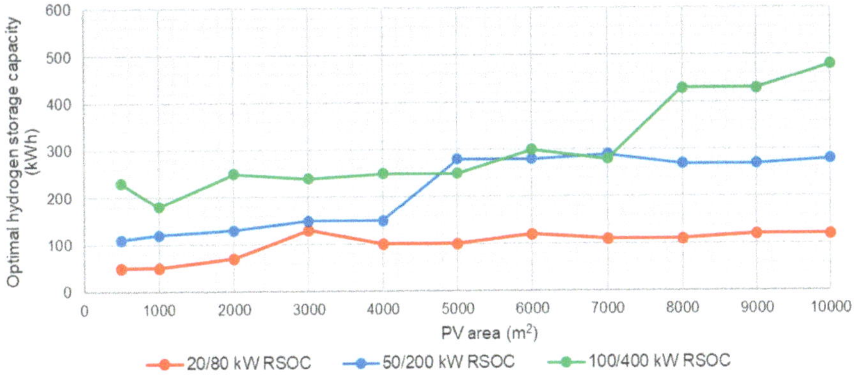

Figure 11. Optimal hydrogen storage size as a function of installed solar PV panel area for the three RSOCHS systems and the reference system.

3.2. *Optimization Model Perfomance*

The model is evaluated by examining the optimization results of various RSOCHS system setups and investigating whether an increased degree of freedom in the energy system can result in lower OPEX values or not. Figure 12 shows the optimal OPEX of the 50/200 kW RSOC system as a function of the hydrogen storage capacity for different solar PV installations. The OPEX should decrease with increasing storage capacity, since more storage entails more relaxed constraints in the optimization problem. This is however not exactly valid for all the results. In the graphs, it can be observed that the trend lines are all decreasing when the storage capacity is increasing, but this is not exactly followed by the individual optimized points. It can be presumed that this small but noticeable inconsistency in the results is caused by some inaccuracy of the optimization model.

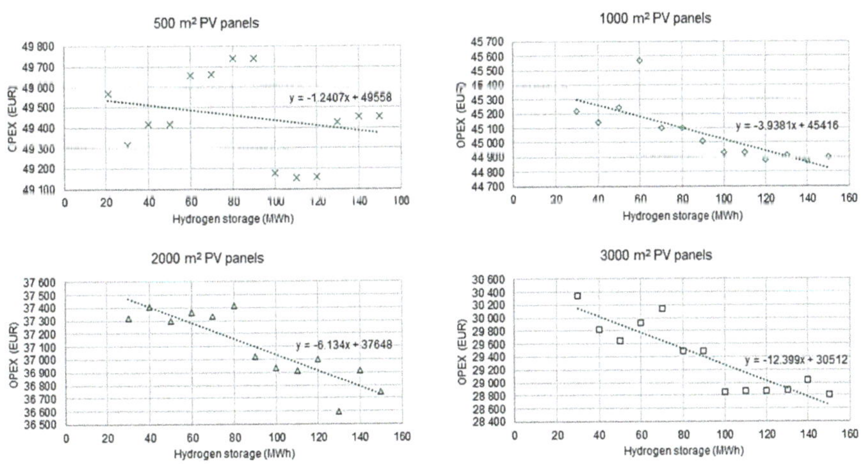

Figure 12. Optimal OPEX of the 50/200 kW RSOC system as a function of the hydrogen storage capacity for different solar PV installations.

One of the main sources of inaccuracy in the method could be the halving approach. Fixing a few points in the first optimizations steps will add more constraints on the following optimization steps which, in turn, will prevent the solution from reaching the global optimum. The peaks in the heat generation graphs in Figures A1–A9 in the Appendix A are visible indications of the inaccuracy caused by the interval halving method. These peaks are most likely a consequence of forced hydrogen combustion due to the additional constraints produced by the interval halving process.

By analyzing deviation from the trend line in the graphs in Figure 12, it can be concluded that the optimization method is still relatively precise. The deviation from the trend line is namely below 2% for each optimized point in Figure 12. We also tried shifting the start of the year by 1400 h (58 days 8 h), but this action did only affect the value of the objective function (the OPEX) by less than 1%. This remark also supports the statement that the optimization method is relatively precise.

An additional drawback of the proposed optimization method is that the system component dimensions cannot be used as design variables in the optimization problem, as the optimization problem is divided into several subproblems with problem-speci0fic variables. Moreover, introducing the RSOC size as a variable would render the problem an MINLP (mixed integer non-linear programming) problem, which cannot be solved by the presented method. Hence, the optimal design of the energy storage system is found by performing several optimizations using different system component dimension combinations. This makes it complicated to use the method for the optimization-based design of energy storage systems with many different components, as it is impossible to include the component investment costs in the objective function.

A more accurate evaluation of the optimization method performance would require a comparative analysis with other optimization methods. For the time being, no other optimization tools are available for solving the type of optimization problems presented in the case study.

4. Conclusions

An interval halving MILP optimization method for seasonal storages was presented and tested by applying it to a case study. The novelty of the method is that it allows integer variables in each time step of the annual energy storage operation. The aim of the case study was to examine the optimal operation and setup of an RSOCHS system and evaluate its suitability as a seasonal energy storage. The operating principle of the RSOCHS system is to use RSOC technology to convert electrical energy to hydrogen gas and to convert hydrogen gas back to electricity while also generating some heat [15].

By analyzing the results of the optimization problem for the investigated case study, it can be noted that the solution of the optimization problem is a relatively accurate estimation of the optimal seasonal storage operation. Moreover, it shall be pointed out that the optimization problem presented in the case study is computationally too expensive to be solved by conventional optimization methods and that no other alternative optimization method suitable for the problem has been found in the literature so far. The model is hence ideal for optimizing MILP seasonal storage problems, but it could also be implemented in other computationally demanding linear scheduling optimization problems with cumulative constraints.

There are, however, a few minor shortcomings that can be discussed regarding the interval halving optimization method. The most significant drawback of the method is caused by the interval halving approach itself. Bisecting the time interval and fixing its boundaries adds constraints to the optimization problem that are not present in a real situation. Consequently, the global optimum of the problem cannot be reached with the method, even if MILP problems can be solved to global optimality. Due to the nature of the interval halving approach, it is also impossible to include the investment cost in the objective function. Hence, it is somewhat inconvenient to use the method as an optimization-based design method for energy storage systems with a large number of design variables.

Further algorithm development could increase the application area of the interval halving optimization method. Other MILP solvers and perhaps even mixed integer non-linear programming (MINLP) solvers could be applied in the model in the subproblem optimization. Thereby, the method could be able to solve a larger variety of optimization problems.

Apart from the restrictions of the optimization model, the mathematical assumptions in the optimization problem formulation also affect the reliability of the case study results. Some mathematical assumptions are made in order to attune the problem to fit the optimization approach, while some assumptions are just made due to the lack of available RSOC operation data. The RSOC operation, for example, is assumed to be linear, even though the RSOC operation is slightly non-linear according to several sources [21,30]. The performance of the RSOC is, in practice, also highly dependent on the operation temperature [30], which is not taken into account in this study, since it would considerably complicate the problem formulation. Uncertain technical factors such as the time required to switch between operational modes and balance of plant energy losses etc. are not taken into account in the model either. It is, however, complicated to create a proper mathematical RSOC model, since ROSC technology is still in an early development phase. A more thorough optimization of the RSOCHS system would therefore have to be postponed until the technology is mature enough to provide sufficient technical and operational data.

The results of the optimization show that the operating cost benefits as well as the self-sufficiency of an RSOCHS system increase when more solar PV panels are installed. The RSOC and hydrogen storage system enables the consumer to utilize more of the generated PV power, which means that less energy has to be imported. An RSOCHS system could hence be an important contributor in achieving a net zero annual energy balance for individual buildings. In building energy systems with big solar PV panel installations, the annual imported energy could be halved by installing an RSOCHS system. This reduction in imported energy would mostly be in the form of a reduction in district heating.

However, installing an RSOCHS system does have a relatively low impact on the OPEX of the energy systems. The maximum capital savings in terms of annual OPEX is only about 5000 EUR, which is only a fraction of the investment cost of the hydrogen gas compressor [15]. The payback time of the RSOCHS system might thus be greater than the total life of the investment.

The optimal hydrogen storage size varies between 50 and 5000 kWh, depending on the RSOC size and the solar PV panel area. The optimal size of the hydrogen storage tends to increase as the RSOC size and the solar PV panel area increases.

The analysis of the ROSCHS system operation optimization showed that the hydrogen storage should be filled during the summer using both electricity generated by the solar PV and electricity imported from the grid. This behavior can be motivated by high PV generation during the summer period and the high district heating prices in the winter period. Because of the high district heating prices in the winter, it is economically feasible to charge the energy storage with electricity from the grid and use it for heating.

The case study in this paper only focuses on the optimal operation of the RSOCHS system and its operational benefits compared to a system without energy storage. To better understand the economic value of the RSOCHS system, a life-cycle cost analysis is required. RSOC technology is still in a development phase and the cost of an ROSCHS system is thus not yet competitive compared with other energy storage technologies [31]. Mass manufacturing of RSOC components is envisaged to bring down the investment cost of the technology, but exactly by how much is still difficult to predict. Hence, RSOC technology might play a significant role in future energy storage and power-to-gas systems.

Author Contributions: Conceptualization, O.L., R.W., F.P., A.H. and J.S.; methodology, O.L.; validation, O.L.; formal analysis, O.L.; investigation, O.L.; writing—original draft preparation, O.L.; writing—review and editing, O.L., A.H., R.W. and F.P.; visualization, O.L.; supervision, A.H., R.W. and F.P. All authors have read and agreed to the published version of the manuscript.

Funding: The authors would like to thank Business Finland. The work was part of Smart Otaniemi innovation ecosystem (Smart Otaniemi Pilot Phase 2, 8194/31/2018) financed by Business Finland.

Conflicts of Interest: The authors declare no conflict of interest.

Nomenclature

a	The beginning of the year
b	The end of the year
$C_{Pex,i}$	Electricity export price at time step i
$C_{Pim,i}$	Electricity import price at time step i
$C_{Qex,i}$	Heat export price at time step i
$C_{Qim,i}$	Heat import price at time step i
CPU	Central processing unit
$E_{H,0}$	Stored hydrogen gas at the beginning and end of the year
$E_{H,cap}$	Hydrogen gas storage capacity
$h_{c,in}$	Specific enthalpy of hydrogen gas at the inlet of the hydrogen compressor
$h_{c,out}$	Specific enthalpy of hydrogen gas at the outlet of the hydrogen compressor
LHV	Lower heating value of hydrogen gas
$m_{i,j}$	Storage load at central point of time interval j in optimization step i
MILP	Mixed integer linear programming
n	The number of the optimization step
N	Number of time steps
OPEX	Operating expense
PV	Photovoltaic
$P_{d,i}$	Electrical power demand of the building at time step i
$P_{E,in,i}$	Electrical power input to the RSOC at time step i
$P_{E,in,max}$	Maximum electrical input of the RSOC
$P_{E,in,min}$	Minimum electrical input of the RSOC
$P_{ex,i}$	Exported electrical power at time step i
$P_{F,out,i}$	Electrical power output of the RSOC at time step i
$P_{F,out,max}$	Maximum electrical output of the RSOC
$P_{F,out,min}$	Minimum electrical output of the RSOC
$P_{im,i}$	Imported electrical power at time step i
$P_{PV,i}$	Solar PV electrical power output at rime step i
$\dot{Q}_{b,i}$	Heat generated by combustion of hydrogen gas at time step i
$\dot{Q}_{d,i}$	Heating demand of the building at time step i
$\dot{Q}_{ex,i}$	Exported heat rate at time step i
$\dot{Q}_{im,i}$	Imported heat rate at time step i
RSOC	Reversible solid oxide cell
RSOCHS	Reversible solid oxide cell and hydrogen gas storage
SA	Simulated annealing
SOEC	Solid oxide electrolysis cell
SOFC	Solid oxide fuel cell
t	Time step size
δ_i	binary operation mode decision variable for the time step i
η_c	Efficiency of the hydrogen gas compressor
η_{is}	Isentropic efficiency of the hydrogen gas compressor
η_m	Electrical motor efficiency of the hydrogen gas compressor

Appendix A

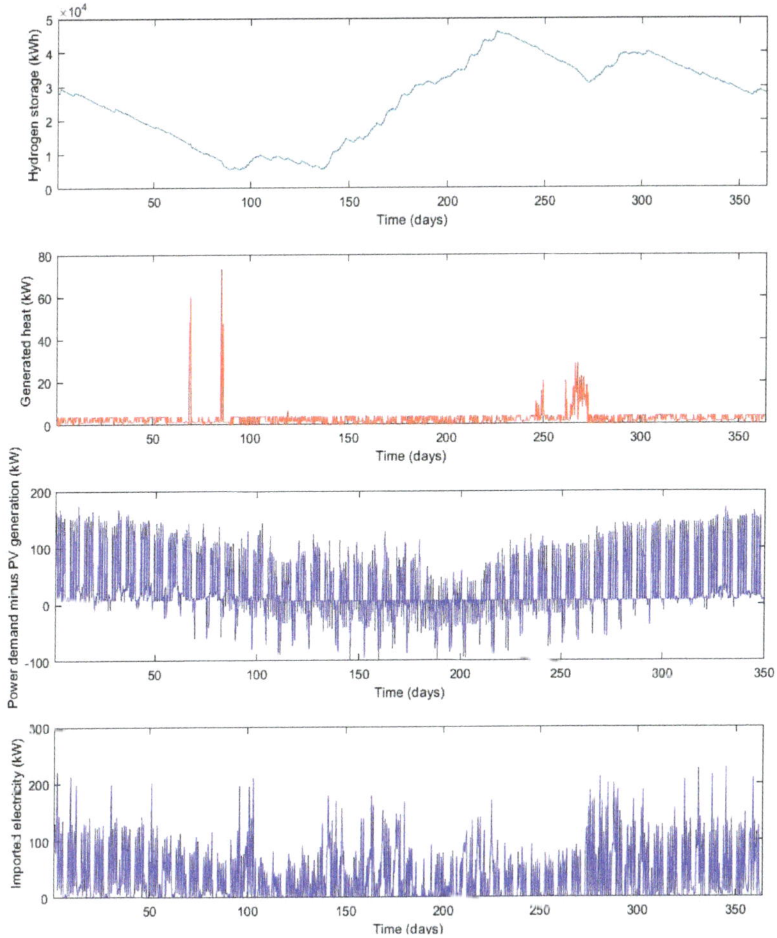

Figure A1. Optimal operation of a 20/80 kW RSOC system with 1000 m^2 solar PV panel.

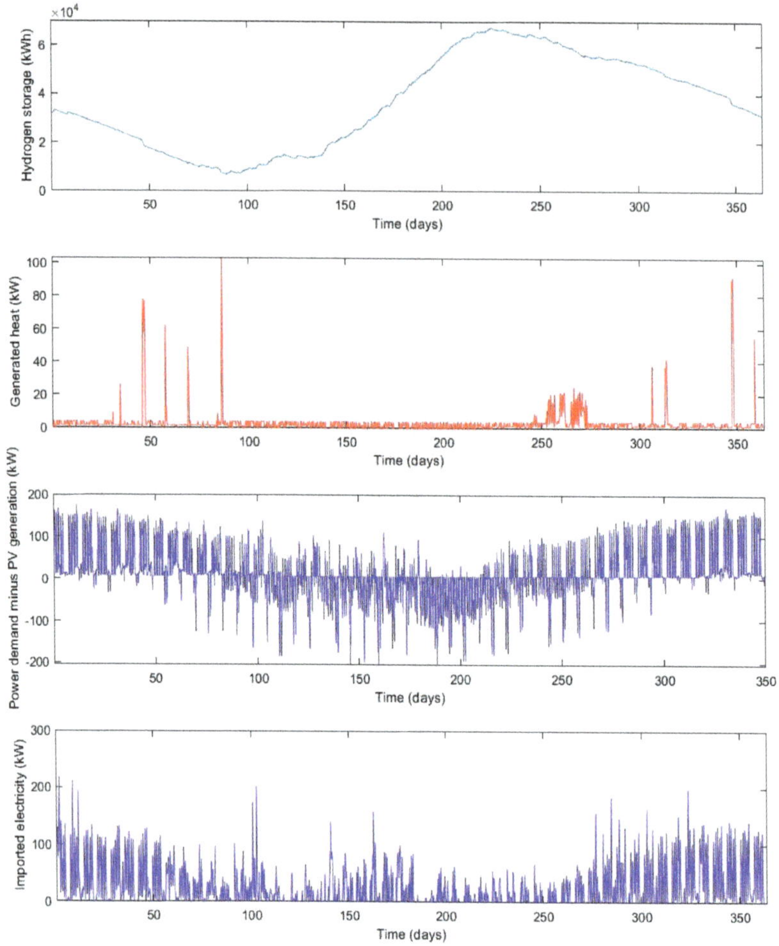

Figure A2. Optimal operation of a 20/80 kW RSOC system with 2000 m^2 solar PV panel.

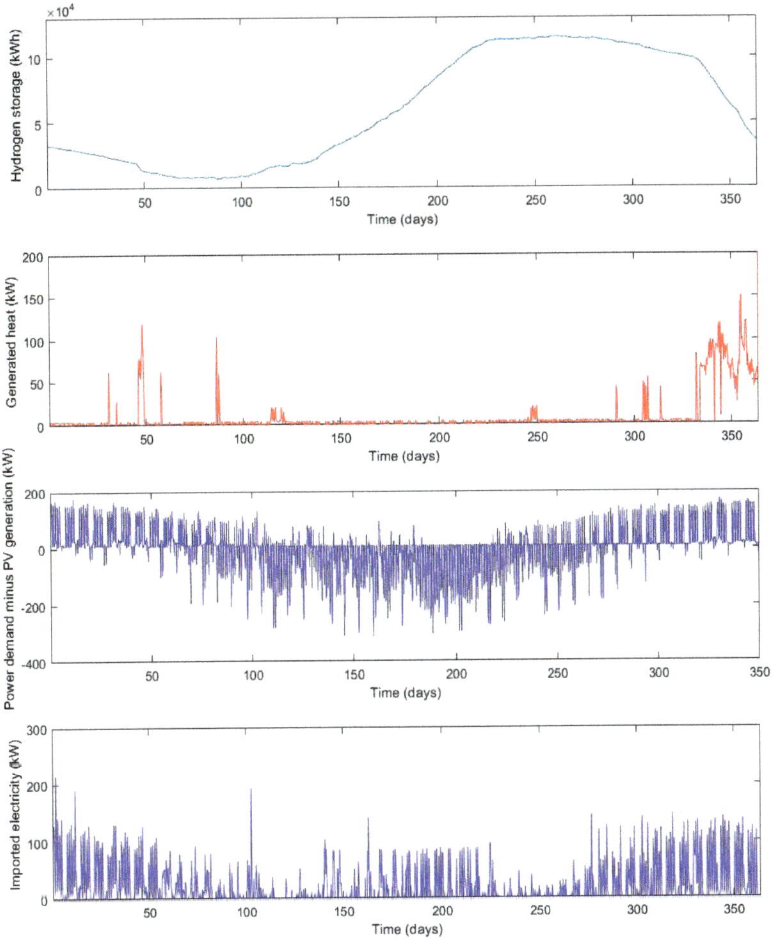

Figure A3. Optimal operation of a 20/80 kW RSOC system with 3000 m^2 solar PV panel.

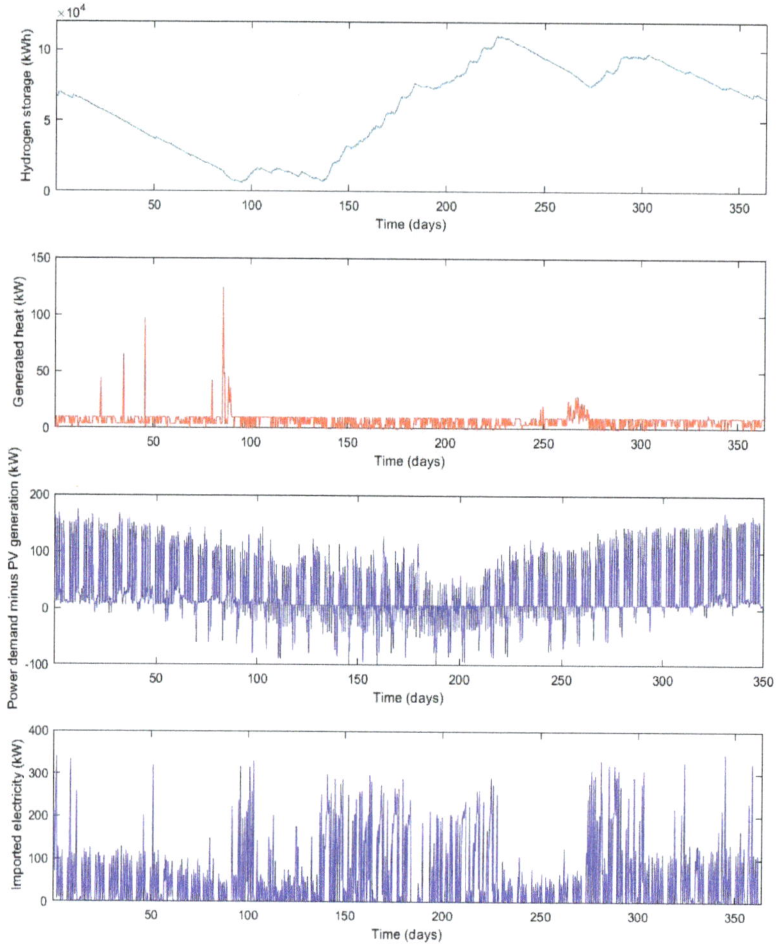

Figure A4. Optimal operation of a 50/200 kW RSOC system with 1000 m^2 solar PV panel.

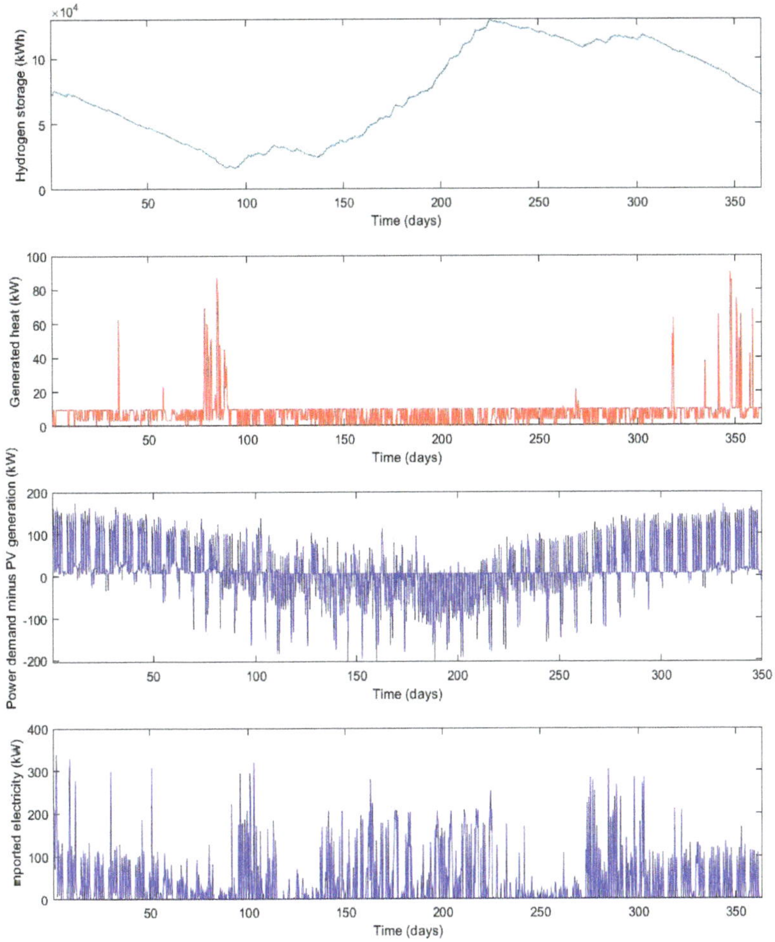

Figure A5. Optimal operation of a 50/200 kW RSOC system with 2000 m^2 solar PV panel.

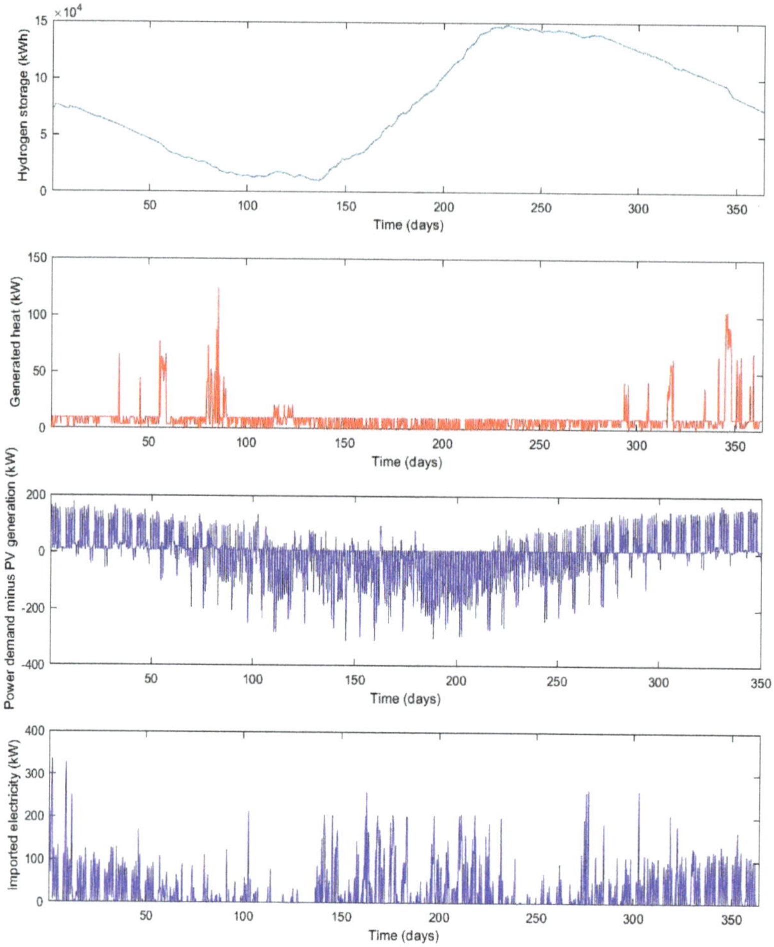

Figure A6. Optimal operation of a 50/200 kW RSOC system with 3000 m^2 solar PV panel.

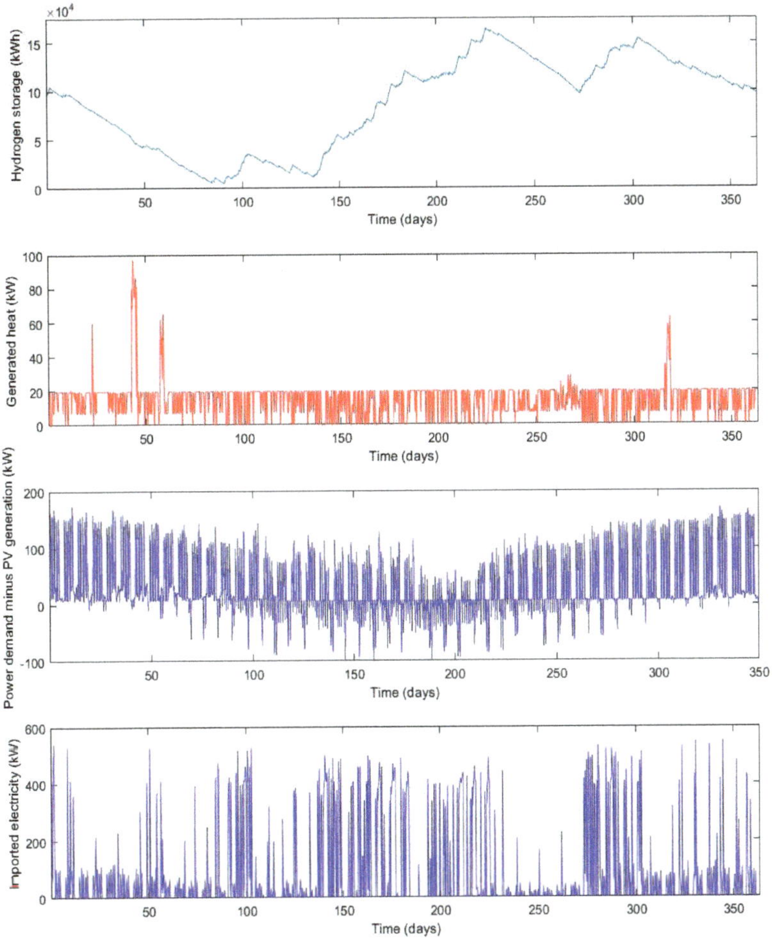

Figure A7. Optimal operation of a 100/400 kW RSOC system with 1000 m^2 solar PV panel.

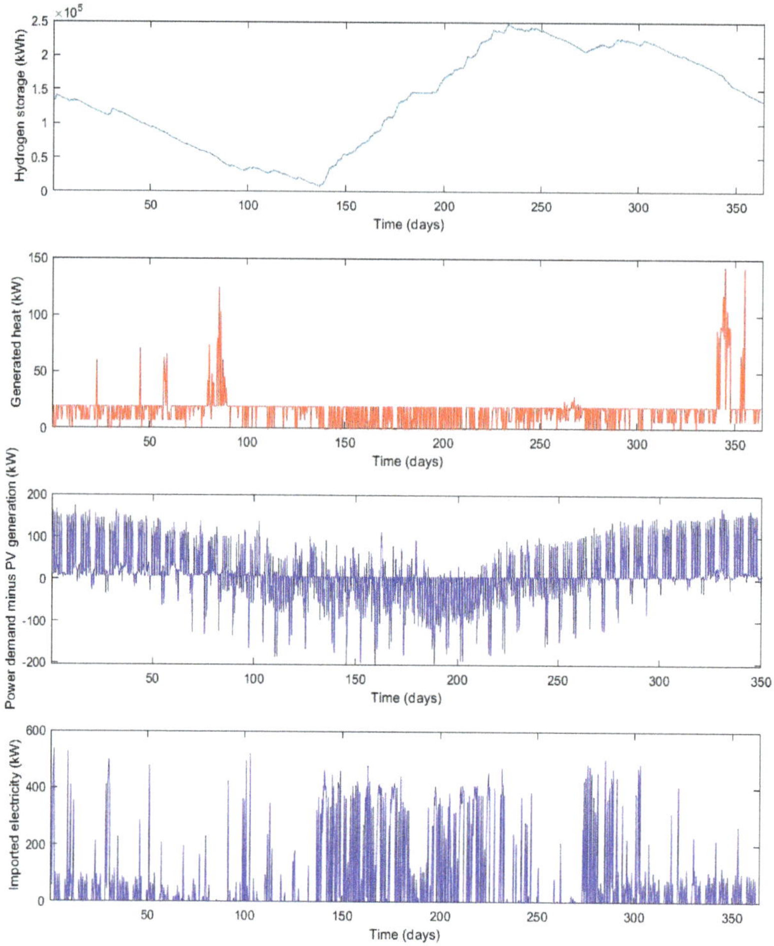

Figure A8. Optimal operation of a 100/400 kW RSOC system with 2000 m^2 solar PV panel.

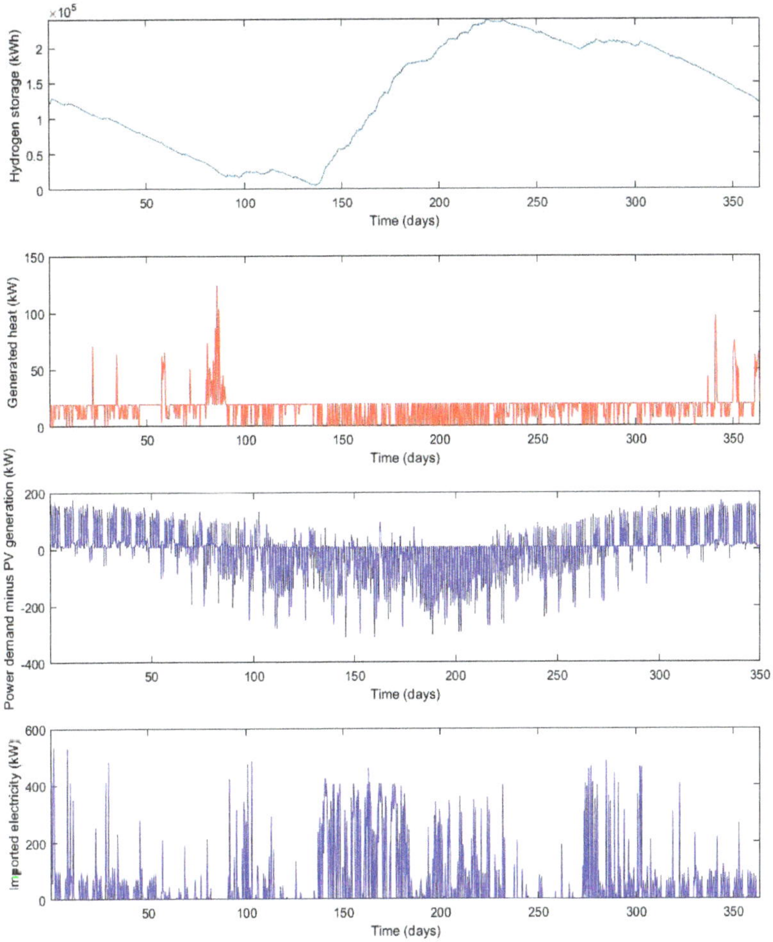

Figure A9. Optimal operation of a 100/400 kW RSOC system with 3000 m^2 solar PV panel.

References

1. Berggren, B.; Wall, M. Two Methods for Normalisation of Measured Energy Performance—Testing of a Net-Zero Energy Building in Sweden. *Buildings* **2017**, *7*, 86. [CrossRef]
2. Teni, M.; Čulo, K.; Krstić, H. Renovation of public buildings towards nZEB: A case study of a nursing home. *Buildings* **2019**, *9*, 153. [CrossRef]
3. Myy Sähkösi Meille - Fortum.fi. Available online: https://www.fortum.fi/kotiasiakkaille/aurinkopaneelit/oman-tuotannon-myynti-lahisahko (accessed on 3 March 2020).
4. Avoin Kaukolämpö Ostohinnat - Fortum.fi. Available online: https://www.fortum.fi/yrityksille-ja-yhteisoille/lammitys-ja-jaahdytys/kaukolampo/avoin-kaukolampo/avoin-kaukolampo-ostohinnat (accessed on 3 March 2020).
5. Kaukolämmön Hinnat Taloyhtiöille ja Yrityksille - Fortum.fi. Available online: https://www.fortum.fi/yrityksille-ja-yhteisoille/lammitys-ja-jaahdytys/kaukolampo/kaukolammon-hinnat-taloyhtioille-ja-yrityksille (accessed on 3 February 2020).
6. International Renewable Energy Agency. *Renewable Power Generation Costs in 2018*; International Renewable Energy Agency: Abu Dhabi, UAE, 2019.

7. Durão, B.; Joyce, A.; Mendes, J.F. Optimization of a seasonal storage solar system using Genetic Algorithms. *Solar Energy* **2014**, *101*, 160–166. [CrossRef]
8. Kerdphol, T.; Fuji, K.; Mitani, Y.; Watanabe, M.; Qudaih, Y. Optimization of a battery energy storage system using particle swarm optimization for stand-alone microgrids. *Int. J. Electr. Power Energy Syst.* **2016**, *81*, 32–39. [CrossRef]
9. Ekren, O.; Ekren, B.Y. Size optimization of a PV/wind hybrid energy conversion system with battery storage using simulated annealing. *Appl. Energy* **2010**, *87*, 592–598. [CrossRef]
10. Zhang, W.; Maleki, A.; Rosen, M.A.; Liu, J. Optimization with a simulated annealing algorithm of a hybrid system for renewable energy including battery and hydrogen storage. *Energy* **2018**, *163*, 160–166. [CrossRef]
11. Kotzur, L.; Markewitz, P.; Robinius, M.; Stolten, D. Time series aggregation for energy system design: Modeling seasonal storage. *Appl. Energy* **2018**, *213*, 123–135. [CrossRef]
12. Steen, D.; Stadler, M.; Cardoso, G.; Groissböck, M.; DeForest, N.; Marnay, C. Modeling of thermal storage systems in MILP distributed energy resource models. *Appl. Energy* **2015**, *137*, 782–792. [CrossRef]
13. Wang, H.; Meng, K.; Dong, Z.; Xu, Z.; Luo, F.; Wong, K. A MILP approach to accommodate more Building Integrated Photovoltaic system in distribution network. In Proceedings of the IEEE Power and Energy Society General Meeting, Denver, CO, USA, 26–30 July 2015.
14. Pinzon, J.; Vergara, P.; Da Silva, L.; Rider, M. An MILP model for optimal management of energy consumption and comfort in smart buildings. In Proceedings of the 2017 IEEE Power and Energy Society Innovative Smart Grid Technologies Conference, Washington, DC, USA, 23–26 April 2017.
15. Lindholm, O. Techno-Economic Analysis of a Reversible Solid Oxide Cell and Hydrogen Gas Storage System. Master's Thesis, Åbo Akademi University, Turku, Finland, 2019.
16. Castañeda, M.; Cano, A.; Jurado, F.; Sánchez, H.; Fernández, L.M. Sizing optimization, dynamic modeling and energy management strategies of a stand-alone PV/hydrogen/battery-based hybrid system. *Int. J. Hydrogen Energy* **2013**, *38*, 3830–3845. [CrossRef]
17. Luta, D.N.; Raji, A.K. Optimal sizing of hybrid fuel cell-supercapacitor storage system for off-grid renewable applications. *Energy* **2019**, *166*, 530–540. [CrossRef]
18. Carapellucci, R.; Giordano, L. Modeling and optimization of an energy generation island based on renewable technologies and hydrogen storage systems. *Int. J. Hydrogen Energy* **2012**, *37*, 2081–2093. [CrossRef]
19. Mixed-Integer Linear Programming (MILP)—MATLAB Intlinprog—MathWorks Nordic. Available online: https://se.mathworks.com/help/optim/ug/intlinprog.html (accessed on 12 March 2020).
20. Wendel, C.H.; Kazempoor, P.; Braun, R.J. Novel electrical energy storage system based on reversible solid oxide cells: System design and operating conditions. *J. Energy Stor.* **2015**, *276*, 133–144. [CrossRef]
21. Reznicek, E.; Braun, R.J. Techno-economic and off-design analysis of stand-alone, distributed-scale reversible solid oxide cell energy storage systems. *Energy Convers. Manag.* **2018**, *175*, 263–277. [CrossRef]
22. Botta, G.; Romeo, M.; Fernandes, A.; Trabucchi, S.; Aravind, P.V. Dynamic modeling of reversible solid oxide cell stack and control strategy development. *Energy Convers. Manag.* **2019**, *185*, 636–653. [CrossRef]
23. Peel, M.; Finlayson, B.; McMahon, T. Updated world map of the Köppen-Geiger climate classification. *Hydrol. Earth Syst. Sci.* **2007**, *11*, 1633–1644. [CrossRef]
24. IDA ICE—Simulation Software - EQUA. Available online: https://www.equa.se/en/ida-ice (accessed on 6 February 2020).
25. H2A Hydrogen Delivery Infrastructure Analysis Models and Conventional Pathway Options Analysis Results—Interim Report. Department of Energy. Available online: https://www.energy.gov/eere/fuelcells/downloads/h2a-hydrogen-delivery-infrastructure-analysis-models-and-conventional (accessed on 3 February 2020).
26. Renewables.ninja. Available online: https://www.renewables.ninja/ (accessed on 7 February 2020).
27. Historical Market Data - Nord Pool. Available online: https://www.nordpoolgroup.com/historical-market-data/ (accessed on 6 February 2020).
28. Electricity Distribution Rates - Caruna. Available online: https://www.caruna.fi/en/our-services/products-and-rates/electricity-distribution-rates (accessed on 3 February 2020).
29. Sähkön ja Eräiden Polttoaineiden vero—Verohallinto. Available online: https://www.vero.fi/yritykset-ja-yhteisot/tietoa-yritysverotuksesta/valmisteverotus/sahko_ja_eraat_polttoaineet/ (accessed on 3 February 2020).

30. Hauck, M.; Herrmann, S.; Spliethoff, H. Simulation of a reversible SOFC with Aspen Plus. *Int. J. Hydrogen Energy* **2017**, *42*, 10329–10340. [CrossRef]
31. Venkataraman, V.; Pérez-Fortes, M.; Wang, L.; Hajimolana, Y.S.; Boigues-Muñoz, C.; Agostini, A.; McPhail, S.J.; Maréchal, F.; Van Herle, J.; Aravind, P.V. Reversible solid oxide systems for energy and chemical applications—Review & perspectives. *J. Energy Stor.* **2019**, *24*, 100782.

buildings

MDPI

Article

Optimal Simulation of Three Peer to Peer (P2P) Business Models for Individual PV Prosumers in a Local Electricity Market Using Agent-Based Modelling

Marco Lovati [1], Xingxing Zhang [1,*], Pei Huang [1], Carl Olsmats [1] and Laura Maturi [2]

1 School of Technology and Business Studies, Dalarna University, SE-79188 Falun, Sweden; mlov@du.se (M.L.); phn@du.se (P.H.); cos@du.se (C.O.)

2 Renewable Energy Institute, EURAC, 39100 Bolzano, Italy; laura.maturi@eurac.edu

* Correspondence: xza@du.se

Received: 22 June 2020; Accepted: 27 July 2020; Published: 29 July 2020

Abstract: Solar photovoltaic (PV) is becoming one of the most significant renewable sources for positive energy district (PED) in Sweden. The lack of innovative business models and financing mechanisms are the main constraints for PV's deployment installed in local communities. This paper therefore proposes a peer-to-peer (P2P) business model for 48 individual building prosumers with PV installed in a Swedish community. It considers energy use behaviour, electricity/financial flows, ownerships and trading rules in a local electricity market. Different local electricity markets are designed and studied using agent-based modelling technique, with different energy demands, cost–benefit schemes and financial hypotheses for an optimal evaluation. This paper provides an early insight into a vast research space, i.e., the operation of an energy system through the constrained interaction of its constituting agents. The agents (48 households) show varying abilities in exploiting the common PV resource, as they achieve very heterogeneous self-sufficiency levels (from ca. 15% to 30%). The lack of demand side management suggests that social and lifestyle differences generate huge impacts on the ability to be self-sufficient with a shared, limited PV resource. Despite the differences in self-sufficiency, the sheer energy amount obtained from the shared PV correlates mainly with annual cumulative demand.

Keywords: microgrid; PV; peer to peer; self-consumption; energy community; local market

1. Introduction

1.1. Background and Literature Review

Positive energy districts (PED) are defined as energy-efficient and energy-flexible building areas with surplus renewable energy production and net zero greenhouse gas emissions [1]. Solar photovoltaic (PV) is ideally a leading renewable source in PEDs due to its easy scalability, simple installation and relatively low maintenance. Distributed PV systems are the main driver in the Swedish PV markets, due to smaller size and distributed ownership, which are better adapted to permeate the urban environment. The installed capacity of PV systems in Sweden is expected to continuously soar in the future, mainly driven by homeowners and private or public companies at relatively small or medium scales, according to its particular market setup and subsidy (e.g., SOLROT deduction, tax reduction, etc.) [2]. However, relying on the subsidy is not sustainable for PV deployment in the long term. At the moment, there is still limited access to capital and appropriate financing mechanisms, resulting in a slow uptake of PV under traditional business models (i.e., power purchase agreements and the net-metering mechanism), which are no longer applicable for small PV systems [3]. The existing business models may need to be further developed to exploit the full potentials generated by distributed energy supply, demand and energy sharing. Thus, in a future without subsidies, prosumers (i.e., small PV owners) will have to sell their excess production at market price back to the grid. This scenario would be unprofitable for PV owners and also strain grid stability and reduce its reliability.

Fortunately, the possibility to form energy communities, where energy can be locally shared, has been regulated at European level in the Clean Energy package presented by the European Commission [4] and at Swedish level under § 22 (a) of the IKN Regulation 2007: 215 [5]. This can be an opportunity for a new business model development within the energy sector, e.g., Peer-to-Peer (P2P) trading. In such business model, consumers and prosumers organise in energy communities, in which the excess production could be sold to other members [6]. The benefits are threefold as the prosumers could make an additional margin on their sale, consumers could buy electricity at a more advantageous price and the grid could be more stable and resilient. This can be a potential solution to promoting PV installation in a sustainable way, while reducing the reliance on subsidies.

To support new regulations, careful design and optimal modelling of P2P business models for PV penetration is necessary by analysing current scenarios and proposing future ways of exchanging energy. Huijben and Verbong [3] summarised three possible ownerships of PV systems: customer-owned (single ownership), community shares (multiple ownership) and third-party ownership. Based on these possibilities, Lettner et al. [7] further described three different system boundaries of a PV prosumer business concept (as illustrated in Figure 1): Group (1) single direct use (one consumer directly uses the generated PV electricity on site); Group (2) local collective use of PV in one building (several consumers share the generated PV electricity with or without the public grid); and Group (3) district power model (PVs are installed in several buildings, where those prosumers directly consume locally generated PV power and the PV electricity is further shared using public or private microgrid). It is possible to have different ownerships in each category of these boundary conditions, resulting in many possibilities and uncertainties in the practical business operation. Learning and mapping (i.e., testing) a wide array of these possible designs and combinations are necessary. There are a few existing regulatory and modelling studies about the P2P PV-electricity trading. Community-owned PV system was surveyed as an innovative business model in Switzerland, where it can seemingly be a successful distribution channel for the further adoption of PV [8]. Roberts et al. tested a range of financial scenarios in Australia, based on the P2P concept, to increase PV self-consumption and electricity self-efficiency by applying PVs to aggregated building loads [9]. Zhang et al. [10] established a four-layer system architecture of P2P energy trading (as shown in Figure 2, i.e., power grid layer, ICT layer, control layer and business layer), during which they focused on the bidding process on business layer using non-cooperative game theory in a microgrid with 10 peers. A price mechanism for the aggregated PV electricity exchange among peer buildings was also developed using either Lagrangian relaxation-based decentralised algorithm [11] or mixed integer linear programming [12]. Jing et al. [13] then applied the non-cooperative game theory to modelling the aggregated energy trading between residential and commercial buildings by considering fair energy pricing mechanism for both PV electricity and thermal energy simultaneously. Lüth et al. [14] designed two local markets for decentralised storage (flexi user market-individually owned batteries) and centralised storage (pool hub market-commonly owned battery), based on a multi-period linear programming. It focused on the evaluation of two different ownerships of batteries and optimised P2P energy trading local markets. They indicated that the end users can save up to 31% electricity bills in the Flexi User Market and 24% in Pool Hub Market. Furthermore, two different ownership structures, namely the third-party owned structure and the user owned structure, were investigated in a P2P energy sharing network with PV and battery storage [15]. These existing studies almost cover all four layers of a P2P network. The impact of other system and market components on the economic performance of PV P2P business models has been investigated, such as EV (Electric Vehicle) batteries [16], gas storage [17], heat pump/hot water storage [18], advanced control [19], energy cost optimisation [20], bidding strategies for local free market [21], double auction market [22], local market designs [23], integration of local electricity market into wholesale multi-market [24], microgrid ICT architecture [25], grid operation [26], etc.

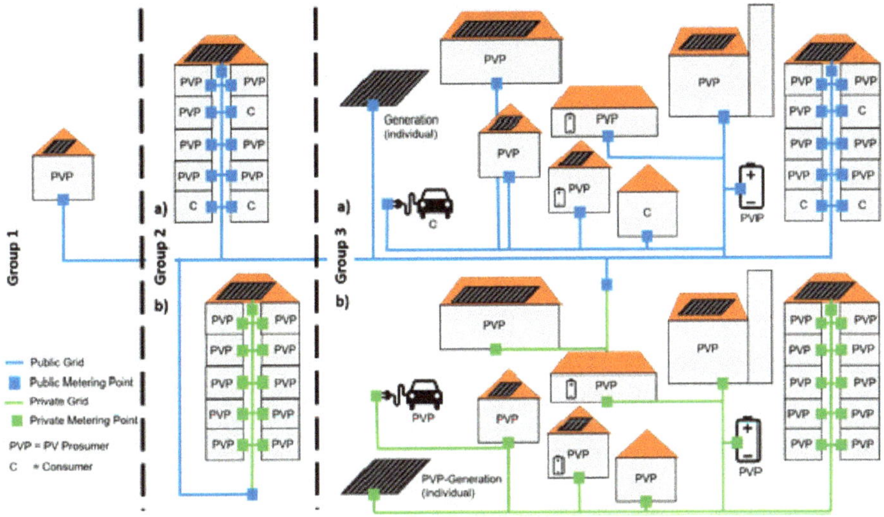

Figure 1. Classification of integration concepts [7].

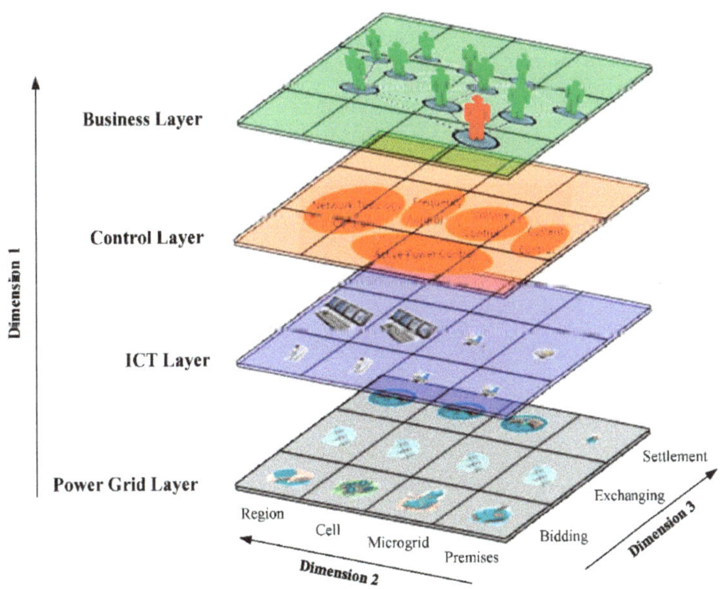

Figure 2. The four-layered system architecture of P2P energy trading from [10].

According to the above studies, a research gap is found in the lack of examination on full P2P energy trading process at the business layer in a local market for individual participant, which, in time sequence, consists of bidding, exchanging and settlement, under different local market conditions with various ownerships of PV systems and market rules. Bidding is often the first process when energy players (generators, consumers and prosumers) agree to trade energy with each other at a certain price for a specific amount of energy. Energy exchanging is the second process, during which energy is generated, transmitted and consumed. Settlement is the last process when bills and transactions are finally settled via settlement arrangements and payment [10], which results in the final economic benefits. In cases of the physical network constraints, due to the varying energy demand and the intermittent generation of PVs, there are always mismatches between sellers and buyers. Such difference between electricity generation and demand are to be evaluated and charged/discharged during settlement stage.

1.2. Novelty and Contribution

Several studies have focused on the technical or economic aspects of the microgrids and shared RES, but the endeavour has been tackled in a segmented way analysing a narrow sample of possibilities among the vast search space of the business models. The existing studies have not yet fully tested the effectiveness and compared the characteristics of various P2P business models, in the case of heterogeneous peer (individual) energy supply/demand and dynamic market rules for the full trading process on the business layer. There is a lack of a concise and efficient method yet to model.

Although this paper only analyses three different setups, it attempts to lay the groundwork for a systematic study of the subject. In other words, the results and the discussion presented in this paper, although not conclusive by themselves, are part of a well-defined search-space. This allows the outcomes to be interpreted from the perspective a larger systematic endeavour.

In summary, the elements of novelty of this paper are described as the following:

(1) The particular result of the study: To the knowledge of the authors, no study has linked the price of the electricity offered within a shared RES to both the risk of economic loss and the potentials for earning among the individual households within the shared microgrid. Furthermore, the dominance of shear annual cumulative consumption over self-sufficiency in determining the earning potential in a shared RES is an unknown phenomenon. It deserves to be further analysed (i.e., tested under different datasets) to be proven.

(2) The examples of business models presented in the study are included in a well-defined search space map (see Figure 3). This facilitates a systematic inquiry and offers a way to organise the results presented in this study and in the follow-ups.

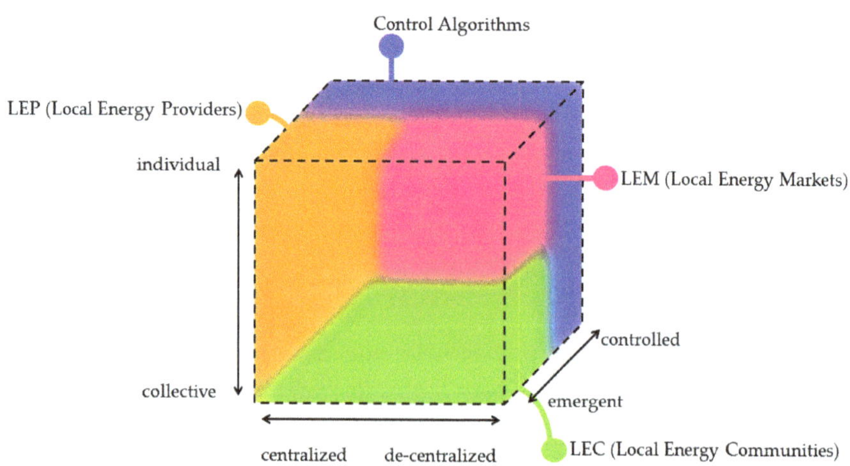

Figure 3. District scale renewable energy systems behaviour map.

This paper studies the P2P business model for 48 individual building prosumers with PV installed in a Swedish community. This paper discovers "latent opportunities" that were previously unknown and optimises the market design and its variables for the best benefit. It has significant influence that integrates energy needs, supply and market rules. This paper is expected to provide knowledge for policymakers to design a fair, effective and economical P2P energy framework. The research results will useful be to optimise PED's three functions (energy efficiency, energy production and flexibility) towards energy surplus and climate neutrality.

2. Materials and Methods

The definition of ownership structures from [3] distinguishes among customers, communities and third parties. In general, a similar distinction could be applied to the behaviour of the local grid instead to the ownership. In this way, the concept of ownership is not associated with the functioning of the grid and it is easier to describe hybrid forms (e.g., some shareholder of an energy provider, or more providers, which form a market although not prosumers). Thinking about the behaviour of the shared system, a space can be defined according to three dimensions (see Figure 3):

(1) The **controlled versus emergent** dimension describes how much there are rules or a controller that directs the exchanges, versus an emergent behaviour from the interactions between agents.
(2) The **centralised versus de-centralised** dimension describes how much the agents are equivalent among each other, versus the presence of few (potentially one) agents that concentrate some functions for a larger number of others.
(3) The **individual versus collective** dimension describes how much each agent controls and directs its own resources (i.e., PV, storage, demand-response resources, etc.), versus having larger pools of agents who share some common resources.

The behaviour map does not refer to any specific levels [10], although the last two (i.e., controls and business) are particularly affected from the volume of the map, in which they are located. In fact, the control of the energy and monetary flows between generation and demand points can be decided by a controller, which can be assigned by the internal rules of a community or emerged as the result of an auction.

2.1. Agent-Based Model

Given the number and nature of the emergent behaviours in the behaviour map (i.e., Figure 3), an agent based model (ABM) simulation was developed to get insight into the energy and economic fluxes exchanged between the different actors in the local grid. Usually, every agent of the simulation represents one household in the local grid (i.e., a consumer or a prosumer), but producers are not excluded. An example of a producer is an energy provider. For instance, companies or investor interacts with the local grid without necessarily being served by it, or the parent grid, i.e., the larger grid in which the local grid is embedded. The local grid could be a microgrid but also a secondary network, where the prosumers are allowed to have a certain level of control of the network.

In an ABM, each agent can interact with all the other agents by trading energy. Thus, it can send energy in exchange for money or vice versa. The movement of energy in the microgrid is an emergent behaviour, which results from the interaction of a number of independent actors. This is opposed to a control algorithm, where the behaviour is set by a series of rules or conditions. Naturally, the freedom of the agents can be limited by the introduction of rules. For instance, a producer could be forced to prioritise the sale of renewable electricity to those consumers that have used the least of it in a given period. If the rules become tighter, the freedom of each individual agent is reduced. While if the rules are as tight as to completely limit any possibility of choice for the agents, the ABM degenerates into a control algorithm.

In the present study, the behaviour of the agents is extremely simplified: the consumers prioritise the purchase of electricity from the cheapest source available at any given time, on the other end the producers have the ability to set the price, and they do so according to the case as explained in the following section (i.e., ownership structures and business models). Figure 4 presents the possible ownership structures arranged into three main families; these are slightly different from those in [3] for the purpose of this study.

1. Local Energy Provider (LEP) (Figure 4a): It occurs when a single agent owns the totality of the production or storage capacity of the entire local network and the other agents are strictly consumers. The owner of the plant can be either a producer or a prosumer.
2. Local Energy Community (LEC) (Figure 4b): It is the case in which a communal plant is shared among all or a group of agents, the shares could be equally distributed or according to other principles such as energy used from the plant or the share of the initial investment.
3. Local Energy Market (LEM) (Figure 4c): It is the most complex and free-form of all the structures; it is characterised by the presence of multiple producers, consumers and prosumers. In this arrangement, the interaction between agents can reach significant complexity and the agents could achieve higher earnings by engaging in intelligent behaviours.

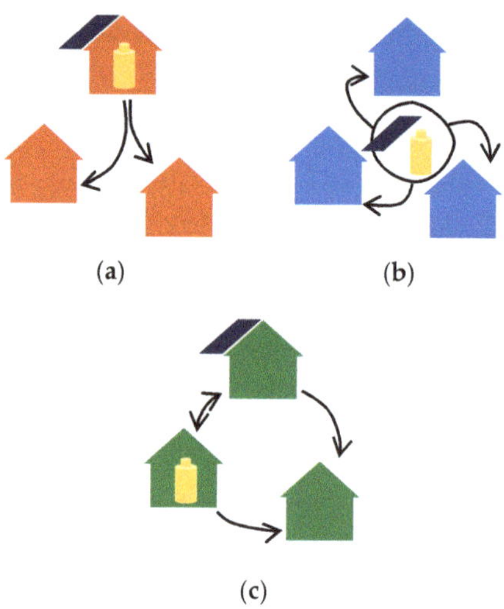

Figure 4. Possible ownership structures organised in three main families: Local Energy Provider (LEP) (**a**); Local Energy Community (LEC) (**b**); and Local Energy Market (**c**).

2.2. Ownership Structures and Business Models

In the case study examined (see Section 2.3), a communal PV plant is shared among the different households in the building. This allows for two of the three basic ownership structures in Figure 4 (i.e., LEP and LEC) to be applied. The ownership structure is intertwined with the business model and the rules of the market. In the following studies, the same communal PV plant is shared between the households in the local grid in three different scenarios:

1. **LEC gratis:** In this arrangement, the electricity from the communal PV plant is given for free when available. All the households participate in the initial investment and in the Operation and Maintenance (O&M) costs of the plant according to equal shares.
2. **LEC LCOE:** In this arrangement, the electricity from the communal PV is given at production cost (i.e., without profit) and the revenues are divided among the shareholders. Although variable shares are possible, in this study, all the households are equal sharers in the LEC (i.e., initial investment and O&M costs, and the revenues are shared equally).

3. **LEP n%:** This arrangement is a pure form of LEP. Thus, the production plant is owned by a single provider who can set the price at its own will. Obviously, the provider cannot set the price higher than that of the parent grid (i.e., the average price for Swedish household consumer as assumed in Section 2.3) as the consumers retain the right to purchase electricity from the cheapest source. In this study, the provider sets the price as half-way between the minimum of the local LCOE and the maximum of the consumer price from the parent grid. More precisely, the provider sets a price at a percentage n so that n = 0 is the LCOE, n = 100 is the price offered by the parent grid and n = 50 is half-way. This set-up is valid under the assumption that the LCOE of the system is lower than the price of the electricity for the consumer. Of course, if this assumption does not hold true, the provider will not be able to charge above market price and will thus operate at the minimum loss.

In all arrangements, the consumer is programmed to buy electricity from the cheapest source. However, by having a single source in the local grid, the choice is only between the local source and the parent grid. This implies that the price of electricity in the local grid must be at any time below the Swedish consumer price. If the local production is absent or insufficient (i.e., local consumption > local production), the demand shall be covered partially or totally by the parent grid. If the local production is not sufficient, at a given point in time, to cover entirely the demand, all the households will be served equally in terms of percentage of their demands as shown in the system of relations in Equation (1).

$$\begin{cases} E_{local} = \eta \times D_{local} \\ E_{house} = \eta \times D_{house} \times \forall house \\ D_{local} = \sum D_{house} \end{cases} \tag{1}$$

where E_{local} and E_{house} are the amount of electricity available in a given time for the aggregated local grid and for a specific household, respectively. η is the self-sufficiency: a number between 0 and 1 that represents the share of the demand covered by locally produced electricity; note that it is the same globally and for each household. D_{local} and D_{house} represent the aggregated demand and the demand of each single household, respectively.

Equation (1) implies that having a larger consumption when the local electricity production is scarce guarantees access to a larger amount of local energy, although equal in percentage. Another consequence of the relation in (1) involves the price of the electricity for each household: the price results from the weighted average (weighted on energy) of the prices from the different sources of electricity purchased. In the specific case of this study, the price can be calculated with the relation Equation (2):

$$P_{house} = P_{local} \cdot \eta + P_{parent} \cdot (1 - \eta) \tag{2}$$

where P_{house}, P_{local} and P_{parent} represent the electricity price for the individual household, the price for the energy produced locally and the price for the energy bought from the parent grid, respectively. η is the self-sufficiency as defined for (1).

Considering that η is the same for every household in the local grid as shown in (1), Equation (2) implies that at any given time there is a unique price of the electricity within the local grid, which depends on the relation between the aggregated energy demand (D_{local}) and the aggregate energy production (E_{local}). Thus, the price for the electricity is solely function of the Hour of the Year (HOY) and is not a function of any given household.

2.3. Case Study

The agent based model is tested on a digital representation of a moderate size residential district (see Figure 5) equipped with a shared PV system + DC microgrid as described in [18]. The group of three buildings with three stories is located in Sunnansjö, Ludvika, Dalarna region, Sweden. The common PV system is formed by the arrays shown in Table 1. In total, there are three arrays on the roof and one on the southern façade (total 65.5 kWp).

Figure 5. Bird view of the small district in the case study [18].

Table 1. Characteristics of the shared PV system.

Block	Facing	Tilt (degree)	Capacity (kW_p)	Production (MWh)
B	South	18	28.4	22
C	East	18	15.9	10.4
A	West	18	15.9	10.3
A	South	90	5.3	3.4

The system capacity and the position of the arrays over the building resulted from an optimisation process, presented in [18], in order to maximise the self-sufficiency while maintaining a positive NPV over the lifetime. In this system, no electric storage was installed. The LCOE (Levelized Cost of Electricity) of the system was calculated to be about 0.83 SEK/kWh (0.077 €/kWh) under the following assumptions:

- Local initial price of the turn-key system without taxation: 10,000 SEK/kWp (935 €/kWp).
- Price of the inverter: 2500 SEK/kWp (234 €/kWp) (changed two times over the lifetime). The number of changes was retrieved as the expected value assuming a lifetime of the inverter between 12 and 15 years.
- Planned lifetime of the system: 30 years.
- Maintenance costs for the system (substitutions, cleaning and inspection): 5109 SEK/year (477 €/year). This value was calculated as the expected value out of 100 stochastic simulations.
- Degradation of the performance of the system: ca. −1.15%/year.

The weather file and the production of the diverse arrays of PV were calculated from PVGIS [27]. The load profile of the 48 households could not be published for privacy concerns. Thus, the study is presented using data generated by LPG (Load Profile Generator) software [28]. Load Profile Generator is a tool that simulates the electric demand for residential light and appliances. The variability of the aggregated curve according to the number of households has been validated against a real low voltage grid consumption [29]. The electric demand is generated by simulating every household component as an agent. Its demand is determined by the power absorption and duration of use of devices among an available selection (see Figure 6). These are chosen by the household components according to a set of activities and needs. The needs are modelled as counters that grow at each time-step: a high counter represents something that is in urgent need of satisfaction. Different needs have different growth rates for each time step, which means that some needs are to be satisfied more often than others.

The parent grid (i.e., the Swedish national grid) was assumed to offer electricity for 1.8 SEK/kWh (0.17 €/kWh) from October to March and 1.2 SEK/kWh (0.11 €/kWh) from March to October. These prices were assumed as a reasonable price for each single household at the annual cumulative level of consumption observed. According to Eurostat (2007–2019), the average price for household electricity in 2019 was 1.39 SEK/kWh (0.1297 €/kWh) for electricity transmission, system services, distribution and other necessary services. If VAT and levies are added, the average price would reach 2.2 SEK/kWh (0.2058 €/kWh) (Eurostat, 2007–2019). It is not clear what taxes can be avoided consuming locally produced electricity, but it is reasonable to believe that VAT can be avoided in both the LEC cases explored as the electricity is offered for free or at a price equal to production cost. Conversely, it is not possible to estimate how much of the base 1.39 SEK can be reduced thanks to the aggregation of the loads. The price of the electricity is not static but is projected to grow linearly over the next 30 years at a rate of +1%/year. This is under the assumption that the national grid will need liquidity to invest in the energy transition. Conversely, the revenues for the energy sold to the grid are set to be worth 0.3 SEK/kWh (0.028 €/kWh), but are assumed to shrink by 1.67%/year under the assumption that the increase in installation of PV will gradually discount the energy during sunny hours.

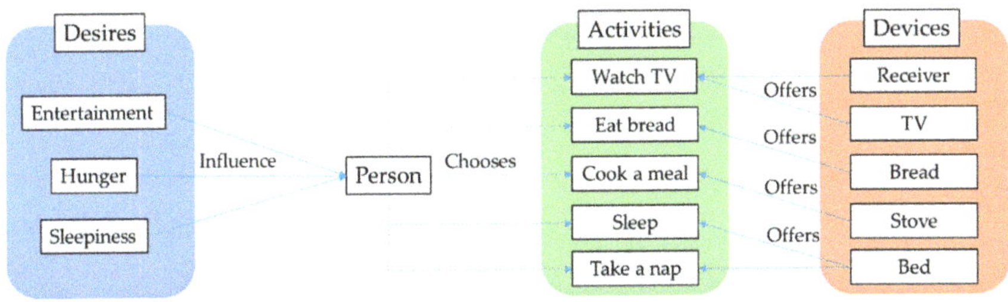

Figure 6. Workflow diagram of the load electricity generation [28].

3. Results

This section begins with a discussion about the self-sufficiency of the different households in the local network. It then proceeds with a techno-economic analysis of each arrangement to establish its features and behaviour (i.e., distribution of risk and profit among stakeholders). Given that the local PV plant is unique, the movement of energy in the network is the same in all the arrangements, thus the self-sufficiency is a static figure throughout the arrangements.

3.1. *Self-Sufficiency of the Households*

PV self-sufficiency is defined as the share of total demand in a household that is being supplied by locally generated electricity from PV system [30]. In this study, the system, as it is designed, allows to cover an estimated 20.2% of the annual cumulative demand of the district. This result is satisfactory for a system without any electric storage (see [31]). The country, with the most electricity production from PV (i.e., Honduras), has an estimate PV self-sufficiency of 14.8% with the EU on average having 4.9%. It has been calculated in references [32] and [18] that the economically optimal self-sufficiency of a conveniently aggregated system, even in absence of electric storage, is comfortably above any penetration level we see today (i.e., often above 20%). The economically optimal self-sufficiency sets a conservative limit of hosting capacity in an electrical system in a regime of self-sufficiency. The P50 (i.e., 50th percentile or median) household has a self-sufficiency of 18.5% as shown in Figure 7a: this value is below the average value of the aggregated district because the slope of the increase is higher to the right of P50 (see Figure 7a). The P50 household has a relatively low self-sufficiency because there is a positive correlation between annual cumulative demand and self-sufficiency (see discussion about Figure 8). In general, the variability in self-sufficiency between the households in the microgrid is high. The most self-sufficient household possesses in fact a value double of the lesser one (14.1% to 28.4%). This strong variability suggests that, even without any deliberate attempt for demand control, some households show habits, or a way of life, that can take the most from the available PV energy.

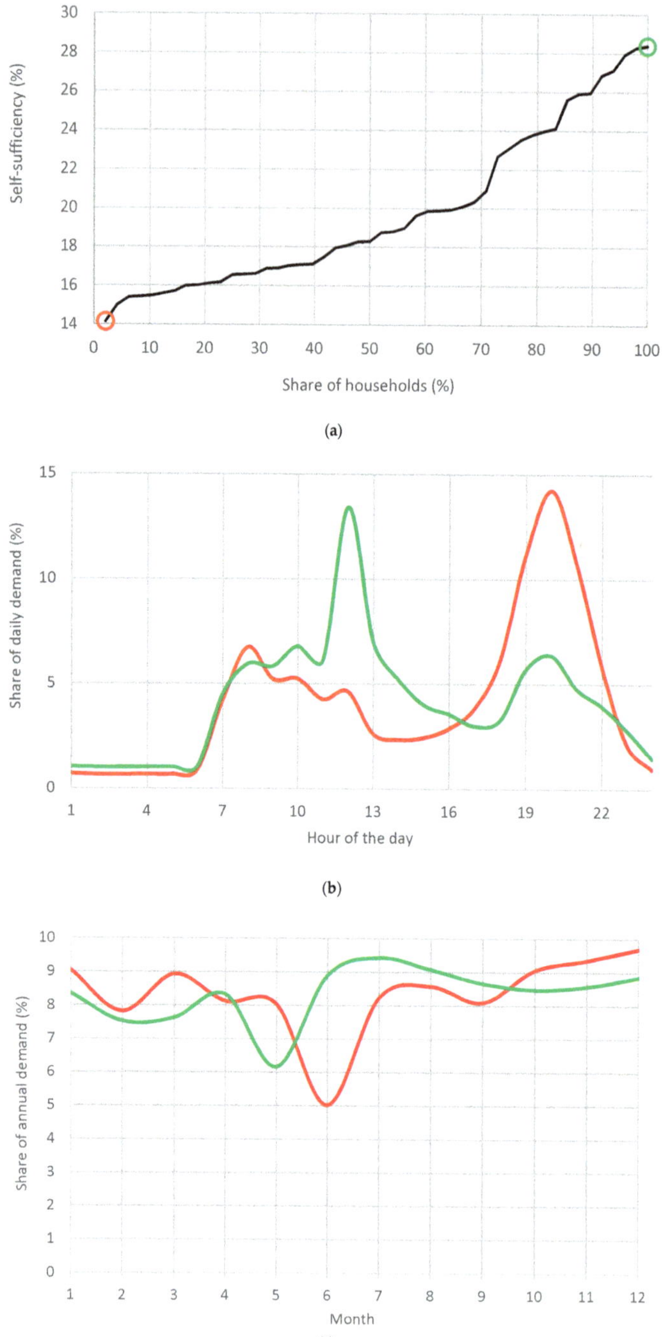

Figure 7. Self-sufficiency of the apartments in the local grid: (**a**) the distribution of self-sufficiencies across the 48 households; (**b**) the hourly average of the extreme households; and (**c**) the monthly average consumption of the extreme households.

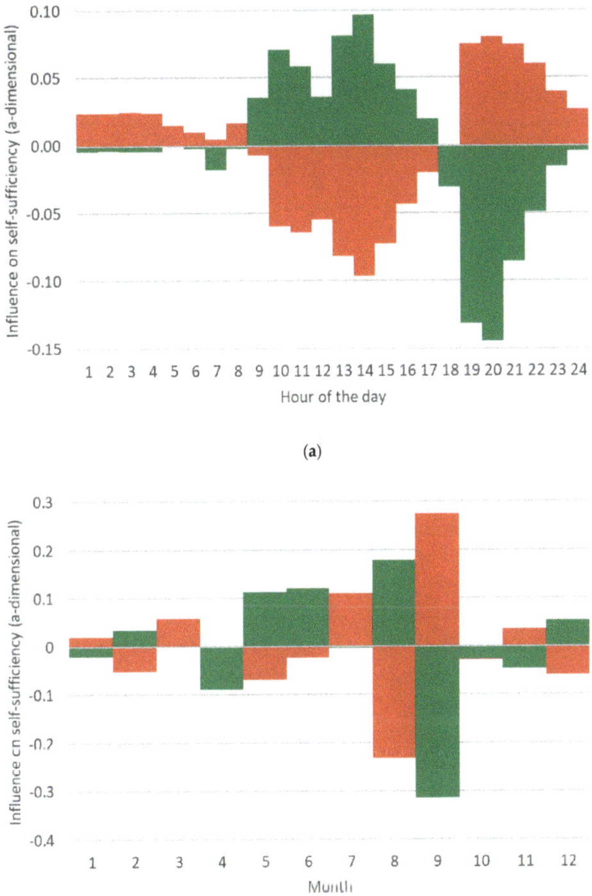

(a)

(b)

Figure 8. Influence on self-sufficiency of high demand in: (**a**) each hour of an average day; and (**b**) each month of the year. The value is a-dimensional but it expresses the positive (or negative) influence of a high electric demand at a given time step compared to all the others (see Equation (3)).

Figure 7b,c shows the share of the annual demand in different hours of the day or month of the year, respectively, i.e., how much of the total annual demand is concentrated during a specific hour of every day or month along the year. In the household with the highest self-sufficiency, the electricity demand around 12:00 is particularly prevalent (see Figure 7b). It indicates that its inhabitants cook at home for lunch. On the other end, the evening peak of the most self-sufficient household is way less prominent than in the lowest one. Looking at the prevalence throughout the months of the year (Figure 7c), the difference is less marked compared to the daily average: both households present a steep drop in sunny months, which seems to indicate an absence due to summer holidays. The most self-sufficient household appears to have had an absence for holidays during May instead of June, as shown in Figure 7c. This might be advantageous as it allows to use more PV electricity when the overall electricity demand of the district is lower and the radiation from the sun is higher. It should be noted that, in general, the best performing household presents a smaller dip in demand for the summer holidays; it is unknown whether it is due to a shorter holiday or the presence of some household components at home.

The examples shown in Figure 7 highlight the two apartments that are extreme in terms of self-sufficiency. To infer more generalised information on the time of high consumption that favours high self-sufficiency (see Figure 8) the following formula was used:

$$ISelfS_{time\ step} = \sum_{HH=1}^{48} \begin{cases} TP_{time\ step,HH} - TP_{time\ step,\ tot} & \forall\ SelfS_{HH} > SelfS_{tot} \\ 0 & \forall\ SelfS_{HH} \leq SelfS_{tot} \end{cases} \quad (3)$$

where $ISelfS_{time\ step}$ is the influence of high energy demand in a given time step (which could be an hour of the day or a month of the year). *HH* stands for Household as the curve results from the sum of all the individual households. $TP_{time\ step,\ \mathbf{HH}}$ and $TP_{time\ step,\ tot}$ are the typical power demand (W) of said time step for the nth household (*HH*) and the whole district (tot), respectively. The sum of all time steps is then rescaled so that it is equal to 1.

In practice, the curve is influenced only by the households that have a self-sufficiency above average. It represents the influence (positive or negative) that the demand in each time step has on the overall self-sufficiency. Unsurprisingly, Figure 8a shows that a lower average demand in the evening and early morning hours is associated with high self-sufficiency. On the contrary, the central hours of the day are generally above average in highly self-sufficient households. It is interesting to notice how the electric demand at 12:00 is in general less beneficial for self-sufficiency than the hours around it: this is somewhat counterintuitive, but it makes sense since at 12:00 the high general consumption due to lunch causes scarcity of renewable energy more often than in the hours immediately before or after. The signal on a monthly basis is not so easy to interpret. It appears to be beneficial to have above-average consumption in August and below-average in September: this is possibly due to a fraction of the households that went on holiday later in any given year. Given the sharp drop in irradiation in the month of September compared to July and August, it seems reasonable that going on holiday in September increases the self-sufficiency over the year.

3.2. *Exploitation of the Common Renewable Resources: Sheer Cumulative Consumption Versus Self-Sufficiency*

Figure 9 shows the relation between the annual cumulative demand and the annual cumulative energy received from the shared PV system. These two variables are strongly correlated (R > 0.9); thus, the quantity of energy consumed from the PV system can be assumed with good confidence from the annual cumulative demand alone (i.e., regardless of the self-sufficiency).

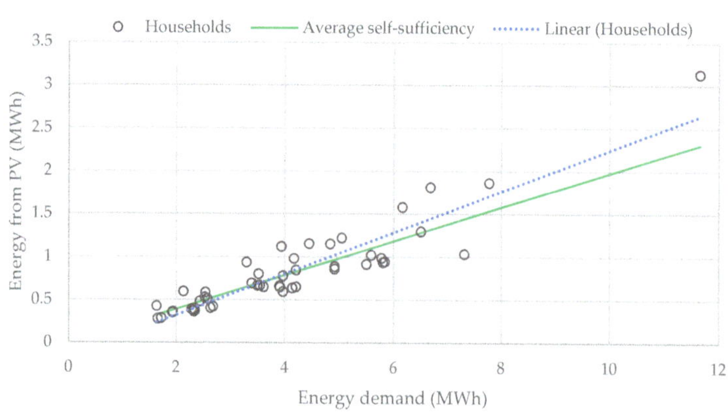

Figure 9. Annual cumulative energy demand and annual cumulative energy used from the PV system for every household in the local grid.

This aspect, although counterintuitive, is a consequence of the highest variability in annual cumulative demand compared to the variability in self-sufficiency: if in fact the highest self-sufficiency is two times the lowest one, the highest cumulative demand is almost five times the lowest one (excluding the highest value as an outlier; otherwise, it is more than seven times). The strong prominence in variability of cumulative demand compared to self-sufficiency reduces the variation in self-sufficiency as a mere noise compared to the other variable (as visible in Figure 9). Furthermore, as self-sufficiency is a share of the demand, it does not have much importance in absolute terms when applied to households with low cumulative demand. This fact represents somewhat a hindrance as it implies that increasing overall consumption works better than improving self-sufficiency to seize larger quantities of scarce local renewable resources. Nevertheless, it is not clear what power an individual household has to change its cumulative energy demand. Further investigation on the aspects that influence the cumulative energy demand (e.g., number of people in the household, cooking habits, holiday habits, etc.) is needed to assess whether it is something that the inhabitants can change. If each household has significant power on the cumulative energy consumption, it is reasonable to fear a sharp increase in the overall consumption after the installation of the communal PV system. It should be acknowledged that the lack of data with respect to other households might focus the attention of the inhabitants on their own energy demand advising them to increase the self-sufficiency. Another interesting aspect, shown in Figure 9, is that the linear interpolation of the household data points has a steeper slope than the average self-sufficiency of the 48 households. This means that the household with the highest annual cumulative consumption also has, on average, a highest self-sufficiency. The highest slope of the interpolation implies that at low consumption the self-sufficiency of a household tends to be lower than average, while at higher consumption tends to be higher. A correlation analysis between annual cumulative consumption and self-sufficiency found a positive, albeit weak, correlation ($R \approx 0.2$). Although it is weak and thus uncertain, the correlation suggests that highly consuming households might have more contemporaneity with the production from PV. This might be due to larger households having some members who stay at home during daytime, or to electric consumption by people who spend daytime at home being larger overall.

3.3. *LEC Gratis*

In this arrangement, the households in the district are shareholders of the system. Thus, they can use the electricity produced by the system for free when available. In this study, the shares of the PV system are equal. Each household will therefore have to pay 13,646 SEK (1275 €) of initial investment plus ca. 342 SEK/year (32 €/year) for maintenance and substitution of the inverter. Different ownership structures are possible, but the business model should be modified to avoid loopholes in the risk–benefit balance. For instance, equal shares could be distributed to a sub-group of the households (i.e., there are consumers who do not hold shares). In this case, an electricity price for non-owners should be established (see Section 2.2 LEP n%).

Figure 10 shows the difference in price between the energy offered by the parent grid and the energy available within the local system. The chart shows monthly values, which refer to the average cost of the electricity that month in the grid. We know from the Section 2.1 "Ownership structures and business models" that at any given time the price of the electricity is unique within the microgrid and depends on the relationship between production of PV and demand (see Equations (1) and (2)). The bars in Figure 10 are the average of all electricity prices of the respective month weighted by the aggregated electric consumption in that month. Obviously, since the energy not met by the local production is bought from the parent-grid, the external price has an influence on the internal one. In simpler terms, the internal price of the electric energy in one month, according to Equation (2) with Plocal = 0, is proportional to the residual demand. Notice that, due to the higher external price, the drop in cost of electricity during the months of March (Month 3) is similar to that in April (Month 4) despite a lower self-sufficiency.

Figure 10. Monthly difference in price between the energy offered by the parent grid and the average paid by the shareholders in a LEC gratis arrangement.

Even if the price of the electricity is the same within the microgrid at any given point in time, the average price paid by each household varies according to the time patterns of consumption. A household will enjoy a lower average price when they consumed a large share of its annual consumption at times when the electricity was free (or at least cheaper). This is to say, a higher self-sufficiency will lower the average price. However, in terms of gross economic benefit (i.e., the sum that can be saved), it is not the average price that matters but the cumulative energy received for free. In this sense, the conclusion from the results in Figure 9 is troublesome as the earnings are not due to the ability to obtain a higher self-sufficiency but simply to the sheer cumulative consumption. In Figure 11, the households in the microgrid are divided into three groups of 16 elements each according to their annual cumulative consumption. As in Figure 9, the correlation of the KPI (Key Performance Indicator) with annual cumulative consumption is evident. In fact, the lifetime economic balance is determined solely by the savings, thus by the sheer quantity of energy that is received by each household. In Figure 11a, it is visible how being in the upper third of the cumulative consumption charts guarantees substantial earnings (IRR: internal rate of return from 1.9% to 6%), in the case of the initial investment about 13,646 SEK (1275 €/household). Conversely, the low-consumption households are doomed to economic losses, which means they are unable to recover the investment itself.

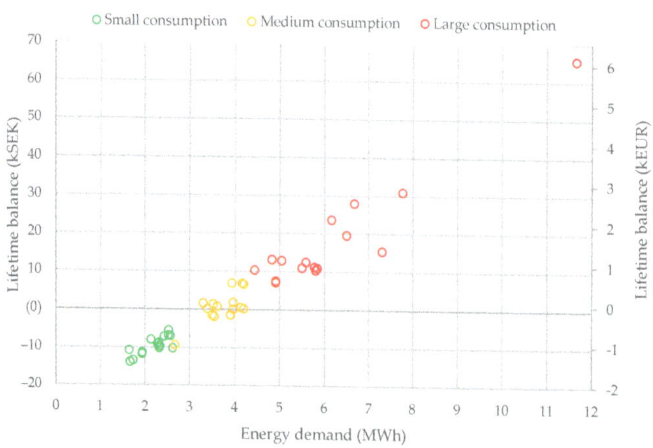

(a)

Figure 11. *Cont.*

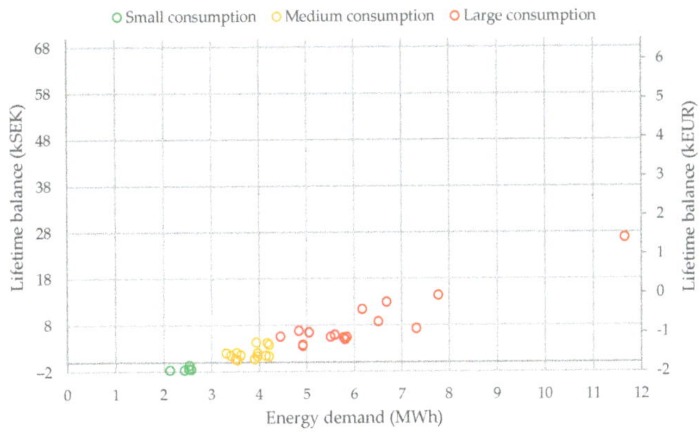

(b)

Figure 11. Cumulative balance over the lifetime of the system against the annual energy demand. The households have been divided in three groups, each of 16 specimens, according to their cumulative consumption: (**a**) LEC Gratis; and (**b**) LEC LCOE.

If the relation between annual cumulative consumption and lifetime earnings would become known by the households in the local grid, there is a risk that there would be a considerable increase of the cumulative demand after the installation of the communal system. This fact, although potentially reducing the risk for those investing in the system (especially in a LEP case), would counteract the purpose of reducing consumption of electricity from the grid.

3.4. *LCOE of LEC*

If the energy is sold at production cost (LCOE), instead of being given for free, the difference in lifetime balance from the different households are greatly reduced, but they persist. In this case, the advantage associated with the use of energy from the system is influenced by the stake of ownership of the system. In general, it can be noted that the lifetime earnings (i.e., Figure 11a,b) follow a linear transformation from the extreme inequality (as in Figure 11a), to a situation of complete equality of earnings (if a LEC grid-price is hypothesised), where no benefit is obtained by the use of on site electricity. In the hypothesis, a benefit for self-consumed electricity would spur increased self-sufficiency. A balance should be found between risk for the low consumption households and reward for the consumption of local renewable energy.

3.5. *LEP N%*

In this arrangement, the PV system is owned by a single provider who has the right to set the price. Obviously, since the parent grid can supply 100% of the demand of the district, the owner cannot set the price higher than the electric grid lest being completely out-bid (e.g., no household would use the owner's energy). In this study, the provider sets the price as half-way between the minimum of the local LCOE and the maximum of the consumer price from the parent grid. More precisely, the provider sets a price at a percentage n so that n = 0 is the LCOE, n = 100 is the price offered by the parent grid and n = 50 is exactly half-way in between.

Table 2 shows how the annual revenues, the balance over the lifetime and the real IRR change according to the price at which the electricity is sold.

Table 2. Annual revenues, lifetime balance and Internal Rate of Return (real) of the investment by different prices set by the owner.

N (%)	Revenues (SEK)	Balance (SEK)	Balance (€)	IRR (%)
0	34,553	−94,058	−8790	−0.5
9.43	37,689	0	0	0.0
25	42,864	155,247	14,509	0.7
50	51,174	404,553	37,809	1.6
75	59,484	653,859	61,108	2.3
100	67,794	903,165	84,408	2.9

Notice how with n = 0% (i.e., the electricity sold at production cost of 0.83 SEK/kWh), the balance and thus the IRR result are negative. This is because the self-consumption of the system is not 100% (it is in fact ca. 85%). In other words, not all the energy produced by the PV system is consumed by the households in the local grid. Therefore, part of the production is sold to the grid below LCOE and results in a moderate loss over the lifetime. The existence of this loss justifies the use of a LCOE adjusted for self-consumption, as described in [18]. This loss also explains why, under LEC LCOE arrangement, some households experience economic losses over the lifetime when the electricity by the communal system is given at price of cost (see Figure 11b). When the electricity is sold at LCOE, the IRR of the PV system is negative, thus holding its shares leads to a loss unless the benefit for cheaper energy outweighs the costs.

Applying an n = 9.43% does not result in any loss or gain over the lifetime of the system. It can be argued that no investor would like to take any risk to have an expected NPV (Net Present Value) of 0 at the end of the lifetime with a discount rate of 0. Nevertheless, there are potential business models for large homeowners such as general contractors or municipalities who could substitute part of the roof and façade cladding with BIPV thus avoiding the cost of an alternative material. Furthermore, this price tag is extremely interesting as price of sale from LEC. It in fact presents the advantage of expected lifetime economic balance in positive ground for each household.

A good business opportunity is finally offered by the n = 100%. This price, while suggesting a real IRR around 3% for the LEP, offers the occupants the opportunity to largely increase their share of renewable energy use without having to pay any upfront cost. In this case, the households have no economic benefit in installing the PV, but they have no risk or upfront investment and could receive information about their own self-sufficiency by the provider, e.g., with a monthly email.

4. Discussion

4.1. Social and Cultural Differences among Households Have a Huge Impact on Self-Sufficiency

In the local grid, if the renewable energy is not enough to cover the electric demand during a specific hour, the aggregated self-sufficiency is assigned to each household regardless of its demand (see Equations (1) and (2)). A large difference in terms of self-sufficiency has been observed within the 48 households, with the individual self-sufficiencies spanning from ca. 14% to more than 28% (see Figure 7a). Considering the absence of active strategies to increase the self-sufficiency in the cluster, such large differences can be attributed only to socio-cultural factors and spontaneous lifestyle choices. In Figure 7b, it appears that the most self-sufficient household has on average the peak of energy consumption at noon (possibly due to home cooking), while the least self-sufficient one has usually its peak consumption at 20:00. Differences are visible also over the different months of the year but their effect is not as clear as in the hours of the day. The large differences observed in self-sufficiency, having no active engagement or use of demand-shifting technologies, invites a deeper analysis and understanding of the existing electric demand and the factors which affect self-sufficiency.

4.2. High Cumulative Energy Demand Is More Effective Than High Self-Sufficiency in Exploiting the Shared Renewable Resource

Despite the large variation in self-sufficiency, it has been observed that the sheer amount of energy used from the system is mainly determined by the annual cumulative demand (see Figure 9). This phenomenon, albeit counterintuitive, is due to the fact that the variability of cumulative demand far outweighs the variability in self-sufficiency (the largest being five or even seven times the smallest one). In other words, the fraction self-consumed is not significant when applied to a group of households whose entire demand is hardly significant compared to others. This fact is problematic because the energy savings (i.e., the main earning mechanism of the investment in some market designs) come from the amount of PV energy consumed and not from the self-sufficiency reached. The relation between annual cumulative consumption and cumulative energy from PV is transposed in the relation between energy consumption and lifetime balance (see Figure 11). The balance in a LEC gratis arrangement (Figure 11a) is almost completely determined by the cumulative consumption, with the self-sufficiency being reduced to a noise in the linear relation. Moreover, if the households are divided into three groups according to their cumulative consumption, the biggest consumers all have positive balance and the smallest consumers all have a negative one. This aspect suggests that, if the communal PV system is installed under a LEC gratis arrangement, the shareholders might increase their electric demand in a bid to outdo each other's energy consumption. This behaviour would possibly defeat the purpose of installing on-site renewables in the first place. It should also be considered that, due to privacy laws and standard practice, each individual household is likely only aware of its own electric demand and self-sufficiency. This lack of data might drive each household to work on improving self-sufficiency instead of annual cumulative demand. It should also be remembered that the earnings are savings, thus increasing the cumulative demand would lead to an increase in the energy bill. In this sense, the increased exploitation of the common electricity through increased cumulative demand would happen only if increased consumption is perceived as a value, for example through the purchase or increased use of energy hungry appliances for cooking or DIY (Do It Yourself) purposes. How easy or difficult it is to change self-sufficiency compared to cumulative demand should also be considered to assess the likelihood of one scenario over the other. For example, cumulative demand might be strongly constrained by working schedule or number of household members. These aspects reiterate the need for a deeper study on the aspect of demand that influence self-sufficiency. From the perspective of the investment in PV, both the changes in behaviour envisioned would increase self consumption, hence earning potential.

4.3. Different Selling Prices Generates Various Business Opportunities

Assuming that the shared PV system is owned by a single entity in a LEP (Local Energy Provider) arrangement, this entity enjoys freedom in setting the price for the sale of electricity. This freedom is nevertheless constrained by the LCOE of the PV system and by the price offered by the parent grid. If the LEP sells electricity at a higher price than the parent-grid it will have no purchaser among the households. This happens because the grid has the capacity to satisfy 100% of the demand of the whole district at any time. For this reason, a coefficient "n" has been devised so that: n = 0 is the LCOE of the local system and n = 100 is the sale of energy at the exact same price as from the parent grid. It has been shown that at n = 0, despite selling at production cost, the lifetime balance is < 0. This is due to the self-consumption being below 100% (i.e., ca 85%), hence ca. 15% of the energy produced being sold at spot price (i.e., 0.3–0.15 SEK/kWh or 3–1.5 € cent/kWh). This loss also explains why in the LEC LCOE arrangement some households still have a negative lifetime balance, as demonstrated in Figure 11b. Another interesting selling price is the one obtained with n = 9.43% because this is the price at which no profit or loss is made from the LEP. This price tag, albeit unattractive as an investment for a third-party PV owner, presents an interesting way for building owners to substitute other claddings on their properties. Using this selling price offers in fact a building material that, contrary to every other, does not cost anything over its lifetime. If applied as common price in a LEC it allows all households to have a positive lifetime economic balance, yet to have individual differences in earnings. It should be said that this price was determined at the end of a previous run when the overall self-consumption was already known. In a real case, to obtain such an equilibrium, the price should be updated at any point in time according to the evolution of self-consumption and energy prices. Selling energy at the price of the parent grid (n = 100) could be an interesting investment as it guarantees the LEP with a real IRR of around 3%; it provides no economic benefits for the household consumers, but it gives them the ability to boost their reliance on renewable without any upfront cost or risk. Furthermore, the possibility for the households to buy voluntarily sized shares of the LEP could kick start a set of tantalising business opportunities.

5. Conclusions

In the study, a newly developed agent-based model was tested on a shared PV system serving a small district comprising 48 apartments in a local community. Different ownership structures were explored. The LEC arrangement was studied both with the electricity given for free to all the equal shareholders or given at a price (in the study the LCOE). For the LEP, because the free offering would make no sense, an array of different prices was tried (see Table 2).

5.1. Key Findings

The main findings of the study are reported as follows and interpreted in the corresponding paragraphs in the discussion Section 4:

- **Social and cultural differences among households have a huge impact on self-sufficiency:** The households were simulated without introducing any demand-response measure or smart control. However, some households achieved a self-sufficiency of almost 30% using the common PV system while others stopped short of 15%.
- **High cumulative energy demand is more effective than high self-sufficiency in exploiting the shared renewable resource:** Despite the large differences observed in self-sufficiency among households, the quantity of energy received from the shared system has been determined almost completely by the annual cumulative demand rather than by self-sufficiency.
- **Different selling prices generates various business opportunities:** Different value of n%, as defined in Section 2.2, generate advantage and interesting features for diverse stakeholders. For instance, a very low n% (i.e., <10%) generates a strong drive for the shareholders to self-consume as much PV energy as possible, but it contains a risk for the least consuming ones. Higher n% (i.e., from ca. 10% to 100%) are interesting for building owners and BIPV solutions and, amid increasing n%, become more and more interesting for third party energy providers.

5.2. Follow-Up Studies

The present study shows a plain set-up and a narrow set of possibilities, but it sets the stage for a broader class of studies. In principle, some of the simplifying assumptions employed in this study should be removed in favour of a higher realism and a more complex modelling; nevertheless, models that are too complex for the level of uncertainty and for the input data available should be avoided.

For instance, it is tempting to change the present model for the prices from the parent grid (i.e., static seasonal price and long-term linear trends for sold and bought electricity) into a spot-price with distribution costs. However, while the change reflects reality better, the long-term modelling of the spot-price would be a daunting task and affected by huge uncertainty. Thus, it might pay off to just maintain a simplified model for the prices (i.e., two seasonal prices for purchase and sale and time of day variation), but to perform a stochastic simulation with variability in the time-evolution of the prices. In other words, any further complexity addition should only be determined by the use case of the model. Furthermore, for this model, the use case is the market design to finance and maintain a fair and remunerative local electric energy system.

On the other end, there are several low hanging fruits that can be easily harvested: for example, while in this study the price was always set by a unique actor (be it a community or a provider), it would be interesting to explore the effect of different prosumer setting each an arbitrary price and explore their interaction. In this sense, one more step could be to endow the agents with some level of intelligence and let them adjust the price reacting to the environment to maximise potential economic gains.

In the present study, there are devices and loads that have not been investigated, such as EVs and electric storages, in the local grid. These features, given a simplified enough model, are extremely easy to be implemented and can constitute a game-changer in the effectiveness of a business model.

Another interesting and potentially prolific research direction would be the study of the demand itself. Given the large variation of self-sufficiency found among the different agents participating in the microgrid, it is possible to find correlation with socioeconomic and lifestyle parameters such as median age, work–home schedules, number of members in an household, etc. This does not constitute information in itself, but it can lead to different results according to the different shared renewable systems. In other words, each social mix might demand a different system (capacity of PV capacity of electric storage).

Regarding the demand, it is of paramount importance to consider how often a house remains vacant due to change or death of the owner. These aspects should be investigated in terms of impact over each business model, but also in terms of risk-mitigating effect of larger local grids. It shall not be forgotten that lower risk can allow lower IRR for the investment, thus unlock wider market niches. The vacancy of the households is also affected by socioeconomic parameters and median age of the households; these aspects likely present spatial variability in different parts of the city and the world.

Author Contributions: Conceptualisation by M.L. and X.Z.; methodology and tool development by M.L.; formal analysis and investigation by M.L. and P.H.; writing—original draft preparation by M.L. and X.Z.; writing—review and editing by C.O.; and Energy Matching Project coordination by L.M. All authors have read and agreed to the published version of the manuscript.

Funding: This research received funding from the European Union's Horizon 2020 research and innovation programme under grant agreement (Energy Matching project No. 768766) and J. Gust. Richert foundation in Sweden (grant number: 2020-00586).

Conflicts of Interest: The authors declare no conflict of interest.

References

1. IEA. IEA EBC Annex 83. Available online: https://annex83.iea-ebc.org/ (accessed on 2 June 2020).
2. Energimyndigheten. Proposed Strategy for Increased Use of Solar Energy. 2016. Available online: https://energimyndigheten.a-w2m.se/Home.mvc?ResourceId=5599 (accessed on 20 June 2020).
3. Huijben, J.; Verbong, G. Breakthrough without subsidies? PV business model experiments in the Netherlands. *Energy Policy* **2013**, *56*, 362–370. [CrossRef]
4. European Commission. Clear Energy Package. Available online: https://ec.europa.eu/energy/topics/energy-strategy/clean-energy-all-europeans_en (accessed on 3 June 2020).
5. Riksdag, S. Regulation (2007: 215) on Exemptions from the Requirement for a Network Concession Pursuant to the Electricity Act (1997: 857). Available online: https://www.riksdagen.se/sv/dokument-lagar/dokument/svensk-forfattningssamling/forordning-2007215-om-undantag-fran-kravet-pa_sfs-2007-215 (accessed on 3 June 2020).
6. Parag, Y.; Sovacool, B.K. Electricity market design for the prosumer era. *Nat. Energy* **2016**, *1*, 16032. [CrossRef]
7. Lettner, G.; Auer, A.H.; Fleischhacker, D.; Schwabeneder, B.D.; Moisl, F. Existing and Future PV Prosumer Concepts. 2008. Available online: https://www.pvp4grid.eu/wp-content/uploads/2018/08 (accessed on 15 June 2020).

8. Stauch, A.; Vuichard, P. Community solar as an innovative business model for building-integrated photovoltaics: An experimental analysis with Swiss electricity consumers. *Energy Build.* **2019**, *204*, 109526. [CrossRef]
9. Roberts, M.B.; Bruce, A.; MacGill, I. A comparison of arrangements for increasing self-consumption and maximising the value of distributed photovoltaics on apartment buildings. *Sol. Energy* **2019**, *193*, 372–386. [CrossRef]
10. Zhang, C.; Wu, J.; Zhou, Y.; Cheng, M.; Long, C. Peer-to-Peer energy trading in a Microgrid. *Appl. Energy* **2018**, *220*, 1–12. [CrossRef]
11. Xu, Z.; Hu, G.; Spanos, C.J. Coordinated optimization of multiple buildings with a fair price mechanism for energy exchange. *Energy Build.* **2017**, *151*, 132–145. [CrossRef]
12. Nguyen, S.; Peng, W.; Sokolowski, P.; Alahakoon, D.; Yu, X. Optimizing rooftop photovoltaic distributed generation with battery storage for peer-to-peer energy trading. *Appl. Energy* **2018**, *228*, 2567–2580. [CrossRef]
13. Jing, R.; Xie, M.N.; Wang, F.X.; Chen, L.-X. Fair P2P energy trading between residential and commercial multi-energy systems enabling integrated demand-side management. *Appl. Energy* **2020**, *262*, 114551. [CrossRef]
14. Lüth, A.; Zepter, J.M.; Del Granado, P.C.; Egging, R. Local electricity market designs for peer-to-peer trading: The role of battery flexibility. *Appl. Energy* **2018**, *229*, 1233–1243. [CrossRef]
15. Rodrigues, D.L.; Ye, X.; Xia, X.; Zhu, B. Battery energy storage sizing optimisation for different ownership structures in a peer-to-peer energy sharing community. *Appl. Energy* **2020**, *262*, 114498. [CrossRef]
16. Tang, Y.; Zhang, Q.; McLellan, B.; Li, H. Study on the impacts of sharing business models on economic performance of distributed PV-Battery systems. *Energy* **2018**, *161*, 544–558. [CrossRef]
17. Basnet, A.; Zhong, J. Integrating gas energy storage system in a peer-to-peer community energy market for enhanced operation. *Int. J. Electr. Power Energy Syst.* **2020**, *118*, 105789. [CrossRef]
18. Huang, P.; Lovati, M.; Zhang, X.; Bales, C.; Hallbeck, S.; Becker, A.; Bergqvist, H.; Hedberg, J.; Maturi, L. Transforming a residential building cluster into electricity prosumers in Sweden: Optimal design of a coupled PV-heat pump-thermal storage-electric vehicle system. *Appl. Energy* **2019**, *255*, 113864. [CrossRef]
19. Thomas, L.; Zhou, Y.; Long, C.; Wu, J.; Jenkins, N. A general form of smart contract for decentralized energy systems management. *Nat. Energy* **2019**, *4*, 140–149. [CrossRef]
20. Alam, M.R.; St-Hilaire, M.; Kunz, T. Peer-to-peer energy trading among smart homes. *Appl. Energy* **2019**, *238*, 1434–1443. [CrossRef]
21. El-Baz, W.; Tzscheutschler, P.; Wagner, U. Integration of energy markets in microgrids: A double-sided auction with device-oriented bidding strategies. *Appl. Energy* **2019**, *241*, 625–639. [CrossRef]
22. Chen, K.; Lin, J.; Song, Y. Trading strategy optimization for a prosumer in continuous double auction-based peer-to-peer market: A prediction-integration model. *Appl. Energy* **2019**, *242*, 1121–1133. [CrossRef]
23. Sousa, T.; Soares, T.; Pinson, P.; Moret, F.; Baroche, T.; Sorin, E. Peer-to-peer and community-based markets: A comprehensive review. *Renew. Sustain. Energy Rev.* **2019**, *104*, 367–378. [CrossRef]
24. Zepter, J.M.; Lüth, A.; Del Granado, P.C.; Egging, R. Prosumer integration in wholesale electricity markets: Synergies of peer-to-peer trade and residential storage. *Energy Build.* **2019**, *184*, 163–176. [CrossRef]
25. Cornélusse, B.; Savelli, I.; Paoletti, S.; Giannitrapani, A.; Vicino, A. A community microgrid architecture with an internal local market. *Appl. Energy* **2019**, *242*, 547–560. [CrossRef]
26. Almasalma, H.; Claeys, S.; Deconinck, G. Peer-to-peer-based integrated grid voltage support function for smart photovoltaic inverters. *Appl. Energy* **2019**, *239*, 1037–1048. [CrossRef]
27. Šúri, M.; Huld, T.; Dunlop, E.D. PV-GIS: A web-based solar radiation database for the calculation of PV potential in Europe. *Int. J. Sustain. Energy* **2005**, *24*, 55–67. [CrossRef]
28. Pflugradt, N.; Muntwyler, U. Synthesizing residential load profiles using behavior simulation. *Energy Procedia* **2017**, *122*, 655–660. [CrossRef]
29. Pflugradt, N.; Teucher, J.; Platzer, B.; Schufft, W. Analysing low-voltage grids using a behaviour based load profile generator. *Renew. Energy Power Qual. J.* **2013**, 361–365. [CrossRef]
30. Luthander, R.; Widén, J.; Nilsson, D.; Palm, J. Photovoltaic self-consumption in buildings: A review. *Appl. Energy* **2015**, *142*, 80–94. [CrossRef]

31. IEA. *Snapshot of Global PV Markets*; IEA: Paris, France, 2020.
32. Lovati, M.; Salvalai, G.; Fratus, G.; Maturi, L.; Albatici, R.; Moser, D. New method for the early design of BIPV with electric storage: A case study in northern Italy. *Sustain. Cities Soc.* **2019**, *48*, 101400. [CrossRef]

buildings

MDPI

Article

Agent Based Modelling of a Local Energy Market: A Study of the Economic Interactions between Autonomous PV Owners within a Micro-Grid

Marco Lovati [1,2], Pei Huang [1,*], Carl Olsmats [1], Da Yan [3] and Xingxing Zhang [1]

1 School of Technology and Business Studies, Dalarna University, 791 31 Falun, Sweden; lova.mrc@gmail.com (M.L.); cos@du.se (C.O.); xza@du.se (X.Z.)
2 Department of Architecture, Aalto University, 02150 Espoo, Finland
3 School of Architecture, Tsinghua University, Beijing 100085, China; yanda@tsinghua.edu.cn
* Correspondence: phn@du.se

Abstract: Urban Photovoltaic (PV) systems can provide large fractions of the residential electric demand at socket parity (i.e., a cost below the household consumer price). This is obtained without necessarily installing electric storage or exploiting tax funded incentives. The benefits of aggregating the electric demand and renewable output of multiple households are known and established; in fact, regulations and pilot energy communities are being implemented worldwide. Financing and managing a shared urban PV system remains an unsolved issue, even when the profitability of the system as a whole is demonstrable. For this reason, an agent-based modelling environment has been developed and is presented in this study. It is assumed that an optimal system (optimized for self-sufficiency) is shared between 48 households in a local grid of a positive energy district. Different scenarios are explored and discussed, each varying in number of owners (agents who own a PV system) and their pricing behaviour. It has been found that a smaller number of investors (i.e., someone refuse to join) provokes an increase of the earnings for the remaining investors (from 8 to 74% of the baseline). Furthermore, the pricing strategy of an agent shows improvement potential without knowledge of the demand of others, and thus it has no privacy violations.

Keywords: urban photovoltaic systems; energy communities; agent based modelling; techno-economic modelling; market design; distributed renewable energy

Citation: Lovati, M.; Huang, P.; Olsmats, C.; Yan, D.; Zhang, X. Agent Based Modelling of a Local Energy Market: A Study of the Economic Interactions between Autonomous PV Owners within a Micro-Grid. *Buildings* **2021**, *11*, 160. https://doi.org/10.3390/buildings11040160

Academic Editor: Ala Hasan and Francesco Reda

Received: 20 March 2021
Accepted: 9 April 2021
Published: 14 April 2021

1. Introduction

1.1. The Problem of Climate Change Has Been Internationally Recognized, The Political Will Is in Place

Climate change is one of the main challenges that threaten the wellbeing or the very existence of human society. This threat cannot be ignored because it can impact a wide range of natural ecosystems and socio-technical systems (e.g., [1–7]). In the last few decades, numerous technologies have been discovered, or improved, that can dramatically reduce our greenhouse gas emissions: renewable or low carbon energy generation facilities, energy storage systems, energy efficiency, and carbon capture devices. The vast majority of countries and international institutions on the planet agree on the danger of climate change and on the need for action [8]. In other words, since the political and social will to build a low carbon economy has been largely achieved, the focus of this study has been put chiefly on practical strategies and effective transition pathways. The subject is how to achieve a transition to a low carbon society in an economically beneficial way and without causing discontent. To reach this goal, it is essential to analyze the impact of different market designs and the choices of different households in creating potentially harmful or unfair economic impacts.

1.2. How Change Can Happen

To transform the will for change to actual change, it is important to understand the causes and the mechanisms that activate the changes. According to Giddens' study [9], an important role in the evolution of technology is played by the interaction between socio-technical regimes: (*i*) the existing dominating technology and the social structure it generated, and technological niches, (*ii*) newer, smaller and dynamic sociotechnical entities that disturb the existing regime. Geels and Schot [10] elaborated different transition pathways (which include transformation, reconfiguration, technological substitution, and de-alignment and re-alignment) elaborating upon previous work and criticisms. In particular, Geels [11] added new elements on the subject introducing the so called 'socio-technical landscape'. The socio-technical landscape is the sum of morals, beliefs, knowledge, and ideas that can spur change in a socio-technical regime. Suarez and Oliva [12] presented different modifications of the socio-technical landscape: regular, hyper-turbulence, specific shock, disruptive, and avalanche. Also Scott [13] speaks about the forces that drive a transition or the conservation of a socio-technical regime, which can therefore be seen as a socio-technical landscape. These forces are divided into three groups: regulative (such as laws and standards), normative (such as values and norms), and cognitive (such as beliefs and search heuristics). The study argues that the stronger of these forces is the cognitive one since it is the most immersive and invisible for the actors under its influence. Other aspects that are fundamental in a transition are the selection pressure and the coordination of resources; these two are deeply interconnected according to Smith et al.'s findings [14]. Specific examples of cognitive forces are the attitudes of end users toward innovative technologies; a large amount of literature exists on this topic (see [15]), but the authors consider obtaining a low carbon economy [16] a rather difficult endeavor.

1.3. Existing Optimization of Urban PV Systems and Research Gap

The coordination of resources, in a practical sense, is of the utmost importance in the case of PV technology. PV technology is currently one of the best candidates to shift the main energy source of our civilization away from fossil fuels. The direct use of the energy from the sun offers an abundant renewable supply with no pollutant emissions during the operational phase, thus producing fewer environmental side effects per unit of energy used compared to the present global energy mix. In general, PV systems can be categorized as urban PV or free-standing utility scale. Urban PV, as the adjective claims, is located in cities on buildings or infrastructure and serves the energy need of its immediate surrounding; this category is the subject of this study. On the other end, free-standing utility scales operate similarly to a traditional power plant: the electricity is generated in a rather large facility, then transmitted and distributed by the wider electric grid.

Both these categories of photovoltaic systems have strengths and weaknesses compared to the other, but both are characterized by a prominence of the initial investment compared to the operational costs. In other words, it takes a long time to repay the initial investment, but after that, savings and revenues can be enjoyed at almost no cost. These characters cause the payback time to be long (over 10 years), even amidst positive Net Present Value (NPV) calculated with discount rate of about 3% (see [17]).

Numerous studies have tackled the economic feasibility of the urban PV systems in the past few years, highlighting different aspects in the results [18]. In particular, a number of studies have optimized the capacity and positions of PV systems according to their lifetime economic performance in self-consumption within the parent building or district (as briefly explained in Section 2.4) [19].

Lovati et al. [17] performed optimization over a school building where PV can be installed both on the roof and on a tilted façade. Most of the electric demand is in winter due to electric heating. Furthermore, summer break reduces demand during summer. The NPV maximization suggests a capacity of PV that can achieve a self-sufficiency above 30%. Moreover, part of the system is installed over the façade because of better self-consumption despite vast residual space on the roof.

Adami et al. [20] considered the differences in price between roof and façade solutions because of the premium paid for aesthetics and multi-functionality and the price reduction due to substitution of a cladding material. It shows that the optimal capacity on the façade is inversely proportional to its unitary price, and it starts to become advantageous when its unitary price falls below 2000 (€/kWp) (considering a unitary price on the roof of 1440 (€/kWp)).

Similarly, Vigna et al. [21] performed optimization over different building clusters (or districts) with varying urban density and functions. The residential function is shown in single family houses or larger high-rise buildings, and the larger buildings are also hypothesized with office or mixed electric demands. The procedure is repeated with two different reward functions: maximize Net Present Value (NPV) and maximize self-sufficiency. When the self-sufficiency is maximized, the office buildings are the best in performance thanks to the good matching between PV power production and building electricity demand (especially in summer). On the contrary, when NPV is maximized, the optimization in purely office buildings is choked by periods of low demand in spring and autumn. A good mixture (i.e., presence of more than one function in a district) is found to be effective to maximize the economic productivity of urban PV.

In Lovati et al.'s study [22] and Bernadette et al.'s work [23], the focus is on the benefit of the aggregation of the demand. The optimization in [22] is performed on a group of 16 houses. The optimization is first performed on each house separately, then on groups of 4 and finally on the whole block of 16. The optimization is repeated with two different reward functions: maximum NPV and minimum Levelized Cost of Energy (LCOE). For maximum NPV the aggregation of the load allowed the optimal capacity, the NPV, and the self-sufficiency to increase simultaneously. For minimal LCOE, to be obtained guaranteeing a self-sufficiency of at least 27%, both the LCOE values were reduced considerably (i.e., the one relative to the self-consumed fraction and the one relative to the whole energy output).

Huang et al. [24] further applied the method to a small residential district in Sweden, in this case a thermal storage was added as a sink for dumping excess PV power through the use of a heat pump. The presence of storage increases the on-site use of PV power output, and thus encourages a larger optimal capacity. Moreover, the effect of PV design at different Electric Vehicles (EV) penetrations in the district is examined. The results show that the self-consumption can reach nearly 80% with a self-sufficiency above 20% even in the baseline case without storage or EVs.

The previous studies have shown the results of optimization, assuming good quality input data; nevertheless, in design practice, the knowledge of the electric demand of the retrofitted building is often not available or is susceptible of high error. Therefore, Lovati et al. [25] tried the optimization process on a static demand obtained from an annual cumulative value. Even in the case of lack of demand data as an input, the method still outperforms other sizing methods and suggests a capacity that is only about 20% larger than the actual optimum. Even if the capacity suggested is almost correct, the values of NPV obtained by the simulation of said capacity with constant demand were considerably higher than those obtained with a realistic demand.

All these studies have focused on the optimal design of an urban PV system; nevertheless they do not discuss the possibility of financing it, nor how the risks and benefits can be shared among the stakeholders in the case that the system is jointly owned. Due to the aforementioned long-payback time of photovoltaic systems compared to other means of electricity production, the financing and lifetime economic performance of urban systems is of utmost importance in order to upscale this technology.

1.4. Relevant Research

As a possible solution for the financing and wider diffusion of renewable energy systems, micro-grids, and in general positive energy districts, are gaining importance in industry and academia. A comprehensive literature review on the subject would perhaps be beyond the scope of this study; nevertheless, the reader might consult the work of

Gjorgievski et al. [26], of which a set of highlights is attempted. The review examines three broad aspects: one is mainly occupied with social arrangements, one is focused on the quality of system design, and one on economic, environmental, and social impacts of the energy communities. The first section contains numerous examples, both proposed and realized, of energy communities of consumers or prosumers. These communities, according to Caramizaru and Uihlein, are already more than 3000 in the EU [27]. Furthermore, the first section of the review describes the role of service providers or initiators such as those described in Capellán-Pérez et al., [28] or Lowitzsch et al. [29]. It also contains a detailed classification of different types of energy communities drawn from categories found in Moroni et al. [30], among others. The section relative to the design of the shared infrastructure offers numerous examples of techno-economic evaluation and optimization studies. A wide variety of technologies and design criteria are presented, most relevant in this context are the works concerned with the overall profitability of rooftop photovoltaic based energy communities. These analysed the operational phase: Roberts et al. [31] and Fina et al. [23], or the design: Novoa et al. [32], Abada [33], Sadeghian and Wang [34], and Awad and Gül [35]. The present study is focused on the operational phase, but it starts from an optimally design shared PV system (see Sections 1.3 and 2.5). In the last section of [26] the economic, environmental, technical and social impacts of micro-grid are discussed by a large set of empirical and theoretical studies.

1.5. Aim and Objectives

The aim is to favour a quick and effortless transition to a higher share of Distributed Renewable Energy Systems (DRES) in residential areas. To achieve this result, different market designs for micro grids of positive energy districts are simulated and the tecno-economic effects of different arrangements are presented and commented on. In particular, the fairness and the resilience of the system is investigated by answering the following research questions/objectives:

1. What is the effect of the price scheme adopted by the prosumers on their savings and revenues?
2. Is the micro-grid of a positive energy district economically feasible when some of the households refuse to invest any money in the shared PV system?
3. Which are the most promising market designs to encourage the adoption of a shared PV system?

In other words, this paper deals with the integration of PV systems in a distributed environment. In particular, it shows the interaction of different owners of a co-owned system in a local energy market (i.e., a loose and free form of energy community). The results show the techno-economic outcome of the different choices that the inhabitants of an energy community can have. It follows a discussion on why such choices could be made, who the potential winners and losers are, and how such an energy community could impact the energy infrastructure and the environment in general.

In order to reach carbon neutrality, it is necessary to integrate suitable onsite renewable energy generation and storage technologies which can offset imported energy from grids. These technologies need to be sustainable; thus, they have to achieve economic, social, and environmental sustainability. The present study is concerned with the path to reach a sustainable positive energy district, and thus it analyses how to finance, build, and run a decentralized renewable energy facility within a residential district. The economic sustainability and the correct policy to avoid excessive inequality in risks and benefits is analysed. Furthermore, the impact of photovoltaic energy sharing in increasing the self-sufficiency of an energy district is simulated and discussed.

2. Research Methodology

This section will provide a brief introduction of the agent-based modelling algorithm used in terms of rules (which define behaviour of the agents), and different scenarios analysed (including various sets of PV distribution within the micro-grid and price schemes).

After that, since the algorithm is applied on a case study, the case study will be described according to its features, how it is modelled, and the determination of the ideal PV system to be installed.

Figure 1 shows the workflow performed in this study, including the geometry of the building, the aggregated demand of the 48 households, a series of techno-economic inputs (see Table 4) and a local weather file. All these inputs are used to obtain the optimal PV system (i.e., the optimal capacity and positions for the modules in the system). After that, the optimal system is tested in a series of scenarios (see following section) using the agent-based modelling.

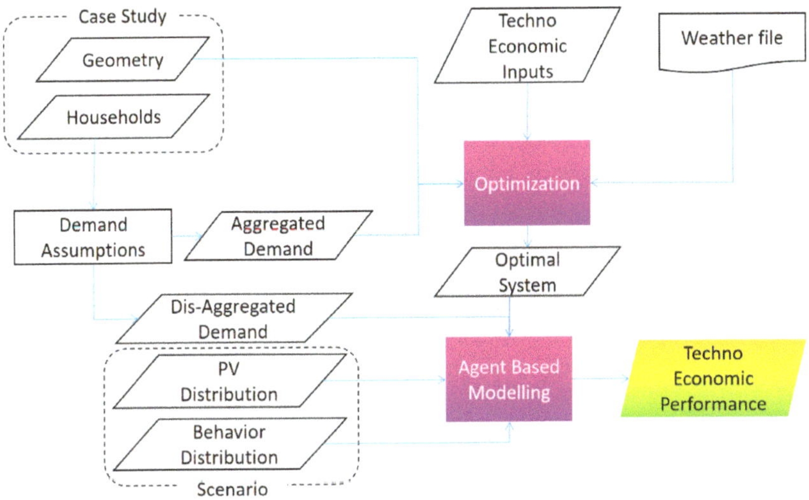

Figure 1. Flow diagram of the data and operations performed for this study.

2.1. Agent Based Modelling and Scenarios

Occupant behaviour in buildings is complicated with different patterns and distribution properties [36–38]. An agent-based model is a model of a complex system where the behaviour is not controlled by a single algorithm, but it comes/emerges from the interaction of a number of sub-systems (i.e., the agents). Lovati et al. [39] proposed a 3D map of the possible algorithms (as shown in Figure 2); the dimensions represent the emergent versus controlled axis, the centralized versus decentralized axis, and the individual versus collective axis. The simulations presented in this study occupy the pink area in the map, and thus are individual, de-centralized systems causing an emergent behaviour in the micro-grid. In fact, in the scenarios described, there are multiple PV owners in the micro-grid and each one can set the price according to its own independent will.

It should be noted that, despite separate ownership, the agent-based model can be described by a simple set of rules:

a. Every household is represented by one independent agent in the simulation.
b. Every agent has an energy balance in each HOY (Hour of the Year). The energy balance is determined by its PV power (if it owns a PV system) minus its power demand in that particular HOY. If the balance is negative, the agent will be a net buyer in that HOY, otherwise it will be a seller. This rule implies that each agent can only sell electric power if it has already satisfied its own demand. Simply, each household can sell only excess PV production.
c. Each seller can set the price for the power he has to export.
d. If the electricity is offered by multiple sellers, the buying agent will buy preferentially from the cheapest source.

e. If the aggregated demand of the district exceeds the offer of the cheapest source, the demand of each household is satisfied proportionally by the cheapest source. If, for example, the cheapest source covers 30% of the aggregated demand in that HOY, each household is provided 30% of its power demand by the cheapest source.
f. If the on-site renewable power exceeds the power demand in a certain HOY, the cheapest sources are consumed preferentially, while the more expensive ones risk being in excess of the demand and sell part (or all) their power to the grid. Those who sell to the grid cannot set the price but are simply valued by the price paid by the grid (which is always way lower than that of the local sellers).

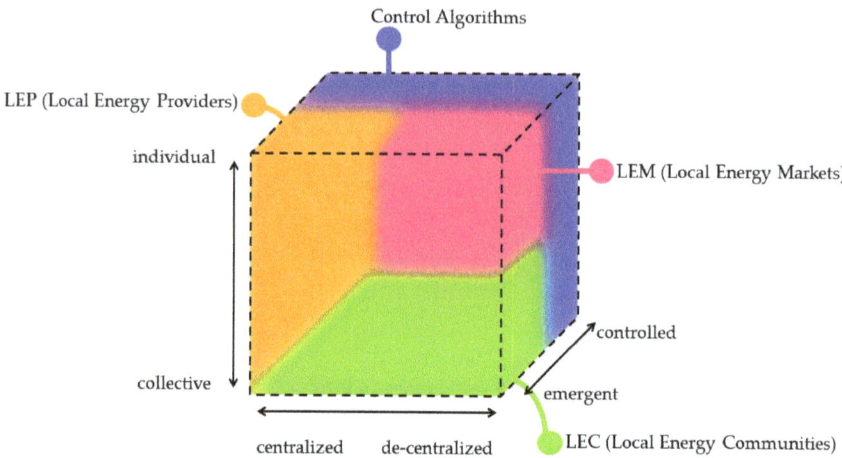

Figure 2. 3D map of the behaviours in a district scale renewable energy system from [39].

The same set of rules is applied on a set of six different scenarios; these are explained in the following paragraphs and summarized in Table 1.

Table 1. PV capacities per household and prices in the 6 different scenarios.

Scenario	PV Capacity (kW/Household)	Electricity Price (at Year 0) (SEK/kWh)
(1)	1.36	1
(2)	1.36	1.19 (summer), 1.78 (winter)
(3)	2.73 or 0	1
(4)	2.73 or 0	1.19 (summer), 1.78 (winter)
(5)	1.36	1 or 1.19 (summer), 1.78 (winter)
(6)	1.36	1 or dynamic

Scenario 1: All residents agree to purchase the PV system; every household purchases an equal share of the total system and has thus the right to 1/48 of the power at any time (i.e., ca. 1.36 kW of capacity each). The price for the sale within the micro-grid is agreed for the long term as the summer grid price/1.2 (thus a static 1 SEK/kWh at the year 0); therefore, whoever buys electricity from another household saves ca. 17% on the electricity cost in summer and 45% in winter.

Scenario 2: All residents agree to purchase the PV system, like in Scenario 1. The price for the sale within the micro-grid is agreed for the long term as 99% of the grid price; therefore whoever buys electricity from another household has almost no savings compared to the grid. In this case it is assumed that using local energy is perceived as a value in itself by the participants in the grid.

Scenario 3: Only 50% of the residents agree to purchase the PV system; every PV equipped household purchases an equal share of the total system and has thus the right to 1/24 of the power at any time (i.e., ca. 2.73 kW each). The price for the sale within the micro-grid is agreed for the long term as the summer grid price/1.2, like in Scenario 1.

Scenario 4: Only 50% of the residents agree to purchase the PV system; every PV equipped household purchases an equal share of the total system like in Scenario 3. The price for the sale within the micro-grid is agreed for the long term as 99% of the grid price, like in Scenario 2.

Scenario 5: All residents agree to purchase the PV system, like in Scenario 1. The price for the sale within the micro-grid is left to the choice of the single household; 50% of the households decide to charge a high price (i.e., 99% of the grid, like case 2 and case 4), the others charge the summer price /1.2 like in Scenario 1 and Scenario 3.

Scenario 6: All residents agree to purchase the PV system. The price for the sale within the micro-grid is left to the choice of the single household like in Scenario 5, and 50% of the households decide to adopt a dynamic price system based on their energy balance in every hour of the year. With this strategy the energy is sold at LCOE whenever the balance is more than double the average balance in that hour of the day. The other 50% charges the 1 SEK per year like in Scenario 1 and Scenario 3.

In general, regardless of the scenario, the behavior of each agent in a time-step can be summarized as shown in Figure 3.

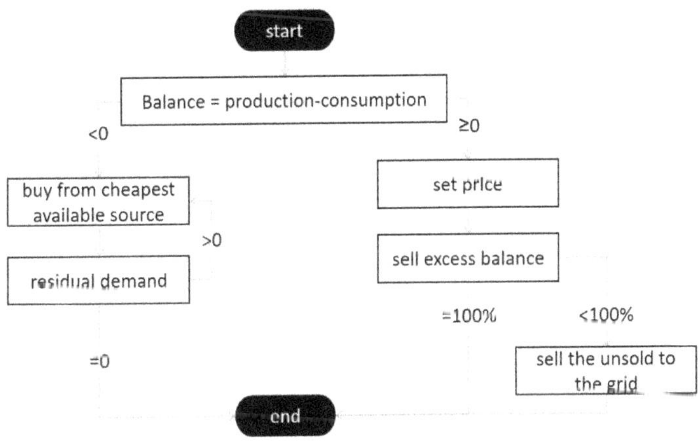

Figure 3. Schematic of the behaviour of each agent in every time-step of the simulation.

2.2. Modelling of the Economic Performance

To understand the profitability and the risks associated to the investment in the shared PV infrastructure the CAGR (Compound Annual Growth Rate) was used as key performance indicator. The CAGR has been considered in its real value i.e., without inflation, and expressed in percentual form according to the following equation:

$$\text{CAGR} = \left[\left(\frac{\text{Inc.} - (\text{CAPEX} + \text{OPEX})}{\text{CAPEX}}\right)^{\frac{1}{\text{lifetime}}} - 1\right] \cdot 100 \tag{1}$$

In Equation (1), real Compound Annual Growth Rate is used to infer the profitability of the investment in the shared PV for each household.

- Inc. represents the cumulative income derived by the ownership of the share of the PV system during its lifetime; it represents the figure before costs (i.e., capital expenditure and operational expenditure) and it is calculated according to Equation (2).

- CAPEX is the capital expenditure; it includes the turn-key cost of the system including design and installation costs, but it assumes no taxation. It can be calculated by multiplying the unitary cost by the installed capacity (see Table 4).
- OPEX is the operational expenditure; it includes a standard annual cost of 80 SEK/kWp year for the substitution and cleaning of the modules, plus substitution of the inverter in case of rupture. The inverters have a cost of 3.5 KSEK/kWp and should be changed at least once in the planned lifetime of the system.
- Lifetime is expressed in years and is assumed as 30 years in this model.

$$\text{Inc.} = \sum_{T=0}^{\text{lifetime}} (\text{Sav.} + \text{Rev.})\cdot(1 - \Delta\eta\cdot T)\cdot(1 + \Delta d\cdot T) \tag{2}$$

Equation (2): cumulative income derived by the ownership of the share of the PV system during its lifetime

- T represents the number of years since the construction of the PV system.
- Sav. Represents the savings due to the avoided purchase of electric power from the external grid, it is calculated according to Equation (3).
- Rev. Represents the revenues obtained by each shareholder by selling excess PV power from their share, it is calculated according to Equation (4).
- $\Delta\eta$ is the variation of the efficiency due to ageing of the PV system. The shared PV is assumed to lose 1% per year (see Table 4).
- Δd is the variation in the price of the electricity for the consumer, it is assumed as +1.5% per year in design stage (see Table 4), but it is then assumed 0 or 2% in the agent based model (see Figure 8 in the results and Figure A1 in the Appendix A).

$$\text{Sav.} = \sum_{Ts=1/1-00:30}^{31/12-23:30} \left(\left(P_{self,Ts}\cdot d_{grid,Ts} \right) + P_{peer,Ts}\cdot \left(d_{grid,Ts} - d_{peer,Ts} \right) \right) \tag{3}$$

Equation (3) savings due to the avoided purchase of electric power from the external grid.

- Ts represents the internal time-step of the model, in this case it is set as 1 h.
- $P_{self,Ts}$ is the power self-consumed in a specific time-step.
- $d_{grid,Ts}$ is the cost of electric power offered by the external grid in a specific time-step.
- $P_{peer,Ts}$ is the power bought from a peer within the local community in a specific time-step.
- $d_{peer,TS}$ is the cost of electric power offered by a peer in a specific time-step.

$$\text{Rev.} = \sum_{Ts=1/1-00:30}^{31/12-23:30} \left(\left(P'_{peer,Ts}\cdot d'_{peer,Ts} \right) + \left(P'_{grid,Ts}\cdot d'_{grid} \right) \right) \tag{4}$$

Equation (4) revenues obtained by each shareholder by selling excess PV power from their share

- $P'_{peer,Ts}$ is the power sold to all peers in a specific time-step
- $d'_{peer,Ts}$ is the price set for selling power to the peers in a specific time-step
- $P'_{grid,Ts}$ is the power sold to the grid in a specific time-step
- d'_{grid} is the price at which the grid purchases power. This price is static, thus is independent by the time-step.

2.3. Case Study Description

A small district (or cluster) of residential buildings was adopted as a case study to test different behaviours of the co owners (see Figure 4). The district is composed of three multi-family apartment blocks and it is located in Sunnansjö (Ludvika), in the Dalarna region of Sweden. The district is currently undergoing energy retrofit within the project

Energy Matching [40] and is already the subject of scientific publication in [24,39,41]. While the geometric and technological properties of the site are known, the demographic information shall not be disclosed to respect the privacy of its inhabitants. For this reason, the electric demand used for the study was generated using LPG (Load Profile Generator) [42] assuming population characteristics as described in the following section (see Section 2.3). In total, there are 48 households in the three multi-family apartment blocks.

Figure 4. View of the (**a**) 3D model used to calculate PV production in the district and (**b**) areas that is possible to dedicate to PV highlighted. The model represents the three multi-family buildings that comprise the 48 households in the study. Please note that each household does not own any physical portion of the roof as it is commonly owned. If a household decides to invest in the shared PV system, it becomes shareholder of the common infrastructure (it does not purchase a physical small system on the roof).

2.4. Electric Demand Assumptions

From previous study, the characteristic of electric demand with different spatial scales possesses various properties, and the aggregated electric use curves of more households indicate weaker randomness [43,44]. The composition of the district, made of 48 households, was assumed to reflect the demographic composition of Sweden as reported in the official statistics [45]. Due to the small sample of households, it was decided to curtail all the households listed in the 'other' category of the official data. These groups are difficult to transfer to a small sample because are numerically small, heterogeneous, and less specified than the main categories. For this reason, the district has been designed to match the main categories of household in Sweden (Table 2) as shown in Table 3. In the tables it is visible that the group 'single + 3 kids' has been removed as in proportion it amounts to almost zero households and would misrepresent the overall number of kids if one household was introduced.

Once the composition of the households was determined, the age and gender of the components have been estimated to match the data available online from the world factbook [46]. The resulting composition of the whole district have been reported in the Appendix A (see Table A1).

Table 2. Number of persons and households in Sweden (excepts those listed in the 'other' category).

-	Male Minor	Female Minor	Male Adult	Female Adult	*n* Households
Single	0	0	921,495	957,910	1,879,405
Single + 1 minor	113,287	84,195	55,871	141,611	197,482
Single + 2 minor	108,963	96,493	25,907	76,821	102,728
Single + 3 minor	65,725	59,947	6146	30,676	36,822
Couple	0	0	1,134,261	1,132,893	1,132,893
Couple + 1 minor	208,795	169,207	377,046	378,958	377,046
Couple + 2 minor	494,165	453,061	472,734	474,492	472,734
Couple + 3 minor	340,621	309,381	194,487	194,905	194,487

Table 3. Number of persons and households in the district under exam as assumed to match the Swedish data [45].

-	Male Minor	Female Minor	Male Adult	Female Adult	n Households
Single	0	0	10	11	21
Single + 1 minor	2	1	1	2	3
Single + 2 minor	1	1	0	1	1
Single + 3 minor	0	0	0	0	0
Couple	0	0	12	12	12
Couple + 1 minor	2	2	4	4	4
Couple + 2 minor	5	5	5	5	5
Couple + 3 minor	3	3	2	2	2

2.5. Calculation of the Optimal PV System

The optimal PV system, used in the micro-grid simulation, is the system that can maximize the self-sufficiency of the district while maintaining a positive NPV. The capacity and positions of such a PV system were found with the same technique and tool from the studies enumerated in the introduction (Section 1.3), which is one result of the H2020 project EnergyMatching. The parameters of the optimization are capacity and position of the PV system, if the capacity is too small the self-sufficiency will be low, but if the capacity is too high the overproduction of electricity will be too frequent, and the system will not be profitable. With the correct capacity (which leads to the largest profitable system) the self-sufficiency can be maximized; furthermore, installing parts of the PV system on the façade instead of on the roof can increase the contemporaneity between production and consumption [17]. In the case study described, considering the assumptions described in Table 4, the optimal system has a PV capacity of ca. 65.5 kWp, and no electric storage. The capacity on the southern slope of the roof (which is the most irradiated) is as high as it can be at 28.4 kWp, while part of the system has been installed on the southern façade (i.e., 5.3 kWp) despite large available spaces on the east and west slopes of the roof. The system can achieve an annual average self-sufficiency of ca. 24.6%. For the detailed design optimization of the PV systems, please refer to [24].

Table 4. Assumption used in the optimization of the PV system.

Parameters	Values
Unitary cost of the PV system	12 KSEK/kWp (ca. 1175 €/kWp)
Unitary cost of electric storage	5.11 KSEK/kWh (ca. 500 €/kWh)
Planned lifetime of the system	30 years
Degradation of the PV system	−1%/year (annual percentual efficiency losses)
Nominal efficiency of the system	16.5%
Performance ratio of the system at standard test conditions	0.9
Price of the electricity from external grid	1.2 SEK/kWh (summer), 1.8 SEK/kWh (winter)
Price of the electricity sold to the external grid	0.3 SEK/kWh
Annual discount rate	3%
Growth of electric price for consumer	+1.5%/year (annual percentual price increases)

The optimization algorithm has a single target: its parameters are the capacity and positions of the PV system, and its reward function is the average annual self-sufficiency of the PV and building system. In the initial stage the building does not feature any PV system (capacity equals 0), then portions of PV area or electric storage are added or moved in an iterative procedure to maximize the reward function. Once it is not possible to further improve the function by adding or moving any component, the system is considered optimal and the algorithm ends.

The assumptions in Table 4 have been used for the dimensioning of the PV system according to the technique described in [17] and in numerous other studies mentioned in

Section 1.3. The lifetime of the system is assumed at 30 years. However, the lifetime of the inverters is considered about 12.5 years and the expenses for its substitution are taken into account among maintenance costs (see [17]). The performance ratio of the system is considered 0.9. This value is not conservative as the average for urban installed systems is generally lower. Nevertheless, it should be considered that this is the performance ratio at standard test condition, which is therefore measured before temperature correction. Several examples in literature are considering the final performance ratio which is either measured experimentally or takes into account the temperature related losses. Several studies regarding urban photovoltaic systems, if controlled for calculated temperature-related losses, show values similar to 0.9 (see [47,48]). Significantly lower performance ratio at standard test condition could be expected in situations of strong partial shading. Furthermore, lower values can also arise in systems characterized by multiple facades if performance ratio is calculated measuring the irradiation in a single point, while production is calculated over the aggregated system. Nevertheless, these situations do not occur in the system under study. In term of efficiency, 16.5% was chosen; this value is conservative respect to the datasheets provided by numerous manufacturers. It should be noted that here, as in [17] and numerous studies mentioned in Section 1.3, the system does not represent the active area or the cumulative area of the modules, but the underlying area dedicated to the PV system. Since this a moderate sized rooftop system, and thus not equipped with empty passages for inspection, a utilization factor of the area of about 80% was assumed. This factor was taken into account by lowering the overall efficiency of the system.

3. Results and Discussion

After the simulation over an entire year in hourly time-step, all the exchanges of money and electric power that have happened in every time step between the different households are recorded in a ledger. The ledger is a Json archive; its size is large as it contains every interaction, both economic and of power, between a every single household with every other single household and the grid. Analysing this large file is possible in order to discover who has earned and who has lost in economic terms. Furthermore, it is possible to track the origin of the power that every household has consumed.

3.1. Self-Sufficiency within the Micro-Grid

Figure 5 shows the distribution of self-sufficiency throughout the district. Every point of the line describes the percentile of households that have a self-sufficiency lower than that described in the y axis. It is visible, for example, that 90% of the district has a self-sufficiency lower than 29%. In other words, most of the district uses at least seventy percent of its energy from the grid. The flip side is that less than 10% of the households in the district has a self-sufficiency below 20%. This implies that the vast majority of households consumes at most 80% of their cumulative demand from non-local electricity. The result is notable because it far exceeds the solar fraction in the energy mix of even the best countries according to [49]. In general, the gradient of the curve seems to be slightly stronger at the extremes. It is rare to have an outlier with high or low self-sufficiency. In fact, about 70% of the households in the district fall within a variance of five percentage points of self-sufficiency (i.e., 22.5% and 27.5%).

3.2. Effects of the Local Energy Market on the Price

Figure 6 describes the average hourly price for Scenario 1 during the different hours of the day within the district. The grey bands represent the variability between different households of the district. If a bar is longer, it means that some households have an average price that is significantly lower than others in that hour of the day. The red ticks represent the average price in that hour for the whole district. It is immediately visible that between 7 P.M. and 4 A.M. the price is almost stable at one point five SEK. The price is stable because there is no electric storage installed in the micro grid and, therefore, when a photovoltaic system is not producing, the price is that of the power provided by the electric grid. Considering that the night-time consumption of the district remains stable throughout the year, and that the electricity is sold at 1.8 SEK/kWh in winter and 1.2 SEK/kWh in summer, the night price paid by the whole district equals almost exactly the average of the two (i.e., 1.5 SEK/kWh). During daytime the price is lower; there are two reasons for this phenomenon. The first reason is that each household owns a share of the local PV system, and therefore all the electricity produced by their own share already belongs to them and is thus free for themselves. The second reason is that they can purchase electricity from their peers, the price of which is consistently lower than that of the grid. For these reasons, those households that are able to use larger share of their own electricity, or at least of the electricity from their peers, can enjoy a lower price for the electricity. It is visible, indeed expected, that the price of electricity is generally lower during the central hours of the day. Furthermore, it can be observed that times of the day of comparatively higher price correspond with moments of high electric demand. For example, it can be seen that the price is comparatively higher at seven and eight AM (when people prepare and consume breakfast) and at noon (when many prepare and consume lunch). This is due to the fact that, despite a large photovoltaic production in that hour, the outlier high demand forces the whole district to supply part of its demand from the external electric grid.

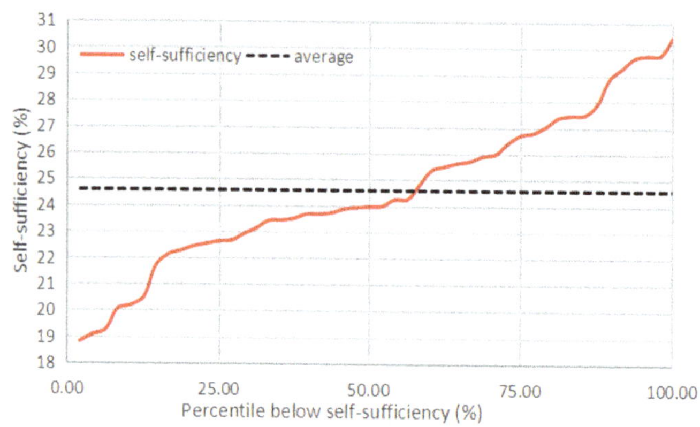

Figure 5. Self-sufficiency within the micro-grid.

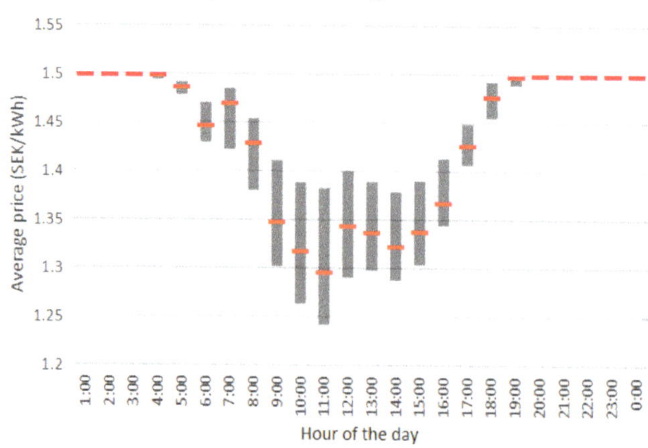

Figure 6. Average price for each hour of the day and its variability within the district (Scenario 1).

3.3. Savings and Revenues

Figure 7 shows the relationship between savings and revenues for each single household within the micro grid. All the sixth scenarios described previously are shown in the chart. In general, the savings are obtained either by using the electricity produced by one's own system, therefore saving 100% of the price from the grid, or else by using the electricity from a peer household at a discounted price. On the other end the revenues are obtained either by selling electricity to the grid, or else by selling electricity to another household, the latter providing a much higher price. In each of the charts, every household is displayed as a circle, its position on the x-axis represents its annual revenues, while its position on the y-axis represents its annual savings. The colour of the circle line represents its belonging to a different category, according to the number of people living in the household. In general, these charts should be studied in their relative difference between each other, rather than in their absolute values. In fact, the absolute value of revenues and savings are not interesting when the cost of initial investment and those for the maintenance of the system are not taken into account. To see the lifetime techno-economic performance of the system, please check the next paragraph, which displays the CAGR. Observing the colour in all the cases it is visible that the smaller households (those which have a smaller electric demand) tend to show lower savings and higher revenues. This phenomenon is quite unsurprising because in the example considered every household purchases an equal capacity to all the others. Therefore, the small households will have a PV capacity that is larger relative to their demand, and will, thus, export and sell a larger fraction of the electricity that they produce.

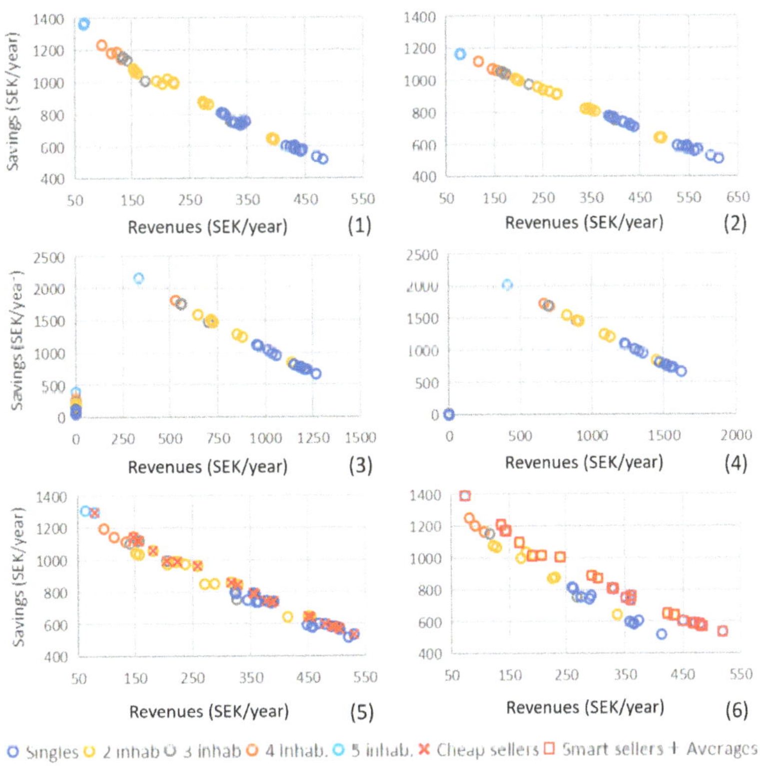

Figure 7. Savings V.S. revenues for each household in the microgrid. Both axes are in SEK/year (**1**) Scenario 1 (**2**) Scenario 2 (**3**) Scenario 3 (**4**) Scenario 4 (**5**) Scenario 5 (**6**) Scenario 6.

3.4. Small Consumers Are 'Sale Oriented', Large Consumers Are 'Savings Oriented' (Scenario 1 vs. Scenario 2)

There is a noticeable difference between Scenario 1 and Scenario 2. In Scenario 1 every household agreed to sell their electricity at a significantly lower price compared to what they agree in Scenario 2 (see Table 1). Therefore, scenario one seems to be particularly advantageous for the largest consumers. This is due to the fact that the largest consumers have high savings but low revenues, and therefore, the sale of electricity at a lower price from others leads to higher savings for them, while it does not impact their revenues significantly. On the contrary, the smallest consumers, that have in general lower savings and higher revenues, will benefit from higher price of electricity. In this case, they will be able to increase their revenues greatly without changing their savings too much. In fact, being smaller, most of their savings comes from their own PV system, which is already over-dimensioned compared to their size. Going back to the largest consumers, a large fraction of their savings implies purchasing electricity from the smaller peers. Another noticeable aspect is that in Scenario 2, all the points are almost linearly correlated. This is due to the high price of the electricity sold, which is almost the same of the cost of electricity from the grid. The smaller remaining differences can be explained by the correlation between the private electric demand and the demand of the whole district. If a household over-produces electricity when the whole district is in over-production, its performances will be slightly lower because it will often sell to the grid. In the opposite case, when a household overproduces at times when there is need from other households, then its performance will be slightly higher.

3.5. When Some Agents Refuse to Invest in the Shared System, The Remaining Investors Have Larger Benefits and Lower Risks (Scenarios 3 and 4 vs. 1 and 2)

Scenarios 3 and 4 reflect Scenarios 1 and 2 in terms of price, but they have the peculiar aspect that only half of the households choose to purchase a PV system. This state of affairs would be very important for any practical application of the micro-grid. This is because, in practice, it is difficult to convince 100% of the tenants in a multi-family dwelling to participate, and especially to invest money, in a PV system. In a realistic setting it is expected that part of the population is unwilling to invest in the system; nevertheless, in the simulation it is assumed that they decided to participate in the micro-grid as simple consumers (i.e., those who do not own any part of the system). The assumption is safe because being a simple consumer only requires to always purchase the electricity from the cheapest source. In this way, to participate as a simple consumer does not have any initial cost nor risks, but it might have a benefit during the lifetime of the system. If Scenario 1 is compared to Scenario 3, it is visible that the points in the latter overwhelmingly outperform those in the former both in terms of revenues and in terms of savings. It is tolerably intuitive that, if only half of the households own a PV system, their revenues will increase. In fact, all the households who do not own PV can only buy the electricity from those who own it, and therefore, the whole local market moves in the direction of a "seller's market". Also, the increase in savings is readily explained. In fact, since the optimal capacity is unchanged in every scenario (see Section 2.4), every PV owner has at his disposal a larger capacity. This fact implies that there is more available electricity for self-consumption in every HOY, even at times of relatively high private electric demand. This spare over-capacity favours an increase in self-sufficiency. Looking at the bottom left corner of the chart for Scenario 3, it can be seen how there is a benefit in terms of savings also for those who do not own a PV system. Of course, these savings are minor compared to those of the other households, and this is due to two specific reasons. The first one is that, by lacking a PV system of their own, these households do not have their own electricity for free, and therefore can only purchase electricity from their peers. The second reason is that every household, before selling electricity, satisfies his own demand (see Section 2.1, in the rule b), and therefore those in the micro-grid who do not own a PV system can only benefit from the left-over electricity from the others. In other words, the household without PV can only purchase electricity when they happen to be in need of power at times when others are in overproduction. Scenario 4, like the Scenario 2, presents a linear correlation between the revenues and the savings of each household. Like Scenario 2 over Scenario 1, Scenario 4 also presents relatively higher revenues and lower savings compared to Scenario 3, thus favouring the smallest consumers. Furthermore, like Scenario 3, it presents a sharp contrast between the PV owners and the other households. It should be noted, though, that this time there is absolutely no benefit in participating in the micro-grid, or at least the benefits are so tiny that cannot be seen by the naked eye. There is nevertheless a benefit for those households: the possibility to increase the share of renewable on-site electricity in their energy consumption. It can be expected that, given the absence of initial investment, most consumers would be willing to increase their renewable energy share. By doing so, they have the possibility to save the planet with a costless, and thus effortless, action.

3.6. Interaction of Competing Sale Strategies within the Micro-Grid (Scenario 5 and Scenario 6)

In the first four scenarios, the price for the sale of PV power and the number of PV owners has been changed. Nevertheless, in all these cases, every PV owner agreed to maintain the same price as everybody else. In Scenario 5 the hypothesis is made that a half of the PV owners prefer to sell at the lower price (i.e., the price of the whole group of PV owners in Scenarios 2 and 4). Meanwhile the other half of the owners prefers to sell at the same price of Scenarios 1 and 3. Observing the revenues and the savings in this arrangement, it can be noted that, in general, the sellers who decide to sell for a lower price ('cheap sellers'), enjoy a higher revenue compared to the others. Savings are not affected by the price of the sale of one's own electricity, but rather by the price of available energy from the other households. For every savings level in the chart, the lowest selling households, which are marked as 'cheap sellers', appear to be on the right side compared to the others, which means that they have managed to obtain higher revenues. There are two factors at play when measuring the revenues in this type of market (where there are two different prices groups). The first factor is the sheer revenues per kWh sold; this acts by lowering the revenues for the so called 'cheap sellers'. The second factor regards the ability to effectively sell your electricity within the micro grid at all. The capacity of not be in over supply, and thus forced to sell most of your power to the grid. If a different price was chosen, the result might have been different, but in this case the increased revenues derived from a higher price scheme are not enough to offset the increased instances in which the electricity cannot be sold due to high price and low demand. Also, in Scenario 6, the last one, the group of households was divided into two sub-groups. As in the previous scenario, some households were selling their electricity at a lower price compared to others. This time, though, the expensive sellers were given the ability to change the price according to a behaviour of their own. The mechanism used to change the price was set as explained in Scenario 6 of Section 2.1. In practice, these households, identified in the chart as 'smart sellers', will sell their power at the LCOE of the system, which is lower than the static price of the cheap sellers, whenever they have an outlier high energy balance in that specific hour of the day. In other words, when a smart seller has an outlier, low power consumption, or an outlier high power production from PV, it will sell its electric power at the lowest possible price. This strategy is extremely simple and is prone to numerous fallacies. In fact, if for example a particular household is on holiday during an unpopular period, its power demand would be unusually low, thus resulting in an outlier high balance. This could cause it to sell at LCOE in a time in which the electricity is indeed in high demand throughout the district. In this example, the household would be selling at the lowest possible price in a time in which the maximum price would still manage to sell to the peers. Conversely, if a household will experience an outlier high demand for its own reasons, it might find itself selling its available power dearly, while there might be plenty of energy available for everyone. It this case it will be forced to sell most its power to the electric grid. Despite the simplicity of this strategy and its obvious flaws, it is visible from the chart that such a simple behaviour is good enough for outsmarting the cheap sellers in the competition for the sale of electric power. Given any savings level, the smart sellers undeniably manage to obtain higher revenues. This is a very important result because it shows that it is possible to create an effective strategy without knowing the consumption of the other agents in the micro grid, thus avoiding privacy issues.

3.7. Effects of the Phenomenon Observed on the Cagr (Compound Annual Growth Rate) for Every Household

Figure 8 represents the real CAGR for the PV owners in the micro-grid; as in the previous set of charts, every circle represents one household.

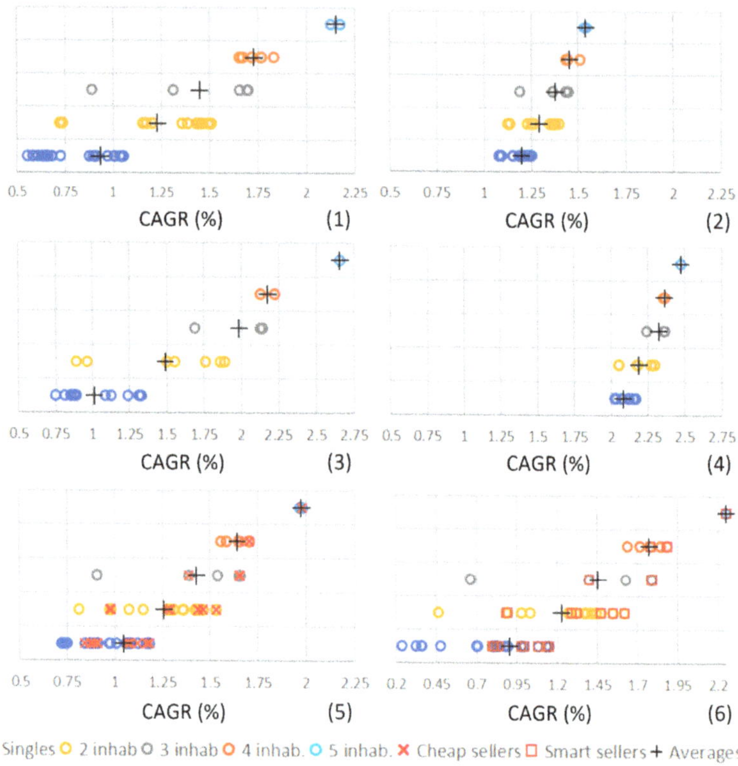

Figure 8. CAGR (Compound Annual Growth Rate) % in the different scenarios (**1–6**), the CAGR are real (i.e., would be higher if adjusted for inflation), and assume a case in which there is no growth in the price of electricity, a chart assuming a 2% linear increase in the price of electricity from Scenario 4 is in the Appendix A.

This time no information is provided on the 'y' axes; the households are simply divided into groups to enhance the readability of the chart. The black crosses represent the average CAGR of each group (according to the number of inhabitants in the household). At first glance these CAGR might look low, but it should be noted that they are produced with two strongly conservative assumptions: the CAGR is real (note it does not consider inflation over a period of 30 years) and the price of electricity for the consumer is considered stable throughout the lifetime of the system (which means it will not grow despite the large investments needed to renew the grid infrastructure and potential future carbon taxation). Some examples of real CAGR considering a 2% linear price growth for the household electricity can be found in Appendix A. In general, it can be seen that larger households, those with higher number of inhabitants and higher electric demand, fare better than the smaller ones in terms of CAGR. This fact must be due to larger households being able to exploit the savings derived by their own PV system more often, and more often benefit from purchasing energy from their peers. Looking at the difference between Scenarios 1 and 2, it is visible how the relative difference in average CAGR between groups is lower in Scenario 2. This means that larger households benefit more from internal sale of electricity compared to smaller ones. This fact is quite intuitive since every household owns an equally sized PV system, and thus having a larger electric demand increases the frequency of over-consumption with respect to one's own electricity. In Scenario 2, where the benefit of buying from peers has been almost completely removed, there is a residual advantage in being a large consumer. This is due to the higher self-consumption of the large consumers. In fact, having a large self-consumption of one's own electricity reduces the risk of not having buyers in a specific HOY; it does so by reducing the very number of hours in which one needs a buyer. Furthermore, though negligible, in case 2 there is also the 1% difference in price between the external and the internal electricity. Scenarios 3 and 4 reflect Scenarios 1 and 2 except that only 50% of the households decide to install PV, and thus are entitled to twice the capacity of before. In Scenario 3 and 4 the CAGR of every group are sensibly increased compared to the respective Scenarios 1 and 2. This at first might sound counter-intuitive for Scenario 3 (i.e., electricity sold for cheap) considering the lower self-consumption given by a larger per capita system. Nevertheless, even though the instances of over-production are more frequent for PV owners, there is a larger number of buyers among the peers because half of the households do not own a PV system, and the price of sale, though low, is still higher than the LCOE of the system. Furthermore, even though self-consumption is lowered by a larger system, self-sufficiency is increased and allows for higher savings from avoided electricity consumption. In other words, having a larger system increases the savings of the owners, and being in a 'seller's market' due to the lower proportion of PV owners, the risk of selling to the grid is also reduced. This explains the sensible increase of CAGR from Scenario 1 to Scenario 3, and the massive increase from Scenario 2 to Scenario 4. In terms of CAGR, Scenario 5 and Scenario 6 reinforce the signal obtained previously in terms of revenues. In fact, in Scenario 5 (when part of the households sells at generally at a lower price) the so called 'cheap sellers' drive the others outside of the market forcing them to sell a large fraction of their electricity to the grid. This is visible by the fact that the cheap sellers consistently occupy the right side of the dispersion in every group (i.e., achieve higher CAGR for every group). Conversely, when the expensive sellers are given the ability to modulate the price according to their own energy balance, despite the simple and flawed nature of the algorithm, they secure the rightmost side of the dispersion in the different groups.

4. Conclusions

4.1. Key Findings

Several previous studies have demonstrated that urban PV systems can be economically feasible while covering a large portion of the residential electric demand (i.e., more than 20% or even 30%); this can be achieved in large parts of Europe without incentives and without necessarily adding any electric storage. It has also been found that the aggregation of the demand strongly increases the optimal capacity of PV, regardless of the reward function used for the optimization.

The optimal design of the PV system does not address the problem of financing and distribution of the risks and benefits of the system, especially if it is shared between multiple owners (which is the most promising case to ramp up urban PV capacity). For this reason, an agent-based modelling environment has been developed. The environment is capable of simulating potential PV production, electric demand, and price setting for each household independently and recording their interactions. The behaviour and PV distribution of the agents has been varied in six different scenarios in which the agents have been free to interact. The annual average self-sufficiency of the whole district reaches 24+%, this is remarkably considering the latitude and the absence of incentives and electric storage.

In general, smaller households (with less components and a lower annual cumulative demand) are characterized by higher revenues and lower savings compared to larger households (see Figure 7); this is due to having a larger system comparative to their size and, thus, a frequent over-production of electricity. The smaller household, being more 'sale oriented' benefitted from scenarios in which the price schemes in the micro-grid were high (see Figure 8 (2) and (4) against (1) and (3)). It seems, therefore, that high prices of the electricity within the micro-grid favour a more equal distribution of the risks and benefits. That said, being the households free to choose their price, they are unlikely to move toward a very high price, which is almost the same as the grid. Even more so, because the lowering of the price appears to be the simplest effective strategy to increase the CAGR (see in Figure 8 (5)).

Considering a real case study, it is very likely that some of the households in a micro-grid (which could be located in a multi-family building or a district) would absolutely refuse to invest money in the shared PV system. Simulating this hypothesis in two scenarios (i.e., 3 and 4) showed increases in both revenues and savings for every group (see Figure 7 (3) against (1), and (4) against (2)). This, unsurprisingly, resulted in improved CAGR for those who own a PV system (see Figure 8 (3) against (1), and (4) against (2)). This result is fortunate in the sense that incentives to invest are automatically generated whenever somebody is unwilling to participate in the investment. In this sense, Scenario 4 is a particularly promising one because a limited fraction of the households could form a consortium and provide renewable electricity at the same price of the one offered by the external grid. This could allow the investor to have safe and interesting CAGR (see case 4 assuming a price growth of 2% in the Appendix A), while the non-investors, albeit without any economic advantage, could increase their renewable use without costs nor risks.

In the last scenario it was hypothesized that part of the households, to out-compete the 'cheap sellers', would adopt a dynamic price scheme based on their own energy balance. In practice, they would sell their excess power at a very low price whenever their balance was above average (compared to that hour of the day). Conversely, they would sell their power at the highest possible price whenever their balance is below the average of that hour. This strategy, despite its simplicity, and despite being unaware of the actual balance of the other households, is shown to easily out-compete the 'cheap sellers'. This is remarkable as it proves that effective strategies are possible without invading other household's privacy, but are solely based on data relative to oneself.

4.2. Future Work

The findings from this technique inevitably lead to new questions; in particular, there are four aspects that should be investigated more deeply:

1. It has been shown that reducing the number of PV owners (leaving unchanged the aggregated PV capacity, which is the optimal one) boosts the CAGRs for those who remain. Nevertheless, this has been done only in two scenarios, in which the percentage of owners was invariably 50% instead of 100%. It would be useful to explore an array of different percentages of PV owners combined in different price schemes, thus understanding the phenomenon more thoroughly. Being a very encouraging aspect of the micro-grid, this advantage of the 'rare owners' might be hiding some effective business models.
2. One feature of this local energy market is represented by rule 'e' from Section 2.1. This rule commands that, in the case of insufficient supply of the cheapest source, the power from the cheapest source should be provided proportionally to every agent's demand. Given the disadvantage experienced by smallest consumers, especially when the overall price of the electricity is low, it would be interesting to explore what happens with a different rule. For example, it would be interesting to provide each agent with equal power instead of satisfying the same proportion. With this difference, having a low consumption would actually boost the self-sufficiency considerably, and perhaps lead to a more balanced share of benefits and risks.

3. The dynamic pricing behaviour with which half of the agents were endowed in Scenario 6 demonstrates the effectiveness of a simple dynamic pricing strategy. While only a proof of concept, this strategy is the first step in exploring a large array of behaviors that the agents could assume. It would be interesting to explore the impact of machine learning driven behaviours varying in complexity and in inputs required (both historical and real-time) [50].
4. This study focuses on economic sustainability and fairness of different ownership and pricing schemes. Thus, it assumes the regulatory aspects as capable to allow a fruitful market structure. However, the regulatory design will be essential to achieve such local market, such as metering, and billing/collection, as well as responsibility allocation etc.
5. The results of this study are obtained in a purely residential district; nevertheless, the presence of commercial, office, or public buildings would increase the contemporaneity of production and load. This effect would generally improve the techno-economic performance of the whole system; this improvement should be quantified to allow for spatial planning of the electric infrastructure.

Author Contributions: Conceptualization, methodology, investigation, M.L.; validation, P.H.; writing—original draft preparation, M.L. and P.H.; writing—review and editing, M.L., P.H., C.O., D.Y. and X.Z.; supervision, X.Z.; project administration and funding acquisition, X.Z. All authors have read and agreed to the published version of the manuscript.

Funding: This research has received funding from the European Union's Horizon 2020 research and in-novation programme under grant agreement (EnergyMatching project N° 768766)., and J. Gust. Richert foundation in Sweden (grant number: 2020-00586).

Institutional Review Board Statement: Not applicable.

Informed Consent Statement: Not applicable.

Data Availability Statement: Not applicable.

Conflicts of Interest: The authors declare no conflict of interest.

Appendix A

The following items do not modify the key findings of the study, yet they provide context and additional information to understand the significance of the work done, for instance:

1. The Section 'Practical issues' below deals with some technical and legislative aspects of the modelling presented and tries to offer a link between the model and its application in the real world.
2. Figure A1 shows the relation between the annual cumulative consumption and the ability to exploit the common electricity, this phenomenon is problematic (see Section 4) and requires further study to be solved.
3. Figure A2 shows the growth rates for Scenario 4 (one of the most promising for implementation) considering a linear growth of the price of electricity of 2% per year.
4. Table A1 shows the composition of the 48 households in the study by gender and age bracket.

Practical Issues

The energy community studied in the present paper relies on the assumption that it is possible to own shares of a common PV system without having an actual physical subsystem. In other words, the ownership structure of the system and the relative rights over the energy produced are decoupled by the physical infrastructure. This could be simply realized by assigning to each shareholder, at any point in time, a share of the aggregated production measured from the whole system. In terms of physical infrastructure, a direct current (DC) loop connected to the parent grid through a single inverter with metering capability, or, alternatively, three DC loops (one for each building) that are connected to a unique inverter would be ideal. If the regulation allows it, the ideal installation of the single inverter would be such that it can modulate its aggregated power to balance the 48 loads of the households, so as to reduce or cancel the voltage drop (or even generate an increase in tension) in the existing low-voltage alternating current (AC) infrastructure. In this arrangement, the 48 households will maintain their existing meter. This arrangement would imply that the 48 households will keep paying the fixed distribution costs to the grid operator. In the simulation presented, the household electricity prices are assumed as varying between 1.2 and 1.8 SEK/kWh; these prices represent the part of the electricity bill that can be avoided by using PV electricity and represent the cost for the electricity, the variable distribution costs, and the levies (i.e., does not include the fixed distribution costs). Of course, this arrangement is only feasible if there is the possibility that the grid operator can obtain data from the inverter of the shared system and charge the electric bill accordingly. The arrangement described would not be different, in terms of physical infrastructure, from the existing grid connected PV systems; in fact there would be no physical connection of the loads to the DC side of the inverter. The only difference would be that the voltage regulation in part of the existing low voltage grid would be distributed to the private inverter rather than managed directly by the grid infrastructure. Other options would be feasible, for example having a single inverter at the interface to the parent grid and one small inverter to connect each household to the DC loop. This arrangement would avoid each household maintaining its own connection and meter to the parent grid but would require each household to purchase an inverter to connect to the DC micro-grid. Furthermore, in this arrangement the micro-grid should be authorized to reduce or shut down the equipment in case of missing payment. This would be essential to protect shareholders to pay for energy from the parent grid or from other peers delivered to insolvent shareholders. Ultimately the issues regarding regulation deserve their own consideration in separate studies, the reader interested in this topic might review [27] for a comprehensive list of types of energy communities substantiated by 24 case studies. For a review of forerunner projects see [51], notable the virtual community from Sonnen (batteries manufacturer). For a comprehensive view of different physical infrastructures [31] can be studied. The present study focusses on the economic aspects of the microgrid, it analyses it in terms of economic sustainability and fairness of different ownership and pricing schemes, and thus, it assumes the regulatory aspects as capable to allow a fruitful market structure, although they might not yet be so.

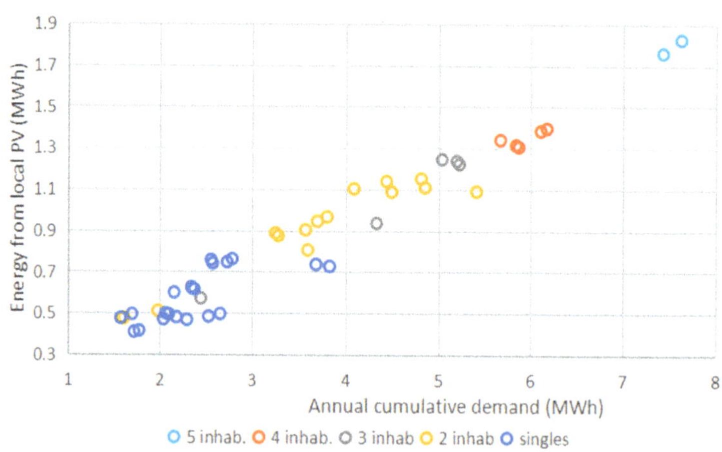

Figure A1. Relation between annual cumulative energy demand and cumulative energy received from the shared system for Scenario 1.

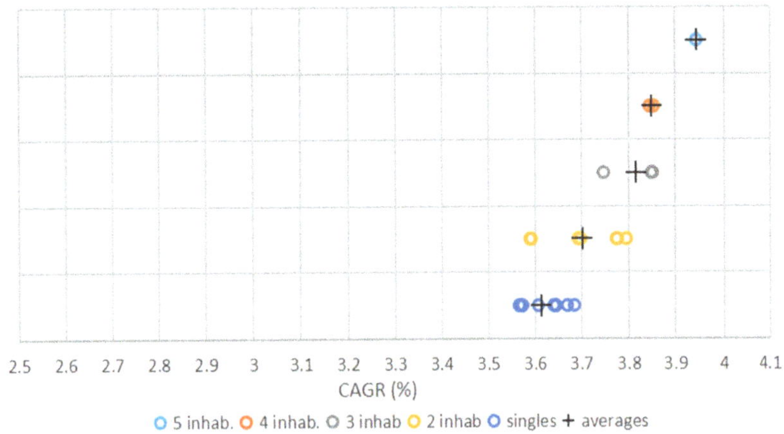

Figure A2. CAGR for Scenario 4 assuming a linear annual growth of the price of electricity for households of 2% of the initial price.

Table A1. Composition by gender and age bracket of the inhabitants in the 48 HH (HouseHolds) in the district.

HH 1	HH 2	HH 3	HH 4	HH 5	HH 6	HH 7	HH 8
Male 25–54 Female 25–54 Male 0–14 Male 0–14 Female 0–14	Male 25–54 Female 25–54 Male 0–14 Female 0–14 Male 0–14	Male 25–54 Female 25–54 Female 0–14 Male 0–14	Male 25–54 Female 25–54 Female 0–14 Male 15–24	Male 25–54 Female 25–54 Female 15–24 Male 0–14	Male 25–54 Female 25–54 Female 0–14 Male 15–24	Male 25–54 Female 25–54 Female 15–24 Male 0–14	Male 25–54 Female 25–54 Female 0–14
HH 9	**HH 10**	**HH 11**	**HH 12**	**HH 13**	**HH 14**	**HH 15**	**HH 16**
Male 25–54 Female 25–54 Male 15–24	Male 25–54 Female 25–54 Female 15–24	Male 25–54 Female >65 Male 0–14	Female 25–54 Female 0–14 Male 15–24	Male >65 Female >65	Male >65 Female >65	Male >65 Female >65	Female 25–54 Male 0–14
HH 17	**HH 18**	**HH 19**	**HH 20**	**HH 21**	**HH 22**	**HH 23**	**HH 24**
Male 25–54 Female 0–14	Male 25–54 Female 25–54	Male 25–54 Female 15–24	Male 55–64 Female 25–54	Male >65 Female 55–64	Male 25–54 Female >65	Male 55–64 Female 25–54	Male >65 Female 55–64
HH 25	**HH 26**	**HH 27**	**HH 28**	**HH 29**	**HH 30**	**HH 31**	**HH 32**
Male 25–54 Female >65	Male 55–64 Female 25–54	Male >65 Female 55–64	Female >65	Male 25–54	Female 25–54	Male 55–64	Female 55–64
HH 33	**HH 34**	**HH 35**	**HH 36**	**HH 37**	**HH 38**	**HH 39**	**HH 40**
Male >65	Female >65	Male 25–54	Female 25–54	Male 55–64	Female 55–64	Male >65	Female >65
HH 41	**HH 42**	**HH 43**	**HH 44**	**HH 45**	**HH 46**	**HH 47**	**HH 48**
Male 15–24	Female 15–24	Male 25–54	Female 25–54	Male 55–64	Female 55–64	Male >65	Female >65

Given the overall socioeconomic conditions of the district where the study is located, it is highly likely that the majority of household will have the ca. 16.3 kSEK (1600 €) to invest for one share (1.36 kWp/household, see Table 1 at the assumed price in Table 4). In case some households are unable or unwilling to pay, the conditions in Scenario 3 or Scenario 4 will happen (see Section 2.1). If, instead, some households are unable yet willing to invest in the system, thus willing to borrow money, the investment will become highly risky whenever the interest rate will be above the 3% used in the optimization phase (see discount rate in Table 4). In areas where the socioeconomic conditions do not allow for investment in the system, the use of public incentives might be the only solution to enable the installation of the system. Since such a system is probably profitable, the incentives do not have to necessarily be direct investments but can simply be borrowings or guarantees to the borrowing institution. These kinds of incentives have the advantage that they only constitute a very limited cost for the public, and primarily mean a social distribution of the risk. For this reason, these types of incentives might be preferred, especially at times of high debt/GDP (Gross Domestic Product).

It so happens that an apartment might change owner or remain vacant. The change of ownership is not modelled in the present framework, and it is unlikely to change the overall profitability of the investment. Anyone, especially if moving in the district, can purchase a share of the system if other households are willing to sell part of their share at an agreed upon price; in this way the residual value of the initial investment is simply bought and exploited by a new owner without changing the work or the profitability of the existing share. Conversely, if the share of the PV system is sold in combination with the property in a change of ownership, it can be resold by the new owner to others if a customer is found at an agreed price. On the other end, if an apartment remaining vacant, or strongly reduces its electricity demand, it can be a threat to the business model. This possibility is strongly affected by the frequency of such events within the community and the size of the community itself, thus it is a subject for parametric study and requires a future work of its own.

References

1. Lovelock, C.E.; Reef, R. Variable Impacts of Climate Change on Blue Carbon. *One Earth* **2020**, *3*, 195–211. [CrossRef]
2. Zhou, P.; Wang, G.; Duan, R. Impacts of long-term climate change on the groundwater flow dynamics in a regional groundwater system: Case modeling study in Alashan, China. *J. Hydrol.* **2020**, *590*, 125557. [CrossRef]
3. Jiang, Q.; Qi, Z.; Tang, F.; Xue, L.; Bukovsky, M. Modeling climate change impact on streamflow as affected by snowmelt in Nicolet River Watershed, Quebec. *Comput. Electron. Agric.* **2020**, *178*, 105756. [CrossRef]
4. Huang, M.; Ding, L.; Wang, J.; Ding, C.; Tao, J. The impacts of climate change on fish growth: A summary of conducted studies and current knowledge. *Ecol. Indic.* **2021**, *121*, 106976. [CrossRef]
5. Leisner, C.P. Review: Climate change impacts on food security- focus on perennial cropping systems and nutritional value. *Plant Sci.* **2020**, *293*, 110412. [CrossRef] [PubMed]
6. He, Q.; Silliman, B.R. Climate Change, Human Impacts, and Coastal Ecosystems in the Anthropocene. *Curr. Biol.* **2019**, *29*, R1021–R1035. [CrossRef]
7. Guo, S.; Yan, D.; Hu, S.; An, J. Global comparison of building energy use data within the context of climate change. *Energy Build.* **2020**, *226*, 110362. [CrossRef]
8. Liu, W.; McKibbin, W.J.; Morris, A.C.; Wilcoxen, P.J. Global economic and environmental outcomes of the Paris Agreement. *Energy Econ.* **2020**, *90*, 104838. [CrossRef]
9. Giddens, A. *The Constitution of Society: Outline of the Theory of Structuration*; University of California Press: Oakland, CA, USA, 1984.
10. Geels, F.W.; Schot, J. Typology of sociotechnical transition pathways. *Res. Policy* **2007**, *36*, 399–417. [CrossRef]
11. Geels, F.W. Technological transitions as evolutionary reconfiguration processes: A multi-level perspective and a case-study. *Res. Policy* **2002**, *31*, 1257–1274. [CrossRef]
12. Suarez, F.F.; Oliva, R. Environmental change and organizational transformation. *Ind. Corp. Chang.* **2005**, *14*, 1017–1041. [CrossRef]
13. Scott, W.R. *Institutions and Organizations: Ideas, Interests, and Identities*; Sage Publications: Thousand Oaks, CA, USA, 2013.
14. Smith, A.; Stirling, A.; Berkhout, F. The governance of sustainable socio-technical transitions. *Res. Policy* **2005**, *34*, 1491–1510. [CrossRef]
15. Bigerna, S.; Bollino, C.A.; Micheli, S.; Polinori, P. A new unified approach to evaluate economic acceptance towards main green technologies using the meta-analysis. *J. Clean. Prod.* **2017**, *167*, 1251–1262. [CrossRef]
16. Atalla, T.; Bigerna, S.; Bollino, C.A.; Polinori, P. An alternative assessment of global climate policies. *J. Policy Modeling* **2018**, *40*, 1272–1289. [CrossRef]
17. Lovati, M.; Salvalai, G.; Fratus, G.; Maturi, L.; Albatici, R.; Moser, D. New method for the early design of BIPV with electric storage: A case study in northern Italy. *Sustain. Cities Soc.* **2019**, *48*, 101400. [CrossRef]
18. Huang, P.; Sun, Y.; Lovati, M.; Zhang, X. Solar-photovoltaic-power-sharing-based design optimization of distributed energy storage systems for performance improvements. *Energy* **2021**, *222*, 119931. [CrossRef]
19. Huang, P.; Zhang, X.; Copertaro, B.; Saini, P.K.; Yan, D.; Wu, Y.; Chen, X. A Technical Review of Modeling Techniques for Urban Solar Mobility: Solar to Buildings, Vehicles, and Storage (S2BVS). *Sustainability* **2020**, *12*, 7035. [CrossRef]
20. Adami, J.; Lovati, M.; Maturi, L.; Moser, D. Evaluation of the impact of multiple PV technologies integrated on roofs and façades as an improvement to a BIPV optimization tool. In Proceedings of the 14th Conference on Advanced Building Skins, Bern, Switzerland, 28–29 October 2021.
21. Vigna, I.; Lovati, M.; Pernetti, R. A modelling approach for maximizing RES harvesting at building cluster scale. In Proceedings of the 4th Building Simulation and Optimization Conference, Cambridge, UK, 11–12 September 2018.
22. Lovati, M.; Adami, J.; Dallapiccola, M.; Maturi, L.; Moser, D. From Solitary Pro Sumers to Energy Community. Quantitative Assesment of the Benefits of Sharing Electricity. *IEEE Trans. Power Syst.* **2019**. [CrossRef]

23. Fina, B.; Auer, H.; Friedl, W. Profitability of PV sharing in energy communities: Use cases for different settlement patterns. *Energy* **2019**, *189*, 116148. [CrossRef]
24. Huang, P.; Lovati, M.; Zhang, X.; Bales, C.; Hallbeck, S.; Becker, A.; Bergqvist, H.; Hedberg, J.; Maturi, L. Transforming a residential building cluster into electricity prosumers in Sweden: Optimal design of a coupled PV-heat pump-thermal storage-electric vehicle system. *Appl. Energy* **2019**, *255*, 113864. [CrossRef]
25. Lovati, M.; Dallapiccola, M.; Adami, J.; Bonato, P.; Zhang, X.; Moser, D. Design of a residential photovoltaic system: The impact of the demand profile and the normative framework. *Renew. Energy* **2020**, *160*, 1458–1467. [CrossRef]
26. Gjorgievski, V.Z.; Cundeva, S.; Georghiou, G.E. Social arrangements, technical designs and impacts of energy communities: A review. *Renew. Energy* **2021**, *169*, 1138–1156. [CrossRef]
27. Caramizaru, A.; Uihlein, A. *Energy Communities: An Overview of Energy and Social Innovation*; Publications Office of the European Union: Luxembourg, 2020.
28. Capellán-Pérez, I.; Johanisova, N.; Young, J.; Kunze, C. Is community energy really non-existent in post-socialist Europe? Examining recent trends in 16 countries. *Energy Res. Soc. Sci.* **2020**, *61*, 101348. [CrossRef]
29. Lowitzsch, J.; Hoicka, C.E.; van Tulder, F.J. Renewable energy communities under the 2019 European Clean Energy Package—Governance model for the energy clusters of the future? *Renew. Sustain. Energy Rev.* **2020**, *122*, 109489. [CrossRef]
30. Moroni, S.; Alberti, V.; Antoniucci, V.; Bisello, A. Energy communities in the transition to a low-carbon future: A taxonomical approach and some policy dilemmas. *J. Environ. Manag.* **2019**, *236*, 45–53. [CrossRef]
31. Roberts, M.B.; Bruce, A.; MacGill, I.J.S.E. A comparison of arrangements for increasing self-consumption and maximising the value of distributed photovoltaics on apartment buildings. *Sol. Energy* **2019**, *193*, 372–386. [CrossRef]
32. Novoa, L.; Flores, R.; Brouwer, J. Optimal renewable generation and battery storage sizing and siting considering local transformer limits. *Appl. Energy* **2019**, *256*, 113926. [CrossRef]
33. Abada, I.; Ehrenmann, A.; Lambin, X. On the viability of energy communities. *Energy J.* **2020**, 41. [CrossRef]
34. Sadeghian, H.; Wang, Z. A novel impact-assessment framework for distributed PV installations in low-voltage secondary networks. *Renew. Energy* **2020**, *147*, 2179–2194. [CrossRef]
35. Awad, H.; Gül, M. Optimisation of community shared solar application in energy efficient communities. *Sustain. Cities Soc.* **2018**, *43*, 221–237. [CrossRef]
36. Qian, M.; Yan, D.; Liu, H.; Berardi, U.; Liu, Y. Power consumption and energy efficiency of VRF system based on large scale monitoring virtual sensors. *Build. Simul.* **2020**, *13*, 1145–1156. [CrossRef]
37. Hu, S.; Yan, D.; Guo, S.; Cui, Y.; Dong, B. A survey on energy consumption and energy usage behavior of households and residential building in urban China. *Energy Build.* **2017**, *148*, 366–378. [CrossRef]
38. Jin, Y.; Xu, J.; Sun, H.; An, J.; Tang, J.; Zhang, R. Appliance use behavior modelling and evaluation in residential buildings. A case study of television energy use. *Build. Simul.* **2018**, *13*, 787–801. [CrossRef]
39. Lovati, M.; Zhang, X.; Huang, P.; Olsmats, C.; Maturi, L. Optimal Simulation of Three Peer to Peer (P2P) Business Models for Individual PV Prosumers in a Local Electricity Market Using Agent-Based Modelling. *Buildings* **2020**, *10*, 138. [CrossRef]
40. EnergyMatching. European Union's Horizon 2020 Research and Innovation Programme under Grant Agreement N° 768766. Available online: https://www.energymatching.eu/project/ (accessed on 20 October 2020).
41. Huang, P.; Lovati, M.; Zhang, X.; Bales, C. A coordinated control to improve performance for a building cluster with energy storage, electric vehicles, and energy sharing considered. *Appl. Energy* **2020**, *268*, 114983. [CrossRef]
42. Pflugradt, N.; Muntwyler, U. Synthesizing residential load profiles using behavior simulation. *Energy Procedia* **2017**, *122*, 655–660. [CrossRef]
43. Xu, J.; Kang, X.; Chen, Z.; Yan, D.; Guo, S.; Jin, Y.; Hao, T.; Jia, R. Clustering-based probability distribution model for monthly residential building electricity consumption analysis. *Build. Simul.* **2021**, *14*, 149–164. [CrossRef]
44. Fan, C.; Yan, D.; Xiao, F.; Li, A.; An, J.; Kang, X. Advanced data analytics for enhancing building performances: From data-driven to big data-driven approaches. *Build. Simul.* **2021**, *14*, 3–24. [CrossRef]
45. Statistics Sweden. Sveriges Officiella Statistik: Personer Efter Hushållstyp, Hushållsställning och kön 31 December 2019 (Excel File). Available online: https://www.scb.se/hitta-statistik/statistik-efter-amne/befolkning/befolkningens-sammansattning/befolkningsstatistik#_Tabellerochdiagram (accessed on 20 October 2020).
46. Central Intelligence Agency. The World Factbook. 2020. Available online: https://www.cia.gov/library/publications/resources/the-world-factbook/index.html (accessed on 16 November 2020).
47. Imenes, A.G. Performance of BIPV and BAPV Installations in Norway. In Proceedings of the IEEE 43rd Photovoltaic Specialists Conference (PVSC), Portland, OR, USA, 5–10 June 2016; pp. 3147–3152.
48. López, C.S.P.; Frontini, F.; Friesen, G.; Friesen, T.J.E.P. Experimental testing under real conditions of different solar building skins when using multifunctional BIPV systems. *Energy Procedia* **2014**, *48*, 1412–1418. [CrossRef]
49. Masson, G.; Kaizuka, I.; Detolleanere, A.; Lindahl, J.; Jäger-Waldau, A. *2019–Snapshot of Global Photovoltaic Markets*; IEA International Energy Agency: Paris, France, 2019.
50. Hu, S.; Yan, D.; Azar, E.; Guo, F. A systematic review of occupant behavior in building energy policy. *Build. Environ.* **2020**, *175*, 106807. [CrossRef]
51. Zhang, C.; Wu, J.; Long, C.; Cheng, M.J.E.P. Review of existing peer-to-peer energy trading projects. *Energy Procedia* **2017**, *105*, 2563–2568. [CrossRef]

buildings

MDPI

Article

Digital Mapping of Techno-Economic Performance of a Water-Based Solar Photovoltaic/Thermal (PVT) System for Buildings over Large Geographical Cities

Santhan Reddy Penaka [1,2], Puneet Kumar Saini [1,3], Xingxing Zhang [1,*] and Alejandro del Amo [4]

1 Energy Technology, Dalarna University, 79188 Falun, Sweden; santhanreddypenaka@gmail.com (S.R.P.); pks@du.se (P.K.S.)
2 Penaka Solar, 78452 Borlänge, Sweden
3 Department of Engineering Science, Uppsala University, 75236 Uppsala, Sweden
4 Abora Solar Company, 50196 Zaragoza, Spain; adelamo@abora-solar.com
* Correspondence: xza@du.se; Tel.: +46-(0)23-77-87-89

Received: 25 June 2020; Accepted: 21 August 2020; Published: 27 August 2020

Abstract: Solar photovoltaic thermal (PVT) is an emerging technology capable of producing electrical and thermal energy using a single collector. However, to achieve larger market penetration of this technology, it is imperative to have an understanding of the energetic performance for different climatic conditions and the economic performance under various financial scenarios. This paper thus presents a techno-economic evaluation of a typical water-based PVT system for a single-family house to generate electricity and domestic hot water applications in 85 locations worldwide. The simulations are performed using a validated tool with one-hour time step for output. The thermal performance of the collector is evaluated using energy utilization ratio and exergy efficiency as key performance indicators, which are further visualized by the digital mapping approach. The economic performance is assessed using net present value and payback period under two financial scenarios: (1) total system cost as a capital investment in the first year; (2) only 25% of total system cost is a capital investment and the remaining 75% investment is considered for a financing period with a certain interest rate. The results show that such a PVT system has better energy and exergy performance for the locations with a low annual ambient temperature and vice versa. Furthermore, it is seen that the system boundaries, such as load profile, hot water storage volume, etc., can have a significant effect on the annual energy production of the system. Economic analysis indicates that the average net present values per unit collector area are 1800 and 2200 EUR, respectively, among the 85 cities for financial model 1 and financial model 2. Nevertheless, from the payback period point of view, financial model 1 is recommended for locations with high interest rate. The study is helpful to set an understanding of general factors influencing the techno-economic performance dynamics of PVT systems for various locations.

Keywords: PVT; water-based PVT; techno-economic analysis; digital mapping

1. Introduction

1.1. Background and Existing Studies

The concept of "electrify everything" considers solar energy as a key renewable technology with an aim of de-carbonization of domestic heating demand [1]. The rapid growth in photovoltaic (PV) installation capacity from the last few years has further strengthened the importance of PV as the main driver of renewable transformation [2]. PV remains an interesting subject area for many researchers, global leaders, and manufacturers because of its reliability, sustainability, ease of installation, and economic feasibility [3]. However, the concurrence of heat/electricity demand and limited roof area in domestic dwellings does require technologies which can generate energy efficiently in both thermal and electrical form. Therefore, there is a huge potential for well-designed systems by combining both solar PV and solar thermal technologies. A relatively new commercialized concept of solar photovoltaic/thermal (PVT) technology can achieve such a goal by generating both electrical and thermal energy together using a single panel [4]. Realizing its importance, the Solar Heating and Cooling Program (SHC) of the International Energy Agency (IEA) has initiated Task 60 for PVT applications and solutions to Heating, Ventilation and Air Conditioning (HVAC) systems in buildings [5]. The task has been active from January 2018 and has built a huge knowledge base around PVT systems for its use in domestic and industrial applications.

PVT systems can be categorized in several ways, however, the most common is based on the heat-transfer medium (air-based/liquid-based) used in the PVT collector [6]. The liquid-based types are dominating the current PVT market in terms of the number of installations due to high efficiency, and ease of integration in existing hydronic systems [7]. In a standard liquid-based PVT collector, the heat carrier is usually water or brine mixture, which is allowed to circulate in a heat exchanger behind the PV cells. The circulation results in a heat transfer through the back sheet of the module, which raises the fluid temperature enough to use for various applications such as, e.g., hot water and swimming pool heating. From a technical perspective, PVT technology is well developed, and it can be coupled with various energy systems. For instance, it can go hand-in-hand with the emerging awareness of heat pump technology with/without borehole storage [8]. However, the current main barriers in PVT development and deployment are lack of testing standards, uncertain financial incentives, and business models across different regions in a niche market. Therefore, the business potential of PVT solution has not been fully explored, although it can be a very efficient solution for domestic and industrial heating requirements.

There are several studies concerning the techno-economic analysis of PVT collectors with a focus on the component and system design [4,9–12]. The most common way is to assess the energetic performance firstly and then carry out an economic evaluation based on dependent variables [4,9,10,12–16]. The prevalent energy performance indexes are energy efficiency and exergy efficiencies [6] while the most popular economic indicators are represented by levelized cost of energy (LCOE), net present value (NPV), and payback period [4]. To name a few studies for technical evaluation, Fudholi et al. [13] investigated electrical and thermal performances on PVT water-based collectors by testing with specific inputs parameters ranging from 500 to 800 W/m^2 solar irradiance and mass flow rate of 0.011 to 0.041 kg/s. The test concluded that absorber performed better at a mass flow rate of 0.041 kg/s and under 800 W/m^2 irradiance, with a measured PV efficiency of 13.8%, thermal efficiency of 54.6%, and overall collector efficiency of 68.4% [13]. Shah and Srinivasa [17] developed a theoretical model using COMSOL multi-physics validation tool with standard test conditions (STC) to measure the PV improved efficiency when it is integrated with hybrid PVT system. Another study performed by Buonomano [18] developed a numerical model to conduct the technical and economic analysis of PVT collectors and compared it with conventional PV collectors installed in Italy. The tool was validated using TRNSYS platform for the energetic and economic performance of systems integrated with PV and PVT collectors together. Yazdanpanahi [19] presented a numerical simulation and experimental validation for evaluation of PVT exergy performance using a one-dimensional steady thermal model and a four-parameter current–voltage model for a PVT water collector. In terms of economic studies, Gu et al. [4] developed an analytical model on basis of combinations of Monte Carlo method to analyze techno-economic performances of solar PVT concentrator for Swedish climates, which considered several essential input uncertainties whereas economic variables were initially assessed. The developed model has expressed results for capital cost range between 4482 and 5378 SEK/m^2 for 10.37 m^2 system cost during the system lifespan of 25 years. The paper results indicated an LCOE of 1.27 SEK/kWh and NPV of 18,812 SEK with a simple payback period of 10 years. It was concluded that the most important sensitivity factor is average daily solar irradiation followed by debt to equity ratio, capital price, regional heating price, and discount rate. Herrando et al. [20] performed techno-economic analysis of hybrid PVT systems for electricity and domestic hot water (DHW) demand for a typical house in London and concluded that such systems can meet 51% of electricity demand and 36% of DHW demand even during low solar global horizontal irradiation (GHI) and ambient temperatures. In the economic aspect, it was also concluded that hybrid PVT technology has better energy yield per unit roof area, which can result in attractive NPV for investor while mitigating the CO_2 emissions. Riggs et al. [10] developed a combined LCOE techno-economic model for different types of hybrid PVT systems applied for process heat application in the United States. The sensitivity analysis of parameters affecting the levelized cost of heat (LCOH) was determined using technical, financial, and site-specific variables. Ahn et al. [21] studied the importance of energy demands, solar energy resources, and economic performances of hybrid PVT systems at different PV penetration levels using Monte Carlo method, whereas the study found that irrespective of PV penetration levels, the uncertainties in energy demands and solar irradiance can influence the energy performance of PVT systems. Heck et al. [22] conducted Monte Carlo method for LCOE based on probability distribution, which concluded that this method provides more realistic information on risk/uncertainty, which triggers more scope of potential investment on electricity generation. However, author defended that the method is slightly complex to use point values.

There is more literature available regarding PVT techno-economic performance than what is presented in this study. However, most of the existing studies focused on a single climate, with a straightforward economic–financial analysis. Furthermore, complicated procedures or individual software (e.g., TRNSYS, Polysun) are used to estimate the performance of PVT collectors, which require detailed modelling skills, and higher computation time. There is a lack of a comprehensive simulation of PVT techno-economic performance through a common tool over a large geographic area, aiming for application feasibility and business potentials. Moreover, many studies have reported the solar energy resource potential of buildings at different spatial scales using digital mapping methods, such as digital numerical maps [23], digital surface model [24], satellite imageries and geographic information systems [25,26], and multi-scale uncertainty-aware ranking of different urban locations [27], which provide direct evaluations for solar application, leading to robust planning decisions. Nevertheless, no study has yet been found for mapping of techno-economic performance of PVT systems.

As a result, this paper aims to fill this research gap by utilizing a validated simulation tool to perform a comprehensive techno-economic performance simulation for a wide range of cities. The results are further analyzed and visualized using a digital numerical mapping approach to establish a comparison among various regions.

1.2. Aim and Objectives

This study aims at simulation and mapping of the energetic and economic indicators of a typical PVT system over different regions to establish a digital performance database for various key performance indicators (KPIs). The economic feasibility of the PVT collector is obtained and compared under various financial scenario models. The data obtained from simulations are used to establish a simple correlation between variables affecting the PVT system.

The main objectives of this paper are to:

(1) Assess the thermal and electrical performance of a typical PVT system [6] in 85 large geographical cities using a validated simulation tool.
(2) Evaluate the economic performance using NPV and payback period using two financial scenarios.
(3) Analysis and visualization of energy and economic performance.

The significance of this paper lies in (1) understanding of typical PVT components behavior at the system level and (2) mapping of the collector energetic and economic performance for different climatic conditions across the world. This research results would reflect the concrete developments in this subject area and help the promotion of potential markets, e.g., discovering the economic feasibility of the PVT system and feasible financial solutions to the PVT system in different regions. This paper evaluates the related business benefits of a typical PVT system, which would help to develop a database as repository of PVT performances in different regions and contexts. The research results will be useful for researchers, planners, and policymakers to further evaluate PVT potentials in a net-zero/positive-energy district towards energy surplus and climate neutrality.

2. System Description and Research Methodology

2.1. Water-Based PVT Collector

Among the different types of PVT technology, the water-based PVT is the most common one that has great possibilities for system integration [28]. This PVT collector type is structured similarly to the typical flat-plate collector, as shown in Figure 1. It is a sandwiched structure comprising several layers, including a glass cover placed on the top, a layer of PV cells or a commercial PV lamination laid beneath the cover with a small air gap in between, heat-exchanging tubes or flowing channels through the absorber and closely adhered to the PV layer, and a thermally insulated layer located right below the flow channels. All layers are fixed into a framed module using adequate clamps and connections. In the heat-exchanging tubes, water is the most commonly used heat carrier medium due to high specific heat capacity and ease of availability. The glass cover is often optional depending on the system design priority for the type of output required (i.e., electricity or heat). The glass cover helps to reduce heat convection losses, but it also causes high solar reflectance losses and thus lowers optical efficiency. In many cases, the glass cover is used when higher heat output is expected, while it is removed when the system is optimized for higher electrical output.

The electrical efficiency of PV cells increases when the pumped cooled water flows across the rigid series or parallel tubes. The flow control is an important factor to achieve overall high performance of the PVT collectors [29]. In addition to electricity production, hot water is generated by absorbing extra heat from the PV layer, which can be used for several applications. The electrical and thermal efficiencies of PVT generally depend on the PV cell type, fluid temperature, fluid flow rate, flow channel size/configuration, and ambient climatic conditions. The collector energetic performance can be measured in terms of energy utilization ratio and exergy efficiency [19].

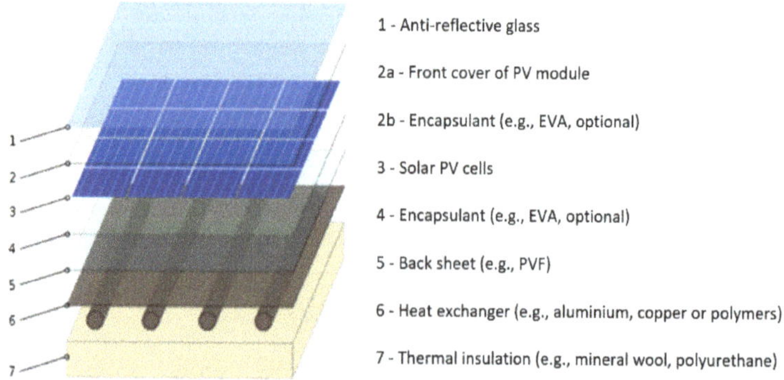

Figure 1. Schematic cross-section of a covered flat-plate photovoltaic thermal (PVT) collector [30].

This paper will focus on a typical PVT collector developed by a Spanish manufacturer named Abora solar. The collector is available on the market, and more than 5700 m^2 of the gross collector is installed for a broad range of applications. The collector is a covered PVT type with an additional layer of glass on the top of the collector (in addition to a glass layer for PV cells) to reduce the heat convection losses. The rated power of the collector is 365 W at standard test conditions (STC) with a collector area of 1.96 m^2 consisting of 72 monocrystalline cells. The main specifications and characteristics of analyzed PVT collector are shown in Table 1.

Table 1. Specifications and characteristics of the modeled PVT collector.

Parameter	Description
Length × width × thickness	1970 mm × 995 mm × 107 mm
Gross collector area	1.96 m^2
Number of PV cells	72
Cell type	Monocrystalline
Rated power	365 Wp
Electric efficiency at STC	17%
Thermal efficiency at STC	70%
Temperature coefficient of PV	−0.41%/°C
Thermal efficiency at zero mean temperature	0.7
Coefficient of thermal losses, a_1	5.98 W/m^2·K
Coefficient of thermal losses, a_2	0.021 W/m^2·K^2
Internal water volume	1.78 L

2.2. Key Performance Indicators

The performance of such PVT collectors is evaluated using standard key performance indicators. The performance of a collector over a specified period can be quantified using the energy utilization ratio (η_e), which is defined as below [31]:

$$\eta_e = \frac{Output\ energy_{electrical}}{GHI \times collector\ area} + \frac{Output\ energy_{thermal}}{GHI \times collector\ area} \tag{1}$$

where GHI is global horizontal irradiation (kWh/m^2), and the collector area is in m^2. However, the exergy value of both electricity and heat is different. Electricity can be regarded as pure exergy whereas heat contains some exergy value. To account for this, "energy" is replaced by "exergy", which has the drawback of being somewhat less intuitive. The overall exergy efficiency takes into account the difference of energy grades between heat and electricity and involves a conversion of low-grade thermal energy into the equivalent high-grade electrical energy using the theory of the Carnot cycle. The overall exergy of the PVT (ε_e) . is defined as following expression:

$$\varepsilon_e = \eta_c\, \eta_{th} + \eta_{el}. \tag{2}$$

Carnot efficiency η_C (%) is defined in the following Equation (3)

$$\eta_C = 1 - \frac{T_{in}}{T_{out}} \tag{3}$$

where η_{th}, η_{el}, T_{out}, and T_{in} are thermal efficiency, electrical efficiency, outlet fluid temperature, and inlet fluid temperature, respectively.

NPV is defined as a measurement of cumulative profit calculated by subtracting the present values of cash outflows (including initial cost) from the present values of cash inflows over the PVT collector's lifetime. In this paper, we use NPV to evaluate a single investment to evaluate the acceptability of the project [4]. A positive NPV indicates that the projected earnings generated by a project or investment, exceed the anticipated costs. In general, an investment with a positive NPV will be a profitable one, and the higher NPV means higher benefits. This concept is the basis for the NPV decision rule, which dictates that the only investments that should be made are those with positive NPV values. NPV is calculated using Equation (4) as below:

$$NPV = \sum_{t=0}^{n-1} \frac{CF_t}{(1+r)^t} - C_0 \tag{4}$$

where, CF_t, r, n, t, and C_0. are the cash flow of particular year (SEK), discount rate, number of years, year of NPV evaluation, and capital cost, respectively.

The payback period is the time for a project to break even or recover its initial investment funds, where the cash flow starts to turn positive and can be given as in Equation (5).

$$PP = T_{(CF_t > 0)} \tag{5}$$

2.3. Research Methodology

The simulation is carried using a validated tool developed by the manufacturer of the studied PVT collector. The Abora hybrid simulation tool [32] was used to map the performance across 85 cities shown in Figure 2. The cities were chosen based on population density and geographical coordinates in different countries to represent a large market potential in these regions. A large number of selected locations for analysis are concentrated within Europe, with limited locations in India, United States, and Australia. The selection of locations is also restricted due to the availability of weather and GHI data in the simulation tool. The simulation tool accepts a wide range of design and financial input parameters, e.g., location and weather resources, electrical and thermal demands, local energy tariffs, specific storage volume, PVT panel and installation parameters, interest rate and financing period, etc. The complete list of various inputs used is shown in Table 2. The performance model used in the tool for evaluation of PVT performance is validated in [24], where a heat pump system integrated with 25 PVT modules was monitored, and measurements were also compared with the dynamic simulation model built in TRNSYS for Zaragoza, Spain. This model has observed thermal and electrical performance of collectors is accurate with measured data (4.2% deviation), however, a slightly higher deviation in heat pump performance was noted due to limitations in the black-box model of the heat pump in the studied energy system.

Figure 2. The simulated locations for techno-economic analysis.

This paper further applies the digital numerical map approach based on heat maps to visualize the performance of various indicators across simulated locations. The simulation results for all locations are exported to Microsoft Excel for calculations of energy and exergy efficiency [33]. After this, the results are visualized using QGIS tool, which provides a heat map rendering to design point layer data with a kernel density estimation processing algorithm [34]. Initially, a parametric study of the components at system level is considered according to the operation flow of the simulation tool indicated in the flow chart shown in Figure 3. Then, the simulations are carried with defined boundary conditions and the results are represented subsequently as monthly electrical and thermal performances, energy savings, economic parameters such as NPV, and payback period.

Table 2. Technical and economic input parameters.

Technical Parameters	Economic Input Parameters
Type of application (domestic/industrial)	Type of mounting structure
Type of demand (hot water/space heating)	Type of inverter
Type of auxiliary system	Material profit margin
Number of bedrooms	Operation and maintenance margin
DHW temperature	Pricing of all system components
Dwellings occupancy	Annual maintenance cost
Number of collectors	Electricity price increment
Collector tilt	Auxiliary fuel price increment
Collector azimuth	Financing period models
Storage tank volume	Interest rate
Meteorological parameters (irradiation/ambient temperature/albedo, etc.)	Opening interest rate
Shadow loss percentage	
Number of additional PV panels	

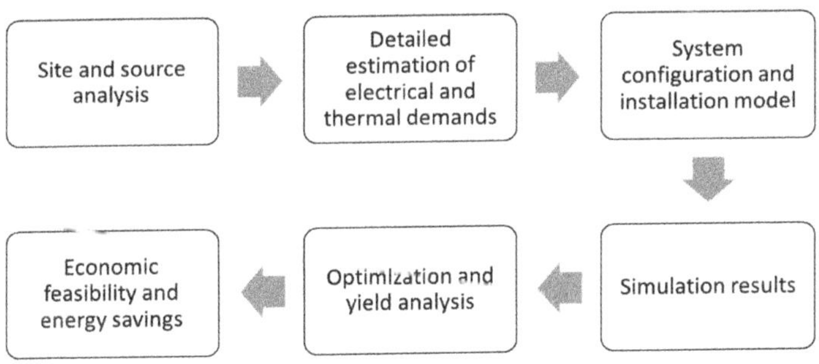

Figure 3. Operation flow of the simulation tool.

This paper also considers the economic performance of the collector in two different financial models, which are described below:

- Model 1: The total system cost is invested in the first year.
- Model 2: Only 25% of total system cost is a capital investment and the remaining 75% investment is considered with the financing period with a certain interest rate.

The economic analysis results highlight the economic parameters, such as NPV and payback period per unit collector area, for all locations. Furthermore, the uncertainty and sensitivity parameters are discussed, and the strategy in decision-making for investing in PVT technology is recommended. The digital mapping method is applied to compile and format the techno-economic performance data into a virtual image, which aims to produce a general map with KPIs of such a PVT system that gives appropriate representations of the dedicated areas.

3. Simulation Tool and Boundary Conditions

3.1. Location and Detailed Demand Analysis

The simulation tool considers the Meteonorm [35] weather database to determine solar and meteorological resources, such as GHI, ambient temp, and wind speed. The thermal and electrical demands change with different categories of buildings, i.e., single and multifamily houses, tertiary buildings (such as hospitals, hotels, and gyms, etc.), and can be selected individually within the tool interface. Specific key parameters are included, such as load profiles, the current auxiliary source of electricity, and energy system details. The simulation engine assesses the total monthly and annual total demand depending on inputs for each application. The monthly energy load (L) needed to raise the temperature of supply water to the desired hot water temperature is calculated using Equation (6):

$$L = m \times C_p \times N \times (T_d - T_s) \tag{6}$$

where 'm' indicates the amount of hot water required per person in a day (in liters), 'C_p' is the specific heat capacity (J/kg·K), 'N' is several days in a month (days), 'T_d' is desired water temperature (°C), and 'T_s' cold supply water temperature in (°C). The monthly demand can also be customized based on consumer utilization in that specific month. For a single-family house, the amount of DHW for one person in a day is considered as 28 L/person/day at 100% occupancy. The demand is kept constant to minimize the variables in the overall system and, thus, to have a fair comparison of collector performance for various locations. The fraction of occupancy can be parameterized to meet the specific thermal demand for the individual location. For tertiary buildings (such as industrial applications), tools consider a different consumption depending on process characteristics.

This simulation tool offers to choose an auxiliary heating system to meet the load demand. This tool also accommodates for the fact that the total collector electricity generation can be utilized for self-consumption or if there is excess electrical energy, it can be sold to the electricity grid in the context of a positive-energy building.

3.2. *System Variables*

This simulation tool consists of several PVT collectors and also recommends the number of collectors that would be required based on optimization of total demand and the storage tank capacity. The specific volume capacity (v/a), which is ratio of tank volume (liter) to collector gross area (m^2) can be changed depending on the number of storage duration hours.

The shading loss fraction on PVT modules can be adjusted manually. There is the provision to integrate PV and PVT collectors in a scenario if the thermal demand is first fully met by PVT modules, and electrical demand is not fully covered.

3.3. *Working Principle of the Simulation Tool*

The simulation tool also optimizes the collector and installation parameters based on the demand, availability, and metrological conditions for a particular location. Simulation results highlight essential parameters such as GHI, irradiation on a tilted surface, thermal demand, thermal production, thermal solar coverage, electrical production, total electric and thermal savings, and environmental impact. The maximum power point P_m (in kW) generated by the PV cells is obtained using Equation (7) depending on the global irradiation on the surface of the module G (W/m^2), ambient temperature T_a (°C), cell temperature T_c (°C), nominal power of photovoltaic collector P_n (kW), G_{STC} irradiance under STC (W/m^2), i.e., 1000 W/m^2, and the temperature variation coefficient of power ($\gamma\gamma$) (%/°C) [36].

$$P_m = P_n \times \frac{G}{G_{STC}}(1 - \gamma(T_c - 25)) \tag{7}$$

The cell temperature T_c is linked to the temperature of the absorber plate, which is dependent on the temperature of fluid going in and out of the module. Cell temperature is calculated for each simulation time step based on inlet and outlet temperatures, and electrical output is then calculated depending on the temperature coefficient of the module.

The instantaneous thermal efficiency of the collector is calculated based on Equation (8)

$$\eta_{th} = \eta_o - a_1\left(\frac{T_m - T_a}{G}\right) - a_2\left(\frac{(T_m - T_a)^2}{G}\right) \tag{8}$$

where η_o is optical efficiency, a_1 is first order heat loss coefficient (W/m^2·K), a_2 is the second order heat loss coefficient (W/m^2·K^2), T_m is the average fluid temperature (°C), and T_a is ambient temperature

(°C). The various characteristics of the simulated module are listed in Table 1 and are validated by real measurements as explained in [25].

The temperature leaving the PVT module T_o is determined using Equation (9)

$$T_o = T_i + \left(\frac{m \cdot C_p}{G \cdot \eta_{th}}\right) \tag{9}$$

where T_i, m, and C_p. represents inlet temperature (°C), fluid mass flow rate (kg/s), and fluid specific heat (kJ/kg·K), respectively. Thermal solar coverage (T_{solar}) is calculated using Equation (10) in this simulation tool

$$T_{solar}\ (\%) = \frac{\text{Total collector thermal production (kWh)}}{\text{Total thermal demand (kWh)}} \times 100. \tag{10}$$

3.4. System Pricing and Optimization

The detailed system cost of the PVT system is defined by customizing each component, such as flat or tilted mounting structure, single-phase or three-phase inverter, material marginal rate, electrical and combustible price escalation rate, annual maintenance cost, etc.

The simulation considers the appropriate dynamic inputs and generates the report of assessment on the key economic performance indicators, i.e., lifetime cash flow with appropriate total annual savings, NPV, and payback period. This simulation tool allows collector economic performance with several financing options shown in Figure 4. For instance:

- The total system cost is invested in the first year as a capital investment.
- The 100% of total system cost can be invested in several years with monthly payment at a certain open and fixed interest rate.
- The 75% of total system cost can be invested in several years with monthly payment at a certain open and fixed interest rate and the remaining 25% of total system cost is to be invested initially as capital investment.

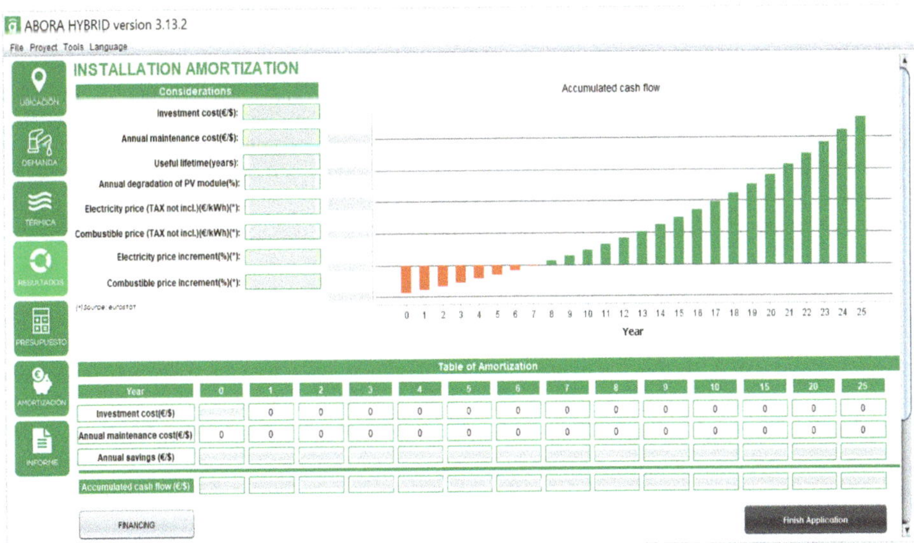

Figure 4. Cost optimization of the PVT system in the simulation tool.

This simulation tool is also flexible in customizing several real-time scenarios, i.e., the number of payments in a single year and the total number of payments in the entire financing period. The early cancellation interest rate can be applied when the system is to be dismantled during the financing period.

3.5. Boundary Conditions

This section pre-determines the boundary conditions for the simulation as shown in Table 3.

Table 3. Boundary conditions for the simulation tool.

Parameter	Description
Type of application	Single-family house
Type of demand	Electricity demand and thermal demand for DHW
Auxiliary system	Electrical heater
Auxiliary system energy price	This is selected individually for each location
No. of people in house	5
DHW temperature	60°
PVT Collector model	aH72SK
No. of collectors	1
Specific volume capacity	80 L/m^2
Inclination	Selected optimally based on a parametric study for maximum energy production
Type of mounting structure	Tilted
Type of inverter	Single-phase inverter
Annual maintenance cost	Assumed that no maintenance is required for a single collector to reduce uncertainties
Electricity and combustible price increment	6% per year is assumed for all the location
System lifetime	25 years
Interest rate	Selected appropriately for each location

Initially, the energy performance of the PVT system is simulated in 85 different locations using the simulation tool. In order to discover and compare the collector energy performance in different locations, the thermal demand is maintained the same in all selected locations. Therefore, the simulated system considers a single PVT collector (1.96 m^2), for a single-family house application with 5 people, for the same demand, and the same tank volume for all locations. These assumptions provide a common system boundary to understand the effect of climatic variables and financing parameters on collector performance. Two types of demands are considered as DHW and electricity use in the building. In the electricity model, no price difference in self-consumed and exported power to the grid is considered. In the thermal system configuration, the auxiliary source for the house is the electricity grid with appropriate energy prices for every location. The generated DHW by the collector is utilized for household purposes using a storage tank connected to the auxiliary system which will deliver demand at the desired temperature of 60 °C, as shown in Figure 5. For each location, the installed tilt and azimuth angles are taken optimally based on higher collector production. The specific volume capacity is assumed 80 L/m^2 for all the locations which is equivalent to total 150 L of storage tank capacity.

In the proposed simplified energy system, PVT collector is directly connected to the tank without any internal or external heat exchanger. The cold water from the tank enters the PVT module, exchanges heat from the absorber, and hot water is fed to the top of the tank. The DHW cold water enters at bottom of the tank, and hot water leaves from top of the tank for DHW supply in the building. The DHW distribution system and associated heat losses are not considered in the analysis. The maximum DHW supply temperature is set at 60 °C, and an electric auxiliary heater is provisioned in the tank for periods when the energy from PVT modules is not enough to meet the DHW load. Electric heater starts and stops at the determined dead band to optimize energy consumption while maintaining the fixed supply DHW temperature. During the periods when tank temperature exceeds the set limit, the energy from PVT modules is fed to a heat sink (air/water heat exchanger), and this spilled energy from the collector is not counted as part of useful energy output.

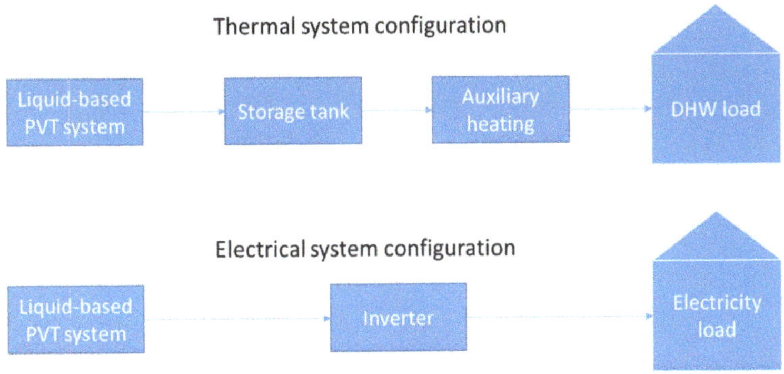

Figure 5. Thermal and electrical system configurations.

In the electrical system configuration, the generated DC power will be converted to AC power using an inverter. Then, it is utilized for household purposes and the remaining will be sent to the electricity grid, whereas the excess electricity demand is taken from the grid connection as shown in Figure 5. As the tilt angle of the PVT collector is a key parameter that will also decide the collector production, a preliminary parametric study is carried for each location to determine the optimal tilt angle for maximum annual collector production.

The total system cost is determined using variables such a module cost, system components cost, annual operation, and maintenance cost. The electricity and auxiliary energy price escalation is assumed to be 6% per year for all the locations. Various parameters considered for economic analysis are shown in Table 4.

Table 4. Parameters considered for economic analysis

Parameter	Value
Abora PVT collector	350 EUR
Cost for Connection kit	128 EUR
Tilted mounting structure	243 EUR
Storage tank	1553 EUR
Valve (servo meter)	127 EUR
Flowmeter	142 EUR
Copper tubes	19 EUR
Isolation tubes	14 EUR
Heat sink	474 EUR
Microinverter	500 EUR
Legal regulations	377 EUR
Electricity price increment	6% annually
System lifetime	25 years
Electricity price	Variable based on each location

The payback time and NPV are estimated by considering a reference system using an electric heater. The price of electricity considered for various locations is shown in Figure 6 below.

The economic performance of the collector in two different financial models is evaluated based on:

- Model 1: The total system cost is invested as initial capital investment in the first year;
- Model 2: 25% of total system cost is capital investment and remaining 75 % is paid within financial period of 7 years with a certain variable interest rate with every location.

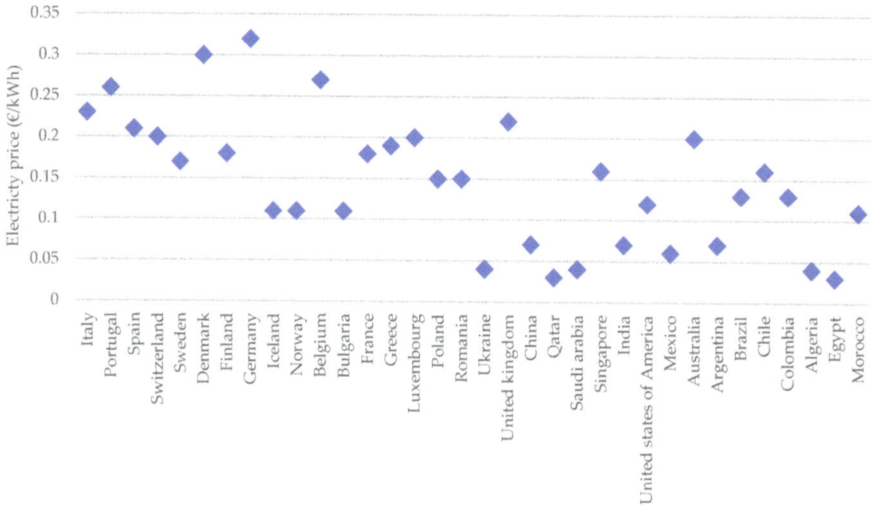

Figure 6. Considered electricity prices in all countries [37].

4. Results and Discussion

This section details the simulation results using the digital mapping approach. Table 5 shows the inputs and results of key performance indicators for all selected locations, and the results are discussed.

4.1. Energy Performance Evaluation of PVT Panel

4.1.1. Collector Thermal Production

The simulated results are visualized using geospatial maps, as they provide clear indication for understanding regional trends for thermal and electrical output even in the case of large datasets. Figure 7 shows the variation in the thermal output of the collector.

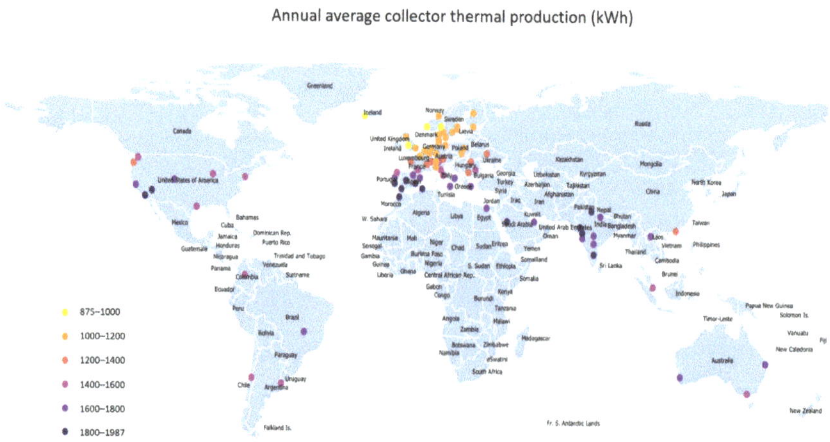

Figure 7. Annual average collector thermal performance.

Table 5. All simulated data of key performance indicators.

Country	City	Latitude	Annual GHI (kWh)	Annual Average Temperature (°C)	Annual Thermal Production (kWh)	Annual Electrical Production (kWh)	NPV per Unit Collector Area for Financial Model 1 (EUR)	NPV per Unit Collector Area for Financial Model 2 (EUR)
Italy	Catania	38	1967	18	1790	487	5140	5541
	Florence	44	1632	16	1520	413	4039	4451
	Milan	45	1233	12	1153	317	2528	2955
	Rome	42	1585	17	1464	401	3797	4211
	Bari	41	1824	17	1679	458	4691	5096
Portugal	Lisbon	39	1939	18	1770	483	4766	5171
	Porto	41	1765	16	1640	447	4246	4657
	Setubal	39	1997	18	1823	495	4966	5368
Spain	Sevilla	37	2134	20	1882	520	4972	5361
	Valencia	39	2043	18	1831	505	4776	5167
	Zaragoza	42	2002	16	1795	498	4649	5041
	Barcelona	41	1904	18	1728	479	4387	4782
	Lugo	43	1567	13	1464	406	3393	3798
	Madrid	40	2019	15	1810	504	4709	5101
Switzerland	Bern	47	1335	10	1270	351	2576	3002
	Davos	47	1612	4	1562	426	2863	3286
	Lausanne	47	1408	12	1329	364	2108	2539
	Zurich	47	1249	10	1186	331	1648	1935
Sweden	Gothenburg	58	1138	10	1073	305	1287	1726
	Linkoping	58	1132	8	1061	304	1257	1697
	Malmo	56	1183	9	1113	316	1424	1863
	Stockholm	59	1179	8	1105	317	1407	1846
	Uppsala	60	1099	8	1024	297	1142	1583
Denmark	Alborg	57	1116	8	1047	298	3041	3463
	Copenhagen	56	1144	10	1079	305	3195	3615
	Odense	55	1102	9	1040	295	2987	3409
Finland	Helsinki	60	1160	6	1086	312	1021	1464
	Oulu	65	1182	4	1112	321	1104	1545

Table 5. *Cont.*

Country	City	Latitude	Annual GHI (kWh)	Annual Average Temperature (°C)	Annual Thermal Production (kWh)	Annual Electrical Production (kWh)	NPV per Unit Collector Area for Financial Model 1 (EUR)	NPV per Unit Collector Area for Financial Model 2 (EUR)
Germany	Berlin	53	1194	10	1128	315	4582	4988
	Dortmund	52	1093	11	1037	291	4034	4446
	Frankfurt	50	1143	11	1078	302	4291	4701
	Hamburg	54	1146	11	1091	306	4363	4772
	Munich	48	1318	11	1257	345	5348	5747
Iceland	Reykjavik	64	968	6	932	266	−145	186
Norway	Bergen	60	926	9	875	253	−576	−163
	Oslo	60	1029	7	962	277	−408	3
	Trondheim	64	1166	7	1107	317	−136	273
Belgium	Brussels	51	1151	12	1094	306	3244	3664
Bulgaria	Sofia	43	1335	13	1264	348	364	813
France	Lyon	46	1422	14	1337	368	1899	2333
	Nantes	47	1408	13	1333	367	1889	2323
	Paris	49	1204	13	1134	315	1279	1718
	Toulouse	44	1522	15	1437	391	2197	2628
Greece	Athinai	38	1915	21	1731	474	3119	3540
Luxembourg	Luxembourg	50	1194	9	1128	318	1661	2096
Poland	Krakow	50	1191	10	1126	315	868	1267
	Warsaw	52	1213	10	1137	320	909	1307
Romania	Bucharest	44	1589	13	1482	406	1841	2153
	Cluj-Napoca	47	1443	11	1365	374	1516	1831
Ukraine	Kyiv	50	1330	10	1242	348	−1287	−1368
United Kingdom	Glasgow	56	1097	10	1045	294	2096	2527
	Liverpool	53	1013	11	965	273	1765	2199
	London	52	1107	13	1048	294	2109	2540
China	Hong Kong	22	1338	24	1251	329	461	725
Qatar	Doha	25	1957	28	1715	462	−1468	−1168
Saudi Arabia	Medina	25	2349	29	1966	540	−828	−401
Singapore	Singapore	1	1618	27	1473	390	1461	1569

Table 5. *Cont.*

Country	City	Latitude	Annual GHI (kWh)	Annual Average Temperature (°C)	Annual Thermal Production (kWh)	Annual Electrical Production (kWh)	NPV per Unit Collector Area for Financial Model 1 (EUR)	NPV per Unit Collector Area for Financial Model 2 (EUR)
India	Bangalore	13	2093	25	1847	489	−12	178
	Bombay	19	1910	28	1687	445	−213	−21
	Hyderabad	17	2005	28	1765	466	−112	79
	Lucknow	27	1921	27	1717	453	−174	17
	New Delhi	29	2157	27	1873	505	35	224
	Surat	21	2168	28	1874	500	26	215
	Wadhwan	23	2159	28	1866	496	17	207
	Yavatmal	20	1938	28	1715	453	−179	13
USA	Chicago	42	1564	11	1475	402	987	1432
	Denver	40	1912	11	1796	483	1695	2133
	Houston	30	1720	21	1582	422	1211	1655
	Las Vegas	36	2278	21	1987	545	2136	2570
	Los Angeles	34	1973	20	1808	489	1722	2161
	New York	41	1597	14	1508	407	1052	1496
	Portland	46	1436	12	1361	374	732	1179
	San Francisco	38	1886	15	1757	478	1616	2056
	Washington	39	1602	15	1510	407	1053	1497
Mexico	Mexico City	20	1848	18	1727	451	−342	−224
Australia	Brisbane	−27	1898	21	1720	452	3940	4339
	Melbourne	−38	1528	15	1426	371	2872	3282
	Perth	−32	1930	19	1731	455	3990	4389
Argentina	Buenos Aires	−35	1703	13	1550	406	65	−2077
Brazil	Brasilia	−16	1928	22	1762	467	1985	2197
Chile	Santiago	−33	1732	15	1570	411	1785	2171
Colombia	Bogota	5	1560	14	1510	394	856	1107
Algeria	Algiers	37	2017	18	1835	495	−1027	−747
Egypt	Cairo	30	2009	22	1791	485	−1551	−1589
Morocco	Rabat	34	2094	18	1907	517	1616	1950

The general trend shows that thermal output is higher in countries with higher irradiation, such as Saudi Arabia, Algeria, Morocco, Brazil, Mexico, India, etc., with annual thermal production above 1800 kWh (area-specific output 918 kWh/m^2) due to high GHI and ambient temperatures. The lower band of average collector production can be seen in Reykjavik, Iceland, and for some locations in Norway, with a specific output of 475 and 500 kWh/m^2, respectively. Similar thermal output is obtained for locations in countries such as Sweden, Finland, United Kingdom, Denmark, etc., with less than 510 kWh/m^2 annual production. The collector shows better performance in countries, such as Spain, Portugal, and Australia, with collector production of above 1600 kWh (816 kWh/ m^2).

Figure 8 shows the correlation of collector thermal production with GHI and ambient temperature. All the simulated data points of these parameters are considered to define the possible trend. Results show that thermal output has a strong linear correlation with GHI with R^2 value close to 0.98. Thus, the location with higher GHI has higher thermal output. In addition, thermal output shows a linear trend with ambient temperature for most of the data points, however, the correlation is not as strong as with GHI. Therefore, ambient temperature cannot be used as a sole indicator to estimate the collector output.

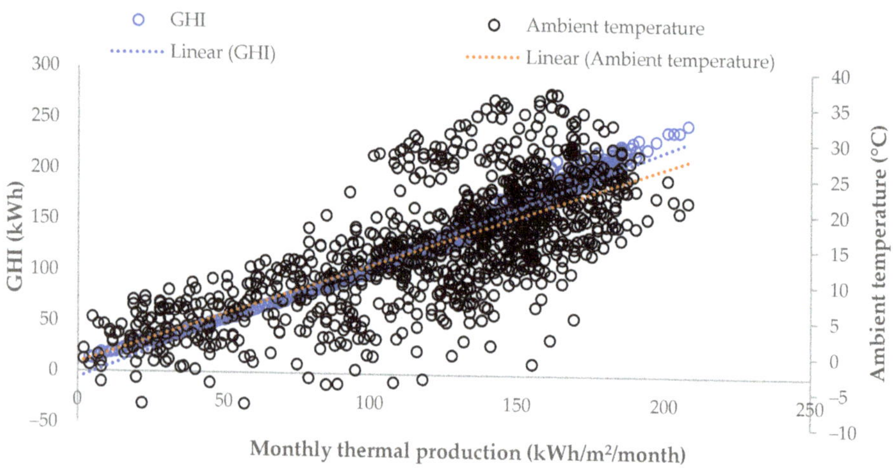

Figure 8. Correlation of collector thermal production with global horizontal irradiation (GHI) and ambient temperature.

4.1.2. Collector Electrical Production

Figure 9 represents the electrical performance of the collector, which shows similar trends as thermal output. For locations in countries with high GHI, such as Saudi Arabia, Algeria, Morocco, Brazil, India, etc., generation is above 500 kWh, and the peak value is in Saudi Arabia with 540 kWh. The electrical production is much less in Iceland with 266 kWh due to less available GHI, and the collector generation is lower than 300 kWh in Sweden, Finland, Denmark, Poland, United Kingdom, etc. The collector performed slightly better in Spain, Portugal, and Australia, with more than 400 kWh annually. However, it shows there is no significant difference in thermal and electrical production trends. Furthermore, a correlation of collector electrical production with GHI and ambient temperature is developed based on all monthly points from all chosen locations and a positive correlation is realized as shown in Figure 10. A large variation in electrical output for similar values of ambient temperature can be observed, which again shows that GHI is the critical parameter governing the electrical output of the collector.

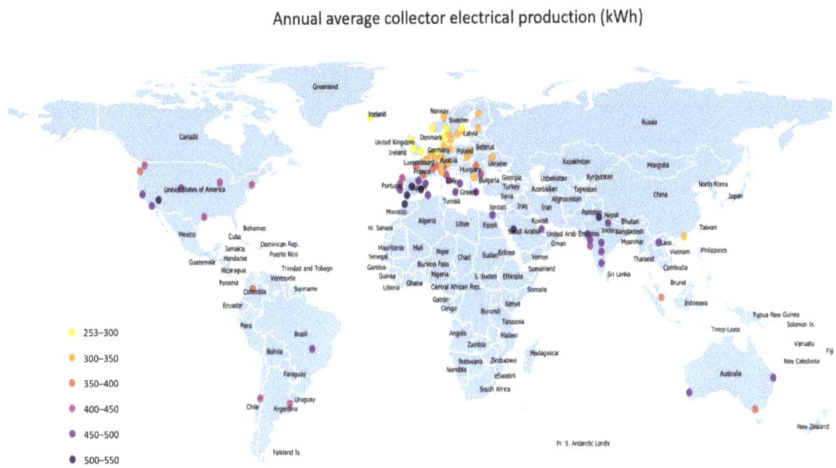

Figure 9. Annual average collector electrical performance.

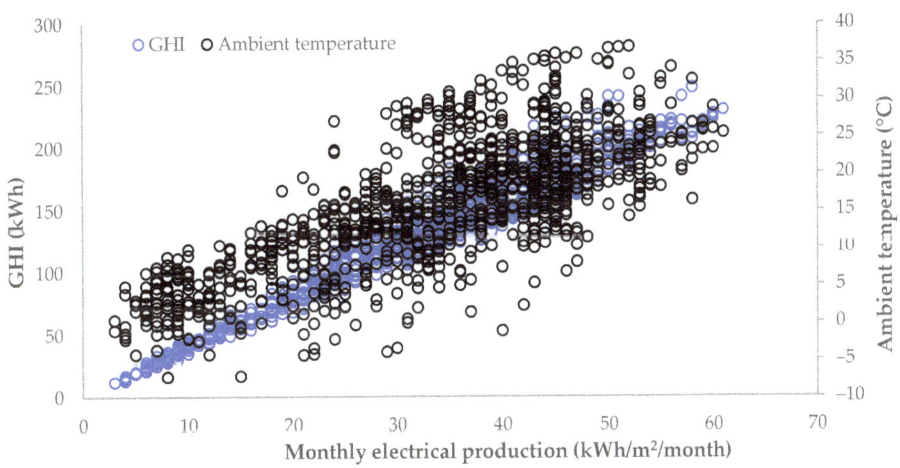

Figure 10. Correlation of collector electrical production with global horizontal irradiation (GHI) and ambient temperature.

A large variation in thermal and electrical output is seen for many countries and is reflected in Figures 7 and 9. The range of collector output with a maximum and minimum value of thermal and electrical production is shown in Figure 11.

The minimum thermal production in blue color represents the minimum production for analyzed location, while the maximum thermal production is indicated with an orange color that represents the highest thermal production of a city in each country. The results show likely high variation in Italy, Spain, United States, and Australia, as many cities were simulated in those countries, and less variation is recorded in countries Denmark, Iceland, United Kingdom, etc., due to the lower number of simulated cities.

In general, PVT collector monthly production is an important key factor in the sizing of a solar system to match the monthly variation of energy consumption. Figures 12 and 13 show the variation in collector monthly thermal and electrical production, respectively. The thermal performance in April and July is relatively higher and less in January and October for the locations in the northern hemisphere, such as Madrid, Stockholm, and Berlin. In Medina, although GHI and ambient temperatures are higher in July, the thermal production is lower compared to in October. This is because the thermal demand in July is less than in October. Therefore, in July, due to high GHI and less thermal demand, the storage tank losses will be higher as the tank temperature increases. Higher tank temperature results in lower thermal and electrical production of collector. As the GHI trend in the southern hemisphere is opposite to the northern hemisphere, the production in January and October is likely higher than the April and July months. In Stockholm, the variation between the months is significant because of seasonal variation in GHI, and the same is lower in Medina, which results in more uniform monthly production.

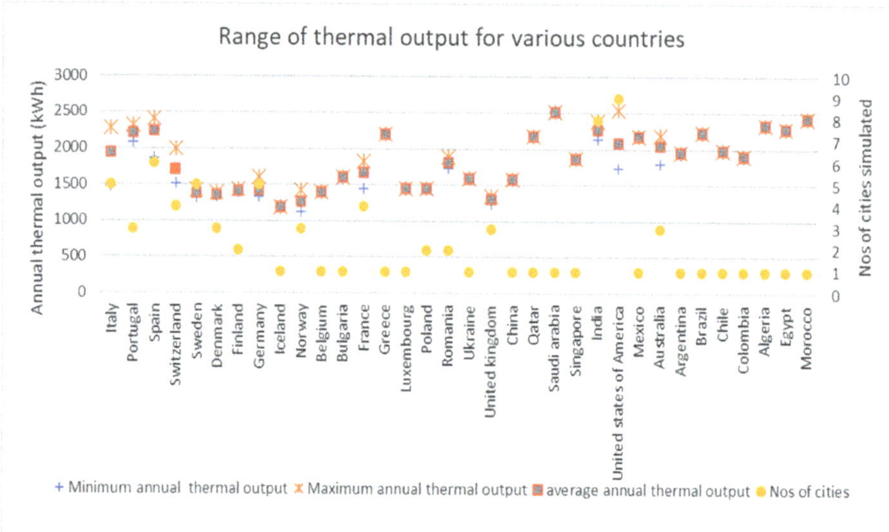

Figure 11. Country-wise collector thermal performance uncertainty.

The trends for monthly electrical production are slightly different than thermal output. For example, in Medina, electrical production is higher in July than in October even though the ambient temperature is maximum in July. This is due to high GHI in July and is in line with findings that the major factor influencing the electrical production is GHI, rather than ambient temperature.

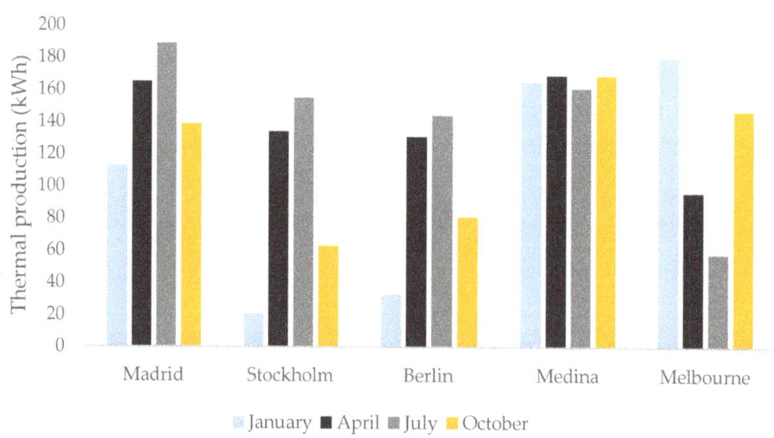

Figure 12. Collector monthly thermal production variation.

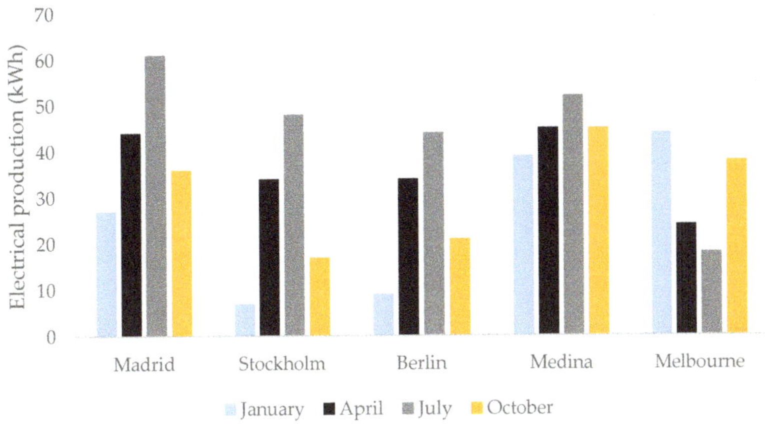

Figure 13. Collector monthly electrical production variation.

4.1.3. Collector Energy Utilization Ratio

The energy utilization ratio of the collector for various locations is shown in Figure 14. The correlation trends between energy utilization ratio and annual average ambient temperature are shown in Figure 15 with consideration of all selected 85 geographical locations to derive a possible trend between the parameters.

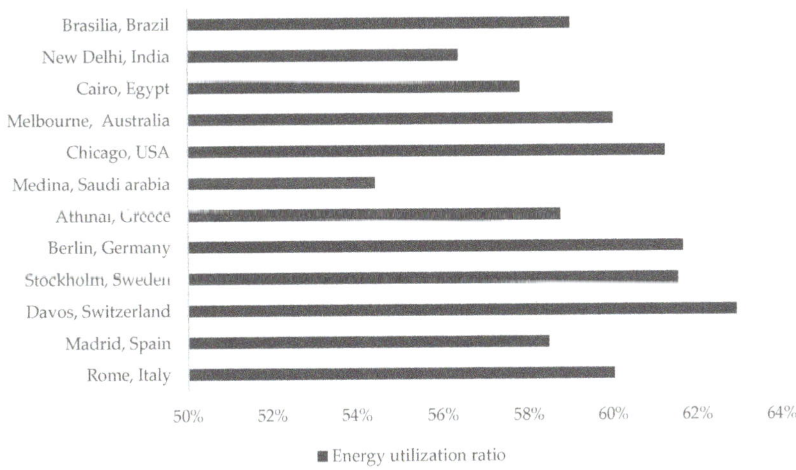

Figure 14. Collector energy utilization ratio.

Some locations show interesting results of system boundaries on PVT collector performance. This can be realized by comparing the energy utilization ratio for Medina (high irradiation) and Davos (low irradiation location). The energy utilization for Davos (63%) is higher compared to Medina (52.5%), even though the absolute value of total energy output is higher for Medina (2506 kWh) compared to Davos (1988 kWh). This is because the load demand for Medina is comparably lower, while the other system design parameters remain the same (collector area, tank volume, etc.), which resulted in higher average tank temp and thus lower collector efficiency for Medina. Results show that the total thermal demand for every location varies depending on the ambient temperature as shown in Figure 16. This is because of the temperature difference between the annual average ambient temperature of each location and desired water temperature (assumed 60 °C), which has to be covered by the collector thermal production.

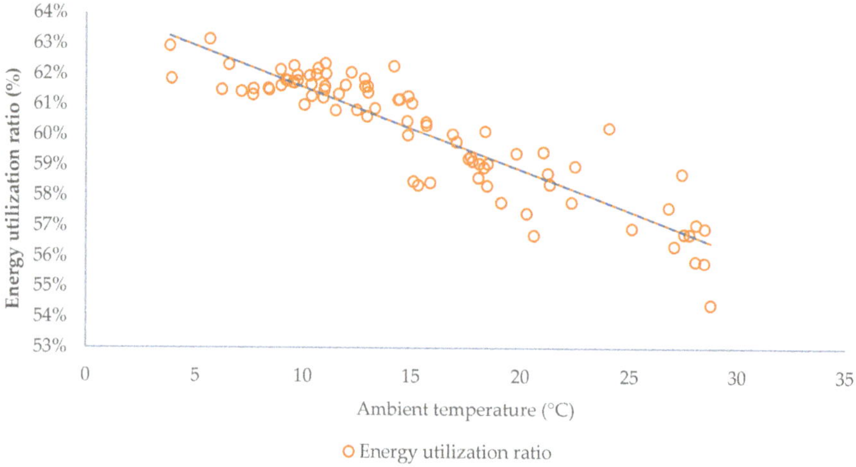

Figure 15. Correlation of energy utilization ratio with the annual average ambient temperature.

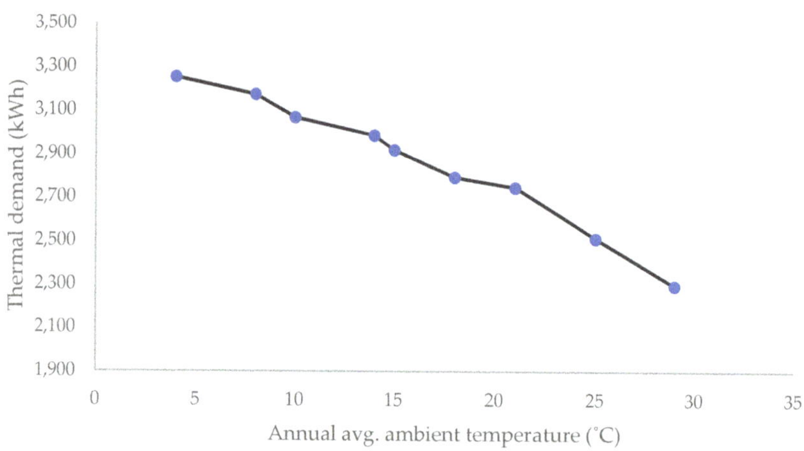

Figure 16. Total thermal demand of single-family house relation with the average ambient temperature.

4.1.4. Collector Exergy Efficiency

From the Carnot efficiency, it can be noted that exergy efficiency is a function of inlet temperature and thermal output of the collector (assumed that the desired output temperature is fixed at 60 °C). Hence, it can be derived that locations with higher ambient temperature will result in less quality of exergy and, thus, lower exergetic efficiency.

Figure 17 shows the correlation of exergetic efficiency with ambient temperature based on all selected 85 geographical locations to derive a possible trend between the parameters. Similar trends can be seen for some specific locations shown in Figure 18. It can be seen that even though the energy efficiency of Madrid is higher compared to Davos, the exergy efficiency of Davos is higher due to lower annual ambient temperature and, thus, higher quality of heat is delivered to the user.

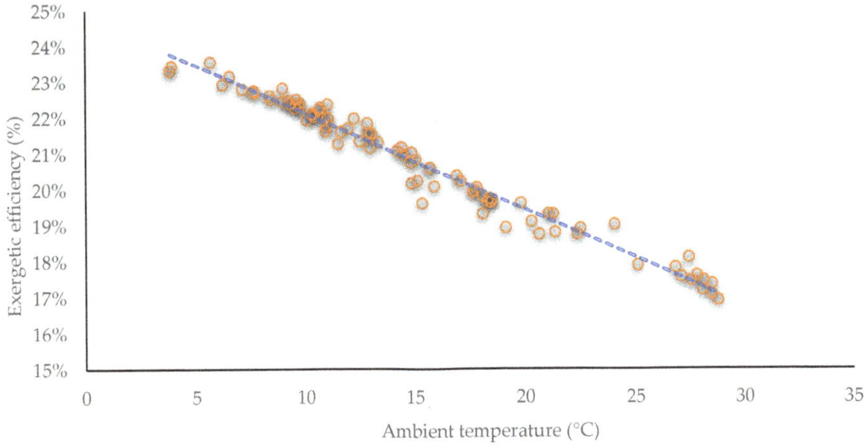

Figure 17. Correlation of exergy efficiency with the annual average ambient temperature.

Figure 18. Collector exegetic efficiency.

4.2. *Economic Performance Evaluation of the PVT Collector*

Based on the above energy performance, the economic performance of such a PVT system is investigated in the 85 different locations. In this section, the NPV per unit collector area is analyzed and represented.

4.2.1. Collector Economic Performance in Financing Model 1

This financing model scenario has assumed that the total cost of the system is invested in the first year of the system period. As the total system cost will be invested in the first year, the interest rate is not considered. Figure 19 is the digital representation of NPV potential per unit collector area with financial model 1 in all 85 geographical cities across the world and Figure 20 shows the NPV potential per unit collector area in geographical cities in the European continent.

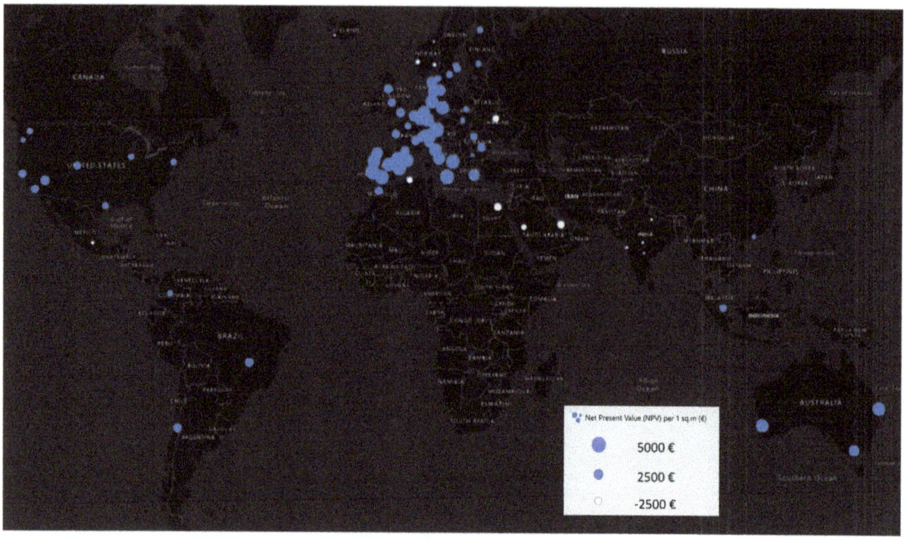

Figure 19. Net present value (NPV) potential per unit collector area for financing model 1.

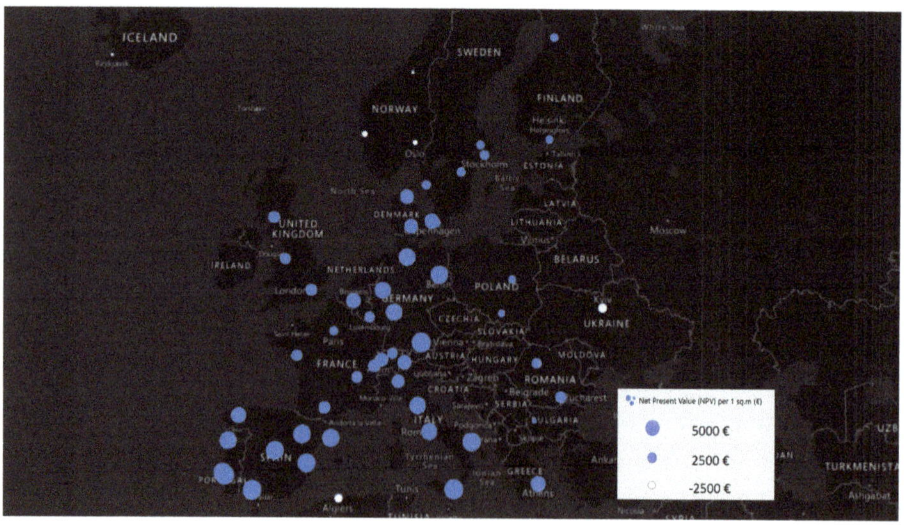

Figure 20. NPV potential per unit collector area in Europe for financing model 1.

The cities with larger dots represent the high NPV potential and cities with smaller dots size represents the least NPV potential. The cities Catania and Munich have the highest potential of 5140 and 5348 EUR, respectively, followed by the cities Bari, Lisbon, Setubal, Sevilla, Valencia, Zaragoza, Madrid, and Berlin, which have potentially more than 4500 EUR per unit collector area. This is due to their high available GHI and electricity grid price, so the energy savings are high in these locations which is reflected in huge NPV potential for this system. Cities such as Oslo, Bergen, Reykjavik, etc., with relatively less electricity grid price resulted in having negative NPV due to lower available GHI. The cities with high collector production such as Medina, Algeria, and Cairo have shown negative NPV potential due to a much lower electricity grid price which eventually showed fewer energy savings.

The NPV potential in all 85 simulated cities has been selected, divided, and segmented for the appropriate countries to define the NPV range per unit collector area of each country as shown in Figure 21. A large variation in NPV can be seen in a few countries, such as Italy and Portugal, due to variability in GHI for simulated locations. However, a smaller variation is identified in countries such as China, Argentina, Brazil, etc., because only one city has been simulated in this paper, which is part of the key uncertainty.

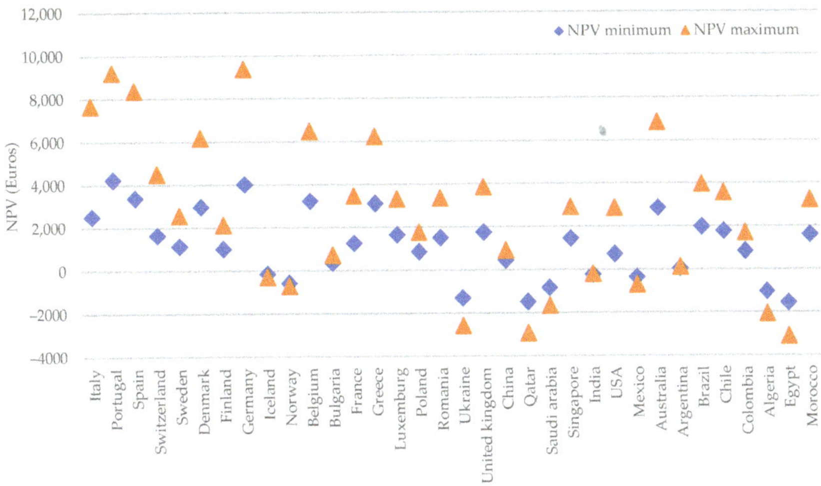

Figure 21. Country-wise NPV potential per unit collector area for financial model 1.

Figure 22 shows the payback period of this PVT system for a single-family house of 5 people in several countries based on financial model 1. The results show that the total system cost will be returned in the first 10 years in countries such as Australia, Belgium, Denmark, Germany, Greece, Italy, Portugal, Spain, Switzerland, etc. This is due to high collector production and high electricity grid price. Although countries such as Algeria, Saudi Arabia, and Egypt have the highest collector production, the grid price is comparatively lower, which reflects the payback period of more than 20 years.

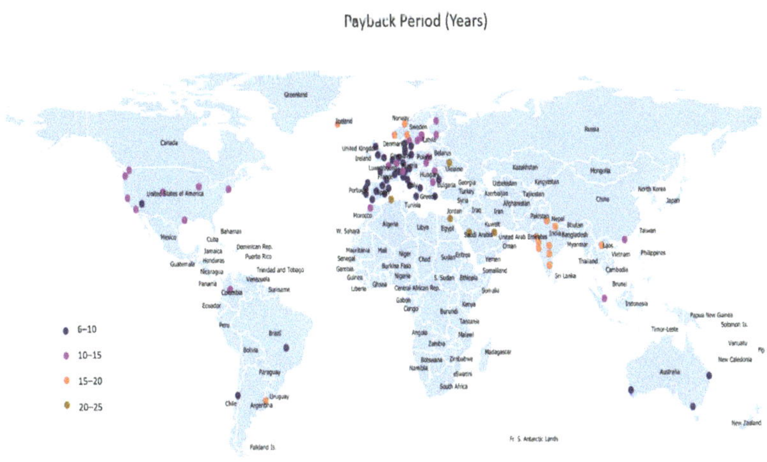

Figure 22. Country-wise average payback period of the PVT collector system.

4.2.2. Collector Economic Performance in Financing Model 2

This financing model has been analyzed by assuming that 75% of total system cost is paid within a financing period of 7 years with a certain interest rate and that the remaining 25% of total system cost is invested in the first year without any interest rate. The NPV potential per unit collector area with financing model 2 in 85 geographical cities across the world is shown in Figure 23, and NPV potential per unit collector area in a specific European continent is shown in Figure 24.

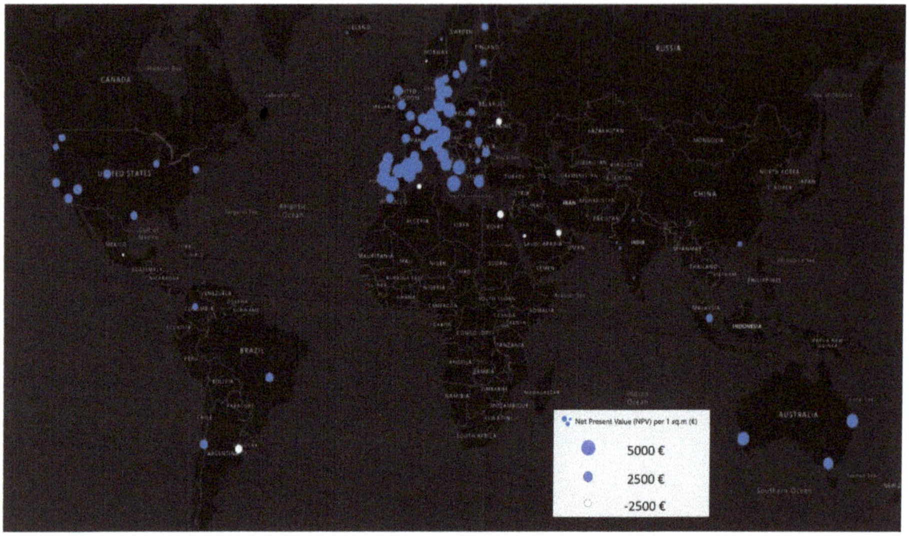

Figure 23. NPV potential per unit collector area for financing model 2.

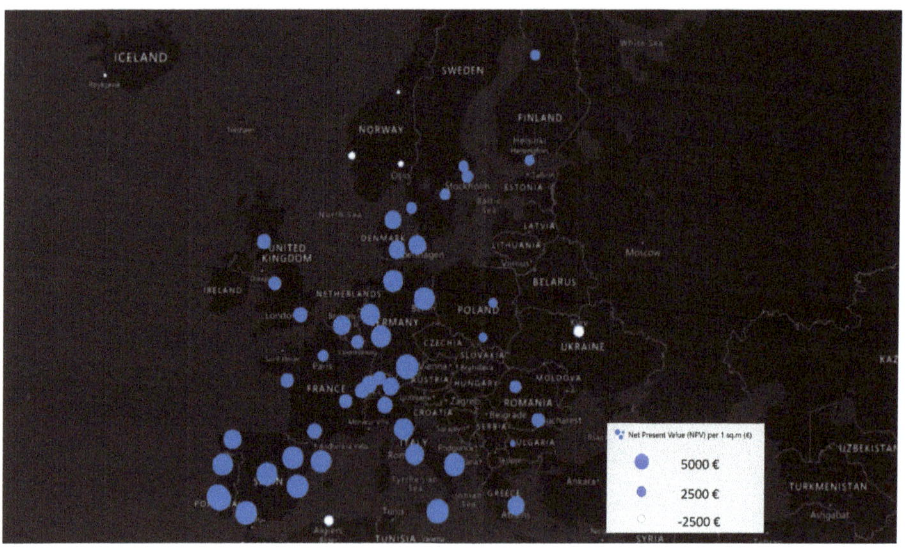

Figure 24. NPV potential per unit collector area in Europe for financing model 2.

The cities with larger dots represent the high NPV potential cities and those with smaller dots represent the lower NPV potential. The cities that showed high NPV potential in financing model 1, such as Catania and Munich, which have shown improved NPV of 5140 and 5348 EUR, respectively, were because of the almost zero interest rates in those countries. This is because if the interest rate is zero, the user needs to pay part of the system cost in later years, and the present value of this investment will be lower due to the time value of money. This will reduce the accumulated investment and thus higher NPV. However, if the interest rate is high, the extra amount paid due to high interest in later years will overweigh the advantage due to the time value of money, and it will decrease the overall NPV. Therefore, financial model 1 is recommended for countries with a high interest rate to maximize the NPV and minimize the payback. Meanwhile, financial model 2 is recommended for countries with zero or lower interest rates to maximize the NPV.

Figure 25 shows the NPV potential per unit collector area in each country for the financing model 2. As compared with financing model 1, there is slightly better performance in NPV in most of the countries. Thus, not much variation has been identified in model 2 compared with model 1.

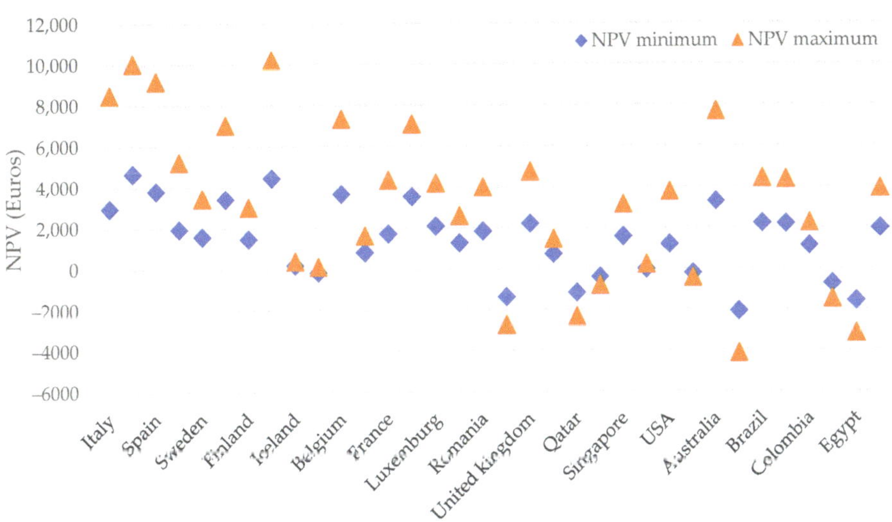

Figure 25. Country-wise NPV potential per unit collector area for financing model 2.

The effect of NPV change due to financial model 2 compared to model 1 is shown in Figure 26. As expected, the countries with high interest rate have shown a negative effect on NPV and countries with less and zero interest rates have shown better NPV potential, such as United States, Australia, and most of the European countries. However, due to the high interest rate of 38% in Argentina, a huge negative impact is identified with financing model 2. Furthermore, a correlation is derived between NPV variations with an interest rate of a specific location in Figure 27.

4.3. Uncertainties

In this paper, the authors acknowledge the possible uncertainties in energy performance analysis. For instance, the delivery water temperature is assumed to be 60 °C and 28 L DHW demand per person for all locations across all cities. In addition, the specific volume ratio (v/a) has been assumed as 80 L/m^2 for all locations, but since it may vary depending on the location and type of application, the resulted collector production would be slightly different in real time, but this approach has been assumed to achieve the goals of this paper.

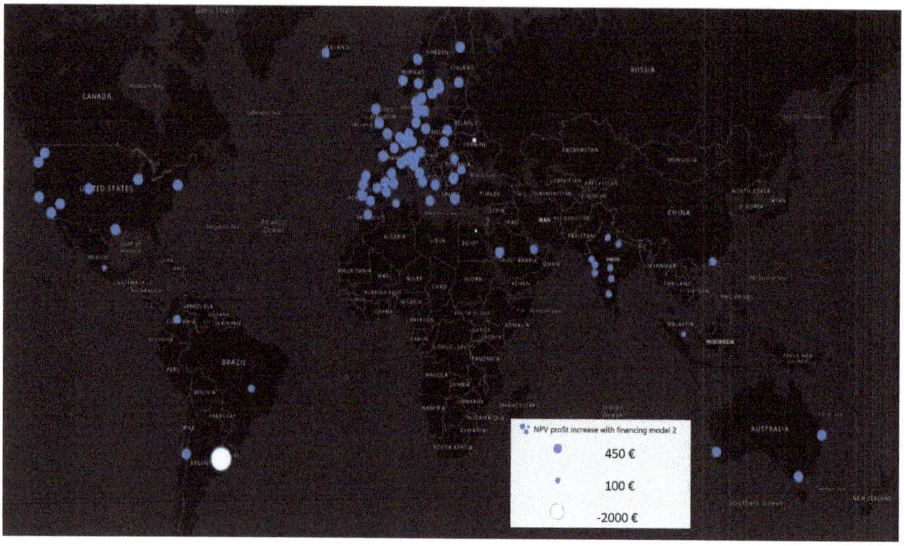

Figure 26. NPV profit increase with financing model 2.

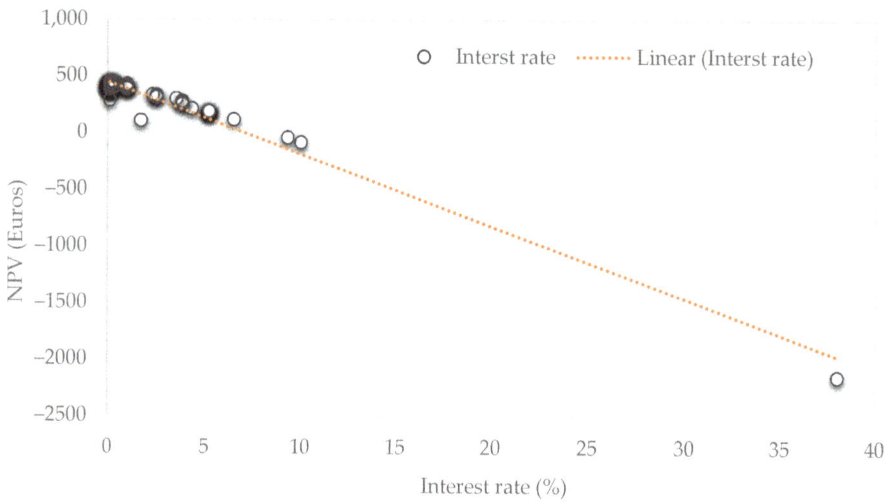

Figure 27. Correlation of NPV potential variation with interest rate.

Furthermore, as the grid price is a key parameter of the total system energy savings, the auxiliary energy price is taken as the generalized price for every specific country, whereas in the real-time case, the energy price would be different for every state/city/municipality depending on localized energy policy. It has been considered because of the unavailability of precise data, which may not be significantly higher. The interest rate is chosen for each country for deriving the NPV potential difference between financing model 1 and model 2. However, only a few countries which have negative and zero interest rate have been assumed as 0.1%, due to the incapability of the simulation tool in accepting negative or null values. However, it has also been realized that the uncertainty of difference between the negative interest rates and assumed interest rates has not been less than 1%, which is not significantly affecting the NPV potential difference. Hence, the assumptions have been considered to achieve the aims in possible optimistic and realistic approaches irrespective of the uncertainties.

5. Conclusions

The performance of a solar PVT consists of PVT collector and storage tank is evaluated for 85 locations across large cities. The optimal tilt angle of the PVT collector, load demand, and electricity prices are chosen appropriately for each simulated location. The results show that the major parameter influencing the PVT performance is GHI, and results derived a strong linear correlation between collector output and GHI. The other factor influencing energetic performance is ambient temperature, source, and load water temperatures. The energetic utilization ratio is dependent on total thermal demand and specific volume ratio (v/a ratio) as it can have a major influence on the fluid temperature in the storage tank and, thus, collector total production. The electrical production by PVT collector is higher in high ambient temperature locations. The highest and lowest energy utilization ratio of the collector is recorded in Reykjavik, Iceland (63%), and Medina, Saudi Arabia (54%), respectively. The highest and lowest exergetic efficiency of the collector has been recorded in Reykjavik, Iceland (23%), and Medina, Saudi Arabia (17%), respectively. Most importantly, the results show that the higher energetic output does not guarantee high economic feasibility. There are several factors such as electricity price, interest rate, and selection of financial model which can highly affect the economic feasibility of PVT collector. The average NPV per unit collector area of 85 geographical cities for financial model 1 and financial model 2 is 1886 and 2221 EUR, respectively. The NPV and payback period analysis of the PVT system has shown positive results for the cities, which have high collector production and high electricity grid price reflecting high energy savings. However, the financing model 1 is highly recommended for the locations with high interest rates and financial model 2 is beneficial for the locations with less interest rates. This paper offers potential insights into the promotion of the PVT market in different regions.

Author Contributions: S.R.P. worked on simulation, analysis, and writing. X.Z. contributed to supervision, concept development, structuring, and writing. P.K.S. contributed to simulation, analysis, and writing. A.d.A. dedicated efforts to simulation and analysis. All authors have read and agreed to the published version of the manuscript.

Funding: This research received funding from the Germany–Sweden joint project: 'Product and process development for the preparing and realization of complete buildings of various types of use using energy efficient, partially energy independent lightweight construction solutions, ENSECO'.

Acknowledgments: The authors acknowledge the useful gains from the IEA SHC Task 60, and the open access support from Dalarna University, Sweden.

Conflicts of Interest: The authors declare no conflict of interest.

References

1. Sommerfeldt, N.; Madani, H. In-depth techno-economic analysis of PV/Thermal plus ground source and heat pump systems for multi-family houses in a heating dominated climate. *Sol. Energy* **2019**, *190*, 44–62. [CrossRef]
2. Joshi, S.; Dhoble, A.S. Photovoltaic -Thermal systems (PVT): Technology review and future trends. *Renew. Sustain. Energy Rev.* **2018**, *92*, 848–882. [CrossRef]
3. Al-Waeli, A.H.; Sopian, K.; Kazem, H.A.; Chaichan, M.T. Photovoltaic/Thermal (PV/T) systems: Status and future prospects. *Renew. Sustain. Energy Rev.* **2017**, *77*, 109–130. [CrossRef]
4. Gu, Y.; Zhang, X.; Myhren, J.A.; Han, M.; Chen, X.; Yuan, Y. Techno-economic analysis of a solar photovoltaic/thermal (PV/T) concentrator for building application in Sweden using Monte Carlo method. *Energy Convers. Manag.* **2018**, *165*, 8–24. [CrossRef]
5. PVT Systems IEA SHC 60-Annex 180504.pdf. Available online: http://task60.iea-shc.org/Data/Sites/1/publications/PVT%20systems%20IEA%20SHC%2060%20-%20Annex%20180504.pdf (accessed on 25 November 2019).

6. Zhang, X.; Zhao, X.; Smith, S.; Xu, J.; Yu, X. Review of R&D progress and practical application of the solar photovoltaic/thermal (PV/T) technologies. *Renew. Sustain. Energy Rev.* **2012**, *16*, 599–617. [CrossRef]
7. Ramschak, T. IEA SHC Task 60. 2020. Available online: https://task60.iea-shc.org/Data/Sites/1/publications/IEA-SHC-Task60-A1-Existing-PVT-Systems-and-Solutions.pdf (accessed on 29 July 2020).
8. IEA-SHC-Task60-Highlights-2019.pdf. Available online: https://task60.iea-shc.org/Data/Sites/1/publications/IEA-SHC-Task60-Highlights-2019.pdf (accessed on 10 June 2020).
9. Buonomano, A.; De Luca, G.; Figaj, R.; Vanoli, L. Dynamic simulation and thermo-economic analysis of a PhotoVoltaic/Thermal collector heating system for an indoor–outdoor swimming pool. *Energy Convers. Manag.* **2015**, *99*, 176–192. [CrossRef]
10. Riggs, B.C.; Biedenharn, R.; Dougher, C.; Ji, Y.V.; Xu, Q.; Romanin, V.; Codd, D.S.; Zahler, J.M.; Escarra, M.D. Techno-economic analysis of hybrid PV/T systems for process heat using electricity to subsidize the cost of heat. *Appl. Energy* **2017**, *208*, 1370–1378. [CrossRef]
11. Wang, K.; Herrando, M.; Pantaleo, A.M.; Markides, C.N. Technoeconomic assessments of hybrid photovoltaic-thermal vs. conventional solar-energy systems: Case studies in heat and power provision to sports centres. *Appl. Energy* **2019**, *254*, 113657. [CrossRef]
12. Kazem, H.A. Evaluation and analysis of water-based photovoltaic/thermal (PV/T) system. *Case Stud. Therm. Eng.* **2019**, *13*, 100401. [CrossRef]
13. Fudholi, A.; Sopian, K.; Yazdi, M.H.; Ruslan, M.H.; Ibrahim, A.; Kazem, H.A. Performance analysis of photovoltaic thermal (PVT) water collectors. *Energy Convers. Manag.* **2014**, *78*, 641–651. [CrossRef]
14. Khelifa, A.; Touafek, K.; Ben Moussa, H.; Tabet, I.; Hocine, H.B.C.E.; Haloui, H. Analysis of a Hybrid Solar Collector Photovoltaic Thermal (PVT). *Energy Procedia* **2015**, *74*, 835–843. [CrossRef]
15. Khelifa, A.; Touafek, K.; Ben Moussa, H.; Tabet, I. Modeling and detailed study of hybrid photovoltaic thermal (PV/T) solar collector. *Sol. Energy* **2016**, *135*, 169–176. [CrossRef]
16. Gagliano, A.; Tina, G.M.; Nocera, F.; Grasso, A.D.; Aneli, S. Description and performance analysis of a flexible photovoltaic/thermal (PV/T) solar system. *Renew. Energy* **2019**, *137*, 144–156. [CrossRef]
17. Shah, R.; Srinivasan, P. Hybrid Photovoltaic and Solar Thermal Systems (PVT): Performance Simulation and Experimental Validation. *Mater. Today Proc.* **2018**, *5*, 22998–23006. [CrossRef]
18. Buonomano, A.; Calise, F.; Vicidomini, M. Design, Simulation and Experimental Investigation of a Solar System Based on PV Panels and PVT Collectors. *Energies* **2016**, *9*, 497. [CrossRef]
19. Yazdanpanahi, J.; Sarhaddi, F.; Adeli, M.M. Experimental investigation of exergy efficiency of a solar photovoltaic thermal (PVT) water collector based on exergy losses. *Sol. Energy* **2015**, *118*, 197–208. [CrossRef]
20. Herrando, M.; Markides, C.N. Hybrid PV and solar-thermal systems for domestic heat and power provision in the UK: Techno-economic considerations. *Appl. Energy* **2016**, *161*, 512–532. [CrossRef]
21. Ahn, H.; Rim, D.; Pavlak, G.S.; Freihaut, J.D. Uncertainty analysis of energy and economic performances of hybrid solar photovoltaic and combined cooling, heating, and power (CCHP + PV) systems using a Monte-Carlo method. *Appl. Energy* **2019**, *255*, 113753. [CrossRef]
22. Heck, N.; Smith, C.; Hittinger, E. A Monte Carlo approach to integrating uncertainty into the levelized cost of electricity. *Electr. J.* **2016**, *29*, 21–30. [CrossRef]
23. Jung, J.; Han, S.; Kim, B. Digital numerical map-oriented estimation of solar energy potential for site selection of photovoltaic solar panels on national highway slopes. *Appl. Energy* **2019**, *242*, 57–68. [CrossRef]
24. Oh, M.; Park, H.-D. A new algorithm using a pyramid dataset for calculating shadowing in solar potential mapping. *Renew. Energy* **2018**, *126*, 465–474. [CrossRef]
25. Mishra, T.; Rabha, A.; Kumar, U.; Arunanchalam, K.; Sridhar, V. Assessment of solar power potential in a hill state of India using remote sensing and Geographic Information System. *Remote Sens. Appl. Soc. Environ.* **2020**, 100370. [CrossRef]
26. Bremer, M.; Mayr, A.; Wichmann, V.; Schmidtner, K.; Rutzinger, M. A new multi-scale 3D-GIS-approach for the assessment and dissemination of solar income of digital city models. *Comput. Environ. Urban Syst.* **2016**, *57*, 144–154. [CrossRef]
27. Peronato, G.; Rastogi, P.; Rey, E.; Andersen, M. A toolkit for multi-scale mapping of the solar energy-generation potential of buildings in urban environments under uncertainty. *Sol. Energy* **2018**, *173*, 861–874. [CrossRef]
28. Chow, T.-T. A review on photovoltaic/thermal hybrid solar technology. *Appl. Energy* **2010**, *87*, 365–379. [CrossRef]

29. Joshi, A.S.; Tiwari, A. Energy and exergy efficiencies of a hybrid photovoltaic–thermal (PV/T) air collector. *Renew. Energy* **2007**, *32*, 2223–2241. [CrossRef]
30. IEA-SHC-Task60-B2-Design-Guidelines-for-PVT-Collectors.pdf. Available online: https://www.iea-shc.org/Data/Sites/1/publications/IEA-SHC-Task60-B2-Design-Guidelines-for-PVT-Collectors.pdf (accessed on 27 July 2020).
31. IEA SHC||Task 60||IEA SHC||Task 60. Available online: http://task60.iea-shc.org/ (accessed on 21 November 2019).
32. 'Contact', Abora Solar. Available online: https://abora-solar.com/en/contact/ (accessed on 29 November 2019).
33. Microsoft Excel, Spreadsheet Software, Excel Free Trial. Available online: https://www.microsoft.com/en-ww/microsoft-365/excel (accessed on 29 July 2020).
34. Welcome to the QGIS Project! Available online: https://qgis.org/en/site/ (accessed on 29 July 2020).
35. 'Intro', Meteonorm (de). Available online: https://meteonorm.com/ (accessed on 22 July 2020).
36. Al-Ghussain, L.; Taylan, O.; Baker, D.K. An investigation of optimum PV and wind energy system capacities for alternate short and long-term energy storage sizing methodologies. *Int. J. Energy Res.* **2018**, *43*, 204–218. [CrossRef]
37. 'India Electricity Prices, June 2019|GlobalPetrolPrices.com', GlobalPetrpPrices.com. Available online: https://www.globalpetrolprices.com/India/electricity_prices/ (accessed on 15 March 2020).

buildings

MDPI

Article

Waste Incineration Heat and Seasonal Thermal Energy Storage for Promoting Economically Optimal Net-Zero Energy Districts in Finland

Janne Hirvonen [1,*] and Risto Kosonen [1,2]

1 Department of Mechanical Engineering, Aalto University, 00076 Aalto, Finland; risto.kosonen@aalto.fi
2 College of Urban Construction, Nanjing Tech University, Nanjing 210000, China
* Correspondence: janne.p.hirvonen@aalto.fi; Tel.: +358-50-431-5780

Received: 15 October 2020; Accepted: 11 November 2020; Published: 17 November 2020

Abstract: In countries with high heating demand, waste heat from industrial processes should be carefully utilized in buildings. Finland already has an extensive district heating grid and large amounts of combined heat and power generation. However, despite the average climate, there is little use for excess heat in summer. Waste incineration plants need to be running regardless of weather, so long-term storage of heat requires consideration. However, no seasonal energy storage systems are currently in operation in connection with Finnish waste incineration plants. This study used dynamic energy simulation performed with the TRNSYS 17 software to analyze the case of utilizing excess heat from waste incineration to supplement conventional district heating of a new residential area. Seasonal energy storage was utilized through a borehole thermal energy storage (BTES) system. Parametric runs using 36 different storage configurations were performed to find out the cost and performance range of such plans. Annual energy storage efficiencies from 48% to 69% were obtained for the BTES. Waste heat could generate 37–89% of the annual heat demand. Cost estimations of waste heat storage using BTES are not available in the literature. As an important finding in this study, a levelized cost of heat of 10.5–23.5 €/MWh was obtained for various BTES configurations used for incineration waste heat storage. In the three most effective cases, the stored heat reduced annual CO_2 emissions of the residential area by 42%, 64% and 86%. Thus, the solution shows great potential for reducing carbon emissions of district heating in grids connected to waste incineration plants

Keywords: seasonal thermal energy storage; waste incineration; district heating; waste heat

1. Introduction

In the heating dominated climate of Nordic countries, heating demand surpasses electricity demand. In Finland, Sweden and Denmark, the combined heating demand is about 400 TWh and is mostly met through district heating [1]. However, district heating is typically produced through combustion of fossil fuels, biomass or solid waste, all of which produce carbon emissions. Thus, there is great potential for reducing carbon emissions in the district heating sector. The reductions could be realized through improved system coupling of waste incineration plants and district heating.

Currently, there are many waste incineration plants in Sweden and Finland, which are used for generating electricity and heat. In Finland, 93% of non-sorted mixed waste is utilized for energy generation [2]. The energy sector in Finland is dominated by co-generation plants. Unfortunately, while solid waste is produced all through the year, the demand for heat falls during summer and part of the heat produced by incineration is wasted. In addition, due to the already high energy conversion of waste, energy generated from waste cannot really be increased. Thus, the available energy should be used more effectively.

One solution is seasonal underground thermal energy storage. This could be done using aquifer thermal energy storage (ATES), where heat is moved between a hot and cold water reservoir. It has been successfully utilized in the Swedish Arlanda airport, where thermal energy storage has been reported to supply 22 GWh/a of cooling and low temperature heat. In a Finnish study, district cooling was generated using heat pumps and waste heat was stored in the hot reservoir of an ATES system. Heat was produced at the cost of 41.5 €/MWh, while the seasonal storage efficiency was 73–83%.

Unfortunately, appropriate aquifers are not available everywhere. Thus, a more general solution could be obtained by using borehole thermal energy storage (BTES) technology. This means storing heat into the ground itself using boreholes fitted with heat transfer pipes. In many studies, the heat to be stored comes from solar energy [3]. Finnish studies on seasonal solar energy storage have shown that it is technically feasible, but economically problematic. The most expensive components of the system are the solar thermal collectors [4] or the solar photovoltaic panels [5], not the actual storage system. Thus, a system with a free source of heat would be a potential breakthrough.

In Emmaboda, Sweden, a high temperature BTES system was built to store waste heat from industrial processes, such as metalworking [6]. The storage volume had a high thermal conductivity of 3.0 W/mK. There were 140 boreholes, each 150 m deep and equipped with a coaxial double tube borehole heat exchanger (BHE). Seasonal storage efficiency of 68% was expected. However, technical problems prevented heat extraction during the initial test period. Heat was first extracted after the fourth year of operation and design performance was expected to be achieved after the fifth year [7]. Further study of the facility revealed that the industrial waste heat was of lower temperature than expected, which reduced the output from the BTES system [8]. Only 19% storage efficiency was achieved. Total construction and consultancy cost of €1.25 M was reported, but not the cost of produced heat. Other examples on storage of industrial waste heat are scarce.

A 50,000 m^3 BTES located in China was designed to utilize both industrial waste heat and solar energy [9]. The supply temperature in the system varied according to the specific process that was supplying the heat at any given moment. However, the maximum temperature for the processes varied from 60 °C to 90 °C. The storage volume had a low thermal conductivity of only 1.1 W/mK. The ground in the core of the BTES was heated up to 40 °C, though this was expected to rise with continued operation. This was purely a technical study with no cost information reported.

One experimental BTES system was connected to a combined heat and power plant in the Czech Republic [10]. The core temperature of the facility reached up to 78.5 °C. This temperature was measured in a dedicated monitoring borehole, so as not to be disturbed by operational heat flows. The average measured thermal conductivity in the ground was a moderate 2.2 W/mK. It was estimated that 65% of injected heat could be sustainably extracted. This is promising for systems intending to utilize heat from waste incineration plants. However, no estimation of the cost of heat was presented.

The main goal of the present study was to show the technical and economic feasibility of borehole thermal energy storage as a supplementary heating system in Finnish district heating grids, where large enough waste heat streams are available. The study also aimed to explain why certain solutions are more cost-effective than others, taking into account the cold Finnish climate and the local ground properties. The novelty of the paper was in the integration of a waste heat process into the district heating system through seasonal storage. No such system yet exists in Finland, but due to the widespread use of district heating and waste-to-energy processes, the potential for reducing carbon dioxide (CO_2) emissions is great. Previous studies on the topic did not report the cost of heat produced through the seasonal waste heat storage system. In this study, we explicitly report the levelized cost of heat obtained from a BTES system. Both the emission reduction potential and its cost are novel and useful information for Finnish and Nordic district heating operators. Paste experimental studies analyze just a single storage system configuration, which may or may not be optimal. The parametric runs performed in the study show the performance range and tradeoffs available for optimal waste heat storage systems, which was not presented in the previous studies.

2. Materials and Methods

This study on incineration waste heat recovery was performed using dynamic energy simulation. A flowchart of the whole calculation process is presented in Figure 1.

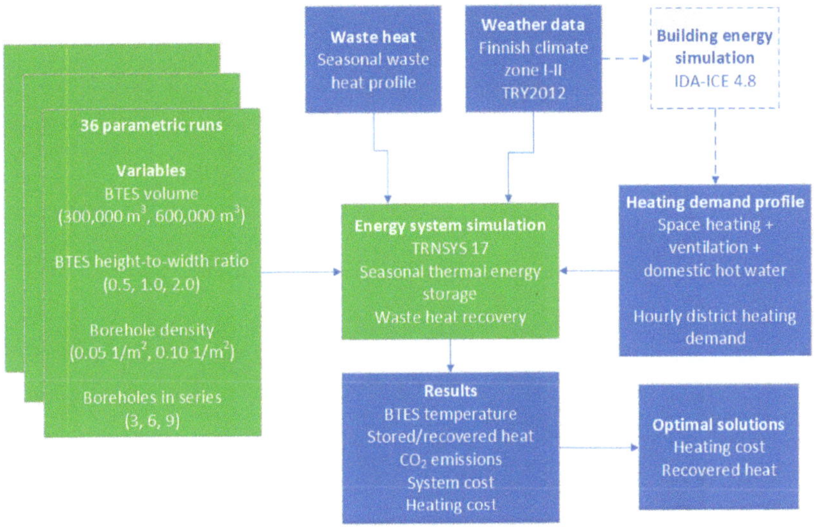

Figure 1. Flowchart of the calculation process.

Heating demand profiles from a previous simulation study were used as input data for energy storage calculations. Various system configurations were tested and out of all the combinations, a few optimal solutions were chosen based on the cost of heat and the amount of heat recovered from the storage.

2.1. District Information

The goal of the study was to analyze the utilization of waste heat from an incineration plant in a new planned residential district in the Finnish city of Vaasa. Vaasa is located in a subarctic climate of Finnish climate zone II, which is represented by the Finnish Test Reference Year 2012 (TRY2012) weather data of Vantaa [11]. The average ambient air temperature in TRY2012 is 5.6 °C. The heating degree day value (S17) for Vaasa in the reference period (1980–2010) was 4469 Kd. The district comprised apartment buildings built according to energy performance requirements of the 2010 Finnish building code [12], with a specific annual heating demand of 64.5 kWh/m^2. With 130,000 m^2 of heated floor area, this corresponds to 8.4 GWh total annual heating demand. Space heating, ventilation and domestic hot water (DHW) consumption profiles were obtained from a previous study [13], in which an apartment building was simulated using the IDA-ICE 4.8 simulation tool [14]. The different demands were combined into a single hourly energy demand profile, which was normalized to the building code requirement.

2.2. Energy System

The planned district was supplied with waste incineration heat and conventional district heating (DH). A TRNSYS model was created to simulate the interaction of a seasonal borehole thermal energy storage (BTES) system with the conventional district heating network and the buildings themselves. The BTES system was a field of boreholes drilled into the ground and fitted with U-tube heat transfer pipes. Heat exchangers (HX) between the DH grid and the BTES were ideal (Type 91) with 95%

efficiency. The BTES system was modelled in TRNSYS using Type 557a. The heat capacity of the ground was 2240 kJ/m^{3}K with a high heat conductivity of 3.5 W/mK [15]. The top surface of the BTES system was covered by a 10 cm polystyrene (Styrofoam) insulation layer to reduce heat losses. Thermal conductivity of the Styrofoam was 0.03 W/mK [16]. The sides of the BTES system and the boreholes themselves were not insulated. The starting temperature of the ground was 5.6 °C [15]. The borehole width was 15 cm and the inner diameter of the U-tube pipe was 3.2 cm.

The boreholes in the BTES system were distributed uniformly in a cylindrical volume. Some of the boreholes were connected in series, such that the hot water for charging was fed into the center of the storage and taken out at the outside edge, going through several boreholes in the process. During discharging, the flow direction was reversed, such that cold water was supplied from the edges. A higher number of series-connected boreholes increases heat transfer efficiency, while reducing the total flow rate and power. It also creates a larger radial temperature difference in the storage, reducing heat losses. This is visualized in Figure 2, which shows a borehole field and the principle of series connection between boreholes (in charging mode).

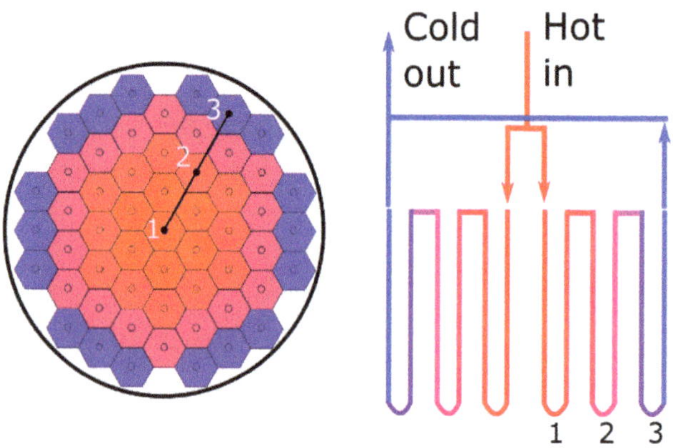

Figure 2. The operational principle of borehole thermal energy storage (BTES). The horizontal BTES cross-section (**left**) shows the borehole grid and the radial heat distribution with a hot core and cooler edges. The series connection of boreholes (**right**) results in a larger temperature change in the heat transfer fluid.

The inlet flow sheds a part of its energy into the central borehole, cooling down in the process. The flow is then redirected to another borehole, cooling down more and finally exiting from the outer borehole at the lowest temperature.

The efficiency of the BTES was defined as, Equation (1):

$$\eta = E_{extracted}/E_{charged} \tag{1}$$

where $E_{extracted}$ is the amount of energy taken out of the storage over a year and $E_{charged}$ is the energy charged into the storage over the same time.

During the heating season, all of the heat in the waste incineration plant was used elsewhere in the city, but in summer there was excess heat available to be used in the new district. According to the local utility, there was 45 GWh of excess heat available from the incineration plant during summer. This was much more than even the annual heating demand of the new district. Thus, all heating demand during summer was covered by the waste heat and any excess could be charged in the seasonal energy storage system. The monthly heating demands are shown in Figure 3.

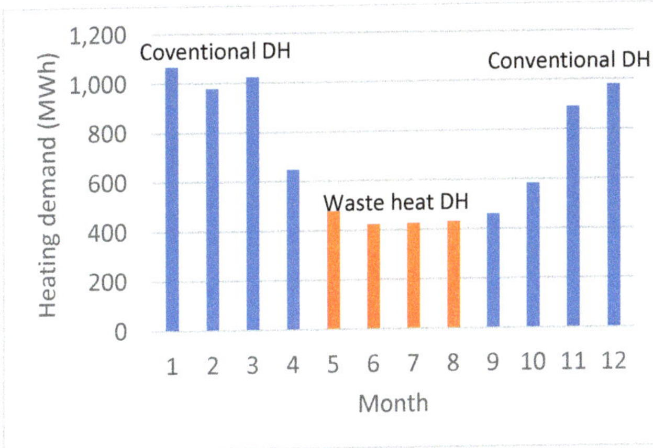

Figure 3. Monthly heating demand in the residential district without seasonal energy storage. The blue color shows when heating is provided by conventional district heating and the orange shows when waste heat is directly utilized.

During the winter months, the monthly heating demand in the district is about 1000 MWh, while during summer the monthly demand remains below 500 MWh.

Waste heat was available from the start of May to the end of August, which limited the charging time of the BTES to four months. For this reason, the charging flowrate to the BTES was 1600 kg/h per borehole loop, while the discharge flowrate was lower at 1200 kg/h per borehole loop.

The new district utilized a low temperature local heating grid with 70 °C supply temperature and 30 °C return temperature from the buildings. The BTES was connected to the conventional district heating system in series, as shown in Figures 4 and 5.

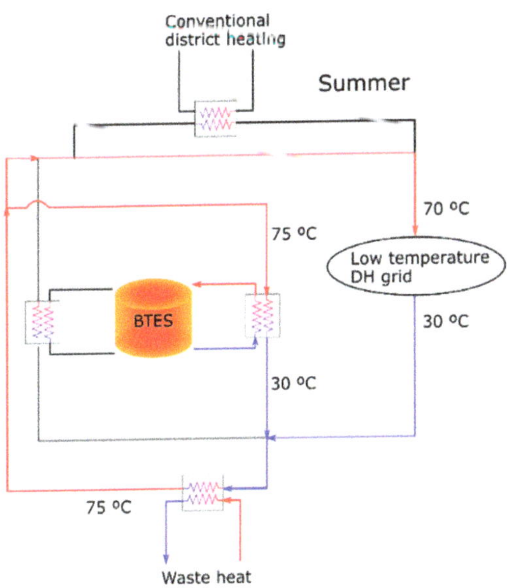

Figure 4. BTES charging mode (summer).

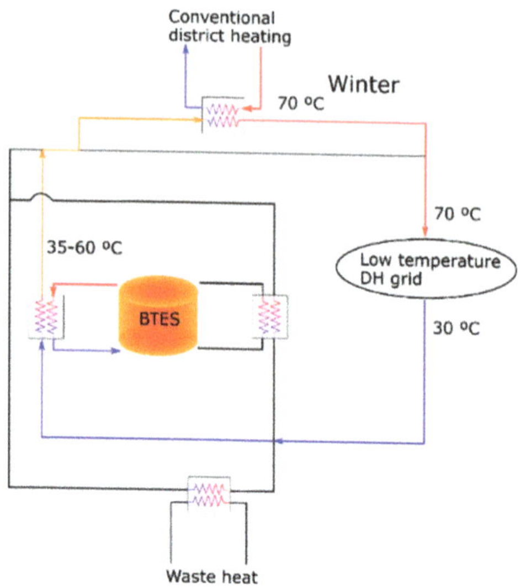

Figure 5. BTES discharging mode (winter).

In summer (Figure 4), waste heat was supplied directly to the low temperature heating grid and to charge the BTES. In winter (Figure 5), the return flow from the local grid was preheated using the BTES and then heated to the required temperature level (70 °C) with the conventional district heating. Heat transfer and losses in the district heating pipes were not modelled, only the connection between the BTES and other grids was modelled. The carbon dioxide emission factor of the conventional district heating was assumed to match the national average of 194 kg-CO_2/MWh [17]. Heat produced by otherwise wasted incineration heat was assumed to have no emissions.

2.3. Cost Calculation

The cost analysis was limited to the direct investment costs of the BTES system, which were the cost of boreholes and piping at 33.5 €/m and the cost of the BTES polystyrene insulation cover at 83.3 €/m^3. The cover was 10 cm thick in every scenario. The district heating grid was assumed to be installed regardless of the BTES system and no extra costs were attached to it. The BTES system was assumed to require annual maintenance valued at 0.5% of the investment cost. The levelized cost of heat (LCOH) was calculated over a 25-year period. Discounting was done using a 3% interest rate and 2% energy price escalation [18].

2.4. Parametric Runs

For the new storage-based heating system to be economically feasible, it needs to be competitive with the conventional heat generation system. According to district heating operators, a levelized cost of heat below 20 €/MWh would be preferable for widespread adoption [19]. For comparison, the after-tax consumer price for district heating in Helsinki in 2020 was 35–66 €/MWh [20]. Parametric runs were performed with various BTES configurations, to see how the economics changed. The variables were the volume of the rock used for storage (V_{BTES}, 300,000 and 600,000 m^3), the height-to-width ratio of the cylindrical storage (h_{ratio}, 0.5, 1.0 and 2.0 m/m), the areal density of boreholes (BH) on the BTES surface, i.e., how many boreholes per area (BH density, 0.05 and 0.10 1/m^2) and the number of boreholes connected in series (N_{series}, 3, 6, 9). The settings of the parametric runs are displayed in Table 1.

Table 1. Parameters of the different borehole thermal energy storage (BTES) configurations tested.

Case	V_{BTES} (m^3)	h_{ratio} (m/m)	BH Density ($1/m^2$)	N_{series}
1	300,000	0.5	0.05	3
2	300,000	0.5	0.1	3
3	300,000	0.5	0.05	6
4	300,000	0.5	0.1	6
5	300,000	0.5	0.05	9
6	300,000	0.5	0.1	9
7	300,000	1	0.05	3
8	300,000	1	0.1	3
9	300,000	1	0.05	6
10	300,000	1	0.1	6
11	300,000	1	0.05	9
12	300,000	1	0.1	9
13	300,000	2	0.05	3
14	300,000	2	0.1	3
15	300,000	2	0.05	6
16	300,000	2	0.1	6
17	300,000	2	0.05	9
18	300,000	2	0.1	9
19	600,000	0.5	0.05	3
20	600,000	0.5	0.1	3
21	600,000	0.5	0.05	6
22	600,000	0.5	0.1	6
23	600,000	0.5	0.05	9
24	600,000	0.5	0.1	9
25	600,000	1	0.05	3
26	600,000	1	0.1	3
27	600,000	1	0.05	6
28	600,000	1	0.1	6
29	600,000	1	0.05	9
30	600,000	1	0.1	9
31	600,000	2	0.05	3
32	600,000	2	0.1	3
33	600,000	2	0.05	6
34	600,000	2	0.1	6
35	600,000	2	0.05	9
36	600,000	2	0.1	9

The parametric runs included all combinations of the mentioned variables. Under these shapes and sizes, the diameter of the BTES varied between 58 and 115 m, while the depth of the BTES varied between 46 and 145 m. The number of boreholes was between 130 and 656 for the smaller BTES size and between 207 and 1042 for the larger BTES size.

The volume of 300,000 m^3 (small) and 600,000 m^3 (large) corresponded to heat storage capacities of 3.5 GWh and 7 GWh, respectively. Before accounting for losses, these small and large BTES systems should be able to store enough heat to cover 50% and 100% of the winter heating demand, respectively. The height-to-width ratio has an effect on heat losses through the uninsulated sides and the insulated top, while the BH density affects the heat transfer power. Having more boreholes connected in series increases the potential temperature change achieved by the underground heat collectors, but reduces the total flow and heat transfer power in the system.

3. Results

3.1. All Cases

The results of the parametric run are presented in Table 2, arranged according to levelized cost of heat.

Table 2. Results of the parametric run arranged by levelized cost of heat (LCOH). Highlighted in bold are three cases determined through the Pareto optimality condition based on maximal heat output and minimal LCOH. See Section 2.4.

Case	V_{BTES}	h_{ratio}	BH Density	N_{series}	BTES Heat Output	BTES Efficiency	Waste Fraction	LCOH
	(m^3)	(m/m)	($1/m^2$)	-	(MWh)	%	%	(€/MWh)
1	**300,000**	**0.5**	**0.05**	**3**	**2761**	**60.1**	**53.9**	**10.5**
7	300,000	1	0.05	3	2652	61.0	52.6	10.5
3	300,000	0.5	0.05	6	2607	60.2	52.0	11.1
13	300,000	2	0.05	3	2368	57.8	49.2	11.6
9	300,000	1	0.05	6	2366	60.3	49.2	11.9
5	300,000	0.5	0.05	9	2385	59.1	49.4	12.2
4	**300,000**	**0.5**	**0.1**	**6**	**4220**	**67.8**	**71.3**	**12.9**
8	300,000	1	0.1	3	4119	67.7	70.1	13.0
2	300,000	0.5	0.1	3	4181	66.3	70.8	13.0
10	300,000	1	0.1	6	4046	69.2	69.2	13.2
6	300,000	0.5	0.1	9	4076	67.9	69.5	13.4
19	600,000	0.5	0.05	3	4186	57.1	70.9	13.6
14	300,000	2	0.1	3	3887	65.9	67.3	13.6
25	600,000	1	0.05	3	4060	58.1	69.4	13.6
21	600,000	0.5	0.05	6	4078	57.9	69.6	14.0
11	300,000	1	0.05	9	2007	57.5	44.9	14.1
12	300,000	1	0.1	9	3783	69.0	66.0	14.2
16	300,000	2	0.1	6	3600	66.8	63.9	14.7
27	600,000	1	0.05	6	3774	58.3	65.9	14.8
15	300,000	2	0.05	6	1857	54.2	43.1	14.8
31	600,000	2	0.05	3	3660	55.0	64.6	14.9
23	600,000	0.5	0.05	9	3804	57.2	66.3	15.1
29	600,000	1	0.05	9	3306	56.5	60.4	16.9
18	300,000	2	0.1	9	3135	65.5	58.3	17.0
33	600,000	2	0.05	6	3082	53.8	57.7	17.8
22	**600,000**	**0.5**	**0.1**	**6**	**5717**	**62.8**	**89.1**	**18.9**
28	600,000	1	0.1	6	5595	64.6	87.7	19.0
24	600,000	0.5	0.1	9	5644	63.3	88.2	19.2
26	600,000	1	0.1	3	5501	62.4	86.5	19.3
20	600,000	0.5	0.1	3	5533	60.7	86.9	19.5
30	600,000	1	0.1	9	5375	64.6	85.0	19.9
32	600,000	2	0.1	3	5277	60.8	83.9	20.0
17	300,000	2	0.05	9	1380	48.1	37.4	20.1
34	600,000	2	0.1	6	5157	62.6	82.4	20.5
36	600,000	2	0.1	9	4684	62.0	76.8	22.8
35	600,000	2	0.05	9	2352	48.7	49.0	23.5

The reported results are the heat extracted from the BTES during the fifth year (BTES heat output), the fraction of stored heat that was utilized during the fifth year (BTES efficiency), the fraction of annual heating demand covered by incineration waste heat during the fifth year and the levelized cost of heat over the whole 25-year calculation period (LCOH).

There was a correlation between higher cost and higher heat output from the storage, but there were also exceptions to the rule. The LCOH in the examined cases ranged from 10.5 to 23.5 €/MWh. Compared with the 56 €/MWh district heating price in Vaasa, the cost of heat was low enough to be of economic interest to utilities. The heat output from BTES was 1380–5717 MWh. The direct utilization of waste heat during summer was 1758 MWh, so the BTES increased waste heat use by 78–325%.

The annual efficiency of the BTES varied between 48.1% and 69.2% and the total waste heat utilization factor (over the whole year) was 37.4–89.1%. The charging temperature in all cases was 75 °C.

Three solutions (cases 1, 4 and 22) were presented in a bold font to show the most cost-effective cases for different energy output levels. These solutions are also shown in Figure 6, alongside all the other examined cases.

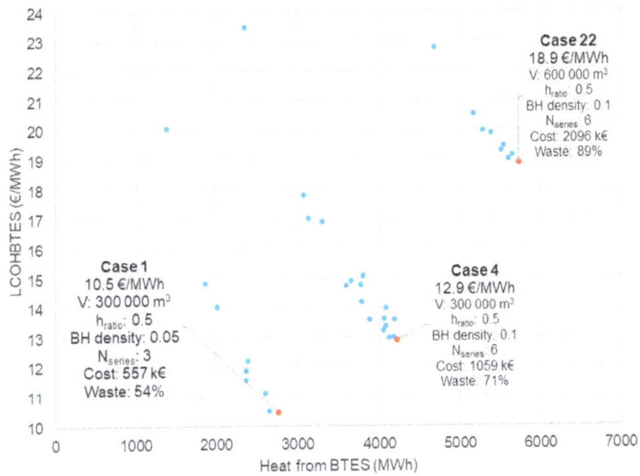

Figure 6. Three highlighted cases based on Pareto optimal selection from all simulated cases.

The solutions were arranged according to LCOH and BTES heat utilization and the three highlighted solutions were selected based on the Pareto optimality concept. A Pareto optimal solution is one where one property of a solution cannot be improved without making another worse. Thus, three points were chosen by trying to minimize LCOH and maximize BTES heat at the same time. Details of the highlighted solutions are reported in Figure 6.

The figure shows three clusters of solutions, which are analyzed further in the following paragraphs. The levelized cost of heat was determined for all cases. Figure 7 shows the LCOH as a function of heat discharged from the BTES during the fifth year of operation, separated by storage size and borehole density.

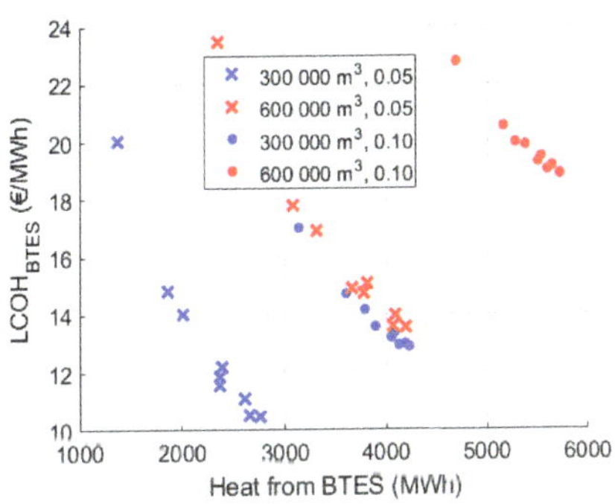

Figure 7. Levelized cost of heat as a function of the fifth-year heat output from the BTES for all examined scenarios. The results are classified according to BTES volume and borehole density.

Three different clusters could be identified along these criteria. The clusters were separated by BTES volume and borehole density. The middle cluster actually combined two different sets of solutions: (i) those with small volume and high borehole density and (ii) those with large volume and low borehole density. Doubling the volume had a similar effect on cost and heat recovery as doubling the borehole density. Both measures increased the number of boreholes, which increased heat transfer capacity and energy provision, but also raised the cost of heating. Increasing the number of boreholes had the same effect under both BTES sizes. Each cluster had a negative linear correlation between heat cost and energy output. The difference in investment costs within each cluster was minimal, which was why increased energy utilization directly reduced the cost of heating.

The same results are presented in Figure 8, but separated by height-to-width ratio and the number of boreholes in series.

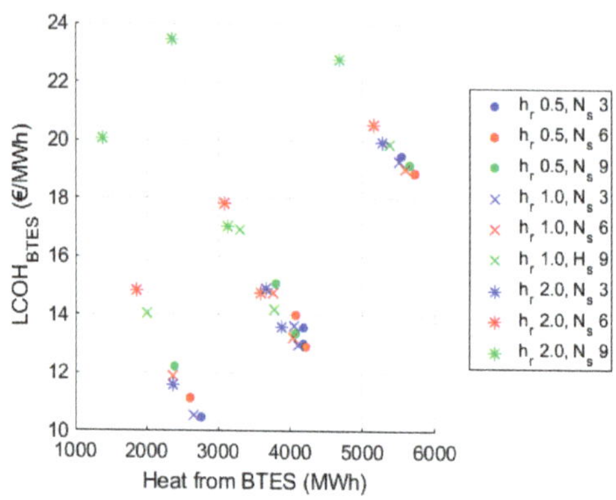

Figure 8. Levelized cost of heat as a function of the fifth-year heat output from the BTES for all examined scenarios. The results are classified according to BTES height-to-width ratio and the number of boreholes connected in series.

In each line of solutions, the most expensive ones were those with a h_{ratio} of 2, meaning BTES systems that were deep and narrow. This could also be seen in Table 2, where the five most expensive solutions had a narrow BTES. If 9 boreholes were connected in series (green), the best shape was a wide one (h_{ratio}, 0.5). Conversely, for a narrow BTES (h_{ratio}, 2), the most cost-effective method was to connect only 3 boreholes in series. When many boreholes were connected in series, the total heat transfer in and out of the storage went down, because the total flow rate was lowered. This might be compensated by increasing the total number of boreholes, which could be done by reducing the height-to-width ratio or by increasing the borehole density. In each solution cluster, the wide BTES shape was the most cost-effective, but the difference to the even shape (h_{ratio}, 1) was not great. There seemed to be a sweet spot of borehole connectivity. Having more boreholes (higher borehole density) favored connecting 6 boreholes in series, instead of 3. More seriality resulted in a higher temperature change of the heat transfer fluid, potentially reducing the need for priming by district heat. However, if the total flow rate, i.e., the number of boreholes was too low, the need for conventional district heating did not change. A balance was required between the output temperature and flow rate on the supply side and power demand and temperature requirement on the demand side. However, the importance of the temperature provided by the BTES would rise if the storage was connected to the DH grid in parallel instead of in series (as in this study).

Figure 9 shows the annual BTES efficiency (fifth year) in all cases as a function of useful heat extraction.

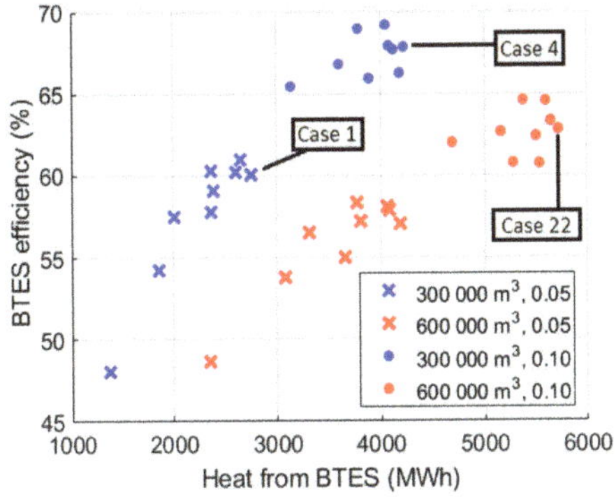

Figure 9. Seasonal storage efficiency as a function of useful heat output from the BTES.

The attained efficiency range in all configurations was 48–69%, with a median of 61%. The highest efficiency was obtained in the small storage. The storage efficiency was generally greater in the small storage than in the large storage, but with a lower energy output. With all other things being equal, a large storage system should have a higher efficiency because of a more favorable surface area-to-volume ratio. However, this is also influenced by the heating demand. At the end of the heating season, the base temperature level remained higher in the large storage than in the small one. This increased the heat losses in the larger storage system. The small storage system was drained more completely, resulting in increased efficiency. It was more cost-effective to size the storage to cover a smaller part of the heating load. On the other hand, a larger storage was required to cover a larger part of heating demand by waste heat.

The BTES configuration influenced the temperature than can be obtained in the storage. Figure 10 shows the storage efficiency vs. the annual maximum temperature of the BTES.

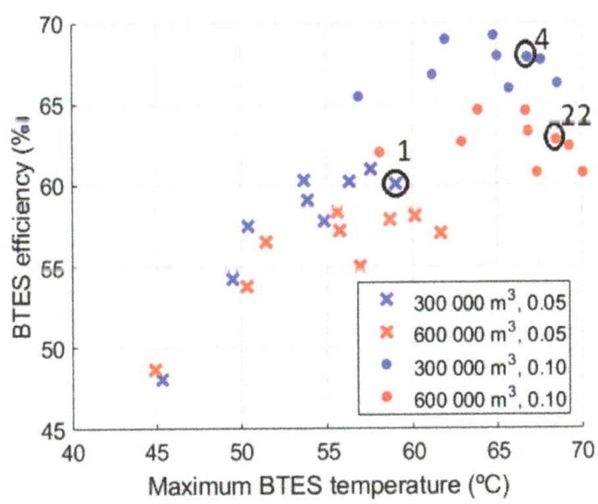

Figure 10. BTES efficiency as a function of maximum storage temperature. The three best cases are highlighted by circles and case numbers.

The results showed a clear correlation between improved efficiency and rising temperature. One could expect the opposite results, where a high temperature increases heat losses and lowers efficiency. However, the temperature of the storage set a maximum limit on the extraction efficiency. Once the demand-side flow was heated up to the BTES temperature, increasing the flow rate on the storage side did not increase heat extraction. This was demonstrated in Figure 10, where the lowest temperature resulted in the lowest efficiency. In all the cases, the size of the seasonal energy storage was so large that the conductive heat losses caused by high temperatures were dominated by the increased ability to meet the heating demands.

3.2. Development of BTES State

Figure 11 shows how the heat storage efficiency of the BTES system developed during the first five years of simulated operation.

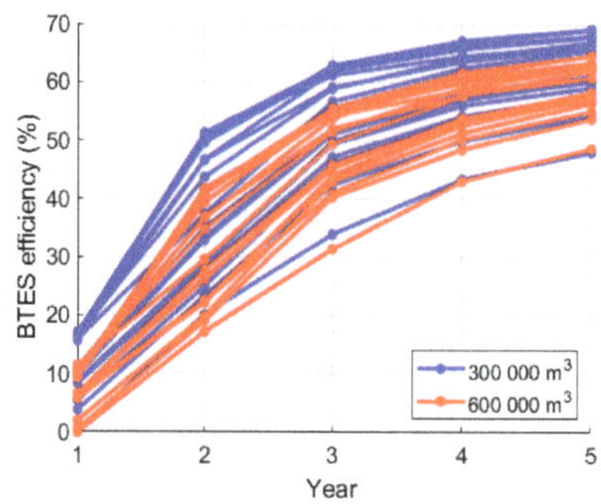

Figure 11. Development of BTES storage efficiency in all the scenarios during the first five years of operation.

During the first year, most of the excess energy went to raising the BTES temperature from the ambient 5.6 °C to the minimum useful temperature of 30 °C, since no heat pumps were included in the system. This resulted in a storage efficiency of less than 20% in every case. The largest rise in efficiency was observed between the first two years, because it was already possible to use the stored heat during the second year. The improvement was smaller every consecutive year as the system started to reach equilibrium. In some cases, the steady-state condition was reached faster than in others. For example, in the highest efficiency case (case 10), the efficiency improved from 63% to 69% between the third and fifth year. In the lowest efficiency case (case 35), the improvement was from 31% to 49%. The latter case involved a large storage system with low borehole density and maximal seriality, which reduced the charging power. With a proper design, i.e., a high enough heat transfer power relative to the storage volume, the BTES system in the highly conductive rocky Finnish ground could be running close to the full efficiency after three years of operation.

Figure 12 shows the development of the waste heat utilization factor over years of operation.

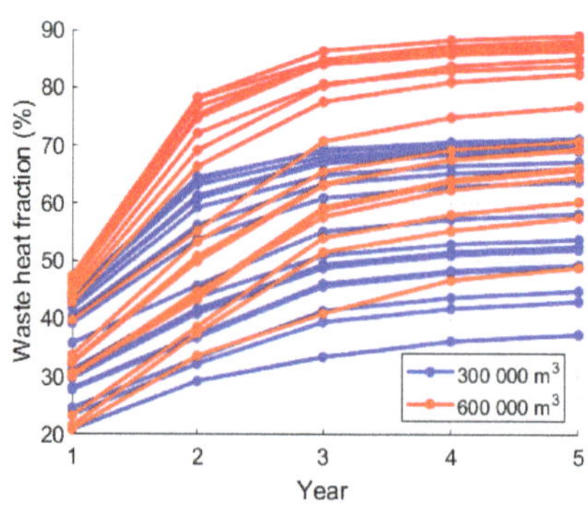

Figure 12. Development of the amount of heat demand covered by waste heat during the first five years of operation.

The cases with a larger storage system tended to have a higher waste heat utilization than those with a smaller storage system. This was opposite to the BTES efficiency, which was increased for smaller storages. In other words, a small storage system allowed a more effective use of the stored waste heat, but reduced the potential to replace conventional district heating. Conversely, in a large BTES, the high temperature levels could be retained for longer, increasing the demand met by waste heat. The minimum first year levels indicated that without seasonal thermal energy storage, less than 20% of the annual heating demand could be met by waste heat.

Figure 13 shows the operational temperature range in the BTES during the last year for all cases.

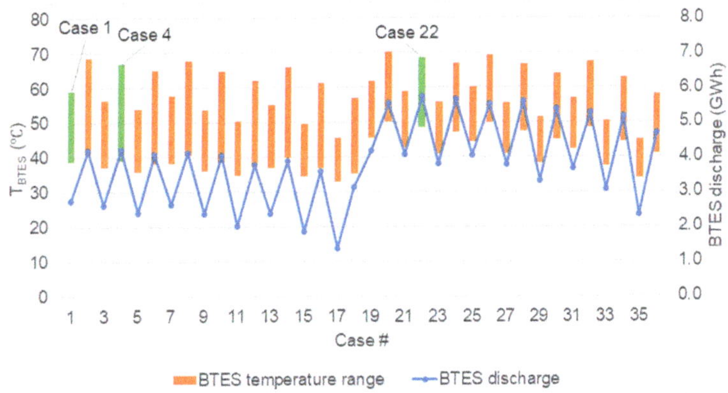

Figure 13. The operational temperature range of the BTES in each studied case during the fifth year. The best cases are identified by green color.

It also showed how much energy was discharged from the storage over the year. Both the maximum temperature and the discharged heat fluctuated up and down systematically. This was caused by the number of boreholes, which is related to the heat transfer capacity of the BTES system. In odd-numbered cases, the borehole density was 0.05 1/m^2, while in even-numbered cases it was 0.1 1/m^2. As it could be assumed, with more boreholes, more energy was charged into the ground and the temperature levels were higher. The annual temperature change in the BTES was also directly related to the number of boreholes, since more boreholes resulted in more heating power. This allowed charging the BTES to higher temperature levels, while also increasing the discharging rate and thus lowering the temperature at the end of the heating season. Cases 31, 33 and 35 demonstrated how increasing borehole seriality from 3 to 6 to 9 in a narrow BTES reduced both the temperature levels and recovered energy. Cases 32, 34 and 36 displayed the same trend, but with a higher number of boreholes. This happened with smaller BTES sizes as well, in cases 13 to 18.

3.3. Performance of the Best Cases

Figure 14 shows the monthly energy flows and the monthly average BTES temperature for the three Pareto optimal system configurations (cases 1, 4 and 22).

It showed how the utilization of stored waste heat increases changed from case 1 to 4 to 22 (waste heat utilization fractions 54%, 71% and 89%, respectively). This was related to the temperature of the BTES. In case 1, the BTES temperature remained below 58 °C and cooled down to 39 °C at the end of the heating season (in April). In case 4, more heat was charged to the ground, raising the peak temperature to 65 °C, while the minimum temperature remained at 40 °C. The increased charging and discharging potentials were caused by the doubling of the number of boreholes, which directly relates to heat transfer power. In case 22, the size of the BTES was doubled compared with the other two cases. Thus, more heat could be charged into the ground. In this case, the peak temperature rose still higher to 67 °C, while the minimum temperature also went up to 49 °C. A larger share of district heating could be displaced because the BTES remained at higher temperatures. In case 22, the energy injected during the first month of charging was more than three times the peak monthly demand in winter. This could be interpreted as there being significant overcapacity in the BTES, but it was mainly because the system was running constantly on full power during the charging period.

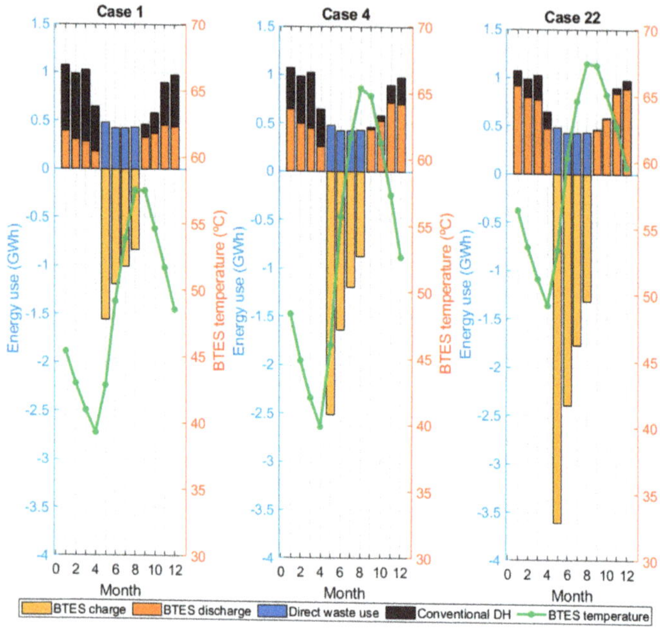

Figure 14. Monthly energy flows and average BTES temperature in the best cases.

In all cases, the charging of the BTES slowed down as summer progressed, because the rising temperature of the BTES reduced the useful fraction of waste heat that could be stored. Injecting more thermal energy to the BTES got harder as the temperature difference between the storage and heat source diminished. The monthly discharge levels showed how the BTES was drained of energy. The stored energy levels at the start of the year were a result of the charging done during the previous year. In each case, the heat discharged from the BTES got lower each month going from December to April. This was because of the lower temperature level in the BTES, which limited the amount of useful heat available. However, at the end of the year (September–December), the discharged energy rose, even though BTES temperature levels were going down. This was due to the increasing heating demand. Higher demand resulted in higher discharge of energy. However, even in the early discharging period, the relative discharge rate, with respect to the demand, was going down.

The heat provided from the storage replaced 2760, 4220 and 5720 MWh of conventional district heating in cases 1, 4 and 22, respectively. This was on top of the energy provided directly by waste heat in the summer. Assuming the Finnish national average for district heating emissions and that heat recovered from waste incineration was emission-free, the BTES system reduced annual CO_2 emissions by 536, 819 and 1109 t-CO_2 in cases 1, 4 and 22, respectively. This corresponded to 42%, 64% and 86% reduction in emissions in cases 1, 4 and 22, respectively, compared with the case using waste incineration heat only in the summer months.

Figure 15 displays the hourly temperature development in the BTES for case 4.

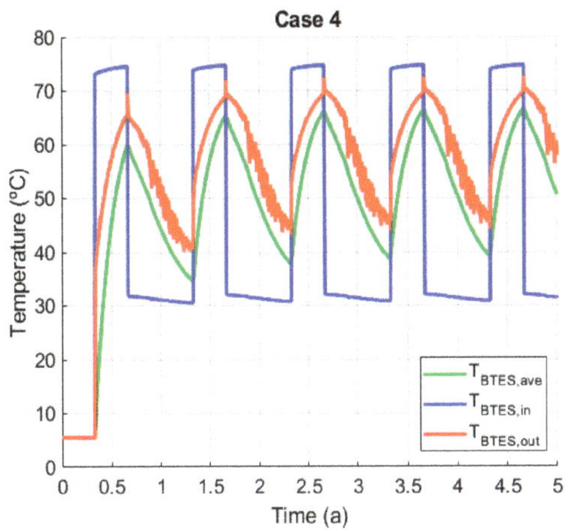

Figure 15. Hourly temperatures in the storage and district heating (DH) system during five years of operation in case 4. The figure shows the average temperature in the storage volume as well as the inlet and outlet temperatures.

It shows the average temperature of the whole storage volume as well as the inlet and outlet temperatures of the charging/discharging flow. The increase in temperature during the first charging period was large and the BTES almost reached the equilibrium temperatures during the second year. At the start of each charging period (excluding the first one), the temperature difference between inlet and outlet was close to 20 °C, but as the BTES heated up, the temperature difference lowered to 6 °C. Similarly, at the start of the discharging period, the BTES was at maximal temperature and the inlet/outlet difference was about 35 °C, but this was lowered to 15 °C by the end of the heating season.

During the later years of operation, the return temperature from the BTES during charging was 55–70 °C. This was still a useful temperature level, which meant that the output from one seasonal energy storage system could be fed to another one. At first, the temperature change of the charging fluid was close to 20 °C, but it was gradually reduced to 6 °C. This revealed an inefficiency in the system, since less and less of the available energy could be utilized. From a global optimization point of view, it would be more useful to heat several BTES systems to lower temperature levels instead of maximizing the utility of a single system. The result might be reduced waste heat utilization in individual communities, but a rise in total efficiency on the city level.

4. Discussion

4.1. Toward a Net-Zero Energy District

The examined district was designed to be powered by external district heating. However, the waste heat provided by the incinerator plant was stored in a local storage system, reducing reliance on the grid during heating season. Strictly speaking, the district is not generating its own energy, but if the concept of net-zero energy district is expanded to cover the whole city, the community as a whole is moved toward net-zero energy when previously wasted heat is utilized. Looking only at the individual district likely results in suboptimal system performance because of the high return temperatures. To properly use the available incineration heat, the return flow from one BTES system should be redirected to another, to create a city-wide charging loop. The optimal temperature levels in such an interconnected system would be worth further study.

As shown in Figure 15, the temperature of the BTES remained above 40 °C at the end of the discharging period, meaning that there was still a lot of energy stored in the ground. With heat pumps, the total energy extraction could be doubled to bring the district close to zero carbon heating (depending on the type of electricity used). Such a district could completely stop utilizing conventional district heating, but to access the waste heat, a DH connection must be maintained regardless. Thus, it may be sensible to use the connection instead of investing in an overlapping heat pump system.

4.2. Implications and Potential

In Finland, only 2.5% of municipal waste goes to landfills [21]. Most of the municipal waste is incinerated for energy. This means that adding to the total waste heat generation is not feasible. Thus, any increased energy use of waste should result from improved utilization rate of the energy generated during low demand. Here, the BTES system shows its importance. District heating in Finland is produced mainly by coal, peat and wood burning, so any shifting of excess waste heat to the heating period has a great potential for emission reduction.

In the European Union (EU-27), 52 Mt of municipal waste is landfilled and 58 Mt is incinerated [22]. The final space heating demand in Europe is estimated to be over 300 TWh, of which the vast majority is produced by natural gas [23]. The share of district heating could be increased significantly, which leaves room for increased use of incineration heat. Most of Europe has increased heating demand in winter, which opens up the potential for storing the summertime excess heat using borehole thermal energy storage.

If the heat storage concept were to be utilized to the maximum potential in as many communities as possible, any individual community could only utilize a fraction of the available excess heat. This would reduce the maximum temperatures possible to reach in the storage systems and create the need for heat pumps to provide more heat. However, the temperature of the ground would still be significantly elevated compared to the natural state, which would improve the coefficient of performance (COP) of the heat pumps. The waste heat could also be used to regenerate the ground to prevent extensive cooling in a scenario with high penetration of ground heat utilization. In such a regenerating role, the waste heat could be distributed widely with little consideration of the temperature.

In a modern floor heating system, the heat distribution temperature is below 40 °C, while the DHW temperature requirement remains at 55 °C, to avoid *Legionella*. If DHW demands were met by building-side heat pumps, the district heating temperature could be lowered. This would reduce losses both in the DH distribution and the seasonal storage. Storage size might also be reduced by draining the stored heat more completely using heat pumps. Of course, the addition of heat pumps would entail costs that would need to be balanced with the obtained benefits. The LCOH of heat generated by the BTES was 10–23 €/MWh, which can be compared to the pre-tax average price of district heating in Finland, 66.5 €/MWh [24]. This leaves plenty of room for profits and payments to the waste heat generator.

4.3. Weaknesses and Reliability

The series connection between the BTES and DH resulted in a high return temperature for the DH flow. This may be a problem for the utility that provides district heating. Another study could present an alternative system, which utilizes a parallel connection, maximizing the cooling in the DH flow.

No actual heating grid was modelled, just the boundary of the residential zone. In practice, there would be some heat losses in the local district heating grid. Heat losses in district heating grids have been estimated to be 8–10% [25], but in a low temperature grid in a cold climate they could be around 5% [26]. Electricity consumption of water pumping was also disregarded. Measurements from a real BTES facility revealed that the electricity consumption of pumping was less than 0.4% of the heat injected into the storage [8]. Thus, the lack of pumping energy should not have a significant influence on the operational cost benefits.

A parametric run does not provide optimal results because many potential system configurations are not tested. Another study could use multi-objective optimization, such as with a genetic algorithm, to further improve the cost-effectiveness of the incineration waste heat storage systems.

5. Conclusions

This study analyzed the potential of a seasonal borehole thermal energy storage system for increasing the utilization of waste heat in a residential community. Heat generated by waste incineration in summer was stored in the ground to preheat district heating flows during the heating season. The main finding was that the storage of waste heat turned out to be a promising low-cost measure to reduce district heating consumption and emissions of heating. The median storage efficiency was 61% in the 36 examined cases, even while the temperature of the storage volume exceeded 60 °C in many cases. This showed that relatively high temperature heat can be stored for long periods of time using large enough storage volumes. The many different simulated storage configurations revealed some guidelines for BTES design.

Compared with only using waste heat from incineration in summer, the use of the BTES increased waste heat use by 78–325% in the 36 examined cases. In the three most cost-effective cases, the levelized cost of heat was 10.5, 12.9 and 18.9 €/MWh for waste heat utilization factors of 54%, 71% and 89%, respectively. Most of the heat required by the community could be provided by the incineration plant. The cost of heat compared to average district heating prices was low enough to be of interest to utilities. A more detailed study considering all the integration costs should be performed.

A BTES system in Finland could be heated up to nearly the final operational temperature level in three years. The density of boreholes in the BTES was essential in determining the energy provided by the storage system. More boreholes meant more heating power and higher temperatures, which directly influenced how much useful heat could be recovered during the heating season. All the best BTES configurations utilized a wide and shallow storage system, which creates a stronger radial temperature gradient and reduces heat losses. When the number of boreholes was increased, it became useful to connect more boreholes in series, to increase the output temperature of the BTES at the cost of a lower flow rate.

In many cases, the return temperature from the storage was high, which reveals a potential for further utilization of the same waste heat stream in other districts. The optimization of a network of storage systems to best utilize a singular waste heat source on the city level would be a good avenue for further study.

In the best three cases, the emissions of district heating in the community could be reduced by 536, 819 and 1109 t-CO_2/a through the use of the BTES, in addition to the direct use of the waste heat in summer. Operators of waste incineration plants should consider building seasonal thermal energy storage systems to communities within reach of heat-generating incineration plants.

Author Contributions: Conceptualization, J.H.; methodology, J.H.; software, J.H.; validation, J.H.; formal analysis, J.H.; investigation, J.H.; resources, J.H. and R.K.; data curation, J.H.; writing—original draft preparation, J.H.; writing—review and editing, J.H. and R.K.; visualization, J.H.; supervision, R.K.; project administration, R.K.; funding acquisition, R.K. All authors have read and agreed to the published version of the manuscript.

Funding: This research was funded by the Academy of Finland under the "Optimal transformation pathway towards the 2050 low carbon target: integrated building, grids and national energy system for the case of Finland", grant number 309064.

Acknowledgments: The authors thank Vaasan Sähkö for providing the idea and basic data for the study.

Conflicts of Interest: The authors declare no conflict of interest. The funders had no role in the design of the study; in the collection, analyses, or interpretation of data; in the writing of the manuscript, or in the decision to publish the results.

References

1. Patronen, J.; Kaura, E.; Torvestad, C. *Nordic Heating and Cooling-Nordic Approach to EU's Heating and Cooling Strategy-Nordic Council of Ministers*; TemaNord: Copenhagen, Denmark, 2017.
2. Vertanen, V. Municipal Waste Treatment in Finland. Available online: http://www.stat.fi/til/jate/2016/13/jate_2016_13_2018-01-15_tau_001_en.html (accessed on 2 May 2019).

3. Gao, L.; Zhao, J.; Tang, Z. A Review on Borehole Seasonal Solar Thermal Energy Storage. *Energy Procedia* **2015**, *70*, 209–218. [CrossRef]
4. Hirvonen, J.; Rehman, H.; Sirén, K. Techno-economic optimization and analysis of a high latitude solar district heating system with seasonal storage, considering different community sizes. *Sol. Energy* **2018**, *162*, 472–488. [CrossRef]
5. Hirvonen, J.; Sirén, K. A novel fully electrified solar heating system with a high renewable fraction-Optimal designs for a high latitude community. *Renew. Energy* **2018**, *127*, 298–309. [CrossRef]
6. Nordell, B.; Andersson, O.; Scorp, A.L. The High Temperature Borehole Thermal Energy Storage Plant in Emmaboda-Report from the First Three Years of Operation 2010–2013, Half-Time Report Luleå University of Technology. Available online: https://www.researchgate.net/publication/273899339_The_high_temperature_Borehole_Thermal_Energy_Storage_plant_in_Emmaboda_-_Report_from_the_first_three_years_of_operation_2010-2013_half-time_report (accessed on 2 April 2019).
7. Nordell, B.; Andersson, O.; Rydell, L.; Scorpo, A.L. Long-term performance of the HT-BTES in Emmaboda, Sweden. Presented at Greenstock 2015. In Proceedings of the International Conference on Underground Thermal Energy Storage, Beijing, China, 19 May 2015.
8. Nilsson, E.; Rohdin, P. Performance evaluation of an industrial borehole thermal energy storage (BTES) project—Experiences from the first seven years of operation. *Renew. Energy* **2019**, *143*, 1022–1034. [CrossRef]
9. Guo, F.; Zhu, X.; Zhang, J.; Yang, X. Large-scale living laboratory of seasonal borehole thermal energy storage system for urban district heating. *Appl. Energy* **2020**, *264*, 114763. [CrossRef]
10. Rapantova, N.; Pospisil, P.; Koziorek, J.; Vojcinak, P.; Grycz, D.; Rozehnal, Z. Optimisation of experimental operation of borehole thermal energy storage. *Appl. Energy* **2016**, *181*, 464–476. [CrossRef]
11. Finnish Meteorological Institute. Test Reference Years for Energy Calculations [In Finnish]. Available online: https://ilmatieteenlaitos.fi/energialaskennan-testivuodet-nyky (accessed on 16 November 2020).
12. Ministry of the Environment. Ympäristöministeriön asetus uuden rakennuksen energiatehokkuudesta (1010/2017) (Decree of the Ministry of the Environment on the energy performance of the new building (1010/2017) [In Finnish]. Available online: https://www.finlex.fi/fi/laki/alkup/2017/20171010 (accessed on 16 November 2020).
13. Hirvonen, J.; Jokisalo, J.; Heljo, J.; Kosonen, R. Towards the EU emissions targets of 2050: Optimal energy renovation measures of Finnish apartment buildings. *Int. J. Sustain. Energy* **2018**. [CrossRef]
14. EQUA Simulation AB, Solna, Sweden. IDA ICE—Simulation Software. Available online: https://www.equa.se/en/ida-ice (accessed on 16 November 2020).
15. Flynn, C.; Sirén, K. Influence of location and design on the performance of a solar district heating system equipped with borehole seasonal storage. *Renew. Energy* **2015**, *81*, 377–388. [CrossRef]
16. Engineering Toolbox. Thermal Conductivity of Selected Materials and Gases. Available online: https://www.engineeringtoolbox.com/thermal-conductivity-d_429.html (accessed on 23 September 2020).
17. Statistics Finland. Tilastokeskus-Energia ja päästöt. Available online: https://pxhopea2.stat.fi/sahkoiset_julkaisut/energia2019/html/suom0011.htm (accessed on 25 September 2020).
18. European Commission. Commission Delegated Regulation Supplementing Directive 2010/31/EU of the European Parliament and of the Council on the Energy Performance of Buildings (Recast) by Establishing Energy Performance Requirements for Buildings and Building Elements. Available online: https://ec.europa.eu/transparency/regdoc/rep/3/2011/EN/3-2011-10050-EN-F1-1.Pdf (accessed on 10 September 2020).
19. Pieskä, M. *Discussion about Waste Heat Storage Potential by a Vaasan Sähkö Oy Representative*; Westenergy Oy Ab: Kvevlax, Finland, 2018.
20. Oy, Helen. Kaukolämmön energia- ja vesivirtamaksut 1.10.2020 alkaen [District Heating Energy and Capacity Costs]. Available online: https://www.helen.fi/globalassets/hinnastot-ja-sopimusehdot/lampo-ja-jaahdytys/kotitaloudet/kaukolammon-energia-ja-vesivirtamaksut-01102020.pdf (accessed on 10 September 2020).
21. Statistics Finland. Waste Treatment in 2018. Available online: https://www.stat.fi/til/jate/2018/jate_2018_2020-06-17_tau_002_en.html (accessed on 28 September 2020).
22. Eurostat. Municipal Waste Statistics-Statistics Explained. Available online: https://ec.europa.eu/eurostat/statistics-explained/index.php/Municipal_waste_statistics (accessed on 28 September 2020).
23. Heat Roadmap Europe. Heating and Cooling Energy Demands. Available online: https://heatroadmap.eu/heating-and-cooling-energy-demand-profiles/ (accessed on 28 September 2020).

24. Statistics Finland. Consumer Prices of Heating Energy in March 2020. Available online: https://www.stat.fi/til/ehi/2020/01/ehi_2020_01_2020-06-11_tau_003_en.html (accessed on 25 September 2020).
25. Poredos, A.; Kitanovski, A. Exergy loss as a basis for the price of thermal energy. *Energy Convers. Manag.* **2001**, *43*, 2163–2173. [CrossRef]
26. Arabkoohsar, A.; Alsagri, A.S. Thermodynamic analysis of ultralow-temperature district heating system with shared power heat pumps and triple-pipes. *Energy* **2020**, *194*, 116918. [CrossRef]

Publisher's Note: MDPI stays neutral with regard to jurisdictional claims in published maps and institutional affiliations.

buildings

MDPI

Article

Emission Reduction Potential of Different Types of Finnish Buildings through Energy Retrofits

Janne Hirvonen [1,*], Juha Jokisalo [1], Paula Sankelo [2], Tuomo Niemelä [3] and Risto Kosonen [1,4]

1 Department of Mechanical Engineering, Aalto University, 00076 Aalto, Finland; juha.jokisalo@aalto.fi (J.J.); risto.kosonen@aalto.fi (R.K.)
2 Finnish Environment Institute, 00790 Helsinki, Finland; paula.sankelo@syke.fi
3 Granlund Consulting Oy, 00701 Helsinki, Finland; tuomo.niemela@granlund.fi
4 College of Urban Construction, Nanjing Tech University, Nanjing 211816, China
* Correspondence: janne.p.hirvonen@aalto.fi

Received: 28 October 2020; Accepted: 3 December 2020; Published: 8 December 2020

Abstract: Energy retrofitting of buildings shows great potential in reducing CO_2 emissions. However, most retrofitting studies only focus on a single building type. This paper shows the relative potential in six Finnish building types, to identify possible focus areas for future retrofits in Finland. Data from previous optimization studies was used to provide optimal cases for comparison. Energy demand of the buildings was generated through dynamic simulation with the IDA-ICE software. The cases were compared according to emissions reduction, investment and life cycle cost. It was found that, in all buildings, it was possible to reduce emissions cost-neutrally by 20% to 70% in buildings with district heating and by 70% to 95% using heat pumps. Single-family homes with oil or wood boilers switching to heat pumps had the greatest emission reduction potential. More stringent requirements for energy efficiency could be mandated during building renovation.

Keywords: energy retrofit; heat pump; CO_2 emissions; building stock

1. Introduction

Energy use in buildings causes 36% of the CO_2 emission in the European Union (EU). This is why the Energy Performance of Buildings Directive (EPBD) declares energy efficiency goals for new buildings in the European Union [1]. However, most buildings in the EU have been built before the EPBD regulations, which is also true for Finland. Relying only on new constructions is too slow to impact CO_2 emissions reduction [2]. Thus, the EPBD has been updated with a requirement for each EU member state to create a roadmap for the renovation of existing buildings [3]. More concretely, in a recent renovation strategy, the European Commission calls for a 60% reduction in the carbon emissions of buildings by the year 2030 [4]. Already, many studies have been done to find the optimal deep retrofit designs for various building types, to minimize cost and energy consumption.

1.1. Localization of Retrofit Solutions

Building retrofits are a timely issue in all parts of the world, both in countries in hot climates seeking to reduce cooling demand and countries in cold climates trying to reduce heating demand. Depending on the building type and climate, the optimal technical solutions are different. Some influential environmental factors are ambient temperature, humidity and solar radiation intensity. On the building-side, differences can arise, for example, from electric load patterns, the use of hot water, and occupational schedule. National policies also influence results, as they may determine the framework for solutions by setting, for example, the base efficiency levels of the current building stock and the requirements for energy efficiency of renovation measures.

Optimization and dynamic building energy simulation have been popular tools for building-related research. For example, in a hot and humid Indian climate, simulation-based optimization was used to design a residential building envelope retrofit, based on phase-change materials and insulation layers [5]. Heat gain in the building was reduced by up to 33.5% after optimal retrofits. A review on building façade retrofits found that, in cooling dominated climates, façade retrofits can reduce energy demand by 15% to 53% [6]. In addition to envelope retrofits, a study based on a South African apartment building included solar panel installation as a building retrofit measure [7]. By optimizing net present value, payback period, and energy savings, energy consumption reduction of 36% to 43% was achieved with a payback period of four years or less. Retrofits of residential villas in Dubai were designed in Reference [8]. A two-stage parametric analysis was used, such that each retrofit measure (insulation, windows and air conditioning) was simulated by itself and then a combined retrofit set was formed based on costs found from a National Renewable Energy Laboratory (NREL) database. In the old buildings of Dubai, with low window-to-wall ratios, thermal insulation of walls was found to be beneficial along with the installation of a more efficient air-conditioning system. Replacing relatively new windows with improved ones was not cost-effective. Since no price data from the United Arab Emirates was available, prices from the USA were used instead. This highlights the need to perform national studies on building retrofits, to provide more information to both businesses and individuals, as well as policymakers.

Air infiltration was found to be of low importance in warm Mediterranean climates [9]. This is because of low temperature differences between indoor and outdoor air compared to cold climates, where infiltration is significant. Temperate climates have their own problems and solutions. Minimizing the cost and greenhouse gas emissions of a retrofitted German office building [10] highlighted solutions, such as improved thermal insulation, increased air-tightness, and a low temperature gas boiler. This is an example of a locally optimal solution, as electricity is very expensive in Germany, while natural gas is relatively cheap, preventing the use of heat pumps (HP). In a study about French houses, thermal insulation of external walls was found to have the highest impact on emissions [11]. On the one hand, deep renovation was not feasible using the French 9-year home improvement loans due to the short amortization period. On the other hand, in a regional level study on the French building stock, 35% reductions in greenhouse gas emissions were obtained with negative costs, when a 50-year horizon was utilized [12]. Retrofits to both the building envelope and heating system were needed to mitigate 70% of emissions. The observed costs were less than 50 €/t-CO_2. This highlights the need for long-term financing, which could be provided through the EU.

Many retrofit optimization studies have also been made for cold climates. Optimization of retrofits on an old Swedish multi-family building showed that improvements to the building envelope or ventilation system were not cost-effective [13]. In fact, the only economical retrofit action was the installation of energy-efficient windows. Another retrofit optimization study of 12 historical residential building types in Sweden also revealed window upgrades as a good solution to improve energy efficiency [14]. Thermal insulation of walls and roof was also cost-effective in many cases. Deep energy retrofits of older Finnish detached single-family houses were examined in Reference [15]. Multi-objective optimization was used to minimize costs and emissions in four age categories of buildings with five different heating systems. Air-source heat pumps were used for auxiliary heating in all optimized buildings and switching from a wood or oil boiler to a ground-source heat pump (GSHP) was the most cost-effective retrofit measure to reduce CO_2 emissions. Similarly, in studies on old Finnish apartment buildings, GSHP was also the most effective way to reduce primary energy consumption [16,17]. An opposite view was presented in a Swedish study that accounted for the whole energy generation chain when considering the energy system retrofit of code compliant and passive level single-family houses [18]. Depending on how the grid electricity was generated, heat pumps could be more CO_2 intensive than district heating (DH) produced by combined heat and power (CHP) plants. This effect stems from heat pumps having to use more high emission electricity during peak demand hours. However, another study found that if all Finnish single-family houses were to perform a deep energy retrofit, the reduction in direct electric heating demand in part of the building stock could compensate for increased heat pump electricity demand in other buildings [19]. At a large scale, the total peak electricity demand could even go down. In Canada, which has a cold climate but with more solar energy than the Nordic countries, solar photovoltaic-thermal collector retrofits in the housing stock could reduce greenhouse gas (GHG) emissions by 17% [20]. Similarly, installing air-to-water heat pumps could results in 23% reduction in GHG emissions [21]. A review of façade retrofit measures showed that façade retrofits are most effective in heating dominated climates, especially ones with high heating degree day values [6]. The review found a range of 7% to 62% energy demand reduction in various studies.

The optimal solutions are influenced by the energy markets and national energy generation systems. The generation mix and local policies influence emissions and the relative benefits of one retrofit measure over another. National energy prices and emission factors were reported in a study that analyzed the cost and emission impacts of energy retrofits in European cities in various climates [22]. For example, electricity cost and emissions were low in France, where electricity is mainly generated by nuclear power [23] and high in Germany, where coal and natural gas are major fuels [24]. Similarly, electrified heating using heat pumps has been economically sensible in Finland [15–17], where the emissions and cost of electricity are relatively low. While the EU calls for electrification of heating, a study based on Canada found that electrification could also increase emissions depending on the local energy infrastructure [25]. This shows that large-scale actions need to be determined according to the local conditions. The most effective solutions will not necessarily be the same even for countries with similar climates. Things like the locally typical façade structure or cost of labor can change the best solutions, even if system efficiencies seem similar on the surface. Thus, the results of any optimization should not be directly utilized in a different context, as many influential factors can change the optimal solutions.

1.2. Retrofits in Different Building Types

Most retrofit studies focus on residential buildings, since they form the majority of the building stock. Other building types have also gained attention. For example, a Finnish office building retrofit was optimized in Reference [26]. The optimized variables were cost and emissions, but the study also took into account the thermal comfort. Thermal comfort of workers could account for 75% of building life cycle cost (LCC). Cost-optimal retrofit solution with a GSHP could reduce CO_2 emissions by 63% while also generating cost-savings. In a study done on large USA office buildings [27], simply adjusting the heating and ventilation setpoints had the potential of reducing energy consumption by 60% in moderate climates but not in cooling dominated climates. In Spain, on the other hand, heating and cooling set points adapted according to the temperature of the previous days were successfully used to reduce both heating and cooling energy consumption by a total of up to 45% [28]. An office building in Hong Kong saw emission reduction of 43% after retrofits [29]. A 19% reduction in CO_2 emissions was obtained in a Turkish university campus building using an optimal combination of energy conservation measures [30]. The energy retrofit of an Italian industrial building with a workshop included measures, such as envelope upgrades, ventilation heat recovery, solar energy and active set point controls [31]. The best out of 1320 examined configurations reduced CO_2 emissions by over 70% without government subsidies. The retrofit study of an Italian hospital involved heat recovery, solar shading and envelope upgrades [32]. With various budget limitations, emission reductions of 180 to 1260 t-CO_2 were obtained. Historical buildings can be a difficult target for retrofits, as visible changes could compromise their special cultural value. However, energy retrofits can also be a protective tool, as improved energy efficiency can ensure that historical buildings will be used even in the future [33].

1.3. Other Aspects of Building Retrofits

Building retrofits can be valuable to investors regardless of any environmental benefits. For example, property values of Canadian office buildings were increased by retrofits [34]. Retrofits could result in decreased operating costs, increased occupancy rates and increased effective rental rates. On the other hand, a large survey of 1550 homeowners in Sweden revealed that energy cost reduction is not a typical reason to renovate a building [35]. Instead, a more common reason is to improve the indoor environment. According to the survey, some barriers to energy renovation are the lack of access to low interest loans and the lack of information related to potential renovation projects. While thermal comfort and indoor air quality are harder to evaluate than energy savings, they do have real value. For example, a Swedish study showed that 77% of retrofit investment cost could be realized as increased value of the house [36]. It also revealed that energy savings promised by dynamic building simulation (58%) were quite close to the actual realized savings (53%). This confirms the importance of dynamic simulation and optimization as a tool for designing emission reducing retrofits.

1.4. Contribution of the Current Study

Optimal renovation measures have been found for several building types in previous studies. In many cases, emissions can be reduced by more than 50%. However, local conditions, such as the climate and energy policy, influence the optimal solutions, while the lack of information can hinder concrete actions. There are differences in the optimal retrofit solutions even between the very similar Nordic countries. Retrofit solutions cannot be copied directly from country to another. The key questions are: In which building types in Finland can the retrofits be done at the lowest cost? Which building types have the greatest impact on Finnish emissions? This paper compiles results from several Finnish building retrofit studies. The results from other studies are adjusted so that the same discounting and energy price source is used in every case, allowing for better comparison of the different cases. The main contribution is to show the potential impact of energy retrofits in different kinds of Finnish buildings. The novelty of this study is in revealing where the building retrofits can have the greatest impact. Compiling the results together helps in gaining a better view of the whole. This can aid decision makers to evaluate their cost and impact relative to other emission reducing actions, such as those in the energy production sector, and to choose priorities within and outside the building sector.

2. Materials and Methods

2.1. Finnish Building Stock

The Finnish building stock is composed of buildings of various ages and many different uses, though it is dominated by residential buildings, as shown in Figure 1. In this study, six common building types have been chosen to provide an insight into retrofit potential and to show which building types to focus on: detached houses [15], multi-story apartment buildings [37], elderly care buildings [38], office buildings [26], educational buildings [39], and commercial buildings [40]. Each building type has been studied in prior research papers. The chosen building types cover 79% of the Finnish building stock [41] and provide an instructive view on the emission reduction possibilities of building retrofits. The buildings excluded from this study are very heterogeneous buildings where energy retrofits are unlikely to happen (such as warehouses) or industrial buildings for which the CO_2 emissions are calculated as part of the industrial sector, instead of the building sector. This means understanding the efficiency, impact, and cost-effectiveness of actions in different building types. Decision-makers can use the information when deciding on retrofitting priorities or the allocation of government grants and other support. Since residential buildings dominate the building stock, they were examined in more detail, using several age classes of buildings.

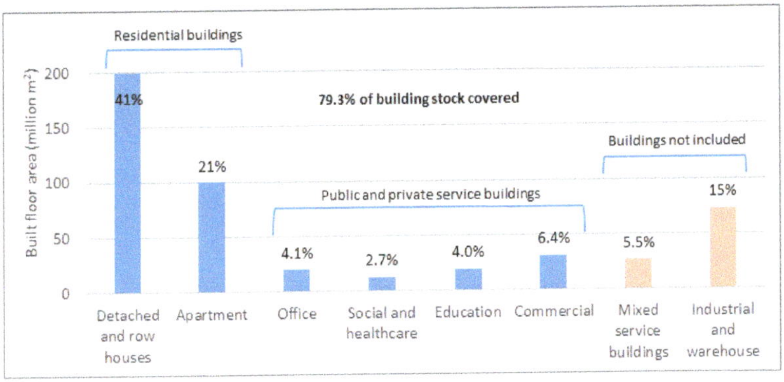

Figure 1. The amount of different building types in the Finnish building stock.

2.2. Simulation and Optimization

Results from dynamic IDA-ICE (IDA Indoor Climate and Energy) [42] simulations from the previous studies were used to determine the energy demand for each building type in this study. IDA-ICE 4.8 (Equa Simulations AB, Stockholm, Sweden) is a comprehensive dynamic building simulation software [43] that takes into account the building geometry, internal structure of the envelope, thermal mass, infiltration, internal gains, heating curves and plant efficiencies etc. It has been validated, for example, in Reference [44,45]. The calculation process is shown in Figure 2. The configuration of each building (envelope properties, solar energy capacity etc.) was determined using the optimized results from previous studies. The optimization in the previous studies was performed using the MOBO (Multi-Objective Building Optimizer) optimization tool [46], which is based on the genetic algorithm NSGA-II (Non-dominated sorting genetic algorithm II) [47]. The use of an optimization algorithm removes the bias of pre-selecting the combinations of retrofit options. Since the original results were obtained using multi-objective optimization, there were many different optimal solutions, each with a unique combination of retrofit options and different cost and emissions. To make the various cases comparable, cost-neutral retrofitting was used by selecting retrofitted cases which had the same life cycle cost as their reference case over a lifetime of 25 years. The selection principle is shown in Figure 3. Finally, the energy costs were recalculated using the same discounting factors for all building types. The annual energy demand in the studied buildings is shown in the Appendix A, in Table A1.

2.3. Building Information—Residential Buildings

2.3.1. Single-Family Houses

The single-family houses (SH) were divided into four age categories, according to the building code in effect at the time of construction (–1975, –2002, –2009, 2010–, for SH1 to SH4, respectively). The basic form of the house was the same for all age categories, a two-story building with a square base and 180 m^2 heated net area. There were several options of main heating systems used in the buildings: district heating (DH), oil/wood boiler, direct electric heating, and ground-source heat pump (GSHP). In the results section of this compilation study, the older buildings SH1 and SH2 have been grouped together, as has been done for the newer building SH3 and SH4.

Heating demand for the single-family houses with on-site boilers were altered from the original source [15] by accounting for oil and pellet boiler efficiencies, which were missing in the original study. The efficiency was 0.81 for the oil boiler and 0.75 for the pellet boiler [48]. This increased the final energy demand, emissions, and cost for these cases.

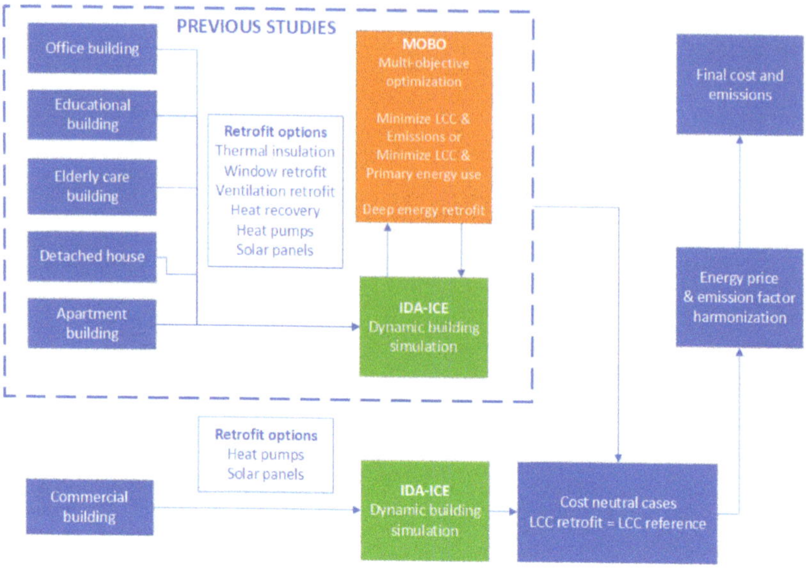

Figure 2. Flowchart of the calculation process.

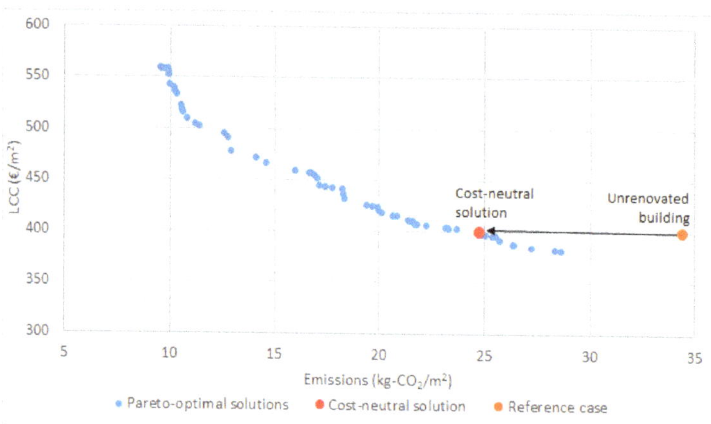

Figure 3. Selection of cases for this study. The cost-neutral cases were chosen from the Pareto fronts generated in previous studies.

When possible, the chosen retrofit level was cost-neutral. In case of buildings with direct electric heating, almost all the solutions were lower cost than the reference case and thus the level B retrofit from Reference [15] was chosen. Similarly, for oil heating replaced by GSHP. In all retrofitted buildings (except the GSHP cases), air-to-air heat pumps (AAHP) were installed due to their cost-effectiveness. Solar thermal and solar electric systems were also utilized in most cases. In houses built before the year 2003, new windows and additional thermal insulation to external walls were installed. This was done in many newer building configurations, as well. The details of the apartment building properties are shown Table 1. The abbreviated ventilation systems mentioned in the table are natural stack ventilation, mechanical exhaust ventilation, and mechanical balanced ventilation with heat recovery. Variable air volume ventilation (VAV) was performed by reducing ventilation flows according to occupation, down to a minimum of 40% during no occupation.

Table 1. Properties of cost-neutrally retrofitted single-family houses (SH). Buildings from four time periods are presented: SH1 (–1975), SH2 (1976–2002), SH3 (2003–2009), and SH4 (2010–).

-	-	Building Envelope				Building Service Systems					
Building	**Heating System**	**Walls**	**Roof**	**Doors**	**Windows**	**Ventilation**	**Radiator**	**GSHP**	**AAHP**	**PV**	**ST**
-	-	W/m^2 K				(HR eff)	°C/°C	kW$_{th}$	kW$_{th}$	kW	m^2
SH1	DH	0.12	0.1	1.4	1	Natural (0%)	70/40	0	6	9	8
SH2	DH	0.12	0.08	1.4	1.5	Exhaust (0%)	70/40	0	3	0	8
SH3	DH	0.17	0.08	1.4	1.4	Balanced, VAV (55%)	45/35	0	5	1	6
SH4	DH	0.17	0.09	1	1	Balanced, VAV (65%)	45/35	0	3	0	4
SH1	Oil → GSHP	0.1	0.09	1	0.5	Natural (0%)	45/35	8	0	6	20
SH1	Wood → GSHP	0.2	0.12	1.4	0.5	Natural (0%)	45/35	7	0	10	0
SH2	Oil → GSHP	0.08	0.09	1	0.5	Exhaust (0%)	45/35	12	0	8	18
SH2	Wood → GSHP	0.19	0.09	1.4	1.5	Exhaust (0%)	45/35	7	0	10	2
SH3	Oil → GSHP	0.08	0.08	1	0.5	Balanced, VAV (75%)	45/35	6	0	8	16
SH3	Wood → GSHP	0.14	0.09	1.4	1.4	Balanced, VAV (75%)	45/35	7	0	9	0
SH4	Oil → GSHP	0.08	0.08	0.8	1	Balanced, VAV (75%)	45/35	14	0	9	10
SH4	Wood → GSHP	0.11	0.09	1	1	Balanced, VAV (65%)	45/35	5	0	10	0
SH1	Wood	0.12	0.09	1	1.8	Natural (0%)	70/40	0	5	0	12
SH2	Wood	0.19	0.1	1.4	1.6	Exhaust (0%)	70/40	0	3	0	4
SH3	Wood	0.25	0.09	1.4	1.4	Balanced, VAV (55%)	45/35	0	3	0	2
SH4	Wood	0.17	0.09	1	1	Balanced, VAV (65%)	45/35	0	1	0	4
SH1	Elec	0.1	0.09	0.8	0.6	Natural (0%)	70/40	0	3	8	18
SH2	Elec	0.08	0.08	0.8	0.6	Exhaust (0%)	70/40	0	4	7	20
SH3	Elec	0.1	0.07	0.8	0.6	Balanced, VAV (75%)	45/35	0	4	8	18
SH4	Elec	0.07	0.08	1	1	Balanced, VAV (75%)	45/35	0	4	8	18

The arrow in the Heating system column indicates a replacement of the heating system with another. Improvements over the reference case have been highlighted in green.

For further reading on the optimization and emission reduction in the single-family houses, please see Reference [15]. On the effects of retrofitting on the heating and electric power demand, see Reference [19].

2.3.2. Apartment Buildings

The apartment buildings (AB) were divided into four age categories, according to the building code in effect at the time of construction (–1975, –2002, –2009, 2010–, for AB1 to AB4, respectively). The buildings of different age were also of different shapes and sizes, ranging from large buildings with 4050 m^2 heated net area to smaller ones at 1585 m^2. District heating was used in each reference case. In the results section of this compilation study, the older buildings, AB1 and AB2, have been grouped together, as has been done for the newer buildings, AB3 and AB4.

The retrofit actions common to all cost-neutral apartment building retrofits included the installation of sewage heat recovery, solar photovoltaic (PV) panels, and solar thermal collectors. In the two older building categories, additional thermal insulation of external walls and roof were also utilized and energy-efficient windows were installed. In new buildings with existing mechanical balanced ventilation, demand-based ventilation was installed during the retrofit in all cases. The list of building properties for all retrofitted cases is shown in Table 2. The ventilation systems mentioned in the table are mechanical exhaust ventilation and mechanical balanced ventilation with heat recovery (HR).

Additional information on the apartment building retrofit optimization can be found in Reference [37]. The impact of retrofits on the hourly power demand of heating and electricity is presented in Reference [49].

2.4. *Building Information—Public and Private Service Buildings*

Table 3 shows the properties of all the retrofitted service buildings. Since their share of the building stock is much smaller than that of residential buildings, these building types were not divided into as many categories. The ventilation controls reported in Table 3 are constant air volume ventilation according to predetermined schedules (CAV, sched) and variable air volume ventilation controlled by CO_2 and temperature sensors (VAV, CO_2 + T). The following sections briefly describe the properties of the retrofitted buildings.

2.4.1. Elderly Care Buildings

The elderly care buildings included an old reference building (pre-1980) and a retrofitted building with either district heating or an air-to-water heat pump. The building size was comparable to a large apartment building at 4709 m^2 heated net area. Retrofit actions in the elderly care building included the installation of additional thermal insulation in the external walls and roof, as well as the installation of energy-efficient windows. Ventilation heat recovery was also installed, along with automated lighting controls, PV panels, and solar thermal collectors. The properties of the retrofitted buildings are shown in Table 3. For additional details on the retrofitting process, please refer to Reference [38].

2.4.2. Office Buildings

The office buildings included an old reference building (pre-1980) and a retrofitted building with either district heating or a ground-source heat pump. It was a large building with a 13,400 m^2 heated net area. The utilized retrofit actions were the installation of ventilation heat recovery and energy-efficient windows, the use of CO_2-controlled VAV ventilation, and the installation of LED lighting and PV panels. Automated lighting control and blinds between the windows were also used in the GSHP case. The properties of the retrofitted buildings are shown in Table 3. More details on the buildings and their optimization can be found in Reference [26].

Table 2. Properties of cost-neutrally retrofitted apartment buildings (AB). Buildings from four time periods are presented: AB1 (–1975), AB2 (1976–2002), AB3 (2003–2009), and AB4 (2010–).

-	-	Building Envelope				Building Service Systems							
Building	Heating System	Walls	Roof	Doors	Windows	Ventilation System and Control		Radiator Temp	GSHP/ EAHP	Backup Heating	Sewage HR	PV	ST
-	-	W/m^{2}K				(HR eff)	-	°C/°C	kW$_{th}$	kW	-	kW	m^2
AB1	DH	0.81	0.08	2.2	0.7	Exhaust (0%)	CAV	70/40	-	-	HP	30	55
AB2	DH	0.34	0.26	0.7	1	Exhaust (0%)	CAV	70/40	-	-	HX	25	100
AB3	DH	0.25	0.07	1.4	1.4	Balanced (60%)	VAV	70/40	-	-	HX	15	50
AB4	DH	0.17	0.09	1	1	Balanced (65%)	VAV	45/35	-	-	HX	15	45
AB1	EAHP	0.23	0.1	1	0.8	Exhaust (0%)	CAV	70/40	35	DH	HP	40	0
AB2	EAHP	0.34	0.26	0.7	0.6	Exhaust (0%)	CAV	70/40	25	DH	HP	45	35
AB1	GSHP	0.36	0.08	0.7	0.7	Exhaust (0%)	CAV	45/35	110	Electric	HP	35	60
AB2	GSHP	0.34	0.26	1.4	0.7	Exhaust (0%)	CAV	65/40	35	Electric	HP	35	25
AB3	GSHP	0.25	0.06	0.7	1.4	Balanced (60%)	VAV	70/40	25	Electric	HX	20	60
AB4	GSHP	0.17	0.09	1	1	Balanced (65%)	VAV	45/35	25	Electric	HX	25	30

Improvements over the reference case have been highlighted in green.

Table 3. Properties of the service buildings.

-	-	Building Envelope			Building Service Systems						
Building	Heating System	Walls	Roof	Windows	Ventilation System & Control		GSHP/AWHP	Backup Heating	PV	ST	Other
-	-	W/m^{2}K			(HR eff)		kW$_{th}$		kW	m^2	
Elderly	DH	0.27	0.08	0.6	Balanced (72%)	CAV, sched	-	-	95	119	automated lights
Elderly	AWHP	0.17	0.08	0.5	Balanced (72%)	CAV, sched	175 (81%)	Electric	153	118	automated lights
Educational	DH	0.54	0.17	1	Balanced (77%)	CAV, sched	-	-	347	168	
Educational	GSHP	0.54	0.09	0.7	Balanced (77%)	CAV, sched	42 (3.3%)	DH	484	0	-
Office	DH	0.35	0.1	0.6	Balanced (77%)	VAV, CO_2 + T	-	-	74	0	LED
Office	GSHP	0.35	0.29	0.7	Balanced (77%)	VAV, CO_2 + T	276 (104%)	DH	76	0	LED, automated lights
Commercial	DH, cost-neutral	0.28	0.22	1.4	Balanced (60%)	CAV, sched	-	-	620	0	-
Commercial	GSHP, cost-neutral	0.28	0.22	1.4	Balanced (60%)	CAV, sched	121 (67%)	DH	650	0	-
Commercial	DH, min cost	0.28	0.22	1.4	Balanced (60%)	CAV, sched	-	-	180	0	-
Commercial	GSHP, min cost	0.28	0.22	1.4	Balanced (50%)	CAV, sched	121 (67%)	DH	195	0	-

Improvements over the reference case have been highlighted in green.

2.4.3. Educational Buildings

The educational buildings included an old reference building (pre-1980) and a retrofitted building with either district heating or a ground-source heat pump. The example building was a large university campus building with 19,000 m^2 of heated floor area. The utilized retrofit actions include the installation of ventilation heat recovery and energy-efficient windows. In the DH case, both solar thermal collectors and PV panels were installed. In the GSHP case, only PV panels were installed, but additional thermal insulation was added in the roof. Because the use profile of the building had sharp changes in demand, the cost-effective heat pump size was very small and district heating provided most of the heating energy even in the GSHP configuration. The properties of the retrofitted buildings are shown in Table 3. A detailed account of the educational building optimization can be found in Reference [39].

2.4.4. Commercial Buildings

The commercial building was a 2554 m^2 hall-like retail space with side offices. The reference building was based on a new commercial building presented in Reference [40], which was then downgraded to match the building code of the 1980s (lower insulation values, no ventilation heat recovery). Unlike the others, this building type was not based on a previous optimization study. With the commercial building, the retrofit consisted only of the installation of solar panels and the installation of GSHP. The cost-neutral level of solar panels was very large, with 80% of power being sold back to the grid. For comparison, a case with a smaller, minimum LCC PV array was also simulated and presented in this compilation. The properties of the retrofitted buildings are shown in Table 3.

2.5. Costs and Emissions

The LCC of the different cases was determined as the discounted sum of the initial investment and lifetime energy use costs. The investment costs were taken directly from the previous optimization studies. The energy costs were recalculated using the same discounting factors for every case. The interest rate was 3%, while the annual energy price escalation rate was 2%. The calculation period was 25 years. The costs of district heating are shown in Table 4. Monthly DH prices were weighted accordingly to get a constant energy cost. The prices of electricity are shown in Table 5. The price consists of the fixed and consumption-based costs, the latter consisting of distribution costs, market-based energy cost, and the national electricity tax. The costs for electricity distribution were influenced by the maximum power demand and season. Since 70% of electricity consumption happens during the colder half of the year, similar weighting was used to get a constant electricity distribution cost. The cost of heating oil was 104 €/MWh and the cost of wood fuel was 56 €/MWh.

Table 4. District heating costs for different building types. P_{DH} is the annual peak district heating power demand. Capacity costs are annual [50,51].

Building	SH	Elder, AB, Office GSHP	Commercial, Office DH	AB1 DH	Educational
Capacity cost (€/kW + €)	$5.15 \times P_{DH} - 6.1$	$53 \times P_{DH} - 60$	$38 \times P_{DH} + 1220$	$36 \times P_{DH} + 3000$	$12 \times P_{DH} + 14{,}020$
Energy cost (€/MWh)	52	52	52	52	52

The emission factors used in the calculations are shown in Table 6. Here, the emissions of wood-burning were given according to actual carbon content, though typically they are assumed to be zero due to absorption in new tree growth. An important local factor on the results of the study is the low emission factor of electricity compared to district heating. Even though the efficiency of heat generation in combustion power plants (with cogeneration or not) is higher than electricity generation in typical power plants, the electricity sector in Finland also includes significant amounts of zero carbon electricity generation, like nuclear, hydro, and wind power. This can be expected to favor electrical heating solutions.

Table 5. Electricity costs for different building types [52,53].

Building	SH Non-Electric	SH Electric	Elder DH, Commercial, AB	Educational, Office, Elder Electric
Fixed cost (€/month)	5.51	17.5	32.24	217
Capacity cost (€/kW/month)	0	1.59	5.58	4.56
Distribution (€/MWh)	40.1	22.2	17.7	14.6
Energy (€/MWh)	56.3	56.3	56.3	56.3
Tax (€/MWh)	27.94	27.94	27.94	27.94

Table 6. Emission factors of energy sources.

Energy Source	Emission Factor (kg-CO_2/MWh)
Oil	263 [54]
Wood	403 [54]
District heating	164 [55]
Electricity	133 [56]

3. Results

Figure 4 shows how much the CO_2 emissions can be reduced in different building types with different heating systems versus what would be the investment cost per heated floor area. The selected cases represent buildings for which a cost-neutral retrofit was done, so that the life cycle cost of the retrofitted case was the same as for the reference case without retrofits. Mostly the cost-neutral measures provided emission reductions of 2 to 40 kg-CO_2/m^2/a, but, in single-family houses with on-site boilers, the reduction potential was as much as 40 to 115 kg-CO_2/m^2/a.

The results in Figure 4 were grouped by a line to high and low-cost cases. On the high cost side were the electrically heated single-family houses, new wood heated single-family houses, commercial buildings and some district heated residential buildings. The position of electrically heated houses is explained by the high cost of electric heating and the resulting high LCC of the reference case. Since the retrofit cases were chosen by cost-neutrality with respect to LCC, this resulted in high investment costs and low cost-efficiency due to the low emission factor of electricity. The cost-neutral commercial building retrofit was also on the high side, because a large number of solar panels were cost-neutral in the long term, but had diminishing emission reducing effects. On the lower cost side, there are the office, educational and elderly care buildings, as well as many of the residential buildings. When considering where to focus retrofit actions and the use of limited funding, buildings on the right side of the dividing line are good candidates. Lower emission reduction unit costs help create more impact for the same investment.

The greatest emission reduction potential was in single-family houses where wood or oil heating was replaced by GSHP (orange squares). This was to be expected, since the emissions of wood and oil are high versus the emissions of electricity. Here, the investment cost was between 200 to 500 €/m^2 and emission reductions between 30 and 120 kg-CO_2/m^2/a. In electrically heated buildings, the chosen measures were costly (300 to 400 €/m^2) with somewhat low emission reduction potential (10 to 20 kg-CO_2/m^2/a).

The other building types are clustered on the lower side of emission reduction and have roughly linear investment cost versus emission reduction relation 50 to 300 €/m^2 and 2 to 40 kg-CO_2/m^2/a. GSHP installation in old apartment buildings (AB1 and AB2) reduced emissions by over 20 kg-CO_2/m^2/a, while, in new buildings (AB3 and AB4), it was between 10 to 15 kg-CO_2/m^2/a. GSHP systems had higher investment costs than district heating systems in the same building type.

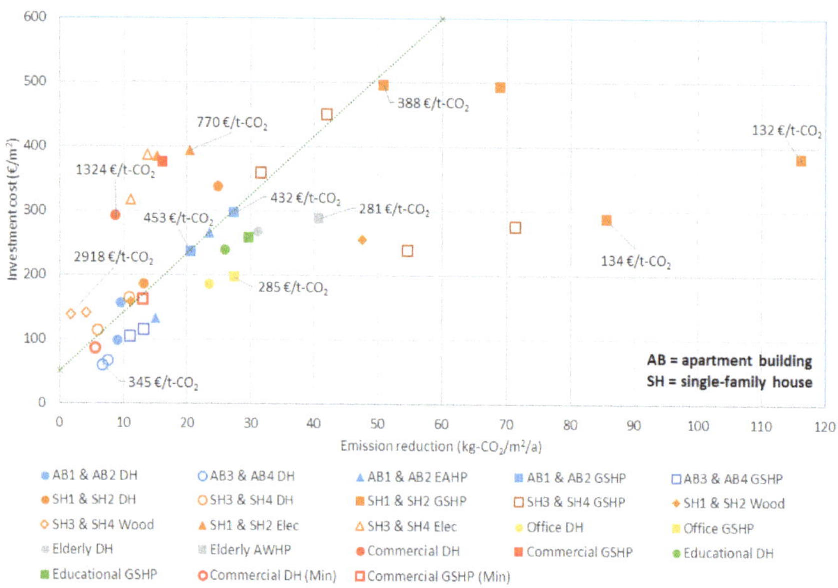

Figure 4. Investment cost for different levels of emission reduction. Different building types are identified by color, while the heating systems are identified by the shape of the symbol. The numbers next to some of the points are examples of emission reduction cost, i.e., what is the investment cost to reduce CO_2 emissions by one ton over 25 years.

Elderly homes (gray) had the largest emission reduction potential of the non-residential buildings. For commercial buildings (red), going from minimum cost PV array to cost-neutral PV array reduced CO_2 emissions only by a little, but significantly increased investment costs (+200 €/m^2). Emissions in office buildings (yellow) and educational buildings (green) could be reduced by about 25 kg-CO_2/m^2/a. Elderly homes and other social or healthcare buildings are especially good candidates for energy retrofits due to their round-the-clock use. Schools are closed in summer and offices during the weekends, but the need for social care buildings remains constant.

Figure 4 also presents some values for emission reduction cost, calculated as investment cost divided by cumulative emission reductions over a 25-year period. These costs ranged from 132 €/t-CO_2 in the most cost-effective case (old single-family house switching from a wood boiler to a GSHP) to 2918 €/t-CO_2 in the least cost-effective case (new single-family house that keeps using a wood boiler). Of course, these investments also result in reduced operation expenses which are not counted in this value. All the cases were cost-neutral with respect to LCC in the original sources, so, in that sense, the lifetime reduction cost was zero.

Figure 5 shows the emission reductions versus life cycle costs. Using LCC, the monetary value of energy savings is taken into account. All examined building retrofits fit between 200 and 600 €/m^2 during the life cycle. The retrofit cases in Figure 5 were grouped to low and high cost cases by a line of constant LCC/emission reduction. All the service buildings were on the low cost side, except for district heated commercial buildings. The priority message in Figure 5 is similar to Figure 4.

In both new and old single-family houses, priority should be given to replace on-site wood and oil boilers with GSHP. Electrically heated houses have a high LCC and average emission reduction because the energy source is expensive but has a low emission factor. In wood heated houses the energy source is cheap, which reduces the economic potential of energy efficiency improvements. In educational and office buildings, the DH and HP cases are very similar in terms of emissions and LCC. Looking at the commercial building, the cost difference between the minimum cost and cost-neutral cases are much smaller using LCC compared to just the investment costs.

Figure 6 shows emission reductions as a percentage value. This reduces the weight of old buildings versus new. The life cycle cost range for the cost-neutral building energy retrofits is 200 to 600 €/m². More specifically, most cases have a life cycle cost of 300 to 500 €/m².

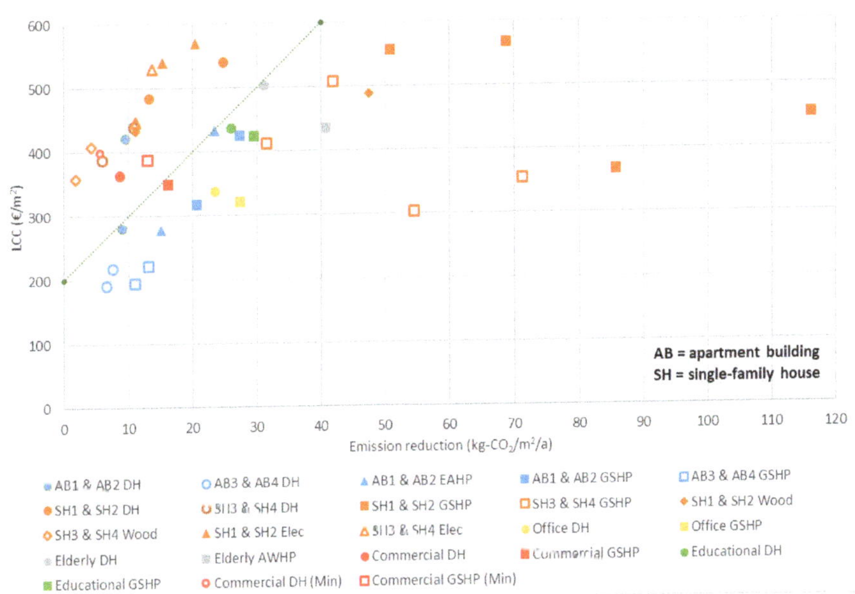

Figure 5. Life cycle cost for different levels of emission reduction.

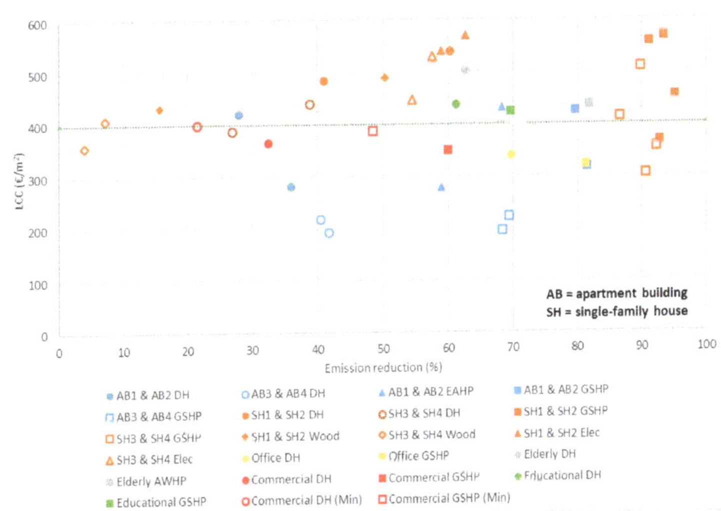

Figure 6. Life cycle cost (LCC) versus emission reduction as a percentage relative to unrenovated reference cases.

GSHP systems provided the biggest relative reductions, 48% to 95%. With district heating, the range for reductions was 20% to 70%. Smallest changes happened in single-family houses that kept using wood boilers. The biggest changes were in single-family houses that switched from on-site boilers to GSHP. Switching from DH to GSHP was also effective for reducing emissions. In most service buildings, the emission reduction potential was over 60% with both DH and heat pumps. This exceeds majorly, for example, the reduction potential in district heated old apartment buildings. In commercial buildings with district heating, simply the addition of solar electricity could reduce emissions by more than 30%, which is a good achievement considering the heating-dominated climate and the minimal solar energy availability during the winter.

Figure 7 shows the absolute and relative emission reductions in the retrofitted buildings. In more than half the cases, it was possible to reduce emissions by more than 50%. Absolute emissions have a roughly linear correlation with relative emissions until about 80% reductions, but there are also cases where the absolute emission reductions between different buildings are the same, even though they have very different relative emission reductions. The reduction potential for district heated buildings was between 20% and 70%, while, in buildings switching to heat pumps, it was 70% to 95%. In the different types of service buildings, the emission reduction potential was between 60% and 85%, and great reductions were possible with both heat pumps and district heating.

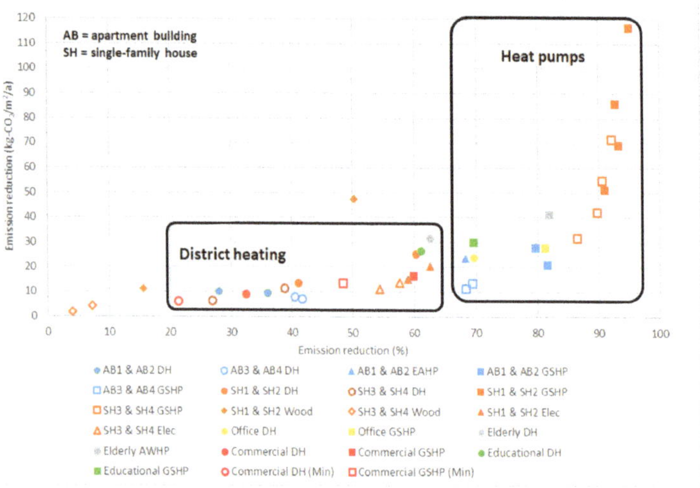

Figure 7. Absolute versus relative emission reduction in all building types.

4. Discussion

4.1. Emission Reduction Potential and Emission Prices

In most of the examined buildings, emission reductions of 30% to 70% could be achieved in a cost-neutral manner and in some buildings as much as 80% to 95%. The investment cost for most buildings was 100 to 400 €/m^2. The emission reduction potential in the majority of examined buildings was 5 to 40 kg-CO_2/m^2/a, but reductions of 40 to 115 kg-CO_2/m^2/a were achievable in single-family homes with oil or wood based boilers, which represent a significant share of the whole building stock. A detailed examination of emission reduction potential in the whole building stock was out of the scope of this paper. However, Finland has 300 million-m^2 of residential buildings and 110 million-m^2 of service buildings [57]. It is estimated that heating of buildings is responsible for 7.8 Mt-CO_2 or 17% of Finnish emissions [58] and that all energy consumption in buildings causes 30% of Finnish emissions [59] or 17 Mt-CO_2. A separate study is forthcoming from Hirvonen et al. to analyze the emission reduction potential in the building stock, taking into account the shares of buildings built in different periods and the distribution of different heating systems. In addition, the economic benefits in the private and public sector should be analyzed as large-scale retrofits also influence the job market and tax revenues.

The minimum investment cost to mitigate 25 years of CO_2 emissions by cost-neutral building renovation was 132 €/t-CO_2. The average cost was 578 €/t-CO_2. This can be compared to the cost of emission reducing measures in the power generation sector and to the cost of carbon taxes or carbon absorption schemes. For example, the Finnish CO_2 tax related to gasoline is 83 €/t-CO_2 [60]. However, if all the taxes on gasoline were accounted as CO_2 tax, the cost would be 336 €/t-CO_2. The cost of compensating for CO_2 emissions by planting trees in Finland over 25 years is 23 €/t-CO_2 [61]. The cost of emission allowances in the European Emission Trading System (EU ETS) is about 25 €/t in the year 2020. This suggests an underpricing of emission allowances, as even cost-effective building energy retrofits are many times more expensive than simply purchasing the right to emit. Of course, these retrofit measures also reduce energy consumption and thus provide monetary savings, as well. In the life cycle point-of-view, the emission reduction cost in this study was roughly zero in all cases. In addition, building retrofits are always long-term affairs that can produce cumulative emission reductions even past the calculation period.

4.2. *Benefits of Building Energy Retrofits*

Building energy retrofits reduce the cost of living and often pay for themselves over time. However, there are also other benefits in doing energy retrofits. For example, energy retrofits increase property values, which can reduce the effective cost of retrofitting. It was estimated that 77% of the investments made into the retrofits of Danish single-family houses was transferred into property values [36]. Reduced heating expenses are valuable even when moving out of the house, as the new owners are willing to pay more for a house with lower upkeep. Increased property values have also been observed in retrofitted office buildings [34]. Improved indoor air quality and thermal comfort are also valued by occupants, but it is hard to determine their value in the market. Lower number of missed work and school days due to lesser respiratory illness impact is a concrete benefit [62]. Indoor air quality has an effect on worker productivity [63], which should be of interest for all businesses. Energy efficiency may also improve brand value, which can result in increased sales [64].

4.3. *Drivers and Obstacles of Change*

Because heat pump systems are cost effective without any government grants, it can be expected that more and more building owners wish to switch away from municipal DH grids into their own heat pump systems. The interests of residents and municipalities—which own the district heating grid—are in conflict because energy saved by residents means less income for the municipality. City-owned utilities can also cause a conflict against the city's own climate goals. For example, utilities can increase the fixed cost of available heating capacity, while reducing the cost of actual consumption to make it more difficult to use heat pumps for support heating. This can be justified through the benefits of the whole district heating system compared to the benefits of individual buildings, as the efficiency of combined heat and power generation is reliant on both the district heating consumption and return temperature [65]. However, discouraging auxiliary heat pumps can result in building owners completely cutting themselves off of the district heating grid. To prevent this, the utilities could also start installing their own heat pump systems either in the client building or in a centralized heat pump facility. This way the use profile of the heat pump could be optimized according to the needs of the heating grid. Centralized heat pumps have the potential to reduce district heating costs, but at the same time they may raise the cost of electricity, which would improve the feasibility of CHP district heating [66]. Several private companies in Finland are already selling their heat pumps as a service, which means that residents do not need to invest their own capital and can get monetary savings from day one.

Lack of capital is a problem for both individual house owners and housing cooperatives. Many housing cooperatives already have significant debts due to plumbing renovations or other unavoidable maintenance. Building owners can be afraid of incurring new debt, even if said debt could be economically profitable. Sometimes energy efficiency can be improved as a side effect of comfort improvements. Comfort has been reported to be a more common reason for wanting to do renovations than cost savings [35]. However, building owners who do not live or work on the premises might not be interested in expending money for the renters' comfort.

It was shown that, in a low interest rate environment, cost-effective building energy renovation is possible in all building types. However, many building owners may be afraid of the investment due to lack of information. New investments into energy retrofits could be encouraged through a public retrofit database. Building owners (companies, housing cooperatives, homeowners) could report their current energy consumption information to an open database, where companies could analyze them and offer solutions. This would lower the threshold of making contact and would automatically allow competing offers. Energy consumption statistics shown before and after retrofits would make previously apprehensive building owners more interested in joining the system, as well. This could be tied to the energy retrofitting subsidies that are in place in 2020–2022 [67]. Completed projects could be required to report their results (such as cost and achieved savings) to the open database in an anonymous way. Municipalities could also encourage energy renovation investments by promising rent discounts for buildings lying on land owned by the municipality, on the condition of achieving improved energy efficiency levels. One solution would be to lower property taxes for energy efficient buildings. However, this would require changes in tax legislation and hurt the most reliable income source that municipalities have. The national government might have to compensate for the income lost this way [68]. Some building owners are afraid of increasing debt loads, as money also needs to be put into mandatory renovations not related to energy. Very low interest renovation loans with long enough amortization periods should be available to encourage building retrofits. The immediate costs could also be externalized through the use of energy-as-a-service provided by retrofitting companies, where the company providing the service provides the initial capital for the improvements and just collects monthly fees as payment.

There are clear societal benefits of energy efficiency, but the sometimes long payback periods can be an obstacle for businesses with high expectations of return on invested capital. Thus, owners of private service buildings might need additional encouragement from the public by, for example, legislation or tax credits. For example, the minimum requirements for energy efficiency improvements during building renovations could be raised according to current economic optimum. Municipalities and the national government could lead the way in building retrofits. They are often both the user and owner of the building, which helps in justifying the investment. The share of public buildings in the Finnish building stock is estimated to be 10% [59], which gives the public sector a major role in advancing building energy retrofitting.

4.4. *Reliability of Results*

A major question in this kind of study is the generalizability of the results. Many assumptions are required regarding the buildings themselves and the cost of energy. For example, the cost of district heating and electricity is reported differently in the Finnish statistics compared to pricing given by the utilities. In the statistics, the cost of DH might be given completely as an energy consumption-based cost, which disregards fixed monthly payments that are determined by the maximum power demand. This can overemphasize the role of energy demand savings. To provide more accurate cost analysis requires knowledge of the monthly or at least annual peak energy demand. Studies often report only the total annual energy demand without any seasonal or short-term variances. In this study, the heating power capacity cost was taken into account through hourly demand profiles, which improves the accuracy of the results, but makes the setup more difficult to understand.

The selected DH consumption-based prices in this study were fixed for the whole year. It is also possible to have a contract with low prices in summer and high prices in winter.

Heat pumps turned out to be the most cost-effective retrofit solution in all buildings. This is mainly due to the lower emission factors of electricity compared to district heating. A national average was used for the district heating emission factor, but in practice there are significant regional differences. Some Finnish cities produce DH mainly with coal or peat, while others utilize wood or waste-based fuels. Thus, additional consideration of the utility of heating system switch is needed, depending on the location of the building. Similarly, the method of electricity generation in the Nordic electricity market (Nord Pool) changes every hour, so even electrified heating is not always equally clean. Decarbonizing the district heating system itself would reduce the relative benefits of electrification. Of course, it would also reduce the absolute impact of any building-side energy saving measures. Thus, the long-term view of building energy retrofits should also consider the feedback between the individual buildings and the energy grid. Buildings with on-site boilers will be good candidates for switching to heat pumps, regardless of changes in district heating plants. When interpreting the results, it is important to take the local context into account. The optimal solutions are based on Finnish cost levels, emission factors and construction practices.

5. Conclusions

In this article, we analyzed the climate impact of cost-neutral building energy retrofits in several different building types. The investment cost and life cycle cost of retrofit actions were also examined. The retrofit configurations were taken from previous studies, each examining a single building type for which cost-optimal retrofit options were found. In this study, the previous results were compiled together. The energy consumption profiles were generated using the IDA-ICE building simulation tool under the test reference weather data for Southern Finland. CO_2 emissions were calculated using the average emission factors for Finnish electricity and district heating, as well as heating fuels. In Finland, the emission factor for electricity is lower than that of district heating or building-side heating fuels. The results apply conditionally to cold climates with relatively clean and low-cost electricity.

In all examined Finnish building types, emissions could be reduced in a cost-neutral way. The potential was different, depending on the heating system and building type. In service buildings where major retrofits were done (office, educational, and elderly care buildings), the emissions could be reduced by 61% to 82% at an investment cost of 180 to 300 €/m^2. In commercial buildings, where only solar electricity or GSHP were installed, the reduction was 33% to 60% for cost-neutral retrofits with an investment cost of 300 to 380 €/m^2. In apartment buildings, cost-neutral retrofits resulted in 28% to 82% reductions in emissions at an investment cost of 60 to 300 €/m^2, with the higher reductions obtained using heat pumps. If the buildings remained in the district heating grid, the emission impact of retrofits was much higher in these service buildings than in apartment buildings. The retrofits were thus more effective per unit in the service building, but the built floor area of apartment buildings is about 20% higher, influencing total impact potential. Both the most and least effective energy retrofits were found in single-family residential buildings, where the range of emission reduction was 4% to 95% at a cost of 100 to 500 €/m^2. The top range of both the reduction potential and investment cost was typically related to GSHP systems. Some residential buildings show significant potential to reduce emissions, but the residents may be lacking in capital. Service buildings have less variance, but there is still a lot of reduction potential. Here, the requirement for high return on investment may slow down deep building retrofits. The government could encourage retrofits by retrofitting public buildings and by giving retrofit grants or long-term low interest loans to both citizens and companies. Social care buildings, such as elderly homes, are good candidates for retrofits because of their high degree of use.

Replacing on-site boilers in single-family houses with ground-source heat pumps was the most effective emission reducing measure both in emission impact and low cost. In houses that kept using wood boilers, the cost-neutral reductions were limited due to the low cost of heating energy. Different types of heat pumps were the most cost-effective energy conservation measure (ECM) in all building types. Other common ECMs were envelope upgrades, such as the installation of new windows or additional thermal insulation to the roof. Solar energy (thermal or electricity) was utilized in all retrofitted buildings, as well. In office buildings, automated and demand-based lighting and ventilation control was cost-effective. In service buildings, such as educational and office buildings, the emission reduction potential was greater than in most apartment buildings. Looking at the total LCC, all non-residential buildings were on the lower cost side. Elderly care buildings had the largest emission reduction potential among the service buildings, which implies that more focus should be given to the other buildings which are always in use regardless of weekends or summer holidays, such as hospitals. However, no single building type could be said to be the best target for emission reduction based on achieved results in the cited studies, as the heating system also influences the result. The important point is that the whole building stock does offer chances for reducing emissions significantly and with reasonable cost. The building code should reflect this through more stringent energy efficiency requirements related to renovation work. To support this, low interest loans with long amortization periods should be available for performing the work.

By replacing district heating or on-site electric/combustion-based heating with heat pumps, emissions could often be reduced by 70% to 95%. Even in buildings which kept using district heating, emissions could be reduced by 20% to 70%. The benefit of heat pumps was dependent on the low emission factor of electricity compared to district heating. Thus, technology switching in district heating generation could alter the balance in favor of district heating. The greatest emission reduction potential was in buildings with on-site oil and wood boilers, which are not participating in the EU ETS. One way to enhance European climate action would be to expand the ETS to include building-side heating systems, as well. This could be done by mandating the purchase of emission allowances for heating fuel suppliers. This would have the largest impact on countries where on-site boilers are dominant. The emission reduction cost in the examined cases was at least 132 €/t-CO_2, and the average cost was 578 €/t-CO_2. This implies that, if building renovation in the EU is to be significantly sped up on an economic basis, the cost of emissions in the EU ETS should go up.

Author Contributions: Conceptualization, J.H., J.J. and R.K.; methodology, J.H., J.J. and R.K.; software, J.H.; validation, J.H., J.J., T.N. and P.S.; formal analysis, J.H.; investigation, J.H.; resources, J.H., T.N. and P.S.; data curation, J.H., T.N. and P.S.; writing—original draft preparation, J.H.; writing—review and editing, J.H., J.J., R.K. and P.S.; visualization, J.H.; supervision, J.J. and R.K.; project administration, R.K.; funding acquisition, R.K. All authors have read and agreed to the published version of the manuscript.

Funding: This research was funded by the Academy of Finland under the project "Optimal transformation pathway towards the 2050 low carbon target: integrated building, grids and national energy system for the case of Finland", grant number 309064.

Conflicts of Interest: The authors declare no conflict of interest.

Abbreviations

AAHP	Air-to-air heat pump
AB	Apartment building
AWHP	Air-to-water heat pump
Balanced	Mechanical balanced ventilation
CAV	Constant air volume ventilation
CHP	Combined heat and power
DH	District heating
EAHP	Exhaust air heat pump
ECM	Energy conservation measure
EPBD	Energy performance of buildings directive
ETS	Emission Trading System
Exhaust	Mechanical exhaust ventilation
GSHP	Ground-source heat pump
HP	Heat pump
HR	Heat recovery

HX	Heat exchanger
LCC	Life cycle cost
LED	Light-emitting diode
Natural	Natural stack ventilation
PV	Solar photovoltaic panel
SH	Single-family house
ST	Solar thermal collector
VAV	Variable air volume

Appendix A

Table A1. Energy demand in the presented buildings before and after retrofit.

-	-	Original		Retrofitted	
Building	**Case**	**DH/Boiler Demand**	**Electricity Demand**	**DH/Boiler Demand**	**Electricity Demand**
-	-	kWh/m^2	kWh/m^2	kWh/m^2	kWh/m^2
AB1 (–1976)	DH	171.9	30.2	118.6	27.6
AB2 (–2002)	DH	124.1	28.1	74.3	22.3
AB3 (–2010)	DH	80.7	36.5	41.7	28.1
AB4 (2010+)	DH	64.8	34.9	31.8	27.1
AB1 (–1976)	EAHP	171.9	30.2	30.8	38.8
AB2 (–2002)	EAHP	124.1	28.1	31.4	34.0
AB1 (–1976)	GSHP	171.9	30.2	0	48.3
AB2 (–2002)	GSHP	124.1	28.1	0	32.2
AB3 (–2010)	GSHP	80.7	36.5	0	40.1
AB4 (2010+)	GSHP	64.8	34.9	0	35.6
SH1 (–1976)	DH	234.3	20.6	76.3	26.2
SH2 (–2002)	DH	177.0	23.2	83.7	36.7
SH3 (–2010)	DH	148.3	28.4	76.0	33.1
SH4 (2010+)	DH	115.8	26.8	71.6	34.2
SH1 (–1976)	Oil → GSHP	280.6	20.6	0	32.8
SH1 (–1976)	Wood → GSHP	303.0	20.6	0	40.4
SH2 (–2002)	Oil → GSHP	212.0	23.2	0	33.4
SH2 (–2002)	Wood → GSHP	228.9	23.2	0	45.0
SH3 (–2010)	Oil → GSHP	177.6	28.4	0	32.2
SH3 (–2010)	Wood → GSHP	191.8	28.4	0	41.1
SH4 (2010+)	Oil → GSHP	138.7	26.8	0	33.3
SH4 (2010+)	Wood → GSHP	149.8	26.8	0	38.7
SH1 (–1976)	Wood → Wood	303.0	20.6	103.9	35.3
SH2 (–2002)	Wood → Wood	228.9	23.2	129.7	38.5
SH3 (–2010)	Wood → Wood	191.8	28.4	123.9	38.7
SH4 (2010+)	Wood → Wood	149.8	26.8	100.3	31.6
SH1 (–1976)	Elec	0.0	233.4	0	81.6
SH2 (–2002)	Elec	0.0	179.5	0	70.7
SH3 (–2010)	Elec	0.0	169.5	0	67.8
SH4 (2010+)	Elec	0.0	140.8	0	62.4
Office	DH	169.6	44.3	33.7	34.1
Office	GSHP	169.6	44.3	0.2	43.9
Elderly	DH	253.9	58.6	75.6	44.2
Elderly	AWHP	253.9	58.6	0	62.5
Commercial	DH, neutral	68.6	118.4	68.6	47.4
Commercial	GSHP, neutral	68.6	118.4	4.5	67.6
Commercial	DH, min cost	68.6	118.4	68.6	69.8
Commercial	GSHP, min cost	68.6	118.4	4.5	91.1
Educational	DH	222.1	46.6	77.7	27.4
Educational	GSHP	222.1	46.6	55.9	26.3

The arrow in the Heating system column indicates a replacement of the heating system with another.

References

1. European Parliament. Directive 2010/31/EU of the European Parliament and of the Council of 19 May 2010 on the energy performance of buildings. *Off. J. Eur. Union* **2010**, *153*, 13–35.
2. Yücel, G. Extent of inertia caused by the existing building stock against an energy transition in the Netherlands. *Energy Build.* **2013**, *56*, 134–145. [CrossRef]
3. European Parliament. Directive (EU) 2018/844 of the European Parliament and of the Council amending Directive 2010/31/EU on the energy performance of buildings and Directive 2012/27/EU on the energy efficiency. *Off. J. Eur. Union* **2018**, *156*, 75–91.
4. European Commission. *A Renovation Wave for Europe—Greening our Buildings, Creating Jobs, Improving Lives*; European Commission: Brussels, Belgium, 14 October 2020.
5. Saikia, P.; Pancholi, M.; Sood, D.; Rakshit, D. Dynamic optimization of multi-retrofit building envelope for enhanced energy performance with a case study in hot Indian climate. *Energy* **2020**, *197*, 117263. [CrossRef]
6. Sarihi, S.; Saradj, F.M.; Faizi, M. A Critical Review of Façade Retrofit Measures for Minimizing Heating and Cooling Demand in Existing Buildings. *Sustain. Cities Soc.* **2021**, *64*, 102525. [CrossRef]
7. Fan, Y.; Xia, X. A multi-objective optimization model for energy-efficiency building envelope retrofitting plan with rooftop PV system installation and maintenance. *Appl. Energy* **2017**, *189*, 327–335. [CrossRef]
8. Rakhshan, K.; Friess, W.A. Effectiveness and viability of residential building energy retrofits in Dubai. *J. Build. Eng.* **2017**, *13*, 116–126. [CrossRef]
9. Guillén-Lambea, S.; Rodríguez-Soria, B.; Marín, J.M. Air infiltrations and energy demand for residential low energy buildings in warm climates. *Renew. Sustain. Energy Rev.* **2019**, *116*, 109469. [CrossRef]
10. Shao, Y.; Geyer, P.; Lang, W. Integrating requirement analysis and multi-objective optimization for office building energy retrofit strategies. *Energy Build.* **2014**, *82*, 356–368. [CrossRef]
11. Foda, E.; El-Hamalawi, A.; Le Dréau, J. Computational analysis of energy and cost efficient retrofitting measures for the French house. *Build. Environ.* **2020**, *175*, 106792. [CrossRef]
12. Rogeau, A.; Girard, R.; Abdelouadoud, Y.; Thorel, M.; Kariniotakis, G. Joint optimization of building-envelope and heating-system retrofits at territory scale to enhance decision-aiding. *Appl. Energy* **2020**, *264*, 114639. [CrossRef]
13. La Fleur, L.; Rohdin, P.; Moshfegh, B. Investigating cost-optimal energy renovation of a multifamily building in Sweden. *Energy Build.* **2019**, *203*, 109438. [CrossRef]
14. Milić, V.; Ekelöw, K.; Andersson, M.; Moshfegh, B. Evaluation of energy renovation strategies for 12 historic building types using LCC optimization. *Energy Build.* **2019**, *197*, 156–170. [CrossRef]
15. Hirvonen, J.; Jokisalo, J.; Heljo, J.; Kosonen, R. Towards the EU Emission Targets of 2050: Cost-Effective Emission Reduction in Finnish Detached Houses. *Energies* **2019**, *12*, 4395. [CrossRef]
16. Niemelä, T.; Kosonen, R.; Jokisalo, J. Cost-effectiveness of energy performance renovation measures in Finnish brick apartment buildings. *Energy Build.* **2017**, *137*, 60–75. [CrossRef]
17. Niemelä, T.; Kosonen, R.; Jokisalo, J. Energy performance and environmental impact analysis of cost-optimal renovation solutions of large panel apartment buildings in Finland. *Sustain. Cities Soc.* **2017**, *32*, 9–30. [CrossRef]
18. Dodoo, A. Techno-economic and environmental performances of heating systems for single-family code-compliant and passive houses. *E3S Web Conf.* **2019**, *111*, 03039. [CrossRef]
19. Hirvonen, J.; Jokisalo, J.; Kosonen, R. The Effect of Deep Energy Retrofit on The Hourly Power Demand of Finnish Detached Houses. *Energies* **2020**, *13*, 1773. [CrossRef]
20. Asaee, S.R.; Nikoofard, S.; Ugursal, V.I.; Beausoleil-Morrison, I. Techno-economic assessment of photovoltaic (PV) and building integrated photovoltaic/thermal (BIPV/T) system retrofits in the Canadian housing stock. *Energy Build.* **2017**, *152*, 667–679. [CrossRef]
21. Asaee, S.R.; Ugursal, V.I.; Beausoleil-Morrison, I. Techno-economic feasibility evaluation of air to water heat pump retrofit in the Canadian housing stock. *Appl. Therm. Eng.* **2017**, *111*, 936–949. [CrossRef]
22. Salata, F.; Ciancio, V.; Dell'Olmo, J.; Golasi, I.; Palusci, O.; Coppi, M. Effects of local conditions on the multi-variable and multi-objective energy optimization of residential buildings using genetic algorithms. *Appl. Energy* **2020**, *260*, 114289. [CrossRef]

23. IEA. France—Countries & Regions. IEA Data Browser, France. 2020. Available online: https://www.iea.org/countries/france (accessed on 15 September 2020).
24. IEA. Germany—Countries & Regions. IEA Data Browser, Germany. 2020. Available online: https://www.iea.org/countries/germany (accessed on 15 September 2020).
25. Prabatha, T.; Hewage, K.; Karunathilake, H.; Sadiq, R. To retrofit or not? Making energy retrofit decisions through life cycle thinking for Canadian residences. *Energy Build.* **2020**, *226*, 110393. [CrossRef]
26. Niemelä, T.; Levy, K.; Kosonen, R.; Jokisalo, J. Cost-optimal renovation solutions to maximize environmental performance, indoor thermal conditions and productivity of office buildings in cold climate. *Sustain. Cities Soc.* **2017**, *32*, 417–434. [CrossRef]
27. Papadopoulos, S.; Kontokosta, C.E.; Vlachokostas, A.; Azar, E. Rethinking HVAC temperature setpoints in commercial buildings: The potential for zero-cost energy savings and comfort improvement in different climates. *Build. Environ.* **2019**, *155*, 350–359. [CrossRef]
28. Bienvenido-Huertas, D.; Sánchez-García, D.; Pérez-Fargallo, A.; Rubio-Bellido, C. Optimization of energy saving with adaptive setpoint temperatures by calculating the prevailing mean outdoor air temperature. *Build. Environ.* **2020**, *170*, 106612. [CrossRef]
29. Zheng, L.; Lai, J.H. Environmental and economic evaluations of building energy retrofits: Case study of a commercial building. *Build. Environ.* **2018**, *145*, 14–23. [CrossRef]
30. Güçyeter, B.; Gunaydin, H.M.; Günaydın, H.M. Optimization of an envelope retrofit strategy for an existing office building. *Energy Build.* **2012**, *55*, 647–659. [CrossRef]
31. Ascione, F.; Bianco, N.; Iovane, T.; Mauro, G.M.; Napolitano, D.F.; Ruggiano, A.; Viscido, L. A real industrial building: Modeling, calibration and Pareto optimization of energy retrofit. *J. Build. Eng.* **2020**, *29*, 101186. [CrossRef]
32. Ascione, F.; Bianco, N.; De Stasio, C.; Mauro, G.M.; Vanoli, G.P. Multi-stage and multi-objective optimization for energy retrofitting a developed hospital reference building: A new approach to assess cost-optimality. *Appl. Energy* **2016**, *174*, 37–68. [CrossRef]
33. Webb, A.L. Energy retrofits in historic and traditional buildings: A review of problems and methods. *Renew. Sustain. Energy Rev.* **2017**, *77*, 748–759. [CrossRef]
34. Carlson, K.; Pressnail, K.D. Value impacts of energy efficiency retrofits on commercial office buildings in Toronto, Canada. *Energy Build.* **2018**, *162*, 154–162. [CrossRef]
35. Azizi, S.; Nair, G.; Olofsson, T. Analysing the house-owners' perceptions on benefits and barriers of energy renovation in Swedish single-family houses. *Energy Build.* **2019**, *198*, 187–196. [CrossRef]
36. Bjørneboe, M.G.; Svendsen, S.; Heller, A. Evaluation of the renovation of a Danish single-family house based on measurements. *Energy Build.* **2017**, *150*, 189–199. [CrossRef]
37. Hirvonen, J.; Jokisalo, J.; Heljo, J.; Kosonen, R. Towards the EU emissions targets of 2050: Optimal energy renovation measures of Finnish apartment buildings. *Int. J. Sustain. Energy* **2019**, *38*, 649–672. [CrossRef]
38. Jokisalo, J.; Sankelo, P.; Vinha, J.; Sirén, K.; Kosonen, R. Cost optimal energy performance renovation measures in a municipal service building in a cold climate. *E3S Web Conf.* **2019**, *111*, 03022. [CrossRef]
39. Niemelä, T.; Kosonen, R.; Jokisalo, J. Cost-optimal energy performance renovation measures of educational buildings in cold climate. *Appl. Energy* **2016**, *183*, 1005–1020. [CrossRef]
40. Saari, A.; Airaksinen, M. *Energiatehokkuutta Koskevien Vähimmäisvaatimusten Kustannusoptimaalisten Tasojen Laskenta*; EU Commission: Espoo, Finland, 2012. (In Finnish)
41. Statistics Finland. Buildings by Area, Heating Fuel/Source of Heat, Intended Use of Building and Information. *PxWeb.* 2017. Available online: http://pxnet2.stat.fi/PXWeb/pxweb/en/StatFin_Passiivi/StatFin_Passiivi__asu__rakke/statfinpas_rakke_pxt_002_201700.px/table/tableViewLayout1/ (accessed on 30 June 2020).
42. EQUA Simulation AB. IDA ICE—Simulation Software. 2019. Available online: https://www.equa.se/en/ida-ice (accessed on 8 August 2019).
43. Vadiee, A.; Dodoo, A.; Gustavsson, L. A Comparison Between Four Dynamic Energy Modeling Tools for Simulation of Space Heating Demand of Buildings. In *Cold Climate HVAC 2018*; Springer: Cham, Switzerland, 2019; pp. 701–711.
44. EQUA Simulation AB. *Validation of IDA Indoor Climate and Energy 4.0 with Respect to CEN Standards EN 15255-2007 and EN 15265-2007*; EQUA Simulation AB: Solna, Sweden, 2010.

45. Loutzenhiser, P.; Manz, H.; Maxwell, G. *Empirical Validations of Shading/Daylighting/Load Interactions in Building Energy Simulation Tools*; IEA—International Energy Agency: Paris, France, 2007.
46. Palonen, M.; Hamdy, M.; Hasan, A. MOBO a new software for multi-objective building performance optimization. In Proceedings of the 13th Conference of International Building Performance Simulation Association, Chambéry, France, 26–28 August 2013.
47. Deb, K.; Pratap, A.; Agarwal, S.; Meayarivan, T. A fast and elitist multiobjective genetic algorithm: NSGA-II. *IEEE Trans. Evolut. Comput.* **2002**, *6*, 182–197. [CrossRef]
48. Lehtinen, T. *Rakennuksen Energiankulutuksen ja Lämmitystehon Tarpeen Laskenta [The Calculation of Energy Consumption and Heating Power Demand]*; Ministry of the Environment: Helsinki, Finland, 20 December 2017. (In Finnish)
49. Hirvonen, J.; Jokisalo, J.; Heljo, V.J.; Kosonen, R. Effect of apartment building energy renovation on hourly power demand. *Int. J. Sustain. Energy* **2019**, *38*, 918–936. [CrossRef]
50. Fortum Oy. Kaukolämmön Hinnat Pientaloille, (District Heating Prices for Detached Houses). Available online: https://www.fortum.fi/kotiasiakkaille/lammitys/kaukolampo/kaukolammon-hinnat-pientaloille (accessed on 23 June 2020).
51. Fortum Oy. Kaukolämmön Hinnat taloyhtiöIlle ja Yrityksille, (District Heating Prices for Companies). 2020. Available online: https://www.fortum.fi/yrityksille-ja-yhteisoille/lammitys-ja-jaahdytys/kaukolampo/kaukolammon-hinnat-taloyhtioille-ja-yrityksille (accessed on 23 June 2020).
52. Helen Oy. Sähkön siirtohinnat Helsingissä (Electricity distribution prices in Helsinki). 2020. Available online: https://www.helensahkoverkko.fi/palvelut/hinnastot (accessed on 23 June 2020).
53. Nord Pool. Historical Electricity Market Data. 2018. Available online: https://www.nordpoolgroup.com/historical-market-data/ (accessed on 14 January 2020).
54. Statistics Finland. *Fuel Classification 2019*; Statistics Finland: Helsinki, Finland, 2019. (In Finnish)
55. Motiva Oy. *CO_2 Emission Factors in Finland*; Motiva Oy: Helsinki, Finland, 2019. (In Finnish)
56. Finnish Energy. *Emission Factors of Finnish Electricity, Email Contact*; Finnish Energy: Helsinki, Finland, 2017.
57. Statistics Finland. Buildings and Gross Floor Area of Finnish Building Stock by Building Type. Tilastokeskuksen PX-Web Tietokannat. 2019. Available online: http://pxnet2.stat.fi/PXWebPXWeb/pxweb/en/StatFin/StatFin__asu__rakke/statfin_rakke_pxt_116g.px/. (accessed on 22 September 2020).
58. Ministry of the Environment. *Pitkän Aikavälin Korjausrakentamisen Strategia 2020–2050*; Ministry of the Environment: Helsinki, Finland, 2020. (In Finnish)
59. Motiva Oy. Rakentaminen ja Rakennukset. Motiva. Available online: https://www.motiva.fi/julkinen_sektori/kestavat_julkiset_hankinnat/tietopankki/rakentaminen_ja_rakennukset (accessed on 22 September 2020).
60. Verohallinto. Nestemäisten Polttoaineiden Verotaulukot. Verohallinto. Available online: https://www.vero.fi/yritykset-ja-yhteisot/tietoa-yritysverotuksesta/valmisteverotus/nestemaiset_polttoaineet/nestemaisten_polttoaineiden_verotaulukk/ (accessed on 10 March 2020).
61. Taimiteko. Taimiteko [The Planting Deed]. Taimiteko. Available online: https://www.taimiteko.fi/info/ (accessed on 10 March 2020).
62. Haverinen-Shaughnessy, U.; Pekkonen, M.; Leivo, V.; Prasauskas, T.; Turunen, M.; Kiviste, M.; Aaltonen, A.; Martuzevicius, D. Occupant satisfaction with indoor environmental quality and health after energy retrofits of multi-family buildings: Results from INSULAtE-project. *Int. J. Hyg. Environ. Health* **2018**, *221*, 921–928. [CrossRef] [PubMed]
63. Lan, L.; Wargocki, P.; Lian, Z. Quantitative measurement of productivity loss due to thermal discomfort. *Energy Build.* **2011**, *43*, 1057–1062. [CrossRef]
64. Hardcastle, J.L. *How to Boost Brand Value, Sales from GHG Management*; Environment + Energy Leader: Fort Collins, CO, USA, 2017. Available online: https://www.environmentalleader.com/2017/02/boost-brand-value-sales-ghg-management/ (accessed on 12 June 2020).
65. Rämä, M.; Niemi, R.; Similä, L. *Exhaust Air Heat Pumps in a District Heating System (Poistoilmalämpöpumput Kaukolämpöjärjestelmässä)*; VTT-CR-00564-15; VTT Technical Research Center of Finland: Espoo, Finland, 2015. (In Finnish)
66. Helin, K.; Syri, S.; Zakeri, B. Improving district heat sustainability and competitiveness with heat pumps in the future Nordic energy system. *Energy Procedia* **2018**, *149*, 455–464. [CrossRef]

67. ARA. Energy Assistance 2020 (Energia-Avustukset 2020). 2019. Available online: https://www.ara.fi/fi-FI/Lainat_ja_avustukset/Energiaavustus (accessed on 22 October 2020). (In Finnish)
68. Ministry of the Environment. *Staggering of the Property Tax of Buildings according to Energy Efficiency and heating System (Rakennusten kiinteistöveron porrastaminen energiatehokkuuden ja lämmitystavan perusteella)*; Ministry of the Environment: Helsinki, Finland, 2009. (In Finnish)

buildings

MDPI

Article

Optimising the Parameters of a Building Envelope in the East Mediterranean Saharan, Cool Climate Zone

Aiman Albatayneh

Energy Engineering Department, School of Natural Resources Engineering and Management, German Jordanian University, Amman 11180, Jordan; aiman.albatayneh@gju.edu.jo

Abstract: Enhancing the energy efficiency and environmental sustainability of buildings is a significant global aim. New construction regulations are, therefore, geared specifically towards low-emission and energy-efficient projects. However, there are numerous and typically competitive priorities, such as making the most of energy usage in residential buildings. This leads to the complex topic of multi-objective optimisation. The primary aim of this research was to reduce the energy consumed for heating and cooling loads in residential buildings in Ma'an City, which is located in the Jordanian Saharan Mediterranean, a cool climate zone. This was achieved by optimising various design variables (window to wall percent, ground floor construction, local shading type, infiltration rate (ac/h), glazing type, flat roof construction, natural ventilation rate, window blind type, window shading control schedule, partition construction, site orientation and external wall construction) of the building envelope. DesignBuilder software (version 6.1) was utilised to run a sensitivity analysis (SA) for 12 design variables to evaluate their influence on both heating and cooling loads simultaneously using a regression method. The variables were divided into two groups according to their importance and a genetic algorithm (GA) was then applied to both groups. The optimum solution selected for the high-importance variables was based on minimising the heating and cooling loads. The optimum solution selected for the low-importance variables was based on the lowest summation of the heating and cooling loads. Finally, a scenario was devised (using the combined design variables of the two solutions) and simulated. The results indicate that the total energy consumption was 1186.21 kWh/year, divided into 353.03 kWh/year for the cooling load and 833.18 kWh/year for the heating load. This was compared with 9969.38 kWh/year of energy, divided into 3878.37 kWh/year for the heating load and 6091.01 kWh/year for the cooling load for the baseline building. Thus, the amount of energy saved was 88.1%, 94.2% and 78.5% for total energy consumption, cooling load and heating load, respectively. However, implementing the modifications suggested by the optimisation of the low-importance variables was not cost-effective, especially the external wall construction and partition construction, and therefore these design variables can be neglected in future studies.

Keywords: building optimisation design; Saharan; cool climate; genetic algorithm; low energy buildings

Citation: Albatayneh, A. Optimising the Parameters of a Building Envelope in the East Mediterranean Saharan, Cool Climate Zone. *Buildings* **2021**, *11*, 43. https://doi.org/10.3390/buildings11020043

Academic Editor: Ala Hasan and Francesco Reda

Received: 18 December 2020
Accepted: 21 January 2021
Published: 27 January 2021

1. Introduction

There is now a global movement towards ensuring more efficient energy consumption in all sectors to mitigate the massive increase in demand for energy and the depletion of natural resources. The residential sector accounts for a significant share of the energy consumed; especially electrical energy, which incurs substantial losses through generation, transmission and distribution. In Jordan, for instance, the residential sector consumes 45.4% of the electricity generated, and most of this energy is spent on heating and cooling loads. Jordan is, therefore, finding it increasingly difficult to meet the demand for energy. This makes it essential to implement energy-efficient technologies in building envelopes to minimise the energy wasted and ensure future sustainable development. Consequently, reducing the energy consumed in the residential sector has been a primary aim of governmental organisations and institutions in Jordan. This has resulted in the creation and dissemination of numerous public advertisements on the use of energy, several campaigns to replace lights in residences with more efficient alternatives, and the provision of incentives to install solar water heaters and solar panels. In addition, several codes and books have been published that describe the procedures for making buildings (both new and existing buildings) more energy-efficient. However, these instructions are not obligatory [1].

Researchers are thus seeking to determine the best ways to build such buildings or increase energy efficiency in existing buildings. They have ceaselessly strived to develop better construction materials, reduce costs, increase the energy-efficiency of buildings and improve thermal comfort. In particular, they have worked towards developing a novel solution to tackle the problem of energy consumption by residential buildings and the impact of this on the environment [1].

Ensuring buildings are more energy-efficient is imperative if the future use of energy is to be sustainable. For optimal thermal performance, there needs to be new, innovative designs for the construction of buildings. To achieve this, effective and accurate software programs are required that can correctly predict the performance of buildings in disparate conditions. However, current software is inadequate. To enhance the prediction of energy usage within buildings and lower their running costs, it is vital to address all relevant physical, social and environmental factors during construction [2].

With a focus on multi-objective optimisation, a group of researchers employed both MATLAB and Energy Plus to enhance the energy design of the casing of a building. The principal objectives were to evaluate overall energy-related global costs, primary energy consumption and discomfort hours. The variables measured were the thermo-physical properties of envelope components, the radiative properties of plasters, set point temperatures, window type and building orientation. A Pareto optimisation was performed using a genetic algorithm (GA) that yielded two optimal solutions: a nearly zero energy building (NZEB) that minimises PEC and a cost-optimal solution that minimises energy-related global cost (GC) [3].

The buildings in question were standard newly constructed residential buildings in four climate zones within the country of Italy:

Zone B: Palermo, located in Sicily (Southern Italy);
Zone C: Naples, located in Campania (Southern Italy);
Zone D: Florence, located in Toscana (Central Italy);
Zone E: Milan, located in Lombardy (Northern Italy).

Climatic Zone A was not considered because it includes only two municipalities; thus, its outcomes were not representative. The results indicate that set point temperatures for space conditioning have a substantial effect on the objective functions irrespective of climate zone, and that the solar absorbance of the external surfaces of roofs and vertical walls rises substantially from warmer to colder climate zones [3].

To identify the most practical and inexpensive design, another set of researchers applied passive design methods. Their objective was to combine external shading devices with self-shading envelopes in a 10-storey hotel building located in an extreme hot–humid zone in Saudi Arabia. DesignBuilder was utilised to perform modelling and energy simulations. Even though the baseline building did not have a solar shade, fibre glass insulation was installed in the walls and roofs while the windows were double-glazed in accordance with Saudi guidelines. The results indicate that the proposed shading reduced annual energy usage by 20.5%. The researchers estimated that two years would be needed to pay back the extra costs incurred by the use of passive shading [4].

Another study constructed and then validated a simulation model called the MEEDI, and created a graphic that depicted the results of a parametric sensitivity analysis of energy efficiency in diverse climates across Chile. The MEEDI measures the monthly energy usage of buildings, a process known as ISO 13790, and utilises two revised techniques to measure solar heat gains and the flow of heat through the floor. The economic viability of this method for construction envelopes was then evaluated. The results indicate that variations in building design can have a direct impact on overall energy efficiency. Worthwhile efficiency benefits and a reduction in carbon emissions can be achieved at reasonable costs through simple changes to the envelope materials. Depending on where in Chile they were located and the overall level of efficiency, payback times ranged from 5 to 27 years. The researchers acknowledged that while the MEEDI model can be employed to measure the precise energy demand within a typical Chilean home, it is sufficiently flexible to enable a breadth of scenarios to be tested. Thus, it can provide measures of efficiency for a disparate range of temperature conditions, envelope fabrics and energy systems [5].

Using DesignBuilder, three drawing halls were simulated on campus in a study conducted at Mansoura University in Egypt. The purpose was to enhance the energy efficiency within these halls according to three criteria: single- or multi-objective optimisation, intuitive or computerised optimisation and parametric or whole-design optimisation. Ignoring daylight requirements, the results indicated that ventilation system functionality, optimising the shade overhangs and louvres of windows, and low glazing transmittance reduced cooling loads by from 26% to 31% in comparison to the base case [6].

Another study employed DesignBuilder to simulate the thermal performance of a 145 square metre rural house in Beijing, North China. The objective was to implement passive energy modifications of the building envelope to decrease the energy consumed annually for heating purposes. The most salient modifications were replacing the traditional windows with double-glazed windows and installing thermal insulation within the building. The results indicated that these modifications saved 60% of the energy consumed for heating purposes [7].

Harkouss, Fardoun and Biwole (2018) conducted a comprehensive study to determine the optimal passive design for residential buildings. Utilising Köppen–Geiger climate classifications, TRNSYS (TRaNsient SYStem) software was employed to simulate the yearly energy usage, costs and comfort of a building in one city in each climate zone for twenty five different climates. The objective was to determine the best way to lower energy demands (for cooling and heating) within the building and also the life-cycle cost (LCC). Effective passive cooling techniques such as blinds and natural ventilation were also introduced to increase the simulated levels of thermal comfort. The study comprised five phases: building energy simulation, optimisation, Multi-Criteria Decision Making (MCDM), a sensitivity study and an adaptive comfort analysis. The ideal passive solution reduced cooling requirements and heating requirements by 52%, 54% and 87%, respectively, compared to the base building. Furthermore, the passive cooling techniques were effective in substantially decreasing overheating and increasing levels of adaptive thermal comfort for residents [8].

Another study evaluated the effect of various climate zones in Australia (representing the primary global climates) on the thermal performance of a building. Design methods were developed to align with each climate zone in order to enhance thermal performance. AccuRate (rating software tools accredited by the Nationwide House Energy Rating Scheme in Australia) was then utilised to assess the thermal performance of the same building in different climate zones. The aim was to ensure that the right design would be employed for a building based on its surrounding climate. The optimal designs for each climate zone were then assessed against a variety of climate variables (wind speed and direction, solar radiation, rainfall and humidity) to develop a sustainable building that will substantially reduce energy consumption whilst maintaining the thermal comfort of residents [9].

Numerous studies have aimed to determine how to improve the energy consumption of buildings; they have found that energy estimation, from the design stage through scientific or practical calculations or simulation programs, was useful, particularly if followed by high-quality construction. The paper now reviews the findings of several studies that explore the relationship between climate factors and the basic design parameters affecting the thermal performance of buildings. Site analysis is the first stage of design. Orientation impacts the building's overall thermal performance and the design of heating and cooling to maintain the thermal comfort of occupants [9].

One set of researchers examined the relationship between building orientation on-site and its thermal performance using Design builder as a simulation program that utilised climatic data on the selected location and data on the building envelope entered by the user. The researchers selected, for their sample, a house that matched the characteristics of 60% of the buildings in Jordan. The results showed that orienting the long axis of the building on the north–south axis had a clear impact on thermal performance. They found that positioning the largest glazing area to the south allowed the building to gain the desired heat in winter. The study also demonstrates the effect of orientations on the airflow inside the building crossing from west to east. Appropriate orientations helped minimise heat loss in the winter months and reduced heating energy by almost 35% compared to the Base Case building, indicating a marked improvement in overall thermal performance [10].

Regarding the windows and glazing of the building envelope, it will not be possible to apply any of the designed solar passive strategies unless a certain amount of solar radiation reaches the building facades. An investigation was conducted on an office building in Turkey to assess the influence of window designs and types on energy performance with respect to wall-to-windows area ratio, shading, orientation, geometrical shape and thermo-physical properties. Simulations of two common types of office room models were conducted with changes in parameters such as the window-to-wall ratio of the façade, total solar energy transmittance of the glazing and shading levels regarding orientations. The results indicated that the most influential parameter on annual energy consumption is the window-to-wall ratio. North-facing office units generate a tremendous energy demand while south-facing office units have the lowest energy demand during the heating season. The optimal configuration has a WWR between 40% and 60% during the heating season. There is no positive impact on increasing windows to wall ratio for the north façade in all locations. The primary recommendation is, therefore, to reduce the window size to decrease greenhouse gas emissions and achieve conditions of thermal comfort for the occupants [11].

Another study found that increasing the shading level of windows results in an incremental increase in annual heating consumption and a decrease in annual cooling consumption. The architectural glazing façade may result in massive energy consumption [12].

The thermal energy performance of a building is influenced by structural materials and slab types. The building structure is limited by multiple factors such as building shape, construction materials types. Several studies have shown that buildings with a high thermal mass require more time and energy to heat up and cool down. A study exploring the impact of concrete structures in Australia on energy performance used simulation programs to identify a benchmarking method for measuring the thermal energy performance of a building using two forms of construction (flat slab and waffle slab) and two types of concrete (normal conventional weight and novel Ultra-lightweight). This classification depends on the properties of the material, its ability to absorb and store heat, and its transmission [13].

The correct use of thermal mass in different climatic conditions depends on the region's prevailing climate. Hot, humid weather (for example, tropical and sub-tropical) is most challenging in this environment, as night temperatures remain high. Its use is primarily as a temporary heat sink. However, it must occupy a strategic position to prevent overheating. It should be placed in an area that is not directly exposed to solar energy and allow adequate ventilation at night to move the stored energy away without further increasing internal temperatures. The results indicated that a higher concrete mass (thermal mass) stored more heat, which then reduced the peak indoor air temperatures, whereas ultra-lightweight concrete generated indoor temperatures that were more sensitive to fluctuations in external air temperatures. Hence, the building required more energy to achieve the desired indoor temperature range. Choosing the appropriate type of concrete and form of construction could reduce the annual cooling energy demand of a highly glazed office building by 14% in colder climate zones and 3% in warmer and hot climates [13].

Wall insulation with a scientific and efficient thermal insulation technology is an essential measure for energy conservation in buildings. The standard method employed to improve the performance of wall insulation is to integrate the insulation layer into the built foundation wall; however, this increases the construction cost and delays construction time due to the construction of both the foundation wall and the insulation layer. The selection of insulation materials is critically important and has a pronounced effect on building energy performance [14].

One study investigated climate responsive solutions in the vernacular architecture of Bushehr city and clarified the design features in the city. Notably, sustainability and adaptation were consistent in using the lowest costing materials and applications from the surrounding environment to create a comfortable environment. This depended on adapting to the climate as it is (passive techniques) rather than artificial building alternatives that support climate change (as is currently the case in the modern era). Vernacular buildings are architectural products that emerged as a response to the requirements of societies before the industrial period and to the insurmountable limits of artificial building created by the region and climate, because of the unique interaction between the human mind and experience gathered by observing natural phenomena in the vernacular buildings [15].

The effectiveness of the thermal performance of the design of this city has been achieved through urban planning concepts that take the climatic characteristics of the site, the relations between the orientation of buildings and the sun, direction and flow of air movement, whether distances between buildings are well organised and the existence of shaded crossing corridors that serve pedestrians into consideration. Designers also need to use the small spaces that separate the buildings to direct the airflow between blocks of buildings. The design for the comprehensive system within Bushehr city was completed by the unique features of individual unit design. This is the result of hundreds of years of optimisation to provide comfortable shelter in a local climate using available materials and known construction technologies [16].

The relationship between the climate zoning and energy performance of buildings, which was investigated during hot summer periods and their impact on a building's thermal-energy behaviour, shows that in an urban area, it is essential to conduct a statistical analysis of microclimate variation during hot periods [17].

Building performance simulation can provide many measures in which an integrated design solution satisfies the design requirements and objectives. In sustainable buildings, it is beneficial to identify the most critical design parameters to develop efficient alternative design solutions or generate optimised design solutions. A sensitivity analysis makes it possible to identify the most critical parameters with respect to building performance, and to focus the design and optimisation of sustainable buildings on these extremely important parameters. Such an analysis will typically be performed at a reasonably early stage of the building design process, where it is still possible to influence such parameters [18].

Research by Zero Carbon Hub of 16 housing developments in the U.K. indicated that all the dwellings presented a higher measured heat loss than predicted. This mismatch between the energy performance predicted at the design stage and once the building is in operation is known as the energy performance gap. To overcome this gap, the researchers investigated the relation between thermal performance and defects, utilising a defects classification system that provides a methodology for defect identification and collection. The construction community needs optimal solutions to comprehensively deal with the design through all stages of construction, starting with good design, simulation of energy assessments, highly efficient construction methods and an awareness of users' behaviour [19].

Other researchers compared different solutions to identify retrofit strategies for a historic building envelope using energy plus and the Pareto method. The main result of the multi-objective optimisation was the generation of eight optimal solutions for the sensitivity analysis with no significant change in energy consumption for cold and hot climates. However, for a hot climate, natural ventilation was necessary in addition to thermal insulation to guarantee human comfort with different costs for these solutions [20].

Another study was conducted to create an effective balance between investment costs, energy consumption and indoor environment quality in residential high-rise buildings. This involved evaluating the effects of varying parameters of energy usage for heating and cooling in each Chinese climate zone. WUFI®Plus (Wärme Und Feuchte Instationär—which, translated, means heat and moisture transiency developed by Fraunhofer Institute for Building Physics in Stuttgart, Germany), simulations were conducted for a base building that complied with energy-efficiency codes for Chinese buildings and for buildings where design parameters differed according to the climate zone. The three most sensitive parameters were then identified and the reductions in overall energy demand for each climate zone estimated, as presented in Table 1 [21]:

Table 1. Sensitive design parameters and energy reduction for Chinese climate zones.

Climate Zone	Sensitive Design Parameters	Maximum Energy Saving kWh/(m^2 per Year)
Very cold and cold zones	Air tightness, thermal transmittance (U-value) of external windows/glass door and insulation thickness	75
Warm winter and hot summer zone	Insulation thickness, solar protection and solar heat gain coefficient (SHGC) of external windows/glass door	50
Cold winter and hot summer zone	air tightness, insulation thickness, and solar protection	40
Mild climate zone	Insulation thickness, air tightness and U-value of external windows/glass door	35

Even though more suitable indoor temperatures result in higher heating and cooling loads, the energy demand associated with the additional heating and cooling loads can be reduced by enhancing the design of the building envelope. Table 2 summarises research on the different design variables and their influence on the thermal performance and comfort of the building.

DesignBuilder software was employed in another study to compare wall and roof thermal insulation with no heat insulation in a standard 100 square metre residential apartment on the 3rd floor of a three-storey building in Egypt. The sensible cooling load, air-conditioner energy usage rate, energy usage charge, and the CO_2 emitted by the power station were measured in terms of energy consumed where the room temperature was set at 24, 22 and 20 °C, respectively. The results indicated that thermal insulation (glass wool blanket), 0.05 m thick in the walls and roof of the floor, meant that 40% of the electricity consumed by the air conditioner significantly reduced the operational costs. Moreover, the energy saved reduced the CO_2 generated by the power plant as a result of increased energy use in the residential building by 30% [37].

The present research aimed to minimise the energy consumed by heating and cooling loads in residential buildings in Ma'an City, which is situated in the Jordanian Saharan Mediterranean, with a cool climate zone. This was achieved by optimising various design variables (window to wall percent, ground floor construction, local shading type, infiltration rate (ac/h), glazing type, flat roof construction, natural ventilation rate, window blind type, window shading control schedule, partition construction, site orientation and external wall construction) of the building envelope.

Table 2. Research on the different design variables.

Input Variables	Output Variables	Type of Study	Aim of Research	Ref.
70 input parameters, for instance: thermostat set points, ground reflectivity, surface convection coefficients, thermal conductivity of materials, solar radiation, heat gains.	Energy consumption, Peak power demand.	Sensitivity analysis: differential sensitivity analysis (DSA), Monte Carlo analysis (MCA), and stochastic sensitivity analysis (SSA)	Passive solar, single zone buildings tested for application of different sensitivity analysis methods	[22]
24 variables, including: building height, length, width, rotation, view factor of the ground, ground reflectivity, insulation thickness, glazing ratio, weather parameters, setpoint temperature, ventilation rates, variables concerning occupancy and building environment	Yearly heating loads, Heating load per m^3, Heating power, Summer comfort factor.	Sensitivity analysis (Morris method and the function analysis system technique (FAST))	Thermal comfort and heating/cooling loads assessment	[23]
Typical parameters: weather conditions physical properties of building materials, internal heat gains (occupancy, metabolic rate, equipment, lighting, set point temperature), interventions in building fabric and systems, infiltration rate, chiller coefficient of performance (COP), boilers efficiency.	Annual heating energy, Annual cooling energy, Carbon emissions	Sensitivity analysis standardized regression coefficients (SRC) method; Uncertainty analysis	University building renovation scenarios assessed regarding changing climate conditions	[24]
Sky temperature, wind effect	Internal air temperature	Computational fluid dynamics (CFD)	Find ways to simulate sky temperature and wind effect	[25,26]
U-values of building components, ACH, heat recovery, thermal bridges	Annual heating energy per m^2 floor area, Cost-benefit analysis based on sampling results.	Sensitivity analysis (Sobol sampling) and Uncertainty analysis	3 case studies combining cost–benefit calculation and bps	[27]
Orientation, U-Wall, U-Roof, U-Floor, U-Windows, solar heat gain coefficient, window frame thickness, Thermo control, ventilation and infiltration rate, occupation, metabolic rate, clothing level	Annual heating Energy consumption, Predicted mean vote index.	Sensitivity analysis	Sensitivity analysis of Building parameters and occupancy parameters effects in residential buildings	[28]
Window shading size, window width, window length, windows infiltration rate, doors infiltration rate, AF-Ventilation, U-Ext Wall, U-Int Wall, U-First Floor, U-Roof, α-Ext Wall, α-Roof, α-Floors, thickness of floor slab	Heating/cooling/fans energy consumption, Total degree hours, Global objective, Cooling degree hours, Heating degree hours.	Sensitivity analysis	Building design optimisation using Sensitivity analysis and genetic algorithm	[29]
Occupants behaviour and thermal comfort model	Amount of energy required to sustain occupant's thermal comfort.	Actual data measurements	-	[30–33]
Orientation, window to wall ratio, skylight to roof ratio, U-Wall, U-Roof, U-Skylight, U-Win, U-Floor, C-Floor, C-Roof, C-Wall, α-Ext Wall, SHGC, T-vis, Infiltration, glazing, floor, parapet, corner linear transmittance	Hourly discomfort index, Cooling energy, Annual electricity energy.	Sensitivity analysis	Sensitivity analysis and optimisation of building design in subtropical regions	[34]
Orientation, window to wall ratio, overhang depth, SHGC, insulation, infiltration rate, insulation thickness, window resistance, T-slab Floor, infiltration, sensible heat recovery effectiveness, air Distribution effectiveness, heating, and cooling setpoints, T-cool air, T-heat air	Energy consumption, Predicted mean vote index.	Sensitivity analysis	Sensitivity Analysis throughout the building design process	[35]
Window to wall ratio, window type, space aspect ratio, insulation, T-Shading, α-front shade, α-back shade	Useful daylight illuminance (UDI), Lighting energy consumption, Heating/cooling energy.	Sensitivity analysis, Uncertainty analysis	Performance of office room with automated shading	[36]

The research consisted of four phases:

- Using DesignBuilder simulation software to develop a three-dimensional model of a typical Jordanian building that complies with local codes and instructions;
- Establishing the design parameters and associated values on which to perform the optimisation;
- Conducting a sensitivity analysis to ascertain the importance of the design variables and then group them accordingly;

- Conducting an optimisation by performing multiple random simulations and selecting an optimised solution from the vast number of design parameter configurations that were generated.

2. Materials and Methods

This section explains the process of optimising several design parameters of a building envelope in the Jordanian Saharan Mediterranean cool climate zone. This involved utilising DesignBuilder software to simulate the thermal performance of a baseline building, evaluating the sensitivity of the design parameters on its heating and cooling loads, and devising an optimised solution for these design parameters.

2.1. Location and Climate

Jordan is located between the Arabian Desert and the eastern Mediterranean area; hence, its climatic conditions are characterized by long, hot and dry summers and short and cold winters. December, January, and February are the coldest months, with average maximum/minimum temperatures of 10 and 5 °C, respectively, while July, August, and September are the hottest months, with average maximum/minimum temperatures of 35 and 20 °C. During the summer, daily temperatures can, in some cases, reach or even exceed 40 °C, especially when a hot, dry southeastern wind blows. In winter, significant amounts of precipitation of between 200 and 400 mm fall, and these decrease/stop during the summer season. Within Jordan, there are different climate zones, some of which are similar to the Mediterranean climate while others are almost identical to a desert climate. The Jordanian weather comprises four distinct seasons in which autumn and spring are within the ideal comfort range for humans.

Situated in the Saharan Mediterranean cool climate zone, the city of Ma'an is the largest governorate in Jordan, with an area of 32,832 km^2 (37% of Jordan's overall land area). It is characterised by a cold Saharan climate with hot summers and mild to cold winters. Precipitation is rare, with an annual average of 40 mm [38].

2.2. Building Baseline Specifications and Shape

This typical building comprised a living room, saloon, kitchen, storage room, master bedroom and two other bedrooms, and two bathrooms with a total area of around 186 m^2. Every room was identified as a separate zone, as shown in Figures 1 and 2. The building is oriented where the long axis faces the south–north axis. The building does not have external shading or wind hurdles, which results in a significant amount of solar radiation entering the building.

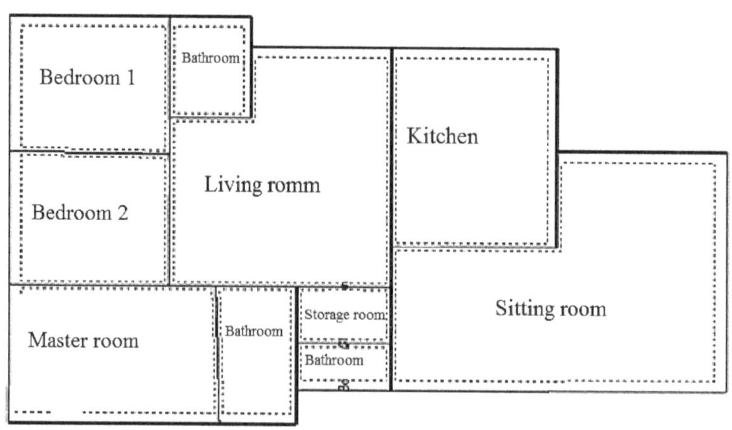

Figure 1. Layout of the building.

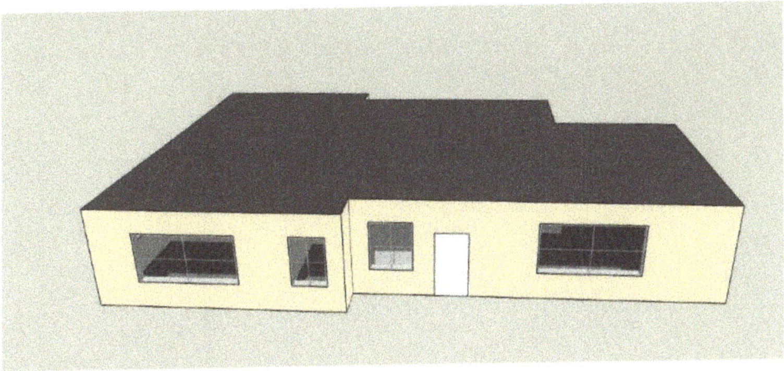

Figure 2. 3D model of the building.

The building envelope materials and its specifications were developed based on typical Jordanian architectural design and adhere to Jordanian building codes. These were applied to the DesignBuilder software as follows.

2.2.1. Walls, Floor and Roof

The building external, internal, roof and floor materials specification and configuration illustrated in Table 3, where the U-Value is the thermal transmittance.

Table 3. Thicknesses and thermal conductivity for walls, roof and floor.

Layer Name	Thermal Conductivity (W/m·K)	Thickness (mm)
	External wall	
Stone	2.2	50
Concrete Reinforced	2.5	100
Extruded polystyrene	0.03	50
Concrete Block	1.6	100
Cement Plaster	1.2	10
U-Value (W/m^2·k)	0.563	
	Internal wall	
Cement Plaster	1.2	30
Concrete block	1.6	100
Cement Plaster	1.2	30
U-Value (W/m^2·k)	2.5	
	Floor	
Cement Plaster	1.2	30
Concrete block	1.6	100
Cement Plaster	1.2	30
U-Value (W/m^2·k)	2.5	
	Roof	
Asphalt	0.7	20
Extruded polystyrene	0.03	50
Miscellaneous materials-aggregate	1.3	100
Concrete Reinforced	2.5	320
Cement Plaster	1.2	20
U-Value (W/m^2·K)	0.535	

2.2.2. External Windows Glazing and Frame

The external windows glassing materials' specification and configuration are illustrated in Table 4.

Table 4. Glazing layer configuration and their thicknesses.

Layer Name	Layer Thickness (mm)
Generic BLUE	6
100% Air Gap	6
Generic CLEAR	6
U-value (W/m^2·K)	3.1
Total Solar transmission (SHGC)	0.5
Light transmission (%)	0.51

External windows glassing surrounded with thermally broken aluminium frame (polyvinyl chloride (PVC)) with U-value = 5.01 W/m^2·K.

2.3. *Simulation Software*

DesignBuilder software (version 6.1) was used to stimulate the energy performance of the building. This is an EnergyPlus-based software tool that analyses building energy, including heating, cooling, lighting, and ventilation. It is also employed to control and measure building carbon, cost, and daylighting performance. DesignBuilder provides a simple and user-friendly way to simulate and optimise the building envelope and environment through rapid comparison of the performance of different building elements. DesignBuilder allows the user to examine the effect of changing each building envelope material on energy performance and consumption in the building.

2.3.1. Energy Retrofit Strategies

To enhance the overall thermal performance of each Jordanian climate, several design variables were considered.

2.3.2. Temperature Set Point of Heating and Cooling

Temperature set points of heating and cooling systems have a significant effect on energy consumption in the building. However, these were fixed at 19 °C for the heating system and 24 °C for the cooling system, as the aim was to reduce heating and cooling loads while maintaining existing levels of thermal comfort.

2.3.3. Building Orientation

An optimum building orientation may reduce energy consumption in the building. To select the most energy-efficient orientation, the base case design was rotated from 0° to 360° in steps of 5° while noting that the base case has a rotation of 90°.

2.3.4. Thickness of External Wall Insulation

The thermal transmittance (U-value) of external walls has a tangible effect on the overall heating and cooling load in the building. Therefore, the effect of increasing insulation thickness was investigated. Table 5 shows how the U-value of the external wall changes as the insulation thickness increases.

Table 5. External wall insulation thickness and its related U-values.

Walls Insulation Thickness (cm)	U-Value ($W/m^2 \cdot K$)
5 (baseline)	0.56
7.5	0.4
10	0.31
12.5	0.25
15	0.21

2.3.5. Thickness of Roof Insulation

The thermal transmittance (U-value) of roof is one of the main elements of the building envelope. Therefore, the effect of increasing the thickness of the thermal insulation layer was tested. Table 6 illustrates how the U-value of the roof changes as the insulation thickness increases.

Table 6. Roof insulation thickness and its related U-values.

Roof Insulation Thickness (cm)	U-Value ($W/m^2 \cdot K$)
5 (baseline)	0.54
7.5 cm	0.39
10 cm	0.3
12.5 cm	0.25
15 cm	0.21

2.3.6. Thermal Mass of Internal Walls

The effect of changing the thermal mass of the internal walls was also tested. Table 7 shows the specifications of the thermal masses that were tested for the construction of the internal walls.

Table 7. Specifications of various thermal masses for internal walls.

Thickness (cm)	Material	Conductivity ($W/m \cdot K$)	Specific Heat ($J/kg \cdot K$)	Density (kg/m^3)
5	Heavyweight concrete block	1.31	840	2240
10				
15				
20				
10	Heavyweight dry concrete cast	1.3	840	2000
20				
30				
10	Lightweight dry concrete cast	0.22	840	720
20				
30				
10	Very lightweight concrete	0.16	840	470
20				
30				
10	High-density concrete	2	1000	2400
20				
30				

2.3.7. Floor Insulation

The effect of insulating the floor with different thicknesses of thermal insulation was tested. Table 8 shows how the U-value of the floor changes as the insulation thickness increases.

Table 8. Floor insulation thickness and its related U-values.

Floor Insulation Thickness (cm)	U-Value ($W/m^2{\cdot}K$)
0 (baseline)	1.88
5	0.5
7.5	0.37
10	0.29
12.5	0.24
15	0.2

2.3.8. Glassing Type

The features and specification of glassing materials have a significant effect on energy performance. Therefore, several glassing types were investigated to examine their effect on heating and cooling loads in buildings. Table 9 presents the specification for different types of glassing.

Table 9. Specification of glassing types.

Type	Total Solar Transmission SHGC	Light Transmission	U-Value ($W/m^2{\cdot}K$)
Double blue 6/6 mm Air (baseline)	0.486	0.505	3.157
Double blue 6/13 mm Air	0.481	0.505	2.708
Double Elec ABS bleached 6/13 mm Air	0.739	0.752	1.772
Dbl Elec Abs coloured 6/13 mm Air	0.155	0.114	1.772
Double LoE Abs bleached 6/13 mm Arg	0.478	0.657	1.322
SageGlass Climatop (triple)	0.248	0.357	0.687
Dbl Elec Ref coloured 6/6 mm Air	0.162	0.137	2.4
Dbl Elec Ref coloured 6/13 mm Air	0.142	0.137	1.772

2.3.9. Window-to-Wall Ratio

Windows are an important element in the building envelope and have a significant impact on heating and cooling loads. This is because they facilitate energy gain from the sun radiation while increasing the infiltration rate and thermal conductivity between the building and the outside environment. Window-to-wall ratios were calculated by dividing total building glazed area by its external wall area.

Therefore, window-to-wall ratios from 0% to 100% were investigated in steps of 1%.

2.3.10. Infiltration Rate (Air Leakage)

Infiltration rate is defined as air leaking into the building from the external environment through cracks, openings, and doors. It is measured in air change per hour (ac/h). Different infiltration rates from 0.25 to 1.5 (ac/h) were tested.

2.3.11. Local (External) Shading

Local shading consists of models built on the outer surface of the window in the building envelope such as overhangs, louvres and sidefins (Figure 3). Different scenarios were tested using each device individually or in combination (see Table 10).

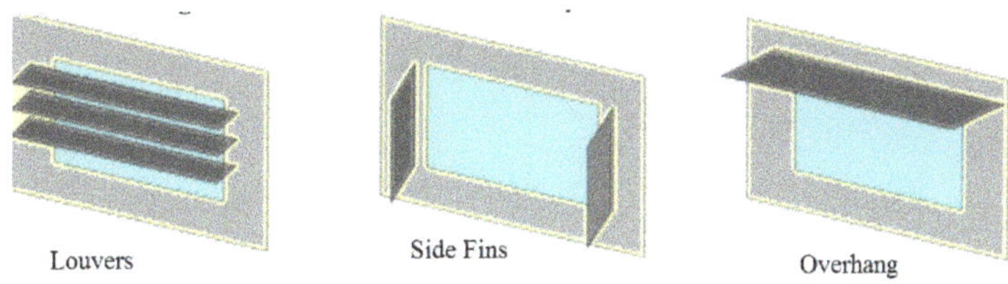

Figure 3. Local shading techniques (overhangs, louvers, and side fins).

Table 10. External window shading scenarios.

Scenarios	Characteristic
Baseline	No shading
1	Overhangs (0.5 m)
2	Overhangs (1 m)
3	Overhangs (1.5 m)
4	Overhangs (2 m)
5	Louvres (0.5 m)
6	Louvres (1 m)
7	Louvres (1.5 m)
8	Overhang (0.5 m) + Louvre (0.5 m) + Side fins
9	Overhang (1 m) + Louvre (1 m) + Side fins
10	Overhang (0.5 m) + Side fins
11	Overhang (1 m) + Side fins

2.3.12. Window Shading Type and Schedule

Window shading devices are tools that cover the window-glassing surface which can be opened or closed to allow or prevent the radiation, or a portion of it, from entering the building, depending on their reflectance and transmittance properties.

Therefore, different types of diffusing shades with different light properties were tested to determine their influence on heating and cooling loads (see Table 11). Additionally, 11 schedules controlling the opening of the window shading were tested.

Table 11. Types of diffusing shades.

Scenarios	Characteristic
Baseline	No shading
1	High reflectance—Low transmittance shade
2	Medium reflectance—Low transmittance shade
3	Medium reflectance—Medium transmittance shade
4	Low reflectance—Low transmittance shade
5	Low reflectance—Medium transmittance shade
6	Low reflectance—High transmittance shade

2.3.13. Natural Ventilation Rate and Schedule

If managed effectively, natural ventilation is extremely useful for lowering the heating and cooling loads by raising the level of heat exchange between the inside and outside of the building according to the difference in temperature.

With a schedule based on room occupancy, the rate of natural ventilation was, therefore, measured in steps of 0.1 (ac/h) from 0 to 5 (ac/h).

2.4. Sensitivity Analysis

It may not be possible to find an optimum solution using a local sensitivity analysis method or by testing one variable at a time, as the design variables might exert an interaction effect on heating and cooling loads. Therefore, an optimisation process for all such variables needs to be performed. However, this could be extremely time-consuming as there may be a large number of variables generating vast combinations of simulations.

To accelerate the optimisation, both global uncertainty and a sensitivity analysis need to be performed to identify those variables that have the largest and smallest effects on the required outcomes and reduce the number of variables involved.

There are several forms of sensitivity analysis (Fourier amplitude sensitivity testing (FAST), Morris, Regression) that can be employed to determine the relationship between the design variables and the required outcomes.

Regression was applied in this study as it is an existing feature in DesignBuilder and can be applied to obtain detailed data for complex models. In this research, the objectives were heating and cooling load, the key indicator that identifies the most and least important variables was the Standardised Regression Coefficient (SRC), and the design variables were those listed in Energy Retrofit Strategies section. To yield accurate results, the following points needed to be addressed:

- Although the number of random simulations is typically 1.5 to 10 times greater than the number of variables, DesignBuilder proposes a number equal to or greater than 10 times this number, with 12 main variables, some of which could have at least 16 possible values, the sample size in this study was 1200 random simulations (100 times the number of variables);
- To define the variables, DesignBuilder orders values in ascending or descending order; for instance, increasing or decreasing U-values for external wall insulation;
- When performing a sensitivity analysis, DesignBuilder suggests avoiding choices with similar values. For example, the U-value of the wall will not change if the insulation is placed at a different location within the wall.

2.5. Optimisation

Having performed the sensitivity analysis, the design variables are divided into two groups according to their effect on heating and cooling loads: a high-importance group and a low-importance group.

Optimisation is then performed, which involves searching for and selecting the design options that are likely to be most efficient in achieving the required objectives.

As such, optimisation is similar to a process in which the performance of objectives changes in line with alterations in the design variables of the building.

To search for optimal design options, DesignBuilder employs genetic algorithms (GA). This is a superior method to parametric regression as a greater number of variables can be included (up to 10), along with two objectives, which, in this research, are "minimising heating load" and "minimising cooling load".

The process is iterative in that first a population of random values are generated (known as a "generation"). Each generation measures the fitness of variables in this population according to the value of the objective function. The fittest individuals are then selected on a stochastic basis and, through recombination and spontaneous mutation, their genomes are employed to create a new generation. Subsequent iterations of the algorithm then utilise the most recent generation of potential solutions. The algorithm usually terminates when the population has achieved a sufficient level of fitness or a specified number of generations have been produced.

The results of the simulated output of each combination of variables are then depicted on a graph with heating on one side and cooling on the other. The minimum values form the "Pareto Front" of optimal designs on the left bottom of the cloud datapoint.

Optimisation was performed with a random sample of 6000 simulations for the high-importance group and 2000 simulations for the low-importance group.

The solutions with the lowest summation of heating and cooling loads for both groups were then selected and combined and a simulation performed to determine whether any further reduction in heating and cooling loads could be achieved.

3. Results and Discussion

To optimise several design parameters of a building envelope, a baseline residential building was designed, and its thermal performance evaluated using DesignBuilder software. This was extremely important as it enabled the thermal performance of various designs to be compared, to assess which have a positive or negative effect on the thermal performance of the building.

3.1. Baseline Building

The simulation indicated that, for Saharan Mediterranean, cool climate zones (Ma'an, Jordan), the annual consumption of energy in a baseline building is 6091.01 and 3878.37 kWh/year for the cooling load and the heating load, respectively.

3.2. Sensitivity Analysis

Using the impeded feature in DesignBuilder, a sensitivity analysis was conducted on 12 design variables. The particular method applied was regression where the indicating factor was the Standardised Regression Coefficient (SRC) and the number of random simulations was chosen to be 100 times the number of design variables.

As presented in Figure 4, the SRC of the cooling load indicates that cooling load is most strongly influenced by window-to-wall percent. The input and output are directly related. Increasing window-to-wall percent, therefore, increases cooling load. Cooling load is also strongly influenced by ground floor construction and local shading type, and moderately influenced by flat roof construction, infiltration (ac/h), natural ventilation rate, window blind type, glazing type, and window shading control schedule. External wall construction, partition construction, and site orientation do not have any notable influence on cooling load and can, therefore, be ignored in further analysis of the cooling load for this model.

As depicted in Figure 5, the SRC of the heating load indicates that it is most strongly influenced by infiltration (ac/h) The input and output are directly related. Increasing infiltration (ac/h), therefore, increases heating load. Heating load is strongly influenced by glazing type and window-to-wall percent and moderately influenced by local shading type. Site orientation, flat roof construction, window blind type, partition construction, external wall construction, window shading control schedule, ground floor construction, and natural ventilation rate do not have any notable influence on heating load and can, therefore, be ignored in further analysis of the heating load for this model.

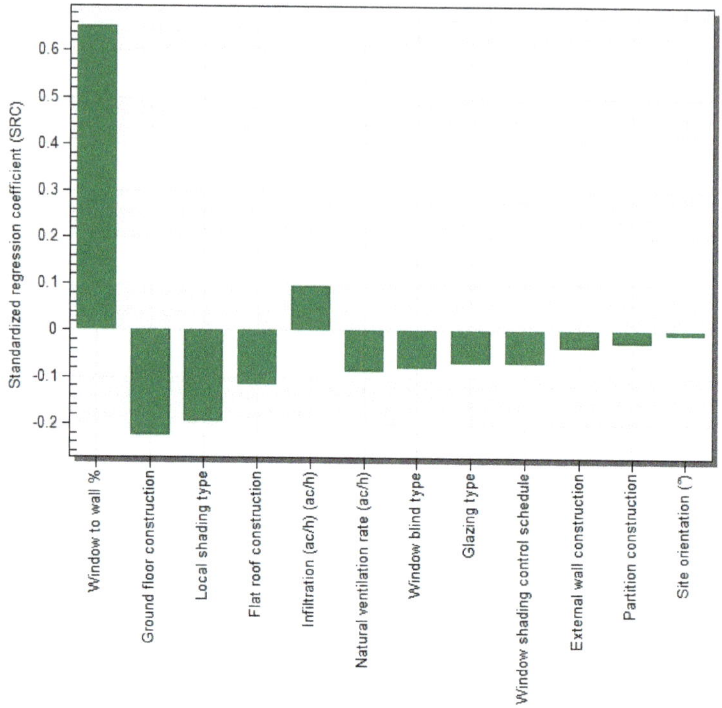

Figure 4. Cooling load sensitivity analysis.

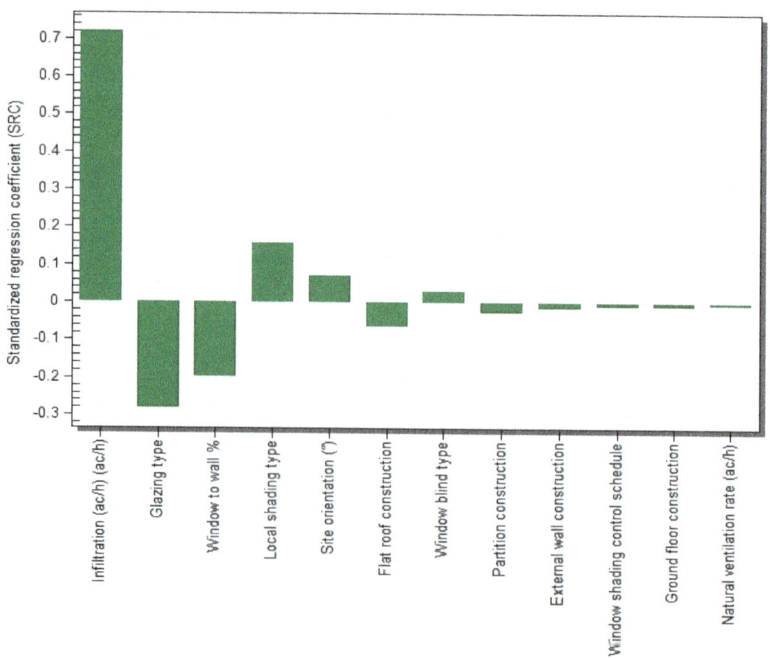

Figure 5. Heating load sensitivity analysis.

The design variables were then categorised according to their importance for heating and cooling loads, as presented in Table 12. This made it easier to determine which variables to include/exclude in the optimisation process.

Table 12. Sensitivity analysis.

-	-		**Cooling Load**	
-	**Importance**	High	Medium	Low
Heating Load	High	Window-to-wall percent	Infiltration (ac/h) Glazing type	-
	Medium	Local shading type	-	-
	Low	Ground floor construction	Natural ventilation rate Flat roof construction Window blind type Window shading control schedule	Partition construction Site orientation External wall construction

Thus, two groups of design variables were identified.

A high-importance group (window-to-wall per cent, ground floor construction, local shading type, infiltration rate (ac/h), glazing type, flat roof construction, natural ventilation rate, window blind type, window shading control schedule) was identified. Based on Table 12 and the SRC graphs, because the cooling load is the dominant, window-to-wall ratio, glazing type, and local shading type will most likely be optimised to prevent the solar gain in summer rather than allowing it to reduce the heating load. Decreasing the U-Value of the flat roof construction and ground floor construction is too expensive but can reduce both heating and cooling loads, albeit with different levels of importance, as increasing the insulation thickness will prevent heat exchange through the large area between the internal space and the outside environment and ground. This will be most effective in preventing heat exchange with the cold air in the summer season, as the cooling load has the highest annual energy consumption. Conversely, the infiltration rate has the highest negative impact on the heating load as it will leak the heat contained in the building to the outside environment while having a slight impact on the cooling load. The infiltration rate can be noticeably reduced at a relatively low cost by applying insulation material on windows edges, wall cracks and doors frames.

A low-importance group (partition construction, site orientation, external wall construction) was identified. These variables have a low influence on heating and cooling loads and, in most cases, increasing their insulation thickness or using better insulation material is extremely expensive. Thus, they will most likely be kept, as they meet the standards, but will still be considered in the optimisation to determine their most suitable values. Site orientation, by contrast, is usually restricted by the land shape and location, but this research considers no such restrictions when determining the optimal orientation. Window blind type and its control schedule depend on the resident's awareness; if they were installed efficiently, they will slightly reduce heating and cooling loads at a low cost.

3.3. Optimisation

The GA technique embedded in DesignBuilder was deployed optimise the two groups of design variables by setting the objectives of the optimisation to minimise heating and cooling loads using 6000 and 2000 random simulations for the high-importance group and low-importance group, respectively.

3.3.1. High Importance Variables

The 6000 solutions from the random simulations are presented in Figure 6, along with their related heating and cooling loads. The Pareto Front (the most optimal solutions) is highlighted in orange.

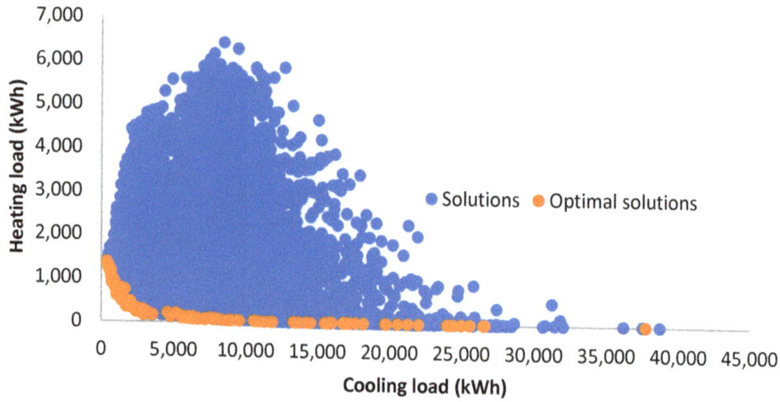

Figure 6. Optimisation of high-importance variables—optimisation analysis results—minimise heating load and cooling load.

Although the original aim was to investigate three solutions (minimising cooling load, minimising heating load, minimising the summation of heating and cooling loads); the fact that the cooling load was dominant in the Saharan Mediterranean cool climate zone meant that the arrangement of solutions based on cooling load (from the smallest to the highest) and the summation of heating and cooling loads were almost identical. Thus, two solutions were considered (minimising heating load, minimising heating and cooling load).

3.3.2. Minimising Heating Load

Because the heating load of the baseline building was extremely low, it was easy to overcome by optimising the high-importance design variables. This was clearly demonstrated in the results of the random simulations, where two of the Pareto solutions had a heating load close to zero and, in 21 solutions, the heating load was less than 10 kWh/year. For these 21 solutions, the cooling load ranged from 14,522.51 to 37,718.28 kWh/year.

Thus, the optimal solution for optimising the heating load regardless of the cooling load will lead to a total energy consumption of 14,532.11 kWh/year. This is higher than the energy consumption of 9969.38 kWh/year for the baseline building. Hence, this optimisation is neglected.

3.3.3. Minimising Both Heating and Cooling Loads

The solution with the lowest summation of heating and cooling was chosen as the total energy consumption for this solution was 1624.34 kWh/year, divided into 718.70 kWh/year for the cooling load and 905.64 kWh/year for the heating load. The design variables for this solution are presented in Table 13.

Table 13. Optimum solution for high importance variables.

Design Variable	Value
Flat roof construction	Flat roof with 15 cm layer of insulation
Glazing type	SageGlass Climatop Blue No Tint (triple)
Window blind type	High reflectance—low transmittance shade
Window shading control schedule	Summer cooling workdays, the shades in the period between 15/05 and 30/09 for 24 h
Infiltration rate	0.25 (ac/h)
Natural ventilation rate	4 (ac/h)
Local shading type	Overhang + side fins (1 m projection)
Window-to-wall ratio	39%
Ground floor construction	Typical ground floor (baseline)

The optimal solution of optimising both heating and cooling load simultaneously was, therefore, chosen for further analysis.

3.3.4. Low Importance Variables

Figure 7 presents the cloud of random simulations for the low-importance group of variables and the Pareto Front (the most optimal solutions).

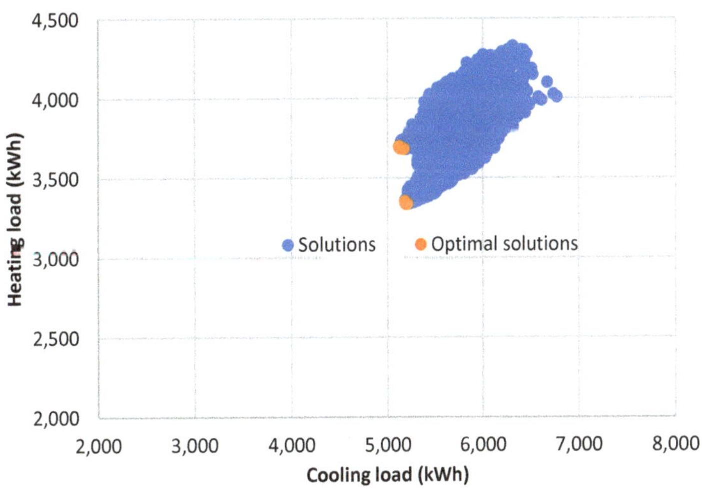

Figure 7. Optimisation of low-importance variables, minimise cooling load and heating load.

Among the 2000 solutions generated, the Pareto Front (optimal solutions) indicates that there only 10 optimum solutions in which the cooling load ranged from 5116.30 to 5205.97 kWh/year and the heating load ranged between 3336.62 and 3697.54 kWh/year.

The small range for these loads was predicted, as the modified design variables were shown to have low importance by the sensitivity analysis. Thus, only the solution with the lowest summation of heating and cooling was chosen as the total energy consumption for this solution is 8525.06 kWh/year, divided into 5182.44 kWh/year for the cooling load and 3342.62 kWh/year for the heating load. The design variables for this solution are presented in Table 14.

Table 14. Optimum solution for low-importance variables.

Design Variable	Value
External wall construction	External wall with 15 cm layer of insulation
Partition construction	Partition wall the concrete blocks replaced with 30 cm very lightweight dry concrete cast
Site orientation	85° (rotating the building 5° counter-clockwise)

3.3.5. Optimal Solution

After choosing the optimum solution for both high- and low-importance variables, the design variables of the two solutions were combined and a simulation for heating and cooling load of the building was created. Total energy consumption was then further reduced to 1186.21 kWh/year, divided into 353.03 kWh/year for the cooling load and 833.18 kWh/year for the heating load. However, the reduction in energy consumed by implementing the modification suggested by the optimisation of the low-importance design variables did not appear to be cost-effective, particularly the external wall construction and partition construction.

4. Conclusions

Buildings consume almost 40% of the world's energy and emit one-fifth of the total CO_2 emissions. These numbers can be reduced by the application of passive design strategies, such as a design for climate. This research conducted an optimisation for the main design variables in a building in Ma'an, which is situated in the Jordanian Saharan Mediterranean, cool climate zone. The aim was to reduce the consumption of heating and cooling loads while maintaining the current levels of thermal comfort provided by existing heating and cooling systems.

A simulation of the thermal performance of a baseline building was conducted, which indicated that the baseline building will consume 9969.38 kWh/year of energy, divided into 3878.37 kWh/year for the heating load and 6091.01 kWh/year for the cooling load.

A sensitivity analysis (SA) was then conducted for 12 design variables to evaluate their influence on both heating and cooling loads simultaneously using a regression method with 1200 random runs. Based on the results, the variables were divided into two groups according to their importance: a high-importance group (window-to-wall percent, ground floor construction, local shading type, infiltration rate (ac/h), glazing type, flat roof construction, natural ventilation rate, window blind type, window shading control schedule) and a low-importance group (partition construction, site orientation, external wall construction). A genetic algorithm (GA) was employed for both the high-importance and low-importance groups. The final results indicated a reduction in energy consumption of 88.1%, 94.2% and 78.5% for total energy consumption, cooling load and heating load, respectively. The total amount of energy saved as a result of the simulation was 1186.21 kWh/year, divided into 353.03 kWh/year for the cooling load and 833.18 kWh/year for the heating load.

These results will help to enhance the codes and the role of public authorities and provide them with an effective evaluation for buildings. This will enable them to develop a plan to spread awareness about the environment and its relationship with buildings and to minimise the footprint of the residential sector. It is important to focus on all types of energy consumption and introduce renewable energy techniques to achieve a net zero energy building.

Numerous priorities, such as maximising energy usage, lowering financial costs, and minimising environmental effects, should be taken into consideration in the optimum design of residential buildings. This makes the complex topic of multi-objective optimisation an important factor to consider in future research.

Funding: This research received seed grant fund (SNERM 04/2018) provided by the deanship of graduate studies and research at the German Jordanian University.

Institutional Review Board Statement: Not applicable.

Informed Consent Statement: Not applicable.

Data Availability Statement: No new data were created or analyzed in this study. Data sharing is not applicable to this article.

Acknowledgments: the author thankful for the seed grant fund (SNERM 04/2018) provided by the deanship of graduate studies and research at the German Jordanian University.

Conflicts of Interest: The author declares no conflict of interest.

References

1. National Electric Power Company. NEPCO Annual Report. 2017. Available online: http://www.nepco.com.jo/store/docs/web/2017_en.pdf (accessed on 14 June 2020).
2. Al-Addous, M.; Albatayneh, A. Knowledge gap with the existing building energy assessment systems. *Energy Explor. Exploit.* **2020**, *38*, 783–794. [CrossRef]
3. Ascione, F.; Bianco, N.; Mauro, G.M.; Napolitano, D.F. Building envelope design: Multi-objective optimization to minimize energy consumption, global cost and thermal discomfort. Application to different Italian climatic zones. *Energy* **2019**, *174*, 359–374. [CrossRef]
4. Alhuwayil, W.K.; Mujeebu, M.A.; Algarny, A.M.M. Impact of external shading strategy on energy performance of multi-story hotel building in hot-humid climate. *Energy* **2019**, *169*, 1166–1174. [CrossRef]
5. Simon, F.; Ordoñez, J.; Girard, A.; Parrado, C. Modelling energy use in residential buildings: How design deci-sions influence final energy performance in various Chilean climates. *Indoor Built Environ.* **2019**, *28*, 533–551. [CrossRef]
6. Samaan, M.M.; Farag, O.; Khalil, M. Using simulation tools for optimizing cooling loads and daylighting levels in Egyptian campus buildings. *HBRC J.* **2018**, *14*, 79–92. [CrossRef]
7. Liang, X.; Wang, Y.; Zhang, Y.; Jiang, J.; Chen, H.; Zhang, X.; Roskilly, T. Analysis and optimization on energy per-formance of a rural house in northern China using passive retrofitting. *Energy Procedia* **2017**, *105*, 3023–3030. [CrossRef]
8. Harkouss, F.; Fardoun, F.; Biwole, P.H. Passive design optimization of low energy buildings in different climates. *Energy* **2018**, *165*, 591–613. [CrossRef]
9. Albatayneh, A.; Alterman, D.; Page, A.; Moghtaderi, B. The significance of building design for the cli-mate. *Environ. Clim. Technol.* **2018**, *22*, 165–178. [CrossRef]
10. Alshboul, A.A.; Alkurdi, N.Y. Enhancing the Strategies of Climate Responsive Architecture, The Study of Solar Accessibility for Buildings Standing on Sloped Sites. *Mod. Appl. Sci.* **2019**, *13*, p69. [CrossRef]
11. Altun, A.F.; Kılıç, M. Influence of Window Parameters on the Thermal Performance of Office Rooms in Different Climate Zones of Turkey. *Int. J. Renew. Energy Res.* **2019**, *9*, 226–243.
12. Luo, Y.; Zhang, L.; Wang, X.; Xie, L.; Liu, Z.; Wu, J.; He, X. A comparative study on thermal performance evalua-tion of a new double skin façade system integrated with photovoltaic blinds. *Appl. Energy* **2017**, *199*, 281–293. [CrossRef]
13. Robati, M.; Kokogiannakis, G.; McCarthy, T.J. Impact of structural design solutions on the energy and thermal performance of an Australian office building. *Build. Environ.* **2017**, *124*, 258–282. [CrossRef]
14. Li, J.; Meng, X.; Gao, Y.; Mao, W.; Luo, T.; Zhang, L. Effect of the insulation materials filling on the thermal per-formance of sintered hollow bricks. *Case Stud. Therm. Eng.* **2018**, *11*, 62–70. [CrossRef]
15. Motealleh, P.; Zolfaghari, M.; Parsaee, M. Investigating climate responsive solutions in vernacular architecture of Bushehr city. *HBRC J.* **2018**, *14*, 215–223. [CrossRef]
16. Shi, L.; Zhang, H.; Li, Z.; Luo, Z.; Liu, J. Optimizing the thermal performance of building envelopes for energy saving in underground office buildings in various climates of China. *Tunn. Undergr. Space Technol.* **2018**, *77*, 26–35. [CrossRef]
17. Pyrgou, A.; Castaldo, V.L.; Pisello, A.L.; Cotana, F.; Santamouris, M. On the effect of summer heatwaves and ur-ban overheating on building thermal-energy performance in central Italy. *Sustain. Cities Soc.* **2017**, *28*, 187–200. [CrossRef]
18. Heiselberg, P.; Brohus, H.; Hesselholt, A.; Rasmussen, H.; Seinre, E.; Thomas, S. Application of sensitivity analysis in design of sustainable buildings. *Renew. Energy* **2009**, *34*, 2030–2036. [CrossRef]
19. Alencastro, J.; Fuertes, A.; De Wilde, P. The relationship between quality defects and the thermal performance of buildings. *Renew. Sustain. Energy Rev.* **2018**, *81*, 883–894. [CrossRef]
20. Roberti, F.; Oberegger, U.F.; Lucchi, E.; Gasparella, A. Energy retrofit and conservation of built heritage using multi-objective optimization: Demonstration on a medieval building. In Proceedings of the Building Simulation Applications, Bolzano, Italy, 4–6 February 2015; pp. 4–6.
21. Zhao, M.; Kunzel, H.M.; Antretter, F. Parameters influencing the energy performance of residential buildings in different Chinese climate zones. *Energy Build.* **2015**, *96*, 64–75. [CrossRef]

22. Lomas, K.J.; Eppel, H. Sensitivity analysis techniques for building thermal simulation programs. *Energy Build.* **1992**, *19*, 21–44. [CrossRef]
23. Sanchez, D.G.; Lacarrière, B.; Musy, M.; Bourges, B. Application of sensitivity analysis in building energy simula-tions: Combining first-and second-order elementary effects methods. *Energy Build.* **2014**, *68*, 741–750. [CrossRef]
24. Tian, W.; De Wilde, P. Uncertainty and sensitivity analysis of building performance using probabilistic climate projections: A UK case study. *Autom. Constr.* **2011**, *20*, 1096–1109. [CrossRef]
25. Albatayneh, A.; Alterman, D.; Page, A.; Moghtaderi, B. The Significance of Sky Temperature in the Assessment of the Thermal Performance of Buildings. *Appl. Sci.* **2020**, *10*, 8057. [CrossRef]
26. Albatayneh, A.; Alterman, D.; Page, A.; Moghtaderi, B. Alternative Method to the Replication of Wind Effects into the Buildings Thermal Simulation. *Buildings* **2020**, *10*, 237. [CrossRef]
27. Burhenne, S. *Monte Carlo Based Uncertainty and Sensitivity Analysis for Building Performance Simulation*; Shaker Verlag: Herzogenrath, Germany, 2013.
28. Ioannou, A.; Itard, L.C. Energy performance and comfort in residential buildings: Sensitivity for building param-eters and occupancy. *Energy Build.* **2015**, *92*, 216–233. [CrossRef]
29. Bre, F.; Silva, A.S.; Ghisi, E.; Fachinotti, V.D. Residential building design optimisation using sensitivity analysis and genetic algorithm. *Energy Build.* **2016**, *133*, 853–866. [CrossRef]
30. Albatayneh, A.; Alterman, D.; Page, A.; Moghtaderi, B. The Significance of the Adaptive Thermal Comfort Limits on the Air-Conditioning Loads in a Temperate Climate. *Sustainability* **2019**, *11*, 328. [CrossRef]
31. Albatayneh, A.; Alterman, D.; Page, A.; Moghtaderi, B. The Significance of Temperature Based Approach Over the Energy Based Approaches in the Buildings Thermal Assessment. *Environ. Clim. Technol.* **2017**, *19*, 39–50. [CrossRef]
32. Albatayneh, A.; Alterman, D.; Page, A.; Moghtaderi, B. The Impact of the Thermal Comfort Models on the Predic-tion of Building Energy Consumption. *Sustainability* **2018**, *10*, 3609. [CrossRef]
33. Albatayneh, A.; Alterman, D.; Page, A.; Moghtaderi, B. Development of a new metric to characterise the buildings thermal per-formance in a temperate climate. *Energy Sustain. Dev.* **2019**, *51*, 1–2. [CrossRef]
34. Li, H.; Wang, S.; Cheung, H. Sensitivity analysis of design parameters and optimal design for zero/low energy buildings in subtropical regions. *Appl. Energy* **2018**, *228*, 1280–1291. [CrossRef]
35. Gagnon, R.; Gosselin, L.; Decker, S. Sensitivity analysis of energy performance and thermal comfort throughout building design process. *Energy Build.* **2018**, *164*, 278–294. [CrossRef]
36. Shen, H.; Tzempelikos, A. Sensitivity analysis on daylighting and energy performance of perimeter offices with automated shading. *Build. Environ.* **2013**, *59*, 303–314. [CrossRef]
37. An-Naggar, A.S.; Ibrahim, M.A.; Khalil, E.E. Energy Performance Simulation in Residential Buildings. *Procedia Eng.* **2017**, *205*, 4187–4194. [CrossRef]
38. Fick, S.E.; Hijmans, R.J. WorldClim 2: New 1-km spatial resolution climate surfaces for global land areas. *Int. J. Climatol.* **2017**, *37*, 4302–4315. [CrossRef]

buildings

MDPI

Article

Strategies of Design Concepts and Energy Systems for Nearly Zero-Energy Container Buildings (NZECBs) in Different Climates

Johannes Koke [1,*], André Schippmann [1], Jingchun Shen [2], Xingxing Zhang [2], Peter Kaufmann [3] and Stefan Krause [3]

1 Faculty of Management, Culture and Technology, Campus Lingen, University of Applied Sciences Osnabrueck, 49809 Lingen, Germany; andre.schippmann@gmail.com
2 Energy and Community Building, Dalarna University, 791 88 Falun, Sweden; jih@du.se (J.S.); xza@du.se (X.Z.)
3 Institut für Strukturleichtbau und Energieeffizienz gGmbH, 09113 Chemnitz, Germany; p.kaufmann@institut-se.de (P.K.); s.krause@institut-se.de (S.K.)
* Correspondence: j.koke@hs-osnabrueck.de; Tel.: +49-591-80098-730

Abstract: Container-based lightweight buildings offer a high ecologic and economic potential when they are designed as nearly zero-energy container buildings (NZECBs). Thus, they are relevant to energy transition in achieving an almost climate-neutral building stock. This paper describes and applies design strategies for suitable building concepts and energy systems to be used in NZECBs for different climates. Therefore, different applications in representative climatic zones were selected. Initially, the global climate zones were characterized and analyzed with regard to their potential for self-sufficiency and renewable energies in buildings. The design strategies were further developed and demonstrated for three cases: a single-family house in Sweden, a multi-family house in Germany, and a small school building in rural Ethiopia. For each case, design guidelines were derived and building concepts were developed. On the basis of these input data, various energy concepts were developed in which solar and wind energy, as well as biomass, were integrated as renewable energy sources. All the concepts were simulated and analyzed with the Polysun® software. The various approaches were compared and evaluated, particularly with regard to energy self-sufficiency. Self-sufficiency rates up to 80% were achieved. Finally, the influence of different climate zones on the energy efficiency of the single-family house was studied as well as the influence of the size of battery storage and insulation.

Keywords: nearly zero-energy building; renewable energy integration; container building; building design; energy simulation; climate study; renewable energy

Citation: Koke, J.; Schippmann, A.; Shen, J.; Zhang, X.; Kaufmann, P.; Krause, S. Strategies of Design Concepts and Energy Systems for Nearly Zero-Energy Container Buildings (NZECBs) in Different Climates. *Buildings* **2021**, *11*, 364. https://doi.org/10.3390/buildings11080364

Academic Editor: Isaac Guedi Capeluto

Received: 24 June 2021
Accepted: 12 August 2021
Published: 18 August 2021

1. Introduction

1.1. Background

Over the past few decades, off-site constructed buildings, such as prefabricated buildings, have come into use as an alternative to the on-site method in both emerging and established economies [1]. Off-site manufacturing is commonly associated with affordable housing and student accommodation, but it is becoming more common in other areas such as hotels, residential areas, the private rented sector, airports, healthcare facilities, and defense-related buildings [2]. In particular, due to the COVID-19 pandemic outbreak, healthcare infrastructure is becoming overwhelmed around the world. In response to bed shortages and facility saturation, modular hospitals are growing quickly in the on-going fight against COVID-19 [3,4].

Owing to the advantages of inherent strength, modular construction, and relatively low costs, the role of containers functioning as building modules has gained in popularity over the past years. Thus, they can be one solution to the above challenges in terms of prefabricated or modular buildings. Compared to traditional building materials that have a large footprint, off-site constructed buildings are more eco-friendly and are in line with the concept of upcycled container architecture [5]. Flexible and quickly assembled modular container buildings usually consist of one or more standardized individual containers with size restrictions that depend in particular on what can be safely and economically built and transported. This is how most container manufacturers build modular sizes in various widths from 8 feet to a maximum of 14 feet. Prefabricated modular buildings are still affected by some deficiencies in both design and certain technical aspects, being unable to meet the occupant's requirements for thermal comfort and energy saving throughout the life cycle of the building.

It is currently estimated that there are more than 17 million retired shipping containers in ports around the world [6]. Most of them are just beyond the official age of retirement from 'active service' after approximately 12–15 years. The rest of them are structurally intact, but are no longer in use for their main purpose. The lifespan of containers used as buildings is much longer and can be assumed to be at least 50 years [7] according to lower degree of mechanical and climatic stresses they are subjected to, as compared to sea containers.

Moreover, because of the high expenses incurred to either destroy or transport them back to the original country and their non-degradable construction materials, they occupy lots of space in the ports. However, this does not mean that these containers cannot be used anymore. The sale of used containers can be a business area in itself, which may consist of depot services and container renovation and re-engineering, as well as the purchasing and onward selling of various types of containers [8].

1.2. Literature Review

A nearly zero-energy building (NZEB) refers to a building that has very high energy performance with the use of zero or very little energy. The nearly zero or very low amount of energy required is normally covered significantly by energy from renewable sources. The energy performance of a building can be determined on the basis of the calculated or actual consumed annual energy [9]. Strategies of design concepts and energy systems for NZEBs have been largely investigated in different climates. Rezaee et al. [10] proposed a parametric framework for a feasibility study of zero-energy residential buildings for the design stage. Their results showed that wall insulation, infiltration, and lighting load are the most significant parameters affecting the region's energy performance. Rey-Hernández et al. [11] performed an analysis of a hybrid ventilation system in an NZEB to meet the heating and cooling demand and IAQ levels. They concluded that the ventilation system with an energy recovery function has great potential. Medved et al. [12] assessed the contribution of energy storage to an NZEB, including heat and cold storage and batteries, which are significant in the transition of NZEBs towards ZEB. Guarda et al. [13] reported the influence of climate change on renewable energy systems designed for NZEBs with a case study in the Brazilian savannah. They indicated that energy demand will increase over the years, while the photovoltaic (PV) system alone will not be able to meet this new demand. As a result, they suggested that the bioclimatic building guidelines must be updated to promote adequate strategies to guide designers to construct resilient buildings when considering the impacts of climate change. Li and Wang [14] proposed an optimal design method to identify the global optimal design solutions for the NZEBs by considering uncertainties. The design process included robust design optimizations of the building envelope and energy systems, which were further tested by a case study in Hong Kong. Lindberg et al. [15] conducted a case study of a German multi-family house to explore the cost-optimal energy system design in an NZEB with grid connection. In the current energy market, they concluded that a heat pump is not a cost-optimal choice. With the same objective of cost-optimal energy performance, Ferrara et al. [16] evaluated a great number of design alternatives for NZEB building envelopes through automated optimization search procedures. They suggested that both envelope design and energy systems should be considered simultaneously at the early design stage in order to achieve a cost-optimal solution.

Several studies have investigated the possibility of achieving an NZEB using containers. Kristiansen et al. [17] investigated the feasibility of an off-grid container unit for industrial construction in China by considering vacuum insulation panels, three-layer glazed windows, natural ventilation, and cooling and heating set points. Additionally, they summarized perspectives on industrialized transportable solar powered zero-energy buildings using PV and battery storage [18]. In addition, this team [19] also performed energy analyses and lifecycle assessments to quantify the lifecycle impacts related to four energy efficiency designs of a container building (conventional, low-energy, net-zero energy, and off-grid). They indicated that the net-zero energy design strategy had the lowest lifecycle impacts in all categories. They also pointed out the transition from a linear approach to a circular approach to building products with the reuse of building structures in the form of containers. This can significantly increase the potential environmental benefits of the building's lifecycle, particularly in designs characterized by a dominant share of the pre-use stage in their total lifecycle impact. Moreover, Vijayalaxmi [20] developed a one-story home with a container structure for the hot–humid tropical climate of Chennai, India. It is claimed that the most effective passive means of achieving thermal comfort in a hot–humid climate is through comfort ventilation. Trancossi et al. proposed thermo-electric and solar heat pumps for self-sufficient container buildings [21]. Taleb et al. [22] developed a container home in a subtropical desert climate with green roofs and green walls as an insulation layer for the container envelope, in which triple glazing windows with low-emissivity film played a key role in reducing the cooling load. Cornaro et al. [23] designed a container-built show house, which had a satisfactory thermal performance with mild and temperate climates by the use of a solar chimney. Bohm [24] undertook the challenge to design a single house for high performance in two very different climates, indicating that it is difficult to find a common design strategy for conservative envelopes of such modular buildings in many diverse climates. It is therefore necessary to develop a category of energy-efficient design for container-based modular buildings under different climate conditions.

1.3. Motivation and Scope

Most of the above studies focused on nearly zero-energy container buildings (NZECBs), where containers are used for single houses or one-story buildings under specific climate conditions. The research gap is thus identified by the fact that there is a lack of a flexible guidance on design strategies for different building archetypes and energy systems in the concept of NZECBs in different climate zones. As a result, this paper proposes passive design concepts for NZECBs through extensive climate analysis, feasible designs for three building archetypes, and in-depth case studies of the related energy system solutions in three representative climate conditions.

One of the main goals is to show how different building types can be built up modularly from easily adaptable container units for different climatic conditions and for different user profiles. The building types are selected according to geographic, social, and cultural aspects in three different countries. Furthermore, the application of a reference building is analyzed and compared for all major climatic zones.

The paper is structured as following: Section 2 illustrates the overall research methodology. Section 3 presents the extensive climate analysis in feasible locations for NZECBs and the detailed climate evaluation in three representative climate conditions for passive design strategies, which are further proposed in Section 4 for three building archetypes as examples. In Section 5, the energy system solutions for NZECBs are studied. The conclusion and future work are finally discussed in Section 6.

2. Research Methodology

The overall research methodology is illustrated in Figure 1 in order to design NZECBs. This study comprises a total of three design phases, from the preliminary (early design) stage, to the scheme design stage, and then to the detailed design stage. The whole study is part of the ENSECO (energy self-sufficient container buildings) project [25], which aims to identify the integrated life-cycle product and process the development of energy self-sufficient buildings in different climate zones using prefabricated affordable containers. The construction design stage with demonstration work is regarded as future work in this paper.

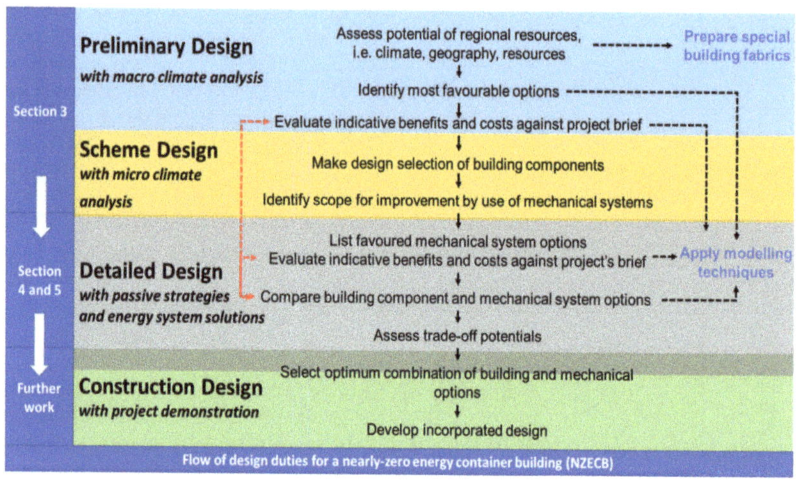

Figure 1. Design stages for a nearly zero-energy container building.

The proposed research method starts from analyzing regional resources, for instance, climate, geography, and resources, to realize the development of the passive potential of buildings under different functional and climatic scenarios. Therefore, all relevant climate zones are characterized, and key values are defined and compared for several representative locations. For the early planning stages of a building project, there are several climatic analysis tools [26]. Among them, the Olgyay Bioclimatic chart is the first bioclimatic chart that includes both temperature and relative humidity so as to construct a thermal comfort zone. However, only limited solutions are available from this chart if the climate is outside of the comfort zone. The other method, the Givoni–Milne Bioclimatic Chart shown in Figure 2, is another tool that is mainly applied for residential-scale construction [27]. There are more alternatives in the area of thermal comfort building design to investigate possible passive techniques for different types of climates around the world, covering passive methods of natural ventilation, evaporative cooling, thermal mass, passive heating, and active methods of conventional air conditioning or dehumidification [27–30].

The scheme design phase stage is based on the identified favorable options. Within this phase, the building components are selected, and it is decided if complementary mechanical ventilation and/or cooling systems are necessary to provide thermal comfort. If the response is positive, the selection of suitable complementary mechanical systems, for instance, air handing units or room unit coolers, takes place in the third stage, the detailed design. Moreover, another consideration in the detailed design stage is the development of outline operation/control strategies to match external/internal conditions. The definition of the ideal energy supply system can be carried out with modelling tools in two steps. The hourly heat and electricity demand is calculated with building simulation tools, e.g., with IDA ICE, and then transferred into the Polysun software in order to design the energy supply system based on renewable energies. Different configurations can be assessed using key indicators such as the degree of self-sufficiency or other economic or ecological parameters. Modelling techniques are used in the preliminary and detailed design phases.

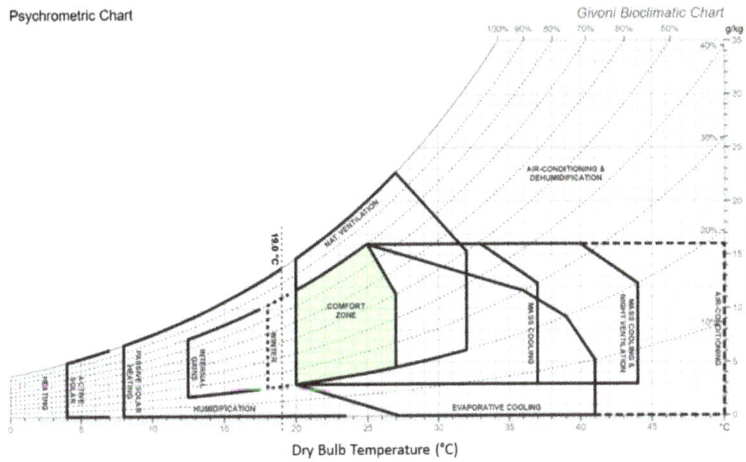

Figure 2. Givoni–Milne Bioclimatic chart.

The proposed design for both passive and mechanical strategies requires iterative calibration progression to avoid unnecessary oversize problems. The results from the detailed design phase are used in the preliminary design phase and vice versa.

The priority should first be given to passive strategies. As a consequence, the key approaches to achieve an NZECB are reflected in the above steps and by maintaining the internal climate of buildings and existing container structures, while minimizing the ecological footprint, as well as maximizing both the use of renewable energy and mechanical systems efficiencies.

3. Climate Analysis

This section starts with a general analysis of the worldwide climate zones at selected representative locations. It aims to describe concise and precise analytical results so that the interpretation together with the conclusions can be used for further building design. Afterwards, three locations are selected for building applications and will be analyzed in more detail.

3.1. General Climate Analysis of Appropriate Regions for NZECBs

There are different definitions of how the earth is divided into climatic zones. According to ASHRAE 169–2020 [31], the Earth's climate is essentially divided into eight different climate zones: Very hot, Hot, Warm, Mixed, Cool, Cold, Very Cold, and Subarctic, which are again differentiated according to different types: Humid, Dry, and Marine. In order to quantify the potentials of each climate zone for renewable energy in buildings, characteristic cities for the different climate zones are defined and characterized as plotted in Figure 3. The corresponding climate zones are listed in Table 1. Although the main focus is on Europe, the selection should also represent areas from the other continents. The subarctic zone is not considered further here. Berlin was selected as a characteristic climate for Central Europe, which is mainly cool–humid. Stockholm, Kiruna, Reykjavik, and Murmansk represent the northern European climate, which varies between a cold and very cold climate, with Kiruna having the lowest annual global radiation and average temperatures. Tashkent belongs to the same climate zone as Ankara, but the Uzbek area is characterized by a particular continental climate with large seasonal temperature fluctuations and a particularly high position of the sun. Mexico City, Melbourne, and Cape Town belong to the warm zone on three different continents. The subtropical Addis Ababa in the hot–humid climate is exposed to particularly high sunlight and is characterized by a so-called diurnal climate in which the mean temperature differences between day and night are greater than the temperature differences between the individual months. Melbourne in Australia and Recife and Mexico City represent a warm climate; Kharga and Abu Dhabi represent a hot and dry climate. Mumbai is characterized by a tropical climate, and is extremely hot and humid with particularly high mean temperatures.

Figure 3. All selected representative locations on a world map. Yellow background color indicates the locations for detailed analysis.

Table 1. The selected regions and the representative climate zones with data from ASHRAE 169–2020 [31].

No.	Location	Country	Climatic Zone
1	Mumbai	India	0A Extremely Hot—Humid
2	Recife	Brazil	0A Extremely Hot—Humid
3	Abu Dhabi	United Arab Emirates	0B Extremely Hot—Dry
4	Kharga	Egypt	1B Very Hot—Dry
5	Addis Ababa	Ethiopia	2A Hot—Humid
6	Melbourne	Australia	3A Warm—Humid
7	Mexico City	Mexico	3A Warm—Humid
8	Cape Town	South Africa	3C Warm—Marine
9	Ankara	Turkey	4A Mixed—Humid
10	Tashkent	Uzbekistan	4A Mixed—Humid
11	Berlin	Germany	5A Cool—Humid
12	Stockholm	Sweden	6A Cold—Humid
13	Reykjavik	Iceland	6A Cold—Humid
14	Murmansk	Russia	7 Very Cold
15	Kiruna	Sweden	7 Very Cold

Climatic data were collected for all cities using the Climate Data Center (CDC) [32] and weather data of the EnergyPlus™ [33]. Global radiation is one of the most important key values for renewable energies and is therefore considered in more detail here. Figure 4 shows the frequency distribution with the number of days per year with a certain amount of radiation energy. The warm and hot locations have their maximum values between 5000 and 8000 Wh/m^2 per day. Only Melbourne has a significantly lower maximum value. The cool and cold locations have a much lower radiation potential with their maximum daily values mainly between 1000 and 2000 Wh/m^2 per day. Only Tashkent has a significantly higher radiation potential. These diagrams are useful in order to evaluate the solar potential over the year. A high frequency of low radiation energy requires greater effort to create energy self-sufficient buildings. Of course, the temperatures are also relevant for building designs, and so further data were collected such as average, minimum, and maximum temperatures; their daily fluctuations; the hours of sunshine; and average wind speeds.

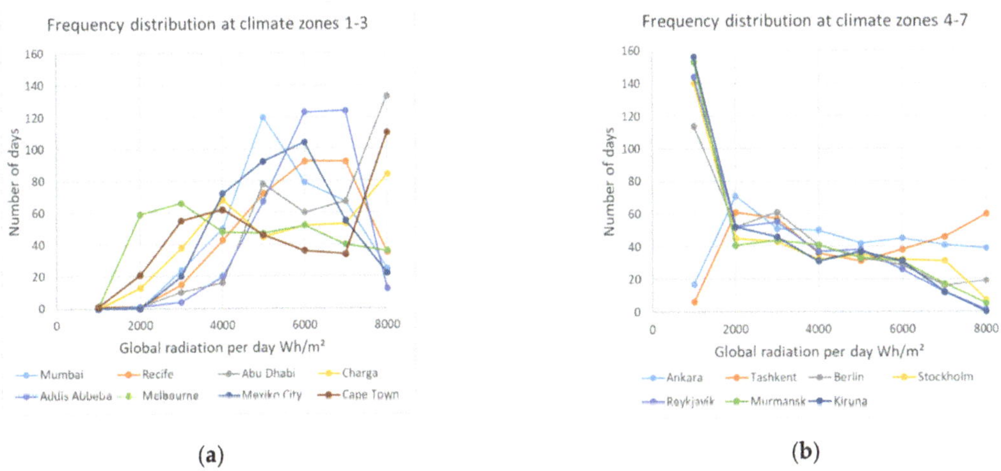

Figure 4. Frequency distribution of daily global radiation for selected cities. (**a**): Hot and warm climate zones 1–3. (**b**): Cool and cold climate zones 4–7.

Table 2 contains all the data, and the cities are sorted in ascending order of their global annual radiation, with Kiruna having a minimum and Abu Dhabi a maximum of solar radiation. The peak values are marked in red and the minimum values in green. High average wind speeds are indicators for the potential of wind turbines as significant energy converters. In this context, wind energy is particularly relevant for Melbourne, Reykjavik, and Cape Town due to their proximity to the coast and the consequent high wind speeds. High average temperatures indicate a need for cooling, which could be potentially compensated by passive measures such as insulation or an increased heat capacity in the building structure if the nighttime temperatures are low enough. The same relationship applies to cold temperatures.

Thus, the heating and cooling requirements for the respective locations can be assessed by a characteristic integrated parameter. Therefore, the temperature–time curves are integrated, once as an integral above the temperature of 23 °C and once below 20 °C. This is based on the assumption that there is a potential cooling requirement (without any building structure) in outside temperatures greater than 23 °C and heating requirements in temperatures below 20 °C. Even if the heating and cooling requirements are particularly dependent on the building and usage conditions, this method is used to derive a building-independent parameter for characterizing the climatic conditions in terms of energy demand. The resulting heating and cooling potentials are plotted in Figure 5. For example, although Addis Ababa and Abu Dhabi have a very high global radiation, the integral predicts an especially high cooling demand, while in Addis Ababa there is predom-

International Climate Zones Category (Zone 6 A) [31]. During the winter, the average temperature drops below freezing (0 °C) from December and reaches the lowest temperature in February. During this period, the sky cover range has an average of around 70%, and the average relative humidity is around 80% to 90% during this period. Summer spans the period from from June to August; it is a mild season with more pleasant days. Extreme hot temperatures are rarely recorded: occasionally the temperature has reached 28 °C within historic records. There is an obvious diurnal temperature variation along with greater windy occasions. Sometimes, it can be very cool or even cold at night since the temperature can drop below 10 °C even in summer.

Higher in the northern latitudes, the days are very short for winter times, and the amount of solar radiation resources is valuable. The winter solar height (December) ranges from 0° at azimuth −40° to 40°, with 7° being the highest at azimuth 0°, so it is clearly low from November to January when the sun is rarely seen. The summer solar height (June) ranges from 0° at azimuth −154° to 154°, with 54° being the highest at azimuth 0°. From around the middle of May to the end of July, the duration of sunshine in Stockholm is about 15 to 17 h per day, when it is not completely dark even at midnight.

In short (according to 18 years' worth of data of meteorological parameters for most stations), the average daily total global horizontal solar radiation fluctuates from 199 to 5278 Wh/m^2; the average monthly air temperature is in the range of −3 to 17 °C, while the average monthly relative humidity varies from 63% to 90%, and the average monthly wind speed has a maximum value of 4 m/s and a minimum value of 2 m/s.

3.2.3. Addis Ababa, Ethiopia, Representing a Hot Region

For the hot climate, Addis Ababa (latitude 9.0° N, longitude 38.76° E) in Ethiopia was selected. From the weather data collected, Addis Ababa is located in the Ethiopian Plateau, which is 2300 m above sea level [15]. It belongs to the hot–humid type according to the International Climate Zones Category (Zone 2 A) [31]. Thus, even though it is located in a tropical climate zone, the climate is mild. It has pleasantly warm daytimes, around 23/25 °C, most of time outside of summer, and cool nights when the temperature drops below 10 °C. In terms of precipitation, it has an annual 1200 mm of rainfall, and the notable rainy seasons occur in June to September and from March to May.

The amount of solar radiation resources is high. The summer solar height (June) ranges from 0° N at azimuth −72° to 72°, with 74° being the highest at azimuth 0°, while the winter solar height (December) ranges from 0° at azimuth −115° to 115°, with 58° being the highest at azimuth 0°. The amount of sunshine in Addis Ababa is high from October to May. In other words, solar resources in Addis Ababa are moderate with average sky cover in the range of around 56%.

In short (according to 18 years' worth of data of meteorological parameters for most stations), the average daily total global horizontal solar radiation fluctuates from 4638 to 6835 Wh/m^2. The average monthly air temperature is in the range of 15 to 17 °C, while the average monthly relative humidity varies from 54% to 83%, and the average monthly wind speed has a maximum value of 4 m/s and a minimum value of 2 m/s.

4. Passive Design Strategies and Concepts for Three Building Archetypes in the Represented Climates

4.1. Design Strategies in Berlin, Germany and a Case Design of Multi-Family House

The moderate temperatures, light winds, and four distinctively marked seasons provide more flexibility in the selection of passive design strategy in terms of architectural engineering. Figure 6 illustrates the historical hourly weather data under a Givoni Bioclimatic Chart overlay. When analyzing the current climate condition over a whole year period, there are a total of 8760 h; 8.1% of these already meet the comfort conditions, while around 57.7% of them require an additional active system design to improve the comfort conditions. The comfort conditions can be satisfied through passive design strategies within the remaining 34.3% of annual time.

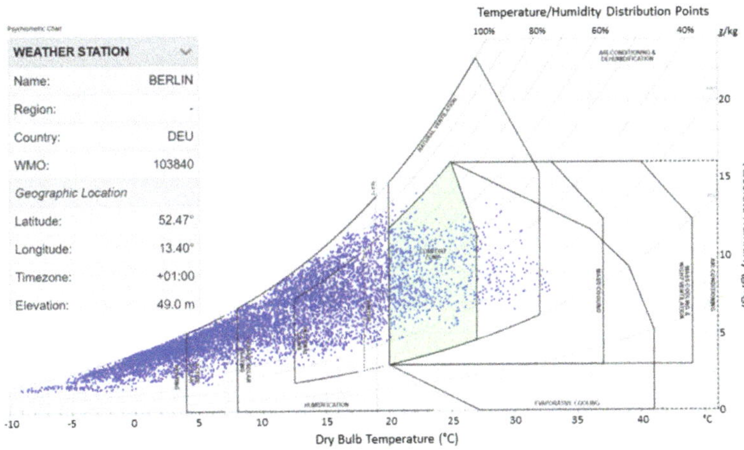

Figure 6. Hourly psychometric chart with Givoni Bioclimatic Chart overlay, Berlin, Germany.

With the help of the Givoni Bioclimatic Chart, the majority of hourly data are located toward the heating process line in Figure 6. This implies that passive heating should be the main concern in passive design strategies, which can be principally improved by building siting and orientation. Firstly, it is always beneficial to learn the traditions from local buildings with common compact floorplans and a central heat source, south facing windows, vestibule entries (working as air locks) to reduce unnecessary infiltration and uncomfortable drafts, and a pitched roof for wind protection. A building's plan geometry is usually quantified based on its aspect ratio. The optimal building aspect ratio varies with orientation, climatic condition, latitude, longitude, and altitude. Here, the approximate aspect ratio of 1:1.6 (which is the proportion of the short side to the long side of the floor plan dimensions) was considered to be optimal because it keeps the exposed surface area-to-volume ratio as compact as possible so that it can obtain balanced heat transmission through the envelopes from cool winters to warm summers.

Building orientation is another important factor to improve a building's thermal performance in relation to both the solar position and the prevailing winds. In this case, the optimal orientation is to locate the E–W axis 18° north of east. The floorplan can be organized towards either winter sun or natural daylight, which is able to penetrate into spaces occupied during the daytime with specific functions that coincide with solar orientation. The long south-facing surfaces of the building can capture solar energy for passive heating during the winter. Simultaneously, the structural mass, such as shear wall, and construction components, such as tiles, slates or a stone-faced fireplace, as well as the service core, can be located at the center of the building's volume. Thus, 'inertia' against temperature fluctuations is provided that can keep winter solar gain and summer night cooling from the perimeter where the circulation zone can be located. Additionally, this can facilitate access to high levels of illumination and reduce the energy required for artificial lighting. However, direct sunlight admitted through the southern facade should be controlled by means of either insulating blinds, heavy curtains, or operable window shutters. These can not only help minimize glare and visual discomfort but also reduce winter nighttime heat losses. In addition, more considerations ought to be given to the exterior. If a basement is used, it must be at least 46 cm below the frost line and insulated on the exterior (foam) or on the interior (fiberglass in a furred wall). A garage or storage area can be additionally added as a buffer zone on the side of the building facing the coldest wind. In terms of landscape, more efforts can be made to create a biophilic summer garden and a cozy winter outdoor space. Landscape trees (either conifer or deciduous) are acceptable beyond 45 degrees from each corner of the glazing for passive solar gain. Dense planting of trees (with at least the same height and distance up to twice the height of the building excluding the roof) can even protect entries from cold winter winds. Sunny wind-protected outdoor spaces (seasonal sun rooms, enclosed patios, courtyards, or verandas) can extend interior living areas with the extra help of exterior wind shields (e.g., wing walls, wind breaks, fences, exterior structures, or land forms).

The main purpose of the opening and glazing design is to maximize natural ventilation and daylight. Both are key elements for an adaptive design as they can significantly boost the occupants' comfort and reduce the required energy consumption. Natural ventilation is most effective when the building is shaped and oriented to optimized summer prevailing winds. Narrow floor plans have the advantage of smooth air flow and cross ventilation through the building floor plan so that both air quality and thermal comfort performance are better. There are adequate wind resources with the prevailing winds coming from west during the summer months; thus, window openings can be located on both the windward sides and the leeward north and northeast sides to generate the pressure differences between the inside and the outside so as to facilitate cross ventilation. As the average wind speed is around 11 m/s during the summer time, it would feel more than a strong breeze, so small opening sizes would be adequate to provide fresh air. Additionally, it is necessary to restrict incoming cold gales from the southwest (around 13 m/s for the monthly average, from Figure 7) and the southeast (around 10 m/s for the monthly average, from Figure 7) during the winter months. In order to have passive solar heating, most of the vision glazing should face the south for winter sun exposure, and overhangs should fully shade windows during the overheated period. When performing the window selection, glazing should possess a minimized U-value as the undesired solar radiation gain has less impact in this climate. Additional placement of small well-insulated skylights (less than 3% of floor area for most clear-day conditions, 5% for most overcast-day conditions) is vital to balance the daytime lighting energy and overheating risk.

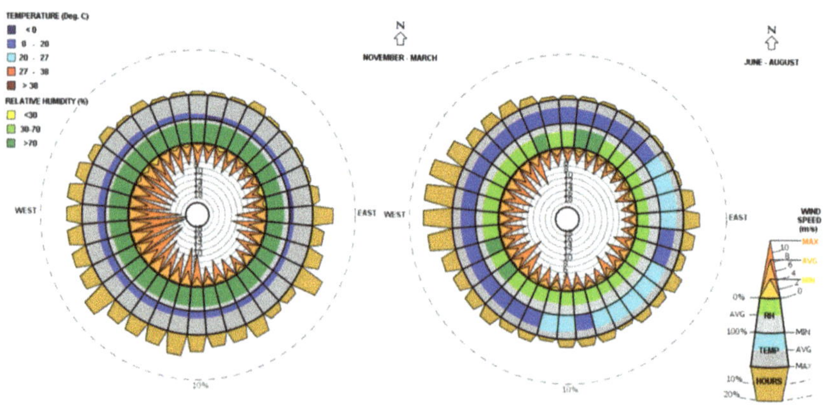

Figure 7. Prevailing winds in winter and summer in Berlin, Germany.

Together with addressing passive heating, the retention of useful internal heat gains is another solution. It is beneficial to borrow the idea from the envelope of traditional local buildings' that use low-mass, tightly sealed, well-insulated construction and a steep pitched roof, which, with a vented attic over a well-insulated ceiling, helps to shed rain and snow and even prevent ice dams. Since we aimed to optimize the passive potential of container house, tightness, as one of the significant features, can be retained and additionally well insulated to gather significant internal heat gains to reduce heating load. Within a super-tight container home, it is also recommended to equip this with a fan-driven HRV or ERV (heat/energy recovery fan) to ensure adequate indoor air quality while conserving energy. Even though the extra insulation (super insulation) and the reduced thermal bridge might prove cost effective, this increases occupant comfort by keeping indoor temperatures more uniform. Meanwhile, it is feasible to lower the indoor comfort temperature at night to reduce heating energy consumption. Thus, a design case is illustrated in Figure 8 as a multi-family house. According to above design strategies, this design case has following characteristics:

- Utilization of structural components in preserving internal heat gains during winter;
- Optimal orientation and window-to-wall ratio so as to strengthen (1) daylighting from multiple side; (2) passive solar gain; (3) flexible natural ventilation strategies (cross ventilation and night flushing);
- Optimization of landscape design, for instance, of green walls and earth sheltering, to achieve both wind protection and direct evaporative cooling;
- Adding overhangs and corridor balconies as fixed sun-shading devices.

Figure 8. A passive design case for a multi-family house in Berlin, Germany.

4.2. Design Strategies in Stockholm, Sweden, and a Design Case of a Single-Family House

In general, the passive design strategies in this climate are similar to those of the previous case, but are characterized by less sunlight, wetter conditions, cooler winter temperatures, and shorter summers. In terms of architectural engineering, Figure 9 shows the historical hourly weather data under the Givoni Bioclimatic Chart. Analyzing the current climate condition over a whole year period, there are in total 8760 h; only 4.7% of these already meet the comfort conditions, while around 66.6% of them require an additional active system design to improve the comfort conditions. The comfort conditions can be achieved through passive design strategies within the remaining 28.7% of the annual time.

With the help of the Givoni Bioclimatic Chart, it was observed that a lot of hourly data are located towards the heating process line in Figure 9. This implies that passive heating should be the main concern in passive design strategies, which can be principally assisted by building siting and orientation. Firstly, it is always beneficial to borrow beneficial ideas from traditional local homes in this cold climate. These suggestions include a common snug and appropriate floorplan with a central fire place, south-facing windows, vestibule entries (air locks) to minimize infiltration and eliminate drafts, and pitched roofs for wind protection. According to design practice, the optimal building dimension ratio is close to 1:1:1 and is partial to multiple stories so as to keep the exposed surface area-to-volume ratio as compact as possible to avoid exposure to the chilly wind environment and significant heat loss through the building's surfaces. An appropriate size can avoid an excessive floor area with an unnecessary heating demand.

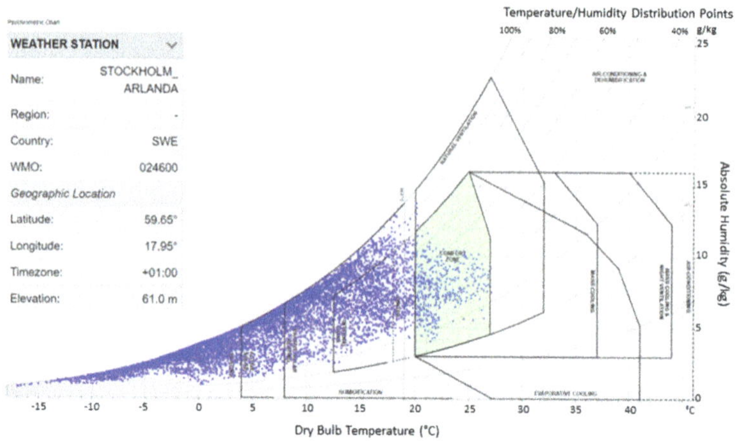

Figure 9. Hourly psychometric chart with Givoni Bioclimatic Chart overlay, Stockholm, Sweden.

In this climate, the floor plan of the building can be organized in such a way that either winter sun or natural daylight penetrate into the day-use spaces with specific functions that correspond to the orientation of the sun. It is recommended to have the long surface of the building oriented toward the south to capture solar energy for passive heating during the winter. The main orientation of the building should be within 30° of south. In this case, the optimal orientation is to locate the E–W axis towards the south. Bedrooms facing southeast benefit from the morning sun, while living rooms orientated west of south will catch the late afternoon sun, which can help delay the evening heating period. Similar to the Berlin case, the structural mass; construction components, such as tiles, slates, or a stone-faced fireplace; and the service core can be located at the center of the building volume. There are two primary reasons for this: one is to provide 'inertia' against temperature fluctuations to store both internal gains and winter day time solar gains from the perimeter; another reason is to have a thermal buffer zone, where the circulation zone can be located on the north side to mitigate lower exterior temperatures.

As sunshine is much more appreciated in this climate, it plays an important role in glazing for passive heating and daylighting as it can significantly reduce the required energy consumption and boost visual comfort with high levels of illumination. Direct sunlight from the south façade can be an excellent source of lighting here. However, it needs to be efficiently distributed throughout the floor plan. If it is a rectangular, a narrow floor plan with the longer sides aligned to the east and west axes can allow a lot of diffuse natural light to enter from the north. In large buildings with a square footprint, access to natural light is restricted to the building periphery and areas near the glazing, so the inner parts of the floor area are dependent on artificial lighting. Introducing daylight from skylights can eliminate this problem (less than 3% of floor area in clear weather, 5% in overcast weather). In order to enable passive solar heating, most of the viewing glass should face south in order to maximize solar radiation in winter. The direct sunlight through the south facade should be controlled either via insulating blinds, heavy curtains, or operable shutters. This can also help minimize glare and visual discomfort as well as reduce heat loss during winter nights.

Even when the climate is cold, passive cooling is still an essential strategy to improve the thermal comfort for around 231 h in a year. Narrow floor plans have the advantage of smooth airflow and cross ventilation through the building floor plan so that both air quality and thermal comfort performance are better. Square floor plans can facilitate passive cooling using the stack effect by means of skylights. There are sufficient wind resources with the prevailing winds coming from the southwest and northeast during the summer months (from Figure 10). Window openings can be located on the both windward southwest facades and the leeward northeast sides to generate pressure differences between the inside and the outside to facilitate cross ventilation. As the average wind speed is around 9 m/s during summer time, it would feel like a strong breeze, so adequate air exchange through small openings is preferable to maintain comfortable interior temperatures and to avoid excessive heat loss. During the cool winter time, the average wind speed is less than a gentle breeze, although the prevailing winds are still mainly from the southwest and south. When performing the window selection, glazing should minimize conductive loss and gain (minimize U-value) because undesired solar radiation gain has less impact in this climate. The additional placement of small, well-insulated skylights (less than 3% of the floor area in a clear climate, 5% in cloudy weather) is crucial to reduce the lighting energy and the risk of overheating during the day.

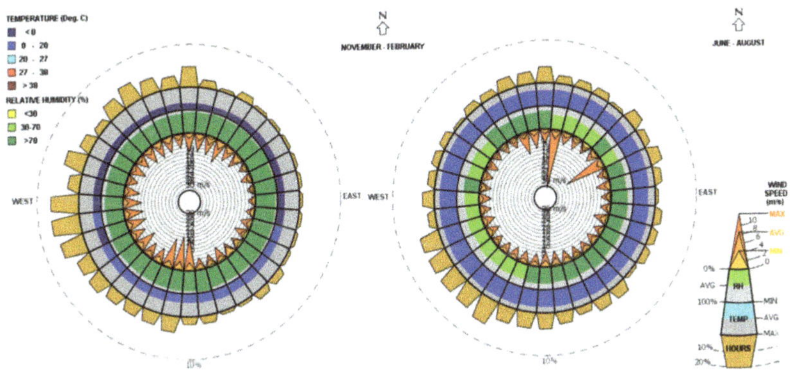

Figure 10. Prevailing winds in winter and summer, Stockholm, Sweden.

In cold climates, traditional buildings prefer to use a low-mass, tightly sealed envelope; thick insulation; and a steep pitched roof with a vented attic over a well-insulated ceiling. The primary reason for this is to maintain a cold roof temperature to avoid risky ice dams, as well as to vent moisture thereby preventing mold and bacteria growth, which usually occurs on or within attics and wall assemblies due to condensation or improper detailing. For the implementation of this, the thermal insulation level of the roof perimeter should be equal or greater to the thermal resistance of the exterior wall. In addition, a 1:300 ventilation ratio is recommended (as specified by most building codes), which is based principally on good historical experience and simple psychometric analysis. Since we planned to optimize the passive potential of a container home, tightness, as one of the significant features, can be kept and additionally well insulated to gather significant internal heat gains to reduce heating load. Within a super-tight container home, it is also recommended to equip a fan-powered heat/energy recovery ventilator to ensure indoor air quality while conserving energy. Even though extra insulation (super insulation) and a reduced thermal bridge might prove cost effective, they increase occupant comfort by keeping indoor temperatures uniform. Additionally, it is feasible to lower the indoor comfort temperature during sleeping time to reduce heating energy consumption. As a result, a design case is displayed in Figure 11 as a single-family house. This involves the main characteristics from the above passive design using the following climate-control strategies:

- An optimal orientation and floor plan so as to strengthen passive solar gain;
- Maximizing daylighting through multiple sides and skylights;
- The utilization of high thermal mass in preserving internal heat gains during winter, assisted using night flushing during summer;
- A hybrid ventilation solution covering both adaptive comfort natural ventilation and fan-forced ventilation;

- The optimization of landscaping design to achieve both wind protection and direct evaporative cooling;
- Movable sun-shading devices;
- The utilization of high thermal mass assisted by night flushing.

Figure 11. A passive design case for single-family house in Stockholm, Sweden.

4.3. Design Strategies in Addis Ababa, Ethiopia

This climate is characterized by high temperatures, constantly high humidity, and great changes in daily temperatures. In terms of architectural engineering, Figure 12 gives the historical hourly weather data according to the Givoni Bioclimatic Chart. Analyzing the current climate condition over a whole year period, there are total 8760 h; 16.6% of these already meet the comfort conditions, while around 15.9% of them require an additional active system design to improve the comfort conditions. The comfort conditions can be achieved through passive design strategies within the remaining 67.5% of annual time. In general, the promising climate control strategies are (arranged by percentage):

- Internal heat-gain preservation;
- Passive solar gain;
- Wind protection of outdoor spaces;
- Using shading elements around windows;
- High thermal mass associated with night flushing;
- Placing direct evaporative features to achieve passive cooling effects;
- Employing ceiling fans or indoor forced-air motion, making the interior feel cooler (to be used on hot days with the windows closed); thus, less air conditioning is needed.

With the help of the Givoni Bioclimatic Chart, there is still a high number of hourly data located toward the heating process line, but generally lower in relative humidity as shown in the Figure 12. This implies that passive heating should be the main concern in passive design strategies, which can be principally assisted by building siting and orientation. The lessons from traditional buildings indicate that high ceilings and tall, operable (French) windows protected by deep overhangs and verandahs are commonly seen in this climatic area. It is useful to create shaded outdoor buffer zones (porch, patio, lanai) oriented to the prevailing breezes, which can extend living and working areas in summer, and use passive solar gain in winter. In terms of floorplans, cross ventilation is maximized by a long, narrow building floorplan as an effective cooling strategy. As there is often a great diurnal temperature variation, night flushing is another strategy with which to remove much of the thermal storage out of envelopes, leading to lower early morning interior temperatures. This is true for the surrounding landscape design and the dense planting of bushes, trees, and ivy-covered walls, especially on the west to minimize heat gain (if summer rains support native plant growth). Furthermore, this climatic area always has wet soil; therefore, raising the building high above the ground is suggested to minimize dampness and maximize natural ventilation underneath the building.

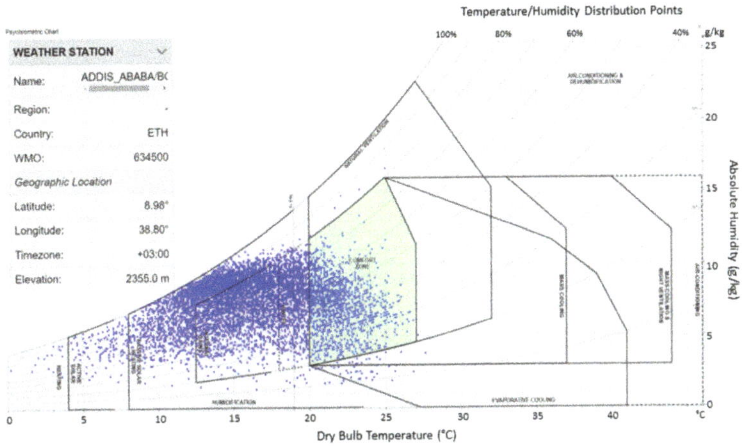

Figure 12. Hourly psychometric chart with Givoni Bioclimatic Chart overlay, Addis Ababa, Ethiopia.

The predominant considerations are given to maximized natural ventilation and reducing or eliminating air conditioning in warm weather if windows are well shaded and oriented to prevailing breezes (shown in Figure 13). As the average wind speed is around 3 m/s, mainly from due north during the summer time, it would feel like a gentle breeze, so adequate air exchange through large openings is preferable to maintain comfortable interior temperatures. In principle, there are some additional suggestions:

- Shade the window well with window overhangs (design dependent on the latitude) or operable shades (e.g., awnings that can be extended in summer) to eliminate cooling load;
- Window openings perpendicularly oriented to the prevailing winds for natural ventilation;
- Locate exterior wing walls and plants around the window direction up to 45 degrees;
- Create the possibility of natural cross ventilation by placing doors and window openings on opposite sides of the building;
- Choose screened porches and patios to provide a longer time of passive comfort cooling without potential insect problems;
- Have a ridge hood, roof monitor, open stairwell, and two-story spaces to enhance the natural indoor buoyancy ventilation.

A design case is proposed in Figure 14 as a school building in Addis Ababa, Ethiopia. In hot and humid climatic zones, local traditional buildings still emphasize natural ventilation in terms of enclosure design. They have a preference to take advantage of lightweight construction with slabs on grade, operable walls and shaded outdoor spaces. Placing enough north glazing is suggested in order to benefit from diffused daylighting and cross ventilation (about 5% of floor area). A whole-house ventilation strategy can be achieved through:

- Maximizing the vertical height between air inlet and outlet (open stairwells, two-story spaces, roof monitors) so as to take advantage of stack ventilation during calm, warm days;
- Using a well-ventilated attic with pitched roofs to guarantee ventilation and rain protection.

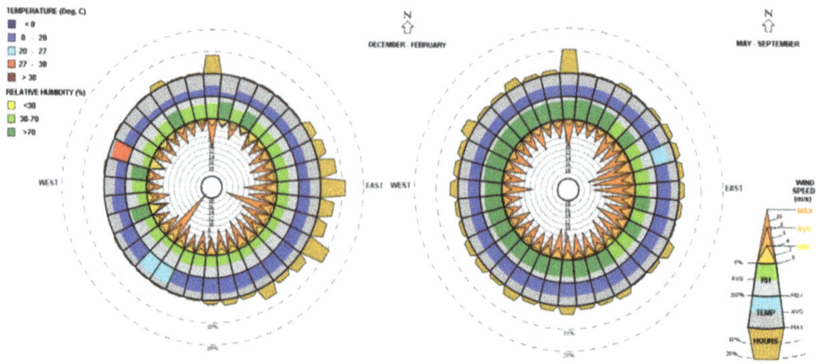

Figure 13. Prevailing winds in winter and summer, Addis Ababa, Ethiopia.

Figure 14. A passive design case for a school in Addis Ababa, Ethiopia.

Figure 12 indicates that the target climatic zone has a significant diurnal temperature fluctuation, which provides sufficient support to the climate-dependent strategy. Taking advantage of diurnal temperature differences, a cooler night-air circulation allows the dissipation of the heat build-up in the whole envelope during the daytime. In order to achieve this passive cooling effect, one simple requirement is to allow the night air to fully circulate the building from the openings and through an open floor plan.

4.4. Development of Multifunctional Components

In order to achieve the project objective of developing partially energy-autonomous lightweight buildings, innovative multifunctional components including the production technology required for their manufacturing were developed. These multifunctional assemblies enable a significant improvement in energy efficiency by using regenerative energies as well as a reduction in the required energy consumption due to improved structural–physical properties (thermal insulation, heat storage). The components developed within the scope of the research project [25] are the roof assembly (Figure 15), the floor assembly (Figure 16), and the side wall assembly (Figure 17).

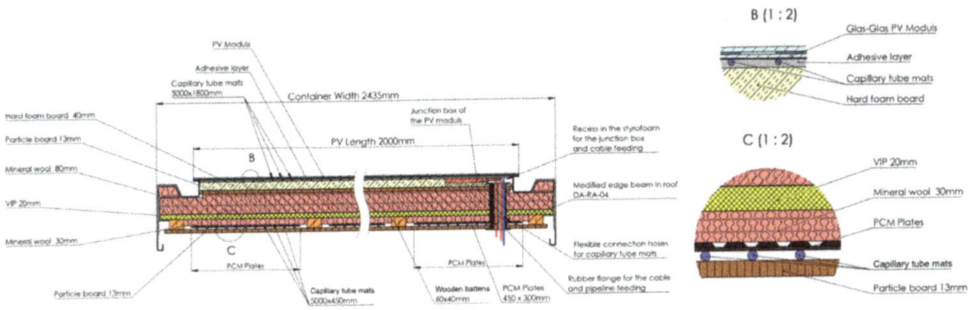

Figure 15. Construction of a multifunctional roof assembly.

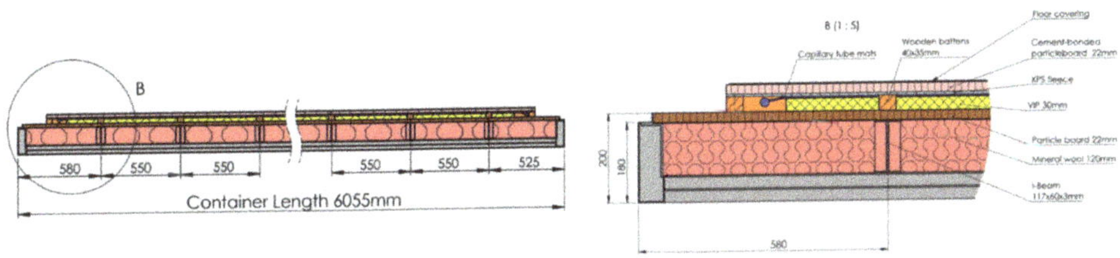

Figure 16. Construction of a multifunctional floor assembly.

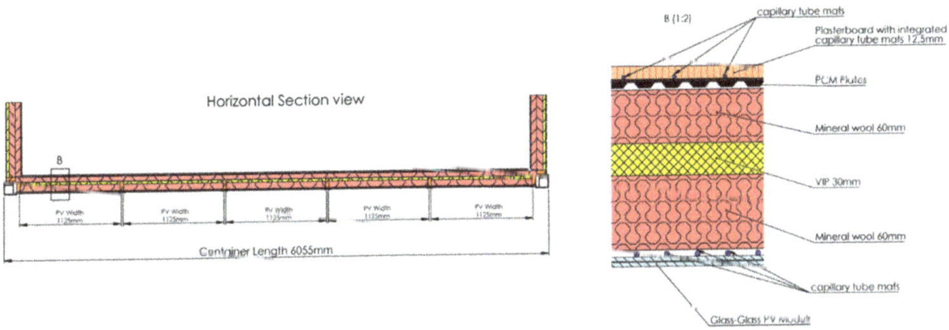

Figure 17. Construction of a side wall assembly.

The use of regenerative energies is achieved by innovative photovoltaic–thermal (PVT) modules integrated into the building's outer shell, which supply the building with electrical and thermal energy. To reduce thermal energy consumption, highly insulating vacuum insulation panels (VIP) were integrated into the assembly constructions. These enable a significant improvement in thermal insulation with simultaneously low insulation thicknesses compared to conventional insulation materials. To improve the low heat-storage capacity of lightweight buildings, phase change materials (PCM) were integrated in macro-encapsulations in the side walls and roof assemblies. This enables a significant increase in heat-storage capacity while maintaining the original wall thicknesses. PCM serves as a thermal buffer that can be useful in regions with large daytime temperatures fluctuations so that the PCM can be regenerate itself regularly.

Environmentally friendly interior climatization is achieved by the integration of capillary tube mats into the wall and roof assemblies, which enable the use of a low-exergy system using renewable energies. The multifunctionality of the assemblies enables a significant improvement in the energy balance and living comfort of container-based lightweight buildings and thus enables the production of lightweight buildings that significantly surpass the state of the art for the fast, cost-effective, and sustainable provision of urgently needed living space.

5. Energy Systems Strategies for Three Different Archetypes in Represented Climates

In this section, the three design cases from the above section are considered for an in-depth energetic analysis. The heating and energy demands are calculated with the building simulation tool IDA ICE. Based on the determined energy requirements, the following energy supply concepts C1-C7 were created, with the emphasis on maximizing the degree of self-sufficiency. They were applied in Polysun for each of the three application cases.

- C1: Air heat pump (AHP) with photovoltaics (PV): An air heat pump for heating and cooling is combined with two water storage tanks, one for heating and one for cooling. The photovoltaic system uses the entire available roof area. A DC battery acts as a power storage device. In order to increase the degree of self-sufficiency, an electric heating rod is integrated into the storage system, which further loads the storage system above the working temperature of the heat pump with excess, self-generated solar electricity.
- C2: Geothermal heat pump (GHP) with PV: A geothermal pump provides heat that is stored in a heat-storage system. A heat exchanger between the earth probe and the underfloor heating cools the building directly. The PV, battery, and heating rod are implemented as in concept C1.
- C3: Combined heat and power plant with PV: A biogas combined heat and power plant (CHP) generates heat and electricity. In order to extend the life of the CHP, an absorption-cooling machine is used to cool the building. One water tank is used for heat and cold. Since the CHP works at a higher temperature level than the heat pump, no electric heating rod is used. PV and battery storage systems are implemented as in concept C1.
- C4: Air heat pump with solar thermal (ST) and photovoltaics: In contrast to concept C1, the area of the photovoltaic system is reduced to provide space for a small solar thermal system (up to eight collectors or 16 m^2).
- C5: Air heat pump with hybrid photovoltaics/thermal collectors (PVT) and PV: The solar thermal collectors from concept C4 are replaced by photovoltaic–thermal hybrid collectors.
- C6: Air heat pump with hybrid collectors: Hybrid collectors cover the entire roof area; otherwise, the approach adopted is similar to that in concept C1. An exception for the multi-family house is that the hybrid collectors cover the balcony balustrade only.
- C7: Air heat pump with photovoltaics and wind turbine (WT): In addition to all components from concept C1, a 3.5 kW wind turbine is integrated.

All concepts were modeled with the software Polysun® for the three application cases. In all cases, the electricity requirements for heat generation, the degree of self-sufficiency, electricity generation by means of PV, electricity supply, and grid feed-in were evaluated.

5.1. Multi-Family House in Berlin, Germany

Due to its design, the multi-family house in Germany has only a small roof area in relation to the usable area. Therefore, the building uses the balcony balustrade as an energy conversion area. Depending on the concept, these were equipped with photovoltaic, solar thermal, or hybrid collectors. The following component specifications apply to the concept C1–C7 for the multi-family home:

- Heat pump: power 5 kW + 2× water storage 1000 L;
- PV (roof): alignment 0°, angle 30°, power 15 kWp (concept C1–C4,C6 + C7)/12 (concept C5), polycrystalline, rated power STC 300 W;
- PV (balcony scaffolding): alignment 0°, angle 90°, power 6 kWp (concept C1–C3, C5 + C7)/12;
- Solar thermal: orientation 0°, angle 90°, gross total area 36 m^2, flat collectors;

- Hybrid collectors (roof): alignment 0°, angle 30°, gross area 17 m^2, power 2.95 kWp;
- Hybrid collectors (balcony scaffolding): alignment 0°, angle 90°, gross total area 34 m^2, power 5.9 kWp;
- Battery: capacity 10 kWh, Li-Ion.

Table 3 shows the results for the apartment building. With the use of the entire available roof area and the balcony balustrade, it is possible to generate 19,233 kWh of electricity, which corresponds to 915 kWh/kWp. The electricity generated is far from sufficient to cover the entire demand. Only concept C3 (CHP) achieves a degree of self-sufficiency of 84.5% under these boundary conditions. Without a CHP, it is currently only possible to achieve a degree of self-sufficiency of about 45%. This value may be increased by making use of the façade for the installation of further photovoltaic modules.

Table 3. Results of the Polysun simulation for the multi-family house in Berlin, Germany.

Multifamily House Berlin	Units	C1: AHP + PV	C2: GHP + PV	C3: CHP + PV	C4: AHP + ST + PV	C5: AHP + PVT + PV	C6: AHP + PVT	C7: AHP + PV + WT
Electricity demand for heat generation	kWh	31,425	11,322	51.0	31,390	32,650	71	28,222
Degree of self-sufficiency	%	30.8	39.7	84.5	25.2	29.7	45.6	38.3
Production PV (AC)	kWh	19,233	19,233	19,233	14,879	18,848	17,888	19,233
Self-sufficiency	kWh	15,272	11,703	15,416	12,604	15,104	8331	13,770
Purchased electricity	kWh	34,397	17,876	2908	37,025	35,789	9995	32,694

5.2. Single-Family House in Stockholm, Sweden

Two consumer scenarios with different power consumptions of 3000 kWh and 6000 kWh for electrical appliances were analyzed for the single-family house. The following specifications were used for concepts C1 to C7 for the single-family house:

- Heat pump: power 8 kW + 2× water storage 1000 L
- PV: orientation 0°, angle of adjustment 45°, power 22.5 kWp (concept C1–C3, C7)/19.2 kW (concept C4 and C5), polycrystalline, rated power STC 300 W
- Solar thermal: orientation 0°, angle of application 45°, gross total area 16 m^2, flat collectors
- Hybrid collectors: alignment 0°, angle 45°, gross total area 17 m^2, power 2.95 kWp (concept C5)/22.1 kWp (concept C6)
- Battery: capacity 20 kWh, Li-Ion.

Table 4 shows the simulation results for the single-family house with low and high power demand. The photovoltaic plant provides a yield of approx. 1000 kWh/kWp so that a degree of self-sufficiency of 80% is possible with low power demand in concept C1. It is striking that the degree of self-sufficiency decreases only by a few percentage points, although the electricity demand doubles in the second scenario. This is mainly due to the very large size of the photovoltaic system and battery storage. With low power consumption, more than 60% of the generated electricity is fed into the grid. With the high consumption, the grid feed is still close to 50%.

Concept C3 (CHP + PV) achieves a high degree of self-sufficiency of 84.4% with low power consumption. However, due to the low electricity consumption, the advantage of CHP, the simultaneous generation of electricity and heat, cannot be used. More than half of the energy generated by CHP has to be fed into the electricity grid. This changes with the higher power demand, so in this case an even higher degree of self-sufficiency of 92.3% is possible. The self-sufficiency is only an electric self-sufficiency, so the biogenic fuel consumption of the CHP is not taken into account.

Table 4. Results of the Polysun simulation for the single-family house in Stockholm, Sweden.

Single-Family House Stockholm	Units	C1: AHP + PV	C2: GHP + PV	C3: CHP + PV	C4: AHP + ST + PV	C5: AHP + PVT + PV	C6: AHP + PVT	C7: AHP + PV + WT
Low electricity demand: 3000 kWh								
Electricity demand for heat generation	kWh	6850	4996	169	5596	6815	332	6561
Degree of self-sufficiency	%	80.2	74.3	84.5	72.8	75.2	72.2	80.3
Production PV (AC)	kWh	22,439	22,439	22,439	19,163	20,532	12,649	22,439
Self-sufficiency	kWh	8794	6908	3776	7288	8380	3366	8684
Purchased electricity	kWh	2210	2504	721	2709	2777	1313	1183
High electricity demand: 6000 kWh								
Electricity demand for heat generation	kWh	6807	4915	169	5463	6729	332	6492
Degree of self sufficiency	%	76.5	64.2	92.3	69.9	71.7	67.5	75.1
Production PV (AC)	kWh	22,439	22,439	22,439	19,163	20,532	12,649	22,439
Self-sufficiency	kWh	10,827	7843	6894	8951	10,093	5191	10,372
Purchased electricity	kWh	3325	4523	635	3868	3992	2506	1875

In large parts of Sweden, the operation of a wind turbine would be possible due to the good wind conditions. Concept C7 (air-WP + PV + wind turbine) achieves a degree of self-sufficiency of more than 80% with the help of a 3.5-kW wind turbine. However, this is only marginally higher than that in concept C1. From an economic point of view, a wind turbine will not be a viable solution in the current situation.

Concept C6, where the hybrid collectors covered the entire roof, is not advantageous for a high degree of self-sufficiency. The low demand does not require the high amount of heat generated by the collectors. The collectors heat up strongly, and the efficiency of the power generation decreases sharply.

5.3. *School Building in Addis Ababa, Ethiopia*

In Ethiopia, the majority of people live in rural areas, and only 4.8% of them have access to the public electricity grid [35]. Thus, energy self-sufficiency is a key issue. The following component specifications apply to the concepts C1–C7 for the school building:

- Heat pump: power 5 kW + 2× water storage 1000 l
- PV: orientation 0°, angle 20°, power 12.6 kWp (concept C1–C3; C7)/9.9 kW (concept C4 + C5), polycrystalline, rated power STC 300 W
- Solar thermal: orientation 0°, angle 20°, gross total area 16 m^2, flat collectors
- Hybrid collectors: orientation 0°, angle 20°, gross total area 17 m^2, power 2.95 kWp (concept C5)/12.39 kWp (concept C6)
- Battery: capacity 10 kWh, Li-Ion.

The roof area of the building is sufficiently large to install a photovoltaic system with 12.6 kWp. In contrast to in the single-family house, hot water is used for hand washing only, so the water temperature at the tapping points can be reduced to 40 °C. Table 5 gives the results of the Polysun simulation for the school in Addis Ababa, Ethiopia.

Table 5. Results of the Polysun simulation for the school in Addis Ababa, Ethiopia.

School Building Addis Ababa	Units	C1: AHP + PV	C2: GHP + PV	C3: CHP + PV	C4: AHP + ST + PV	C5: AHP + PVT + PV	C6: AHP + PVT	C7: AHP + PV + WT
Electricity demand for heat generation	kWh	6903	4910	854	4036	5244	1150	6085
Degree of self-sufficiency	%	84.2	79.4	91.7	75.6	81.7	64.7	85.1
Production PV (AC)	kWh	18,906	18,906	18,906	14,850	17,112	10,609	18,906
Self-sufficiency	kWh	15,146	12,696	10,948	11,432	13,338	7912	14,285
Purchased electricity	kWh	2873	3365	1010	3731	3052	4359	2936

Due to the good weather and climate conditions, PV energy can produce a 50% higher gain in Ethiopia than in Sweden. The gain is approx. 1500 kWh/kWp. Approximately 80% of the photovoltaic electricity can be used directly since the total electricity demand mainly occurs during daytime hours for the school's operation. In addition, the position on the equator clearly favors self-sufficiency. Throughout the whole year, the position of the sun is already high in the early morning hours. Although the photovoltaic energy production is only about 1000 kWh higher than the demand, a remarkably good degree of self-sufficiency of 84.2% is possible.

As a school, the heat requirement is not sufficient for the consumption of the generated heat of the hybrid collectors from concept C6. The concept has the worst degree of self-sufficiency. In rural Ethiopia in particular, a connection to an electricity grid is not common. Therefore, we analyzed the times of the day during which restrictions on energy supply are likely to happen without a grid. The respective deviation from complete energy self-sufficiency is determined in percentage points and summed up for each time of day. Figure 18 shows the annual sum of the percentage points over time. A high deviation means that there is not enough power available for regular school operation. The greatest deviation occurs during the first hour of school operation. However, the deviations remain very small for most of the day. Only in the evening hours does the likelihood of restrictions rise.

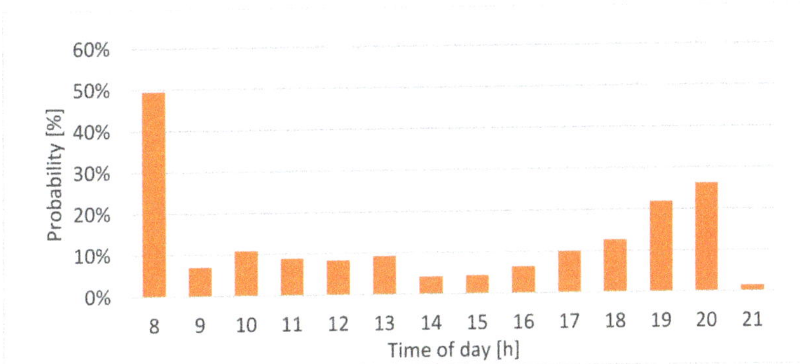

Figure 18. Probability of not meeting the electricity demand. School concept C1 in Addis Ababa.

5.4. Influence of Different Climate Zones on the Energy Self-Sufficiency: A Case Study of a Single-Family House in Different Climates

Finally, the influence of different climate zones on energy self-sufficiency was investigated for several locations (Berlin, Stockholm, Kiruna, Reykjavik, Murmansk, Tashkent, Mumbai, Addis Ababa). For this evaluation, a simplified building model was used in

Polysun, which represents the energetic behavior of a single-family house as described above. The simulation considered a dynamic calculation, which was based on the weather data and the building data. Since the heat requirements for the three application cases were already available at the specified location, the simulation of these specific cases was based on the heating load, so there may have been minor deviations between the simplified model (Polysun) and the detailed model (IDA ICE).

In addition, the parameters of battery size, power consumption, and thermal insulation were varied for the locations. The results for high and low electricity consumption (3000 kWh/a and 6000 kWh/a) as well as for battery storage sizes of 10 kWh and 20 kWh are plotted in Figure 19. Generally, the degree of self-sufficiency should decrease with decreasing storage size. The effect is noticeable in all locations except Reykjavik and less pronounced in locations with little sunlight (Kiruna, Murmansk). Battery storage in these locations can use a smaller design. If the power consumption is particularly high, the degree of self-sufficiency decreases as expected. The effect of the storage size is particularly pronounced in sunny locations (Mumbai, Tashkent).

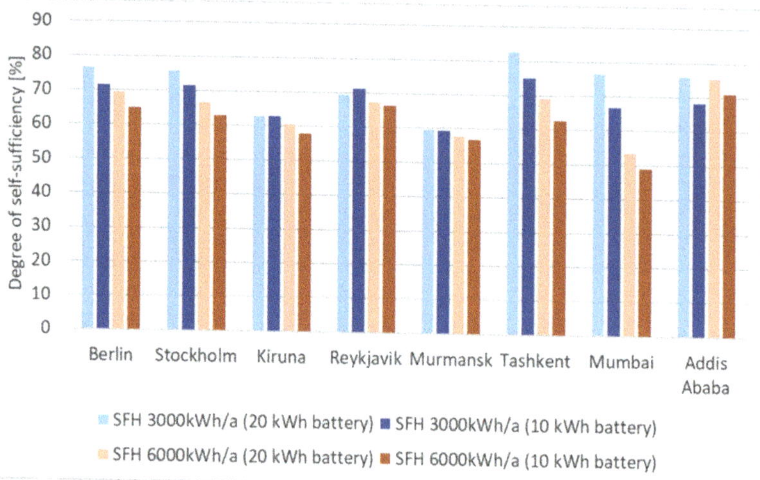

Figure 19. Comparison of the degree of self-sufficiency for different power consumption and battery storage sizes for several locations (single-family house).

The passive house standard, as used in the previous investigations, is relatively cost-intensive due to the complex technology with vacuum insulation panels (VIP). Therefore, the influence of such thermal insulation systems for different climate zones was also simulated and is presented in Figure 20. For this purpose, the container building in the passive house standard (heat transfer coefficient U = 0. 12 W/m^{2}K) was compared to a conventional glass-wool-insulated container building (U = 0.27 W/m^{2}K). The heat capacity of the conventionally insulated building was lower by 30%. In hot regions, the high level of thermal insulation is not advantageous, so the buildings in these locations can be built more economically. The levels of self-sufficiency hardly differed here (Tashkent, Mumbai, and Addis Ababa). In the other cases, the degree of self-sufficiency was reduced moderately by approx. 3–8 percentage points.

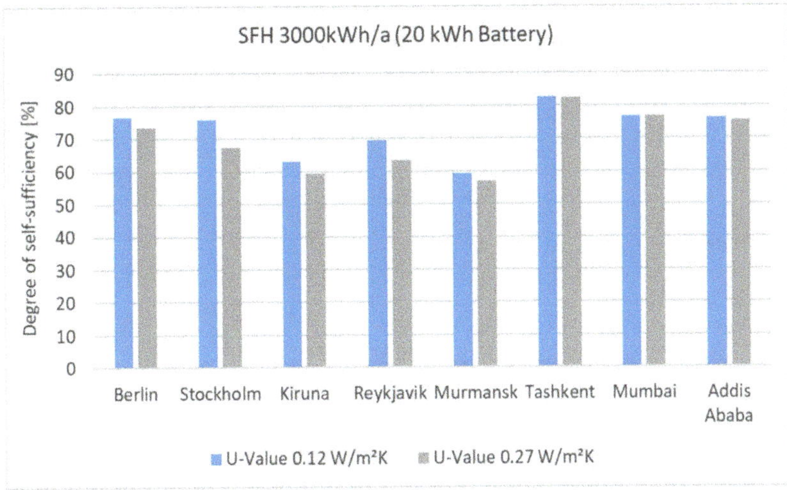

Figure 20. Degree of self-sufficiency for several locations with a changed heat transfer coefficient (U) of the building with a 20-kWh battery (single-family house).

6. Conclusions and Future Work

In this paper, we demonstrated how to systematically design container-based buildings with nearly zero-energy consumption. Owing to their lightweight construction, container buildings have economic and ecological advantages that can be further increased by using recycled containers. Based on a general climate analysis, an architectural design for three building types was derived. The building types were selected according to geographic, social, and cultural aspects in three different countries. The goal of almost-zero energy container buildings was achieved through the use of specific wall constructions with vacuum insulation panels and phase-change materials as well as with the design of various renewable energy systems consisting of photovoltaics, battery storage, solar thermal energy, and air and geothermal heat pumps, as well as combined heat and power plants. The selected locations differ considerably in terms of use and climatic conditions. Therefore, energy supply concepts were systematically developed for all applications, which should enable the highest possible degree of self-sufficiency whereby a degree of real self-sufficiency of approx. 80% can be considered a realistic goal when standard conditions for electricity demand and comfortable temperature are applied. This self-sufficiency goal can be achieved for the SFH and the school building with concept C1 through the use of photovoltaics, batteries, and air heat pumps. The MFH has a comparatively disadvantageous surface-area-specific energy requirement and surface-area-to-volume ratio, so photovoltaics are not sufficient here. Instead, concept C3 with a CHP unit is required for a high degree of self-sufficiency. However, the total annual energy production is often significantly higher than the actual annual energy demand, which makes the buildings on-grid self-sufficient when electricity can be fed into and consumed from the grid. The building solutions need to be adapted individually to the different applications and climates by using the described tools and strategies. The sophisticated wall construction with VIP is particularly beneficial for cool climates, while the PCM serves as a thermal buffer that can be useful in hot regions with large daytime temperature fluctuations so that the PCM can regenerate itself regularly.

In this paper, passive design potentials have been extensively evaluated at the early design stage, while thermal comfort and sensation should be further included in the following stage. Further work will target the lifecycle assessment and economic evaluation of the energy efficiency measures.

Author Contributions: Conceptualization, J.K., J.S. and X.Z.; methodology, A.S. and J.S.; formal analysis, J.K., A.S. and J.S.; writing—original draft preparation, J.K., A.S., J.S., S.K., P.K. and X.Z.; writing—review and editing, J.K., J.S. and X.Z., Funding acquisition, P.K. All authors have read and agreed to the published version of the manuscript.

Funding: This research was funded by Bundesministerium für Wirtschaft und Energie (BMWi), (ZF4574901AT8).

Conflicts of Interest: The authors declare no conflict of interest.

References

1. Kamali, M.; Hewage, K. Life cycle performance of modular buildings: A critical review. *Renew. Sustain. Energy Rev.* **2016**, *62*, 1171–1183. [CrossRef]
2. Elements Europe. Ressidential. Available online: https://elements-europe.com/sectors/residential/ (accessed on 20 July 2020).
3. Harrouk, C. Alternative Healthcare Facilities: Architects Mobilize their Creativity in Fight against COVID-19. Available online: https://www.archdaily.com/937840/alternative-healthcare-facilities-architects-mobilize-their-creativity-in-fight-against-covid-19 (accessed on 11 August 2020).
4. Karmode Prefabicated Technologies. Prefab Modular Hospital Buildings. Available online: https://karmod.com/en/page/prefab-modular-pandemic-hospitals (accessed on 11 August 2020).
5. Shen, J.; Copertaro, B.; Zhang, X.; Koke, J.; Kaufmann, P.; Krause, S. Exploring the potential of climate-adaptive container building design under future climates scenarios in three different climate zones. *Sustainability* **2020**, *12*, 108. [CrossRef]
6. Islam, H.; Zhang, G.; Setunge, S.; Muhammes, A. Life cycle assessment of shipping container home: A sustainable construction. *Energy Build.* **2016**, *128*, 673–685. [CrossRef]
7. Dara, C.; Hachem-Vermette, C.; Assefa, G. Life cycle assessment and life cycle costing of container-based single-family housing in Canada: A case study. *Build. Environ.* **2019**, *163*, 106332. [CrossRef]
8. Sharma, S. Maersk Line. What Happens When Containers Retire? Available online: https://www.supplychaindigital.com/scm/what-happens-when-containers-retire (accessed on 7 August 2019).
9. Hermelink, A.; Schimschar, S.; Boermans, T.; Pagliano, L.; Zangheri, P.; Armani, R.; Voss, K.; Musall, E. Towards nearly zero-energy buildings: Definition of common principles under the EPBD. *Ecofys*. 2012. Available online: https://ec.europa.eu/energy/sites/ener/files/documents/nzeb_full_report.pdf (accessed on 12 September 2020).
10. Rezaee, R.; Vakilinezhad, R.; Haymaker, J. Parametric framework for a feasibility study of zero-energy residential buildings for the design stage. *J. Build. Eng.* **2021**, *35*, 101960. [CrossRef]
11. Rey-Hernández, J.M.; San José-Alonso, J.F.; Velasco-Gómez, E.; Yousif, C.; Rey-Martínez, F.J. Performance analysis of a hybrid ventilation system in a near zero energy building. *Build. Environ.* **2020**, *185*, 107265. [CrossRef]
12. Medved, S.; Domjan, S.; Arkar, C. Contribution of energy storage to the transition from net zero to zero energy buildings. *Energy Build.* **2021**, *236*, 110751. [CrossRef]
13. Aparecida da Guarda, E.L.; Mansuelo, R.; Domingos, A.; Hoffmann Martins Jorge, S.; Durante, L.C.; Machado Sanches, J.C.; Leão, M.; Apolônio Callejas, I.J. The influence of climate change on renewable energy systems designed to achieve zero energy buildings in the present: A case study in the Brazilian Savannah. *Sustain. Cities Soc.* **2020**, *52*, 101843. [CrossRef]
14. Li, H.; Wang, S. Coordinated robust optimal design of building envelope and energy systems for zero/low energy buildings considering uncertainties. *Appl. Energy* **2020**, *265*, 114779. [CrossRef]
15. Lindberg, K.B.; Fischer, D.; Doorman, G.; Korpås, M.; Sartori, I. Cost-optimal energy system design in Zero Energy Buildings with resulting grid impact: A case study of a German multi-family house. *Energy Build.* **2016**, *127*, 830–845. [CrossRef]
16. Ferrara, M.; Fabrizio, E.; Virgone, J.; Filippi, M. Energy systems in cost-optimized design of nearly zero-energy buildings. *Autom. Constr.* **2016**, *70*, 109–127. [CrossRef]
17. Kristiansen, A.B.; Satola, D.; Lee, K.; Zhao, B.; Ma, T.; Wang, R.Z.; Gustavsen, A.; Novakovic, V. Feasibility study of an off-grid container unit for industrial construction. *Sustain. Cities Soc.* **2020**, *61*, 102335. [CrossRef]
18. Kristiansen, A.B.; Ma, T.; Wang, R.Z. Perspectives on industrialized transportable solar powered zero energy buildings. *Renew. Sustain. Energy Rev.* **2019**, *108*, 112–124. [CrossRef]
19. Satola, D.; Kristiansen, A.B.; Houlihan-Wiberg, A.; Gustavsen, A.; Ma, T.; Wang, R.Z. Comparative life cycle assessment of various energy efficiency designs of a container-based housing unit in China: A case study. *Build. Environ.* **2020**, *186*, 107358. [CrossRef]
20. Vijayalaxmi, J. Towards sustainable architecture—A case with Greentainer. *Local Environ.* **2010**, *15*, 245–259. [CrossRef]
21. Trancossi, M.; Cannistraro, G.; Pascoa, J. Thermoelectric and solar heat pump use toward self sufficient buildings: The case of a container house. *Therm. Sci. Eng. Prog.* **2020**, *18*, 100509. [CrossRef]
22. Taleb, H.; Elsebaei, M.; El-Attar, M. Enhancing the sustainability of shipping container homes in a hot arid region: A case study of Aswan in Egypt. *Archit. Eng. Des. Manag.* **2019**, *15*, 459–474. [CrossRef]
23. Cornaro, C.; Rossi, S.; Cordiner, S.; Mulone, V.; Ramazzotti, L.; Rinaldi, Z. Energy performance analysis of STILE house at the Solar Decathlon 2015: Lessons learned. *J. Build. Eng.* **2017**, *13*, 11–27. [CrossRef]
24. Bohm, M. Energy technology and lifestyle: A case study of the University at Buffalo 2015 Solar Decathlon home. *Renew. Energy* **2018**, *123*, 92–103. [CrossRef]

25. *ENSECO Produkt- und Verfahrensentwicklung zur Vorbereitung und Realisierung Kompletter Gebäude Verschiedener Nutzungsarten Mittels Energieautarker Container, Project Funding Code ZF4574901AT8*; BMWi—Federal Ministry for Economic Affairs and Energy: Berlin, Germany, 2018–2020.
26. Matsumoto, H.; Tsuzuki, K.; Susanti, L. Bioclimatic Analysis in Pre-Design Stage of Passive House in Indonesia. *Buildings* **2017**, *2*, 24.
27. Sayigh, A.; Marafia, A.H. Chapter 1—Thermal comfort and the development of bioclimatic concept in building design. *Renew. Sustain. Energy Rev.* **1998**, *2*, 3–24. [CrossRef]
28. Lam, J.C.; Yang, L.; Liu, J. Development of passive design zones in China using Bioclimatic approach. *Energy Conversat. Manag.* **2006**, *47*, 746–762. [CrossRef]
29. Rakoto-Joseph, O.; Garde, F.; David, M.; Adelard, L.; Randriamanantany, Z.A. Development of climatic zones and passive solar design in Madagascar. *Energy Convers. Manag.* **2009**, *50*, 1004–1010. [CrossRef]
30. Singh, M.K.; Mahapatra, S.; Atreya, S.K. Development of bio-climatic zones in north-east India. *Energy Build.* **2007**, *39*, 1250–1257. [CrossRef]
31. Climatic Data for Building Design Standards. *ANSI/ASHRAE Standard 169-2020, American Society of Heating, Refrigerating and Air-Conditioning Engineers*; ASHRAE: Atlanta, GA, USA, 2020; ISSN 1041-233.
32. Deutscher Wetterdienst. Clima Data Center—OpenData-Bereich. Available online: https://www.dwd.de/DE/klimaumwelt/cdc/cdc_node.html (accessed on 5 September 2019).
33. EnergyPlus. Weather Data. Available online: https://energyplus.net/weather (accessed on 5 September 2019).
34. American Society of Heating, Refrigerating and Air-Conditioning Engineers (ASHRAE), International Weather for Energy Calculation (IWEC). Available online: http://www.equaonline.com/ice4user/indexIWEC.html (accessed on 5 September 2019).
35. Länderprofil Äthiopien: Informationen zur Nutzung und Förderung Erneuerbarer Energien für Unternehmen der Deutschen Branche, Stand August 2014. Available online: https://www.dena.de/fileadmin/dena/Dokumente/Pdf/3110_Laenderprofil_AEthiopien_REG.pdf (accessed on 5 September 2019).

buildings

MDPI

Case Report

A Top-Down Digital Mapping of Spatial-Temporal Energy Use for Municipality-Owned Buildings: A Case Study in Borlänge, Sweden

Samer Quintana [1,2,*], Pei Huang [1], Mengjie Han [1] and Xingxing Zhang [1]

1 School of Technology and Business Studies, Dalarna University, 79188 Falun, Sweden; phn@du.se (P.H.); mea@du.se (M.H.); xza@du.se (X.Z.)
2 Department of Engineering Sciences, Uppsala University, 75237 Uppsala, Sweden
* Correspondence: ssq@du.se; Tel.: +46-7309-48879

Abstract: Urban energy mapping plays a crucial role in benchmarking the energy performance of buildings for many stakeholders. This study examined a set of buildings in the city of Borlänge, Sweden, owned by the municipality. The aim was to present a digital spatial map of both electricity use and district heating demand in the spatial–temporal dimension. A toolkit for top-down data processing and analysis was considered based on the energy performance database of municipality-owned buildings. The data were initially cleaned, transformed and geocoded using custom scripts and an application program interface (API) for OpenStreetMap and Google Maps. The dataset consisted of 228 and 105 geocoded addresses for, respectively, electricity and district heating monthly consumption for the year 2018. A number of extra parameters were manually incorporated to this data, i.e., the total floor area, the building year of construction and occupancy ratio. The electricity use and heating demand in the building samples were about 24.47 kWh/m^2 and 268.78 kWh/m^2, respectively, for which great potential for saving heating energy was observed. Compared to the electricity use, the district heating showed a more homogenous pattern following the changes of the seasons. The digital mapping revealed a spatial representation of identifiable hotspots for electricity uses in high-occupancy/density areas and for district heating needs in districts with buildings mostly constructed before 1980. These results provide a comprehensive means of understanding the existing energy distributions for stakeholders and energy advisors. They also facilitate strategy geared towards future energy planning in the city, such as energy benchmarking policies.

Keywords: digital mapping; spatial; temporal; energy use

Citation: Quintana, S.; Huang, P.; Han, M.; Zhang, X. A Top-Down Digital Mapping of Spatial-Temporal Energy Use for Municipality-Owned Buildings: A Case Study in Borlänge, Sweden. *Buildings* **2021**, *11*, 72. https://doi.org/10.3390/buildings11020072

Academic Editor: Ravi Srinivasan

Received: 22 January 2021
Accepted: 15 February 2021
Published: 18 February 2021

Publisher's Note: MDPI stays neutral with regard to jurisdictional claims in published maps and institutional affiliations.

1. Introduction

Buildings represent large energy end-users worldwide. In the EU, buildings currently consume over 40% of total primary energy usage [1]. With its sights set on the new paradigm shift regarding energy production, efficiency and climate change, Sweden will implement strategies to reach national targets for energy efficiency in the building sector by 2050. According to these targets, energy use per square metre should decrease by 20% by 2020 and 50% by 2050, in comparison with use in 1995—this is a national target for energy efficiency in the housing sector [2]. In 2010, over 50% of the world's population were living in urban areas. By 2050, this number is expected to reach 75% [3]. Urban development and the expansion of cities, through the modification of land uses (from natural to artificial), cause a shift in the local energy budget and energy supply/demand patterns. Such a transformation has significantly changed the microenvironment and the related energy usage in urban cities [4]. The mapping of urban building energy plays a crucial role in understanding the multitude of agents that take part in the energy performance of buildings and thus in setting up the benchmarks in different districts for various stakeholders.

In Swan and Ugursal's study, the modelling approaches for energy consumption in a number of buildings were classified into bottom-up or top-down approaches [5]. The bottom-up approach is more appropriate when there is a need to evaluate the energy consumption based on a highly detailed level of data and to model technological systems [6]. Bottom-up models can be divided into two types: deterministic (or engineering) and statistical. The statistical methods search for correlations, utilizing a sample of information from energy bills as a source of data for energy modelling and analysing the link between energy consumption and a range of different variables (e.g., building shape, age and occupant behaviour) [7]. They can also take into account socioeconomic effects in the equations. They calculate reliable consumption based on the available information on the current status of buildings. However, due to their strong dependency on available historical consumption data, these bottom-up statistical methods are restricted to predicting the impact of new technology options and energy saving potential after the application of refurbishment measures [8]. The bottom-up deterministic methods are detailed models which are based on thermodynamic relationships and heat transfer calculations [9]. The main advantage of an engineering-based method is the ability to predict energy saving potential for buildings when some renovation measures are to be implemented [10]. These modelling approaches require a large amount of information about the building structures and parametric input to estimate the energy usage of a set of reference buildings of the stock based on a numerical model. Additionally, the evaluation of urban planning scenarios is computationally extensive, and the availability of construction and geometrical data needed as input for the models is very scarce. The top-down approaches treat the entire residential sector as one energy sink. Unlike the bottom-up approaches, the top-down methods are suitable for a large-scale analysis and not for the identification of the possible improvements to the building at urban and local levels [11]. Compared with the bottom up-approaches, the top-down methods are relatively easy to develop based on the limited information provided by macroeconomic indicators, such as price and income, technology development pace and climate. As summarised by Swan and Ugursal, the top-down approaches have advantages including the capacity for long-term forecasting in the absence of any discontinuity, inclusion of macroeconomic and socioeconomic effects, the simple input information required and the capacity to encompass trends [5].

Both the bottom-up and top-down approaches can assist the spatial–temporal analysis of the energy demand at the district level. For instance, Schneider et al. (2017) developed two bottom-up statistical extrapolation models for spatial–temporal analysis of the geo-dependent heat and electricity demand of a building stock located in Switzerland [12]. They calculated the heat demand using a statistical bottom-up model applied at the building level. Due to the large variability in the electricity usage, they estimated the municipality-level electricity load curve by combining socio-economic indicators with the average consumption per activity and/or electric device. Chen et al. (2019) established a Geographic Information System (GIS) based multi-criteria index system for spatial–temporal analysis of the energy demand in a university located in China [13]. They used the developed system to investigate the characteristics of (i) the temporal dynamic, (ii) the load fluctuation and (iii) the district load spatial distribution as well as the coupling relationships of power loads for heating/cooling between single buildings and the entire university district. They also implemented principal component analysis to identify the buildings which had large impacts on the district power demand. Unlike most of the existing approaches, which estimate the district energy demand at different spatial–temporal levels as functions of the characteristics of either individual buildings or cities and their occupancy levels, Mohammadi and Taylor (2017) connected spatial–temporal heterogeneous human behaviour with the city-level building energy use [14]. They first examined the temporal manifestation of the energy use fluctuations in urban buildings driven by spatial mobility patterns of the population, and then they developed a multivariate auto-regressive model for spatial–temporal analysis of the urban-level building energy demand in the City of Chicago based on a yearly individual positional record. Their study reveals that human mobility can account for the collective energy consumption in urban spaces. As can be seen from the abovementioned literature, spatial–temporal analysis of the geo-dependent urban-level heat and electricity demand is important for urban-scale planning and can bring several benefits: (i) it is beneficial for the construction of a geo-referred database for a specific location; (ii) it enables the estimation of the energy saving potentials that can be achieved by different retrofit programs and thus assists decision making; (iii) it supports the investigation of the influential factors affecting electricity demands and the optimization of the operation and management of district heating/cooling systems and district power dispatches.

Besides the top-down and bottom-up approaches, there is also a typology approach, which is based on the synthetic characteristics of a group of buildings. The European TABULA project defined building typology as "a systematic description of the criteria for the definition of typical buildings as well as a set of exemplary buildings representing the building types" [15]. It takes into account aspects such as climate, period of construction, spatial and housing models, technologies, design rules, building codes, planning regulations, economic constraints, building construction techniques, the organization of construction companies and worksite organizations [16]. Dascalaki et al. has demonstrated that the typology approach is effective in investigating the energy performances of building stocks [17]. By drawing on various building typologies, an energy benchmarking system could be developed as representative of a large percentage of the entire urban building stock. This approach has also been utilised in European Commission energy projects like RePublic_ZEB [18]. An effective tool to support the typology approach in analysing urban building stock patterns and "typologies" is spatial cluster analysis. For instance, Lucchi et al. conducted a spatial cluster analysis using data-mining methods (i.e., an hdb-scan algorithm) and a GIS method to investigate the energy performances in a historic town in Calavino, Italy [19]. Such clustering analysis can overcome the inaccuracies related to the application of the traditional building stock analysis approach. Similarly, Miao et al. proposed a clustering method to automatically extract and identify urban spatial patterns and functional zones based on massive amounts of volunteered geographic information collected in Beijing, China [20]. The study results show that these methods can effectively identify urban spatial patterns and thus can contribute to urban energy simulation.

In the context of sustainable cities, spatial visualization is a very effective approach that can help decision-makers in the urban planning process create future energy transition strategies and implement energy efficiency and renewable energy technologies. The most fundamental energy visualization tools use simple lines, pie charts and bar charts to show the energy usage patterns over time at the individual building level. For instance, the Pulse Dashboard presents trend-line energy consumption data for each commercial building [21]. The Building Dashboard presents energy usage using bars [22]. Other 2D visualization techniques include cluster maps, component planes, spiral displays, time logs and thematic 2D maps [23,24]. However, as the number of analysed buildings increases, these conventional visualization techniques may not perform well due to the limited information that can be presented; most of them can only reveal the temporal characteristics of land but not the spatial characteristics. Compared with 2D visualization, 3D visualization is more realistic and psychologically appealing for the human brain. Geographic information system (GIS) techniques can be used for the visualisation of the energy demand or production in buildings from the urban to the regional scale, or even at a national one. These visualization techniques include "hit maps" (i.e., aggregated data in 3D charts) [24] and 3D city models with semantic objects [25]. There are many studies using GIS techniques to visualise the energy data in building stocks. For instance, Mattinen et al. (2014) developed a method for estimating and visualizing the energy use and greenhouse gas emissions from a residential building stock located in the Kaukajärvi district, Finland [26]. Using such a visualization model, they also analysed the impacts of behavioural and technical changes on the energy performance in the building stock. Finney et al. made a comprehensive mapping of heat sources and sinks in Sheffield City,

UK [27]. Based on the heat source mapping, they linked these smaller systems to create a combined heat and power-based urban-scale network of energy generation and delivery. Huang et al. (2019) used a GIS technique to obtain the roof area in Kowloon district in Hong Kong. Using the obtained roof area, they evaluated the solar power potential that would be available for the whole district by installing rooftop PV panels, which was then used as the input for designing public charging stations. The solar PV potentials were visualised using different colours on the Kowloon district map [28]. Similarly, Ramachandra and Shruthi used the GIS technique to map the wind energy resources of Karnataka state, India. Based on the wind-power mapping, they analysed the variability of these resources, considering spatial and seasonal aspects [29]. Despite the abovementioned literature, until now the utilization of 3D visualization in spatial–temporal analysis of urban-scale energy usage has been very limited.

Although there are existing studies of mapping energy uses in different cities, spatial energy analyses in local municipalities are necessary as they will be different in various city and culture contexts. Specific consideration should be paid to the differences between cities when the aim is to optimise the integration of urban energy systems operated in buildings and promote renovation and renewable energy systems. This is because cities differ from each other at the local, national and international levels from the perspectives of geography, socio-economy, culture, infrastructure, and information platform. The types of cities and districts determine the kinds of users and needs and consequently the nature (qualitative and quantitative) of the policy/regulation schemes and the calibration/adjustment of the energy infrastructures. Citizens' behaviours and needs/preferences with regard to energy may be different from each other in different cities, which can lead to great differences in energy demand. Within the same framework of transforming to a sustainable and liveable city, different areas must not only adopt standardised approaches but also take into account specificities at the local level. Dedicated research into cities and districts at the local scale is therefore of paramount importance to ensure the proper mix between international/national scenarios and local measures.

The urban energy mapping and analysis for Borlänge city have not yet been done. This study therefore aimed to cover this research gap by examining a set of buildings owned by the municipality of Borlänge, Sweden. The first step of the study was to conduct a spatial–temporal analysis of both electricity use and district heating demand. A top-down approach was considered based on the energy consumption data of the municipality-owned buildings. It was expected that this study would be able to provide insights that allow an understanding of the existing local energy distributions. It also facilitates strategy geared towards future energy planning in this city.

This paper is structured as follows: Section 2 indicates the data sources and the methodology used to process the data; results and discussion are presented in Sections 3 and 4; a conclusion is included afterwards.

2. Data Sources and Research Methods

2.1. Data Sources

Acquiring the necessary data to create an urban model can be a difficult endeavour. New general data protection regulation (GDPR) laws instituted by the European Parliament regulate how data can be acquired, handled and stored in order to protect the privacy of individuals [30]. Energy consumption data include sensitive information that falls within the bounds of the new regulation, greatly complicating the data acquisition. Depending on the data resolution, storing the information can be complicated as it may not be kept for long periods of time or may be stored in obsoletes systems, making it difficult to be of use.

The primary source of the data used for this model was Tunabyggen, a municipality-owned company that constructs, manages and rents a set of buildings in the Borlänge municipality. The data were provided in the PDF format, with a total number of 375 pages of monthly data for electricity demand, district heating and hot water flow rate for the year 2018. The geographical information, specifically the vector data for the property information and LIDAR data for the Borlänge municipality, was obtained from the official Swedish surveying institution, Lantmäteriet. Other social statistics and specific data such as building year of construction, percentage of occupation, demographics and typologies were acquired from hitta.se, a Swedish search engine that offers a telephone directory, addresses and maps. To complete and validate the model, it was necessary to use some extra information that was obtained by visual inspection, including the number of floors and the area and shape of the roofs. The flowchart in Figure 1 further describes the processes, databases and validation operations.

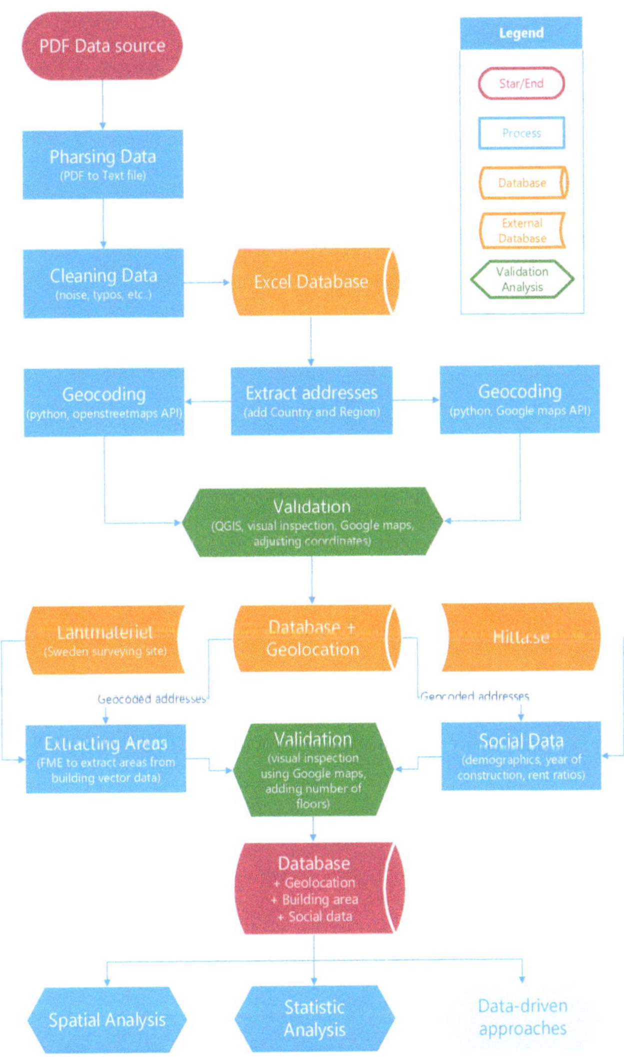

Figure 1. Flowchart for data processing, extraction, geocoding and validation.

2.2. Data Extraction

The first step in the process was to extract the information from the data source provided. The archaic PDF data structure format had to be transformed into a common format that could be used by other applications. In order to extract the data, a custom

Python script was written to parse out the information. Then, the data were further inspected for missing data and error correction. From the 375 pages in PDF format, a total of 262 addresses and 463 entries of monthly data for electricity (kWh), district heating (MWh) and flow rate (m3) for the year 2018 were extracted.

2.3. Geocoding

The addresses extracted from the data source were further expanded to the city and the country. Then, they were processed with a Python script using an application program interface (API) for OpenStreetMap (OSM). Figure 2 shows the script flowchart that was run, which used the pandas and geopy libraries. In parallel, another script was used to connect to the API geocoding services of Google Maps. Two outputs from each geocoding service were obtained with the longitudes and latitudes of the addresses. The output format for the coordinate system was the standard LL-WGS84 [31]. The locations for a total of 222 out of the 262 entry points were found on the first iteration.

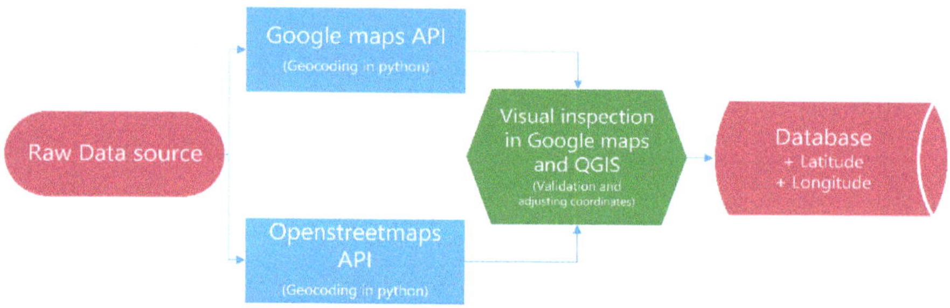

Figure 2. Flowchart for OSM and the Google Maps Python application program interface (API) geocoder.

2.4. Geocoding Validation

The results were plotted and further inspected for validation. During this process, the locations were geocoded and manually centred in the property area, as shown in Figure 3. The red dots were the geocoded locations that were manually centred in the building properties (green polygons). The output amounted to 238 out of the 262 total addresses, leaving a total of 24 addresses and 31 entry points that, due to unspecific naming, we were not able to geocode until manual visual inspection and analysis of the context were undertaken. The preliminary result generated a total of 250 geocoded addresses and 12 unclarified ones.

2.5. Area Merger Code, Area Validation

The next parameters were extracted from the Swedish survey database Lantmäteriet [32]. The building property vector information was provided in a shapefile (.shp) format, a digital vector storage format for storing geometric location and associated attribute information.

Using the Feature Manipulation Engine (FME) tool, it was possible to extract and calculate the areas for the geocoded address points [33]. This information was compared to the visually inspected area in order to analyse its accuracy. The extra information stored in the shapefiles was incorporated into the dataset. This information included a building description, coordinates in the Swedish reference system SWEREF-99-TM and a unique object identity [31].

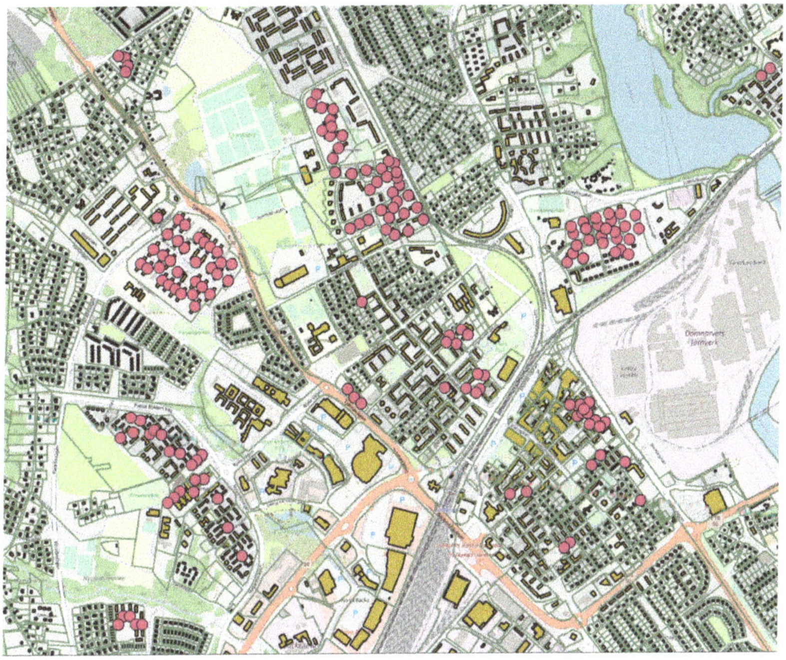

Figure 3. Geodata: building property vectors with adjusted coordinates.

2.6. Data Processing

All the different sources of information were finally combined together and inspected for errors or inconsistencies. The total building area was calculated using the number of floors and the buildings' polygon surface areas. Finally, the results for the energy consumption, electricity and district heating in kWh/m^2 for the year 2018 were obtained. From the 250 total addresses that were geocoded, 28 addresses were excluded from the analysed dataset due to missing, erroneous or abnormal information. The initial dataset contained 236 entries for electricity demand and 108 for district heating demand, which were reduced to 228 and 105, respectively, due to the following reasons: (i) some entries related to utility building samples that had no coherent energy demand on a normalised per metre square surface—for example, the energy demands of some laundry buildings were not representative for this dataset as they were detached from the buildings they provided services to; (ii) the building occupancy ratios for the entries were close to zero—some of the buildings in the sample were unoccupied, so their energy demand was close to zero. The final sample dataset consisted of 228 buildings for the electricity data and 105 buildings for the district heating data.

3. Results

3.1. Statistic Data Analysis

In the considered building samples, all of the buildings were residential buildings and related facility buildings (such as laundries, storage, etc.). The energy use was normalised by dividing it by the heated floor area. The definition of the heated or living floor area has a large impact on the magnitude of the area-specific energy requirement. In Sweden, the heated floor area is defined as the floor area that is heated to more than 10 °C. As a result, in this study, we assumed the heated floor area was on average 87% of the total external floor area for the analysis [34]. In addition, electricity demand was further normalised by considering the occupancy ratio of each building. For heating demand there was no need to consider the occupation ratio, as it is common in Sweden for heating systems to stay on even when a building is unoccupied.

The data were normalised by dividing the total energy consumption by the total living space area. This worked for most cases, i.e., self-sufficient homes and buildings with integrated facilities. In other cases, when the utility facility was in a detached building, it was unclear how many buildings it provided services to. Therefore, the area of the facility in itself was not representative of the total area it served, giving an impression of a very energy-inefficient building. In Sweden, most communal arrangements have a dedicated laundry, garage or storage facility for the community. The utilities room is usually in the basement of apartment buildings, providing an assortment of laundry and/or ironing machinery. In other cases, this facility can be completely detached from the main building. The annual electricity demand for lighting and appliances in the building samples is shown in Figure 4.

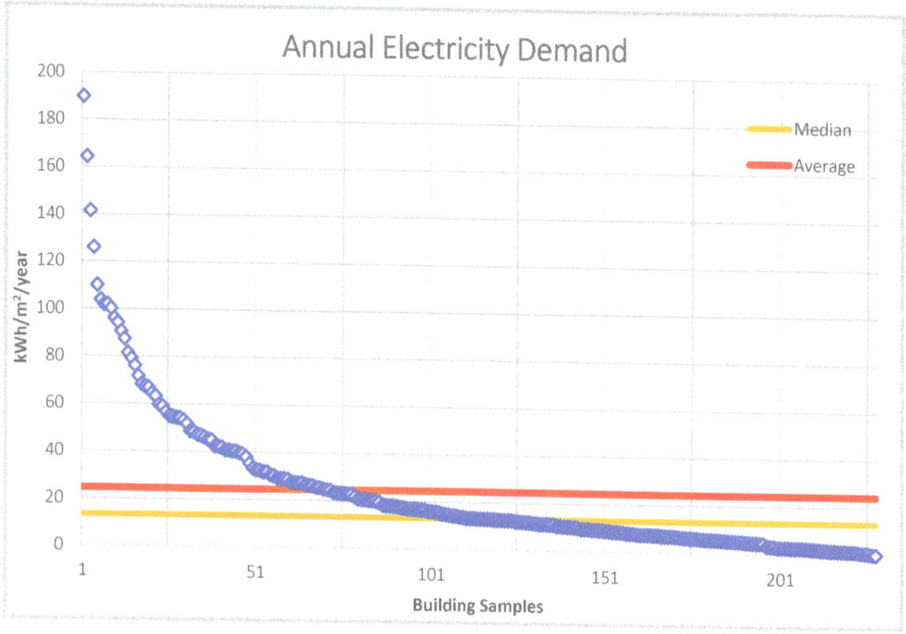

Figure 4. Annual electricity demand for building samples.

The average electricity demand of the 228 building samples was 24.47 kWh/m^2, with a total range from a minimum of 0.16 kWh/m^2 to a maximum of 189.89 kWh/m^2. Compared to the average electricity demand of 30–36 kWh/m^2 in the Swedish context [35], the average electricity demand of the building samples was reasonably low. This corresponded with the build year, zone, occupant background and purpose of the buildings, as most of the occupants in the sampled buildings were life renters, students or had a relatively low income. The median electricity demand was 13.17 kWh/m^2, which means that 50% of the sampled buildings demanded less electricity than this value. Furthermore, over 75% of the sampled buildings achieved electricity use lower than 30 kWh/m^2.

The Swedish Housing Agency's building rules [36] stipulate requirements for the energy performances of buildings depending on their use, end-use heating system and climate zone. The energy performance (heating demand) requirements are given as the specific energy use, comprising the purchased energy for space heating, domestic hot water and electricity for fans and pumps but excluding electricity for household appliances and lighting [37]. The annual heating demand for the building samples is displayed in Figure 5.

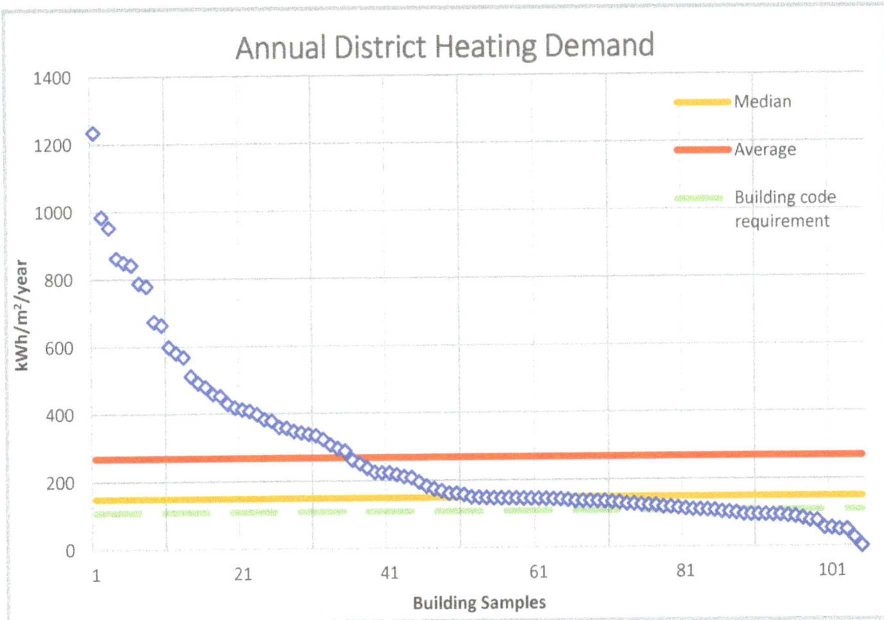

Figure 5. Annual heating demand for building samples.

The average heating demand of the 105 building samples was 268.78 kWh/m^2, with a total range from a minimum of 0.41 kWh/m^2 to a maximum of 1234.52 kWh/m^2. Borlänge city belongs to Climatic Zone II in Sweden, for which the new building code requires an annual energy use of up to 110 kWh/m^2 for non-electrically heated buildings (i.e., heated with district heating). In addition, the criteria for passive houses include even have higher requirements, with a value up to 35% lower compared to the building code [38]. Thus, the average heating demand in the building samples was much higher than either the building code or the passive house standard, about twice of requirement stipulated by the building code and three times the requirement of the passive house standard. The median heating demand was 149.34 kWh/m^2, which means that 50% of the building samples demanded less heating than this value. Approximately 21% of the building samples achieved a lower heating demand than 110 kWh/m^2. The difference between the different municipalities is clear. In Gävleborg, it was found that the average heating demand was about 185 kWh/m^2 in 2010. Across the whole of Sweden, the average annual energy use for heating in one- or two-dwelling buildings was reported to be about 158 kWh/m^2 per year in 2014 [39]. Therefore, the heating use in Borlänge city was found to be at a high level when compared to that of the closest regions and the average figure for the country. However, this high energy demand can be explained by the fact that over 56% of the buildings in the sample were constructed before 1980 and therefore may not be energy-efficient dwellings.

Some of the building samples with a high energy demand corresponded to small districts or clusters of buildings. Even though the total area was aggregated and normalised, it is possible that some heating surfaces for common and utility areas were missing. It is also possible that these nodes required more energy due to some kind of distribution inefficiency.

Annual average heating demand varies considerably depending on the year of construction of a building. For buildings built after 1980, the heating demand was about 97–98 kWh/m^2 in 2004, while for those built before 1980 heating demand was from 120–133 kWh/m^2 per year [40]. For the sampled buildings with a documented year of construction, the average heating demand for buildings constructed before 1980 was about 246.46 kWh/m^2 per year, with these buildings accounting for 98,838 m^2 of the heated floor area, as shown in Table 1. There is, therefore, great potential—amounting to an improvement of about 13,487 MWh per year—for these buildings built before 1980 to improve their energy performance through renovations, such as increasing the thermal insulation of the walls/roofs or upgrading windows and heating radiators. The rest of the buildings in the study case accounted for 132,912 m^2 of the heated floor area, with a heating demand of about 296–297 kWh/m^2. There appeared to be no significant difference between the data for the buildings built after 1980 and those with unclassified years of construction but, due to an even higher heating demand, they still offer great potential for energy saving, around 24,917 MWh per year.

Table 1. Comparison between heating demand in the studied case and the average data for Sweden.

Year/Case	Heating Demand	Area	Potential Savings
2018, case study, 1980>	296.90 kWh/m^2	11,315 m^2	2114 MWh/m^2
2018, case study, <1980	246.46 kWh/m^2	98,838 m^2	13,487 MWh/m^2
2018, case study, N/A	297.53 kWh/m^2	121,597 m^2	22,803 MWh/m^2
2018, Swedish building code, [36]	110 kWh/m^2	Historical heating demands are shown in the left columns according to building code and practice. In comparison to the studied case, great potential savings in heating can be observed.	
2014, Swedish practical average, [39]	154 kWh/m^2		
2010, Gävleborg practical average, [39]	184 kWh/m^2		
2004, Pallardó [40], 1980>	97 to 98 kWh/m^2		
2004, Pallardó [40], <1980	120 to 133 kWh/m^2		

3.2. Spatial Data Analysis

A digital mapping method was applied in this study to compile and format the energy data into a virtual image and thus to produce a general map of energy use in Borlänge city based on the building samples, offering appropriate representations of the dedicated areas and districts.

By using a geographic information system tool—QGIS—it was possible to visualise the sample energy data on a spatial map of Borlänge [41]. Using the yearly electricity and heating demand, as measured in the unit kWh/m^2, as the weight factor, along with the longitudes and latitudes of the addresses, two digital maps were generated, as shown in Figures 6 and 7, for electricity use and heating demand, respectively.

These digital maps provide an interactive and scalable way of visualizing the energy use across the city, which can be used to spot abnormalities or faulty energy data points. They also provide a spatial representation of identifiable hotspots for electricity uses in high-occupancy/density areas. For district heating demands, they show hotspots with buildings mostly constructed before 1980. For instance, some of the hotspots can be easily identified as several student accommodation areas in the northwest quadrant. These highly dense buildings showed high electricity consumption since the occupants remain indoor for most learning and living activities; but, at the same time, these buildings had relatively low heating needs as the buildings are well maintained and insulated. It can be observed from these two maps that electricity use mainly depended on the occupancy density, with higher population per floor area usually resulting in higher electricity use. On the other hand, district heating demand was dependent on the building itself, with poorly-insulated buildings leading to higher heating need. As a result, electricity use and heating demand did not always appear in the same district/area since they were influenced by different parameters. This results offer clear insights for the planning of urban energy infrastructure and distribution, as well as with regard to the potential contributions from local renewable energy source (RES) systems. For instance, more extensive electricity distribution or greater RES power generation are necessary for highly dense residential areas, while better heating should be distributed to those areas with buildings mostly constructed before 1980.

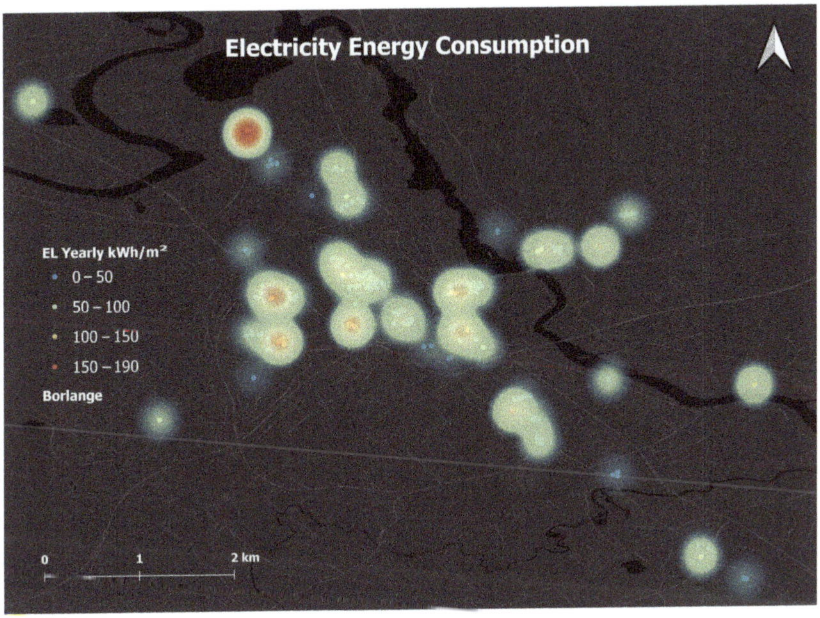

Figure 6. Digital mapping of electricity use in Borlänge city based on building samples.

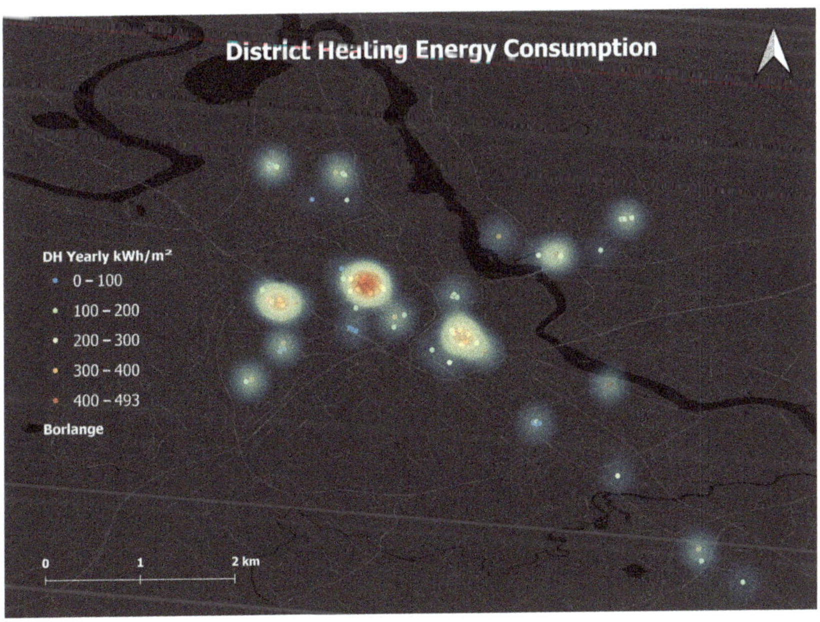

Figure 7. Digital mapping of heating demand in Borlänge city based on building samples.

3.3. Temporal Data Analysis

The yearly energy aggregation, providing a global overview of the data, was analysed in the previous section, but energy demand varies strongly depending on time. Seasonal and daily patterns have been regularly noted in the literature. In this section, we report on the analysis of monthly energy consumption. The same methodology and assumptions as before are applied, in this case at the monthly scale.

The air temperature data for the year 2018 in Borlänge indicated that there were direct correlations with the energy consumption and temperatures. During the winter months, the temperature drops below 0 °C. Afterwards, a short spring rapidly transitions into the summer season, which is accompanied by a pleasant temperature around 20 °C. A relatively smooth transition from autumn to winter occurs between September and November.

There was a slightly negative correlation between the electricity demand and the temperature at the monthly scale. The correlation was stronger for district heating because not only the median values but also the maximum values of district heating energy showed a significant decrease from April and increase from August (Figures 8 and 9).

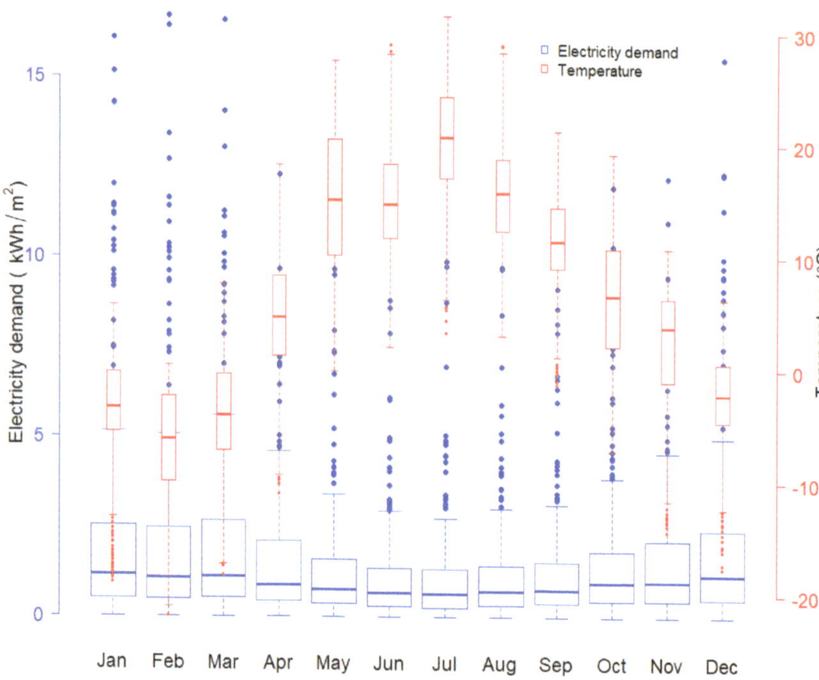

Figure 8. Monthly electricity demand and air temperature in Borlänge in 2018.

Further analysis of the dataset provided a better understanding of the temporal dimension, as the previous analysis was outlined based on a yearly aggregation. On the monthly temporal scale, the seasonal weather impact was more evident. Winter usually refers to December, January, February and March in Borlänge. In this period, the average electricity consumption was 2.65 kWh/m^2 per month. In contrast, the summer season, comprising May, June, July and August, had an average electricity consumption of 1.52 kWh/m^2 per month. The average consumption for the transitional seasons, spring and autumn, was 1.94 kWh/m^2 per month. Table 2 shows the descriptive statistics, such as the mean value, minimum and maximum monthly and yearly values for 2018 and the median and standard deviation calculations. It can also be seen that the variations of electricity demand in winter were higher than in other seasons. One possible reason might have been that additional electricity heaters and more lighting devices were used in winter. The electricity usage varied depending on individual factors and the effect was significant in summer. Figure 8 shows the electricity demand per month in a boxplot. The high monthly electricity demand data series corresponds to laundry, parking and other facilities, with a peak monthly consumption above 15 kWh/m^2 per month. It can be further observed that almost all the "outliers" appear in the upper part of the boxes and that the number of "outliers" is similar for each month. This explains why the mean values are higher than the median. Thus, the influences of these "outliers" should be noticed when evaluating load distribution.

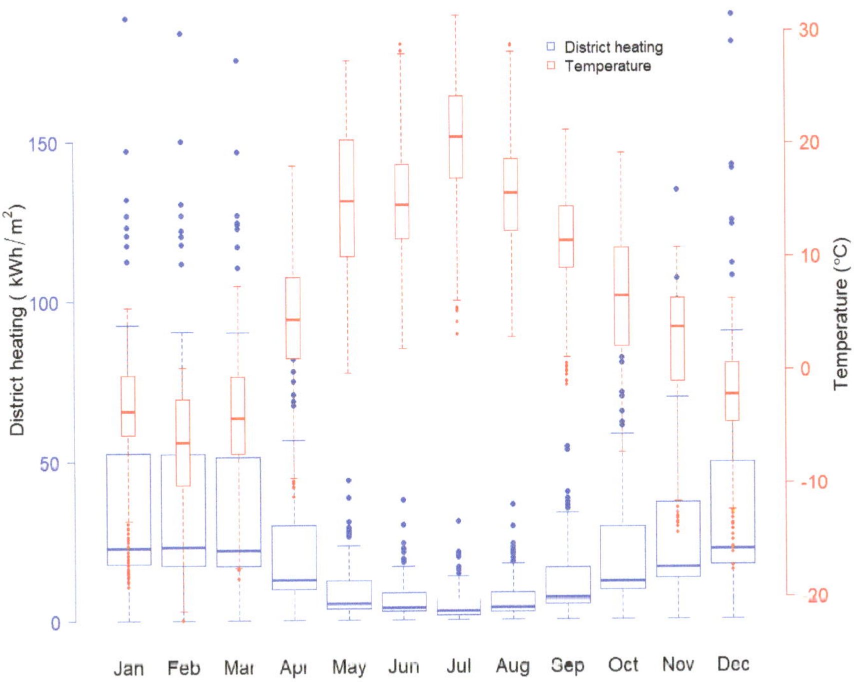

Figure 9. Monthly district heating demand and air temperature in Borlänge in 2018.

Table 2. Electricity demand per month: averages, minimum, maximum and median.

2018 kWh/m^2	Mean	Min	Max	Median	Standard Deviation
January	2.818	0.018	24.966	1.459	3.648
February	2.648	0.014	24.044	1.330	3.472
March	2.621	0.016	22.870	1.331	3.348
April	1.975	0.012	14.970	1.091	2.435
May	1.632	0.010	13.409	0.917	2.068
June	1.468	0.001	13.015	0.813	1.949
July	1.441	0.008	13.020	0.776	2.041
August	1.540	0.007	13.009	0.858	2.020
September	1.666	0.016	13.339	0.937	2.074
October	1.971	0.017	14.229	1.169	2.432
November	2.165	0.017	17.060	1.210	2.640
December	2.524	0.022	22.065	1.324	3.139
Yearly	24.470	0.163	189.886	13.175	29.384

District heating energy demand was significantly more consistent than electricity demand, as shown in Table 3 and Figure 9. For the winter months, there was an average heating energy demand of 38.8 kWh/m^2 per month. In contrast, summer had an average consumption of 7.27 kWh/m^2 per month. The transitional seasons, spring and autumn, had an average of 21.11 kWh/m^2 per month. The intra-difference for each winter and summer seems to be negligible while the inter-difference is obvious.

Table 3. District heating demand per month: averages, minimum, maximum and median.

2018 kWh/m^2	Mean	Min	Max	Median	Standard Deviation
January	39.232	0.037	188.436	22.741	34.686
February	39.204	0.037	183.786	23.068	34.354
March	38.448	0.045	175.248	21.941	33.809
April	22.698	0.023	99.735	12.767	20.404
May	9.603	0.043	43.763	5.304	8.959
June	7.192	0.043	37.452	4.083	7.012
July	5.300	0.017	30.831	2.979	5.586
August	6.987	0.033	36.026	3.876	6.827
September	12.082	0.022	54.051	7.115	10.980
October	21.700	0.048	100.274	11.978	19.515
November	27.981	0.030	134.077	16.269	24.970
December	38.361	0.038	188.769	21.958	36.329
Yearly	268.788	0.414	1234.521	149.347	238.991

Thus, comparing the electricity demand to the district heating monthly energy demand, it is observed that district heating adhered to a more homogenous pattern following the changes of the seasons. For the sampled buildings, the heating was managed by central systems. The variations were determined more by the building envelopes, physical parameters and weather conditions than by the occupant behaviours. This explains the regular pattern and the lower number of "outliers" for each month. The high heating demand values might have been due to poor insulation material or inefficient energy systems.

3.4. *Information Map of Spatial–Temporal Energy Demand*

This part of the study aimed to increase the level of detail of the spatial–temporal energy analyses with the intention of presenting a digital spatial–temporal information map of both electricity use and district heating demand. The initial data consisted of electricity and district heating monthly energy consumption for the year 2018, visualised in graphs and mapped in 2D heatmaps for electricity and district heating, respectively. We expanded the information from the energy use map for benchmarking of large-scale buildings. For this purpose, a new database structure was created by merging the spatial information from Lantmäteriet, the database with the geocoded addresses and the temporal energy demand. The basic workflow for merging these datasets is shown in Figure 10.

The process consisted of using the two databases, in the ".csv" and ".shp" formats, to create a single database with all the features from both files. For the geocoded data, the coordinates were in the standard WGS-84 format and had to be re-projected into the SWEREF-99-TM format to be consistent with the survey data. The re-projected point coordinates were then extracted and used as markers to select the polygons from the survey data. In parallel, the coordinates and areas were extracted from the survey data and used as an underlay, as shown in Figure 11. The enhanced polygons then contained the energy information.

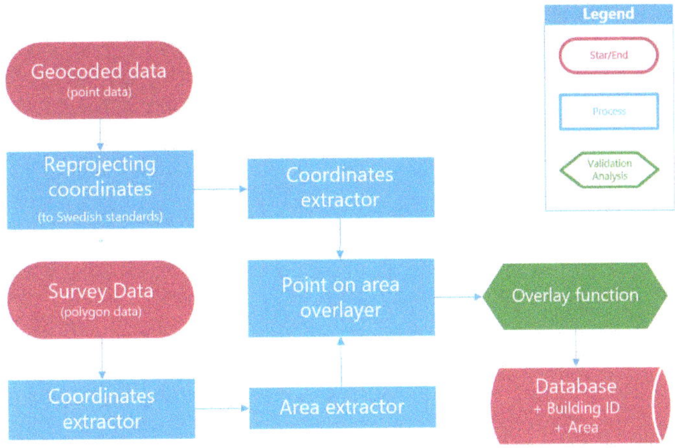

Figure 10. Area merger workflow.

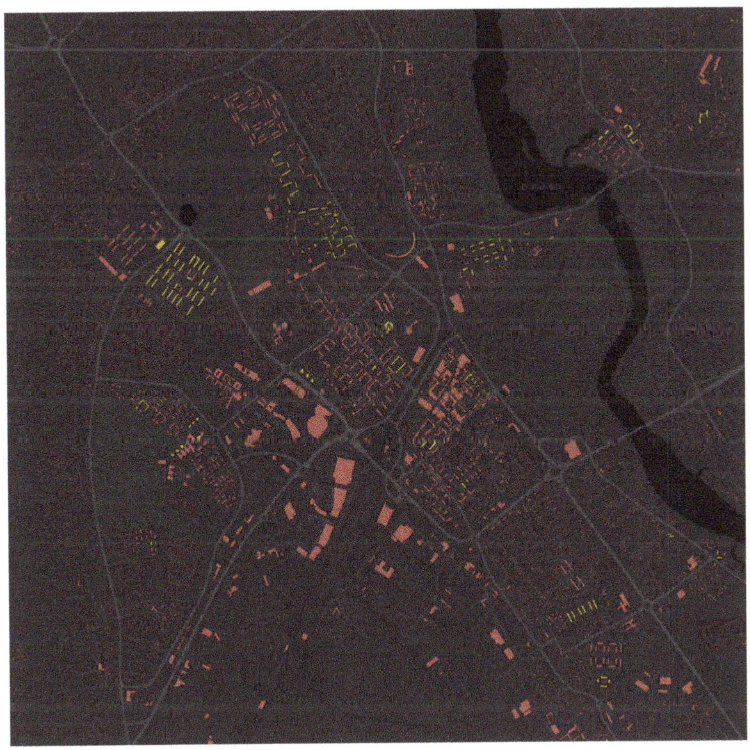

Figure 11. Borlänge building footprints (in red) and this study's building footprints (in yellow).

The merged database, now in the ".shp" file format, contained the areas, identification numbers and general parameters for the buildings, as well as the energy demand information per month in kWh/m^2, as show in Figure 12. Unnecessary data were filtered out of the dataset.

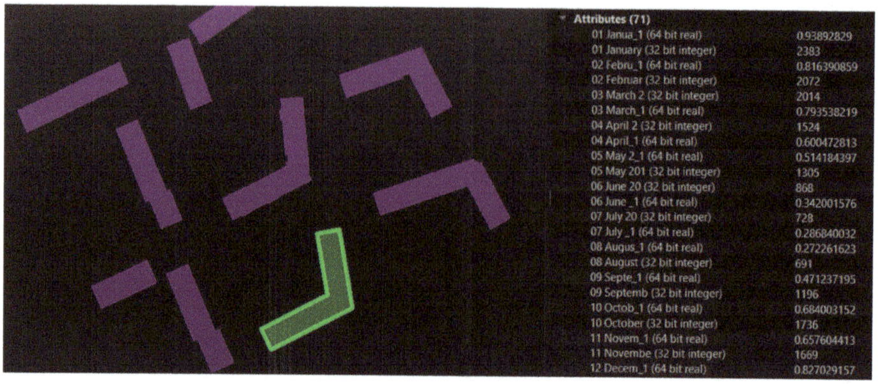

Figure 12. This study's building footprints with energy use information.

3.5. D Visualization Methodology for the Integrated Spatial–Temporal Energy Demand

It was furthermore possible to visualise the integrated spatial–temporal energy demand information in a 3D map. This activity was achieved with a novel digitalization approach. Three data files were required to generate the 3D model, obtained from Lantmäteriet and from the new database shapefile: (i) laser imaging, detection and ranging (LIDAR) data, used for measuring distances by illuminating the target with laser light and measuring the reflection with a sensor; (ii) vector maps of the building properties in the "shapefile" format, which is a geospatial vector data format for geographic information system software; and (iii) an "orthophoto" file, which is an aerial photograph or satellite imagery geometrically corrected to a uniform scale.

The methodology involved a parallel data transformation of the three data files using the Feature Manipulation Engine (FME) tool. The data were deconstructed and reconstructed until the desired outcome was achieved. The building shape data could be used as a clipping mask on both the "orthophoto" and the LIDAR data, defining the building boundaries and creating a unique building identification. The "orthophoto" could be then applied to generate textures for building roofs, as well as for the terrain. The LIDAR data was used to generate the building geometries and heights and the terrain elevation.

Consequently, a 3D city model could be finally generated, as illustrated in Figure 13, with the purpose of complementing the generation of an urban modelling framework/information. The model could further include the spatial–temporal energy use information, as indicated in Section 3.4, and thus assist the master energy planning of the buildings in the city.

Figure 13. Example of 3D model visualization.

4. Discussion

This study aimed to establish the groundwork for further urban-scale energy exploration. The methodology developed here made it possible to pre-process and validate the data for further analyses and could be applied to other study cases. This serves to demonstrate that, even with a limited set of data, meaningful models can be created when the data are manipulated and processed properly, as shown in the geocoding process or in the 3D model creation.

Data analyses were done for different temporal and spatial scales; yearly averages showed correlations for the energy demand, year of construction and typology of the buildings, indicating the potential for future renovation opportunities. The monthly study showed a direct correlation between energy demand and weather patterns. The spatial analysis put into perspective the different zones and clusters of buildings in the urban landscape. It showed that the distribution of district heating demand was greater closer to the centre of the energy production source, while it dispersed in the outskirts; electric energy demand increased in the outskirts, especially during the winter time, explained by the use of electric heaters in the absence of district heating.

Moving from building-level energy modelling towards urban-level modelling presented many challenges. Data were often difficult to acquire and entailed many obstacles, such as privacy laws, the format in which data were stored, data accuracy and, at times, a lack of any data at all. Data pre-processing and processing can also be time consuming and highly demanding on computational power, leading to a compromise between accuracy and simplified assumptions. In this paper, a lot of attention was put into the level of detail of the data, avoiding simplification where possible. In future analyses, the dataset parameters will be submitted to machine learning models to observe the relationships and dependencies between each element, such as between weather conditions, energy demand and building efficiency. The dataset will be expanded to incorporate a large number of extra parameters, from weather data—like air temperature, humidity and irradiance information—to social data and energy data—like energy certification ratings—thus expanding our understanding of the relevance of the specific parameters.

Examining energy patterns at a yearly scale provides a general overview of high-consumption dwellings and helps in categorising them in terms of their function and efficiency. Information about potential energy saving can be obtained and comparisons can be drawn about the overall energy performance of a city in relationship to the country regulations. Examining energy patterns at a monthly scale, on the other hand, specifically shows the relationship between energy demand and seasonal changes. This level of detail makes it possible to see how a building performs under different circumstances. For instance, some buildings might perform well in the winter and in the summer but might require more energy for cooling due to high insulation or limited ventilation. Attention should be paid to the relationship between the energy demand and weather conditions at a larger scale, moving from years to months, days or even minute resolutions. In the spatial context, moving from one dimension (data points) to two dimensions (data on a plane; maps) expands the perception and interpretation of the data analysed, allowing spatial relationships to be understood better. The next step is to expand the spatial data in such a way that leads toward a 3D model, in which the data can be explored in a more direct and realistic manner.

Energy master planning (EMP), at the district and city levels, provides the possibility of untangling the challenges to the dynamics of energy needs and supply. The detailed 3D city information model is an essential digital EMP platform to engage different stakeholders in communication and thus help them to identify their roles in sustainable energy transition. In this model, buildings have shape and volume, the sun casts shadows and vegetation is present. These data explicitly convey a lot of extra information to stakeholders—energy and urban planners—allowing them to explore the data through an interactive platform.

In future work, the main focus will be on creating a 3D model incorporating a high level of detail for buildings, terrain, energy demand and other building characteristics. Visualizing large amounts of information is challenging and for this purpose a graphical user interface (GUI) must be created to enable interaction with the energy information in the 3D model, making it possible to show, hide or filter various information either with numbers or colour codes.

5. Conclusions

A dedicated spatial–temporal analysis of both electricity use and district heating demand in a Swedish local-city context was provided in this study using a toolkit for top-down digital mapping. The average electricity demand in the Borlänge building samples was 24.47 kWh/m^2, which was reasonably lower than the average value in Sweden. The mean value of heating of the building samples was 268.78 kWh/m^2, which was much higher than either the building code or the passive house standard. The heating use in Borlänge city remained at a high level when compared to the closest regions and the average figure over the country. In particular, there was great potential for the improvement of energy performance, amounting to savings of about 13,487 MWh/year for the buildings built before 1980 and around 24,917 MWh/year for the rest of the buildings.

The digital maps provided a spatial representation of the identifiable hotspots for electricity uses in high-occupancy/density areas and for district heating needs in districts with buildings mostly constructed before 1980. Visualizing the energy use across the city also showed that there was some apparent correlation in the electricity use and heating demand hotspot locations, as, for example, in the increase of electricity use and decrease of district heating in the periphery.

Further expanding the temporal scale from yearly to monthly values made it possible to study the electricity use and the district heating pattern in relation to the changes of the seasons and temperatures over the year. As expected, heating demand was increasingly relational to the decrease of temperatures, as was the electricity demand, although at a lower intensity.

The approach to generating the information map for the spatial–temporal energy demand was finally concluded with the three datasets including spatial information from

Lantmäteriet, geocoded addresses and temporal energy demand. This method also expanded the potential to integrate energy information into city information models at a 3D level through parallel data transformation of the three data files with the Feature Manipulation Engine tool. The overall result offers clear insights for the planning of urban energy infrastructure and distributions, as well as a potential contribution for local RES implementation.

Author Contributions: Conceptualization, S.Q. and X.Z.; methodology, S.Q., M.H., and X.Z.; formal analysis, S.Q. and X.Z.; writing—original draft preparation, S.Q., P.H.; writing—review and editing, P.H., M.H. and X.Z.; visualization, S.Q. and M.H.; supervision, X.Z. All authors have read and agreed to the published version of the manuscript.

Funding: The authors would like to note the financial support from the Swedish Energy Agency (UBMEM project: 46068).

Institutional Review Board Statement: Not applicable.

Informed Consent Statement: Not applicable.

Data Availability Statement: The data presented in this study are available on request from the corresponding author.

Acknowledgments: The authors also thank Tina Lidberge for accruing data from Tunabyggen. The masters students, such as Péter Tempfli, Mohsin Raza, Anastasiia An and Mrudula Talari, are appreciated for their support.

Conflicts of Interest: The authors declare no conflict of interest.

References

1. Huang, P.; Lovati, M.; Zhang, X.; Bales, C. A coordinated control to improve performance for a building cluster with energy storage, electric vehicles, and energy sharing considered. *Appl. Energy* **2020**, *268*, 114983. [CrossRef]
2. Ministry of Sustainable Development Sweden. Swedish national roadmap for implementing the European Environ-mental Technologies Action Plan (ETAP). Regeringskansliet. 2006. Available online: www.sweden.gov.se (accessed on 10 May 2020).
3. UN-Habitat. *Global Report on Human Settlements 2009: Planning Sustainable Cities*; UN-Habitat: Nairobi, Kenya, 2009.
4. Moghadam, S.T.; Coccolo, S.; Mutani, G.; Lombardi, P.; Scartezzini, J.-L.; Mauree, D. A new clustering and visualization method to evaluate urban heat energy planning scenarios. *Cities* **2019**, *88*, 19–36. [CrossRef]
5. Swan, L.G.; Ugursal, V.I. Modeling of end-use energy consumption in the residential sector: A review of modeling techniques. *Renew. Sustain. Energy Rev.* **2009**, *13*, 1819–1835. [CrossRef]
6. Kavgic, M.; Mavrogianni, A.; Mumovic, D.; Summerfield, A.; Stevanovic, Z.; Djurovic-Petrovic, M. A review of bottom-up building stock models for energy consumption in the residential sector. *Build. Environ.* **2010**, *45*, 1683–1697. [CrossRef]
7. Nouvel, R.; Mastrucci, A.; Leopold, U.; Baume, O.; Coors, V.; Eicker, U. Combining GIS-based statistical and engineering urban heat consumption models: Towards a new framework for multi-scale policy support. *Energy Build.* **2015**, *107*, 204–212. [CrossRef]
8. Moghadam, S.T.; Toniolo, J.; Mutani, G.; Lombardi, P. A GIS-statistical approach for assessing built environment energy use at urban scale. *Sustain. Cities Soc.* **2018**, *37*, 70–84. [CrossRef]
9. Bruse, M.; Fleer, H. Simulating surface–plant–air interactions inside urban environments with a three dimensional numerical model. *Environ. Model. Softw.* **1998**, *13*, 373–384. [CrossRef]
10. Mauree, D.; Coccolo, S.; Kaempf, J.; Scartezzini, J.-L. Multi-scale modelling to evaluate building energy consumption at the neighbourhood scale. *PLoS ONE* **2017**, *12*, e0183437. [CrossRef] [PubMed]
11. United Nations. Global Initiative for Resources Efficient Cities. United Nations Environment Programme, 2015. Available online: http://unep.org/ (accessed on 10 May 2020).
12. Schneider, S.; Hollmuller, P.; Le Strat, P.; Khoury, J.; Patel, M.; Lachal, B. Spatial–Temporal Analysis of the Heat and Electricity Demand of the Swiss Building Stock. *Front. Built Environ.* **2017**, *3*. [CrossRef]
13. Chen, S.; Zhang, X.; Wei, S.; Yang, T.; Guan, J.; Yang, W.; Qu, L.; Xu, Y. An energy planning oriented method for analyzing spatial-temporal characteristics of electric loads for heating/cooling in district buildings with a case study of one university campus. *Sustain. Cities Soc.* **2019**, *51*, 101629. [CrossRef]
14. Mohammadi, N.; Taylor, J.E. Urban energy flux: Spatiotemporal fluctuations of building energy consumption and human mobility-driven prediction. *Appl. Energy* **2017**, *195*, 810–818. [CrossRef]
15. Loga, T.; Diefenbach, N.; Stein, B.; Dascalaki, E.; Balaras, C.A.; Droutsa, K.; Kontoyiannidis, S.; Villatoro, O.; Wittch-en, K.B. Main Results of the TABULA Project. In *Typology Approach for Building Stock Energy Assessment*; Institut Wohnen und Umwelt: Darmstadt, Germany, 2012.

16. Morganti, M.; Pages-Ramon, A.; Coch, H.; Isalgue, A. Buildingmass and Energy Demand in Conventional Housing Typologies of the Mediterranean City. *Sustainability* **2019**, *11*, 3540. [CrossRef]
17. Dascalaki, E.G.; Droutsa, K.G.; Balaras, C.A.; Kontoyiannidis, S. Building typologies as a tool for assessing the energy performance of residential buildings—A case study for the Hellenic building stock. *Energy Build.* **2011**, *43*, 3400–3409. [CrossRef]
18. RePublic_ZEB. Refurbishment of the Public Building Stock towards NZEB. Available online: http://www.republiczeb.org/filelibrary/WP2/D2-1Public-Building-Stock_final_29-11-2016.pdf (accessed on 9 February 2021).
19. Lucchi, E.; D'Alonzo, V.; Exner, D.; Zambelli, P.; Garegnani, G. A Density-Based Spatial Cluster Analysis Supporting The Building Stock Analysis In Historical Towns. In Proceedings of the Building Simulation 2019: 16th Conference of IBPSA, Rome, Italy, 2–4 September 2020; pp. 3831–3838.
20. Miao, R.; Wang, Y.; Li, S. Analyzing Urban Spatial Patterns and Functional Zones Using Sina Weibo POI Data: A Case Study of Beijing. *Sustainability* **2021**, *13*, 647. [CrossRef]
21. Pulse Dashboard. Pulse Energy, Vancouver, British Columbia. 2014. Available online: http://executivepulsesoftware.com/Wordpress/ (accessed on 16 September 2014).
22. Building Dashboard. Oakland, California: Lucid. Available online: http://www.luciddesigngroup.com/buildingdashboard/index.html (accessed on 16 September 2014).
23. Mhalas, A.; Kassem, M.; Crosbie, T.; Dawood, N. A visual energy performance assessment and decision support tool for dwellings. *Vis. Eng.* **2013**, *1*, 7. [CrossRef]
24. Murugesan, L.K.; Hoda, R.; Salcic, Z. Design criteria for visualization of energy consumption: A systematic literature review. *Sustain. Cities Soc.* **2015**, *18*, 1–12. [CrossRef]
25. Gröger, G.; Plümer, L. CityGML—Interoperable semantic 3D city models. *ISPRS J. Photogramm. Remote Sens.* **2012**, *71*, 12–33. [CrossRef]
26. Mattinen, M.K.; Heljo, J.; Vihola, J.; Kurvinen, A.; Lehtoranta, S.; Nissinen, A. Modeling and visualization of residential sector energy consumption and greenhouse gas emissions. *J. Clean. Prod.* **2014**, *81*, 70–80. [CrossRef]
27. Finney, K.N.; Zhou, J.; Chen, Q.; Zhang, X.; Chan, C.; Sharifi, V.N.; Swithenbank, J.; Nolan, A.; White, S.D.M.; Ogden, S.; et al. Modelling and mapping sustainable heating for cities. *Appl. Therm. Eng.* **2013**, *53*, 246–255. [CrossRef]
28. Huang, P.; Ma, Z.; Xiao, L.; Sun, Y. Geographic Information System-assisted optimal design of renewable powered electric vehicle charging stations in high-density cities. *Appl. Energy* **2019**, *255*, 113855. [CrossRef]
29. Ramachandra, T.V.; Shruthi, B.V. Wind energy potential mapping in Karnataka, India, using GIS. *Energy Convers. Manag.* **2005**, *46*, 1561–1578. [CrossRef]
30. European Parliament. *Regulation (EU) 2016/679 of the European Parliament and of the Council—of 27 April 2016—on the Protection of Natural Persons with Regard to the Processing of Personal Data and on the Free Movement of Such Data, and Repealing Directive 95/46/EC (General Data Protection Regulation)*; The European Parliament and the Council of the European Union: Brussels, Belgium, 2016; p. 88.
31. WGS 84. Lantmateriet.se. Available online: https://www.lantmateriet.se/sv/Kartor-och-geografisk-information/gps-geodesi-och-swepos/Referenssystem/Tredimensionella-system/WGS-84/ (accessed on 21 January 2021).
32. Ås Sivborg, S. Lantmateriet. Available online: https://minkarta.lantmateriet.se/ (accessed on 21 January 2021).
33. *FME Software*, Copyright (c) Safe Software Inc.: Surrey, BC, Canada, 1998.
34. Mata, É.; Kalagasidis, A.S. *Calculation of Energy Use in the Swedish Housing*; Chalmers University of Technology: Göteborg, Sweden, 2009; p. 52.
35. Mata, É.; Kalagasidis, A.S.; Johnsson, F. Energy usage and technical potential for energy saving measures in the Swedish residential building stock. *Energy Policy* **2013**, *55*, 404–414. [CrossRef]
36. Boverket. *Boverket´s Mandatory Provisions and General Recommendations, BBR, BFS 2011:6 with Amendments up to BFS 2018:4*; Code of Statutes of the Swedish National Board of Housing, Building and Planning: Stockholm, Sweden, 2011; p. 154.
37. Dodoo, A.; Gustavsson, L. Effect of Energy Efficiency Requirements for Residential Buildings in Sweden on Lifecycle Primary Energy Use. *Energy Procedia* **2014**, *61*, 1183–1186. [CrossRef]
38. FEBY. *Specification for Zero Energy, Passive and Low-Energy Houses*; Sveriges Centrum för Nollenergihus: Sweden, 2012.
39. Swedish Energy Agency. New Regional Energy Statistics for Single- or Two-Dwelling Buildings. 2015. Available online: http://www.energimyndigheten.se/en/news/2011/new-regional-energy-statistics-for-single--or-two-dwelling-buildings/ (accessed on 10 May 2020).
40. Pallardó, G.G. *Energy Consumption in Tertiary Buildings in Sweden Case-Study: M-building at Lund University—LTH, Lund University*; Lund University: Lund, Sweden, 2011.
41. QGIS Development Team. *QGIS Geographic Information System*; Open-Source Geospatial Foundation Project: Zürich, Switzerland, 2002.

buildings

MDPI

Article

Citizens and Positive Energy Districts: Are Espoo and Leipzig Ready for PEDs?

Zarrin Fatima [1,*], Uta Pollmer [2], Saga-Sofia Santala [3], Kaisa Kontu [4] and Marion Ticklen [5]

1 VTT Technical Research Centre of Finland, PL 1000, 02044 VTT Espoo, Finland
2 Fraunhofer Center for International Management and Knowledge Economy IMW, Neumarkt 9-19, 04109 Leipzig, Germany; uta.pollmer@imw.fraunhofer.de
3 KONE Corporation; Keilasatama 5, 02150 Espoo, Finland; sagasofia.santala@kone.com
4 Citycon Corporation & Citycon Finland, Iso Omena, Piispansilta 9 A, 02230 Espoo, Finland; Kaisa.Kontu@citycon.com
5 City of Espoo, P.O. Box 1, 02070 Espoo, Finland; marion.ticklen@espoo.fi
* Correspondence: Zarrin.fatima@vtt.fi

Abstract: In urban transformation, no solution works without citizen support. With increasing numbers of building technologies and large-scale urban development on its way across cities, it has become vital to keep citizens informed, engaged, and content with the new changes. This paper looks at citizen engagement in Espoo (Finland) and Leipzig (Germany), and it determines whether the cities are ready for developing and implementing positive energy districts (PEDs). The authors studied the cities' operations and current citizen engagement methods to understand how the efforts could be combined and improved. The analysis indicated that the city of Espoo already has a well-established system that continuously promotes citizen engagement at various levels, and combining the available infrastructure with company experts on citizen participation will allow Espoo to seamlessly transition towards PEDs in the near future. The city of Leipzig has a rich experience due to several national projects and participation in an earlier European project, which enabled the city to set clearer goals for the future and modify existing citizen methods. As lighthouse cities, findings from Espoo and Leipzig are also aimed at cities across Europe and beyond to boost development of PEDs together with citizens.

Keywords: positive energy districts; citizen; cities; participation; citizen engagement

Citation: Fatima, Z.; Pollmer, U.; Santala, S.-S.; Kontu, K.; Ticklen, M. Citizens and Positive Energy Districts: Are Espoo and Leipzig Ready for PEDs? *Buildings* **2021**, *11*, 102. https://doi.org/10.3390/buildings11030102

Academic Editor: Geun Young Yun

Received: 10 January 2021
Accepted: 24 February 2021
Published: 6 March 2021

1. Introduction

1.1. *What Is Citizen Engagement?*

Citizen engagement is collectively defined as public participation, stakeholder involvement, co-creation, civic engagement, participatory democracy, or activism [1]. It is also described as individual or collective behavior that focuses on determining the social problems of a community [2–4]. A clear definition of citizen engagement does not exist, but its true nature is the interaction between citizens and government [4].

Citizen engagement has a vital role in service delivery [5]. It is also said to be the 'meaningful involvement of individual citizens in policy or program development.' In other words, citizen engagement requires an active intentional dialogue between citizens and public decision makers [6].

Citizen participation may range from individuals participating in steering committees to partnerships with certain community groups. The Canadian Institutes of Health Research (CIHR) claims that there is no 'one size fits all' approach because each organization, scenario, or audience may require differing engagement practices. Bringing citizen voices into planning, decision-making, implementation, or evaluation processes requires two-way communication. Hence, input must not be pursued for the sake of input, but communication has to be interactive to generate 'informed participation' to reach a common goal [7].

As cities around the world look for ways to involve their citizens in various discussions, citizens themselves are eager to bring their ideas and concerns forward. However, in order to have a high level of inclusive engagement, it is essential to involve a broad and an unbiased selection of society as much as possible in the decision making processes [6].

To be able to make a true impact, citizen engagement has to be embedded in every stage of the decision making process and be conducted with deep commitment, proper allocated time, and co-creation interest. Merely adopting a tick-box approach or having citizen participation at the end of the process will not have the desired result or be beneficial for stakeholders. Citizen engagement has to be added in the 'Bold City Vision' and the overall targets of the project instead of making it an add-on activity. However, it is also a fact that municipal organizations and public servants do not have training to make citizen engagement as effective as desired, and the municipal procedures very often do not cater to involving citizens [6].

Nonetheless, capacity building solutions are now on the rise, e.g., in the Netherlands, a new legal framework for spatial planning called 'Omgevingswet' (Environmental Law) will soon be introduced, and it has participation and co-creation with citizens as one of the main pillars. The new framework does not offer a specific method, but it gives municipalities the freedom to develop their own approaches fitting to their local context [8]. The EU Joint Research Centre leads a community of practice (CoP) on citizen engagement having plans for a manual and online resource catalog aimed at organizations and project needs while also continuing an annual Festival of Citizen Engagement. Additionally, the Citizen Focus Action Cluster at the European Innovation Partnership on Smart Cities and Communities (EIP-SCC) congregates several initiatives and actions on citizen engagement and operates as a mutual learning and matchmaking platform [9].

1.2. The Necessity of Citizen Engagement

The question of the necessity of citizen engagement has been raised many times and may be answered with the fact that continuous innovations do indeed impact the everyday life of citizens while bridging many fields of urban life. A recent booklet published by Smart Cities Information System (SCIS) emphasized that the inclusion of citizens may help to address concerns, increases transparency about plans, and invites diversified and vulnerable groups who might not actively participate otherwise. In parallel, it also strengthens collaborative actions and bottom-up innovations, enhances the sense of trust and community ownership, and develops resource efficiency as unforeseen problems may be avoided; most importantly, citizen views can help explore more sustainable and viable solutions that may function better in the local context [6].

The emerging trend of positive energy blocks and districts (PEBs/PEDs) is a transition towards more energy-conscious behavior that calls for extensive and innovative engagement approaches and co-creation practices, as this will lead to citizens who understand, trust, and use and feel ownership of the measures adopted in their district [6]. Citizen engagement is essentially part of the quadruple helix innovation model that identifies four major actors in the innovation system: science, policy, industry, and society [10], hence placing a further emphasis on how participation can deliver new forms of deliberation and operationalization of the democratic process within the city, eventually leading to higher local impact and building citizen trust [6].

Citizen engagement has significant importance in relation to the success, development, and implementation of PEDs. A recent publication by the European Commission [11] described a PED as having defined borders and an area that:

- is based on open and voluntary participation, is autonomous, and is efficiently controlled by its citizens;
- has the primary purpose to deliver environmental, economic, or social community benefits;
- has an overall energy balance of zero or positive over a year;

- consists of buildings having very high energy performance, complying with minimum energy performance requirements and local building codes;
- consists of buildings that are either nearly zero or have a very low amount of energy demand;
- covers the building demand through renewable energy sources to a large extent; and
- has the possibility to produce renewable energy either onsite or nearby

In parallel to the above characteristics, the concept of 'renewable energy communities' has also emerged, as emphasized by the first and second points. For an energy community to be a success, the interaction and collaboration of all stakeholders is the key to find the best-fit solution for citizens. Nonetheless, municipalities must continuously to play their role in the energy transition and managing the collaboration. This also implies that social innovation, including behavioral change, has to be taken into consideration to deliver PEDs. There may be many non-technical obstacles along the way, but community engagement is vital to ensure the buy-in of community members [11].

1.3. *What Is Happening Worldwide?*

Cities are the melting pot for de-carbonization strategies relating to energy, transport, buildings, industry, and agriculture [12]. Cities have access to large capital and abundant know-how, and so they therefore have the ability to create economies of scale essential for the piloting and scaling up of new ideas [13]. Nonetheless, cities are in need of citizens who are not only political actors but also users, producers, consumers, and owners. A combined effort from these actors may have a huge impact on local urban areas, associations, and homes, thus propelling the climate transition, advancing the economy, and preserving the environment. As evidently said by the Mission Board for Climate-Neutral and Smart Cities, citizens and civil society have to be given more substantial roles, new platforms for action, and better resources [13]. The Mission Board will support 100 European cities in their dynamic transformation towards climate neutrality by 2030, eventually supporting the European Green Deal and becoming climate-neutral by 2050. Citizens are in the center of all action as the mission calls for citizens to be change agents and demands that cities focus on citizens' health and wellbeing, healthier lifestyles, and adopting a 'by and for the citizens' way of thinking and working [13]. Cities are further pushed to put into practice the 'leave no one behind' (just transition) component of the Green Deal [14].

Allen et al. [15] evaluated the relationship between e-participation as a type of co-production and service performance by utilizing multiple large longitudinal datasets from a smart city mobile platform. The study gave evidence that citizen e-participation in co-production can increase the performance of service delivery, a link that is usually believed to be true instead of tested. Feedback and monitoring through the platform led to more issues being resolved, and service delivery had a larger influence on complex problems such as damaged roads, which generally require multiple actors and may need more time to be resolved. Simple issues such as waste management had less participation. However, e-participation may still be limited in general. Few researchers have considered why some citizens engage actively while others do not. A study by Choi and Song [16] in South Korea claimed that people with a stronger social capital—commitment, ownership of the community, and trust in government—have a greater likelihood to become part of e-participation. Emphasis only on technology-driven factors such as usefulness, ease of use, and perceived behavioral control become insignificant when tasks demand greater civic engagement. Attention has to be paid to how to nurture individual social capital (e.g., virtue of good civic norms) through proper procedures, as well as institutional and political reform [16].

Citizen engagement has remained a crucial component of smart cities in recent years. As these projects last for lengths of four-to-five years, cities have ample time to experiment and implement new tools and approaches. Nonetheless, two elements have to be considered for a smart city to be successful. First, citizens must be part of the design so that the smart city answers to the real needs of the people. Second, each city has distinct characteristics that have to be included to create a citizen participation strategy truly tailored and adapted to the local context [17]. Simonofski et al. described five context factors that impact citizen participation strategies in two smart cities (Namur and Linköping): the smart city consideration, the drivers for participation, the degree of centralization, the legal requirements, and the citizens' characteristics. The factors are applicable for any city, however, because even though similar stakeholders and participations methods were used in these two cases (direct interaction, living lab, open data, and online platform), the methods were implemented for varying reasons that led to varying challenges being encountered [17].

With the emerging number of smart cities around the world, there has been strong focus on transactions between citizen and government. The presence of smartphone and smart city technologies have further stimulated micro-transactions between citizen, government, and information broker (for example, tax payment in exchange for services). Johnson et al. [18] explored how the modern smart city includes the citizen as a series of micro-transactions encoded on the real-time landscape of the city. This transactional citizen is monitored by smart city sensors and is integrated into smart city decision-making through certain platforms. The concept is based on four broad modes of transaction—type (intentional contribution), tweet (intermediated by third party), tap (convened or requested transaction), and pass (ambient transaction based on movement)—and enables one to understand how citizens interact and find potential avenues for private sector influence [18].

To provide a few examples of smart cities, the European funded IRIS project (Integrated and Replicable Solutions for Co-Creation in Sustainable Cities) created a planning framework to steer activities around co-creation and citizen engagement. The project created a citizen engagement ladder based on design and system thinking that includes phase 1: awareness-raising; phase 2: mapping; phase 3: scoping; phase 4: co-creation and design scenarios; phase 5: touchpoints and influencers; and phase 6: feedback loops [19]. Similarly, the CityxChange project developed the Citizen Participation Playbook to support local communities in PEBs and PEDs [20]. This project developed a roadmap of four distinctive citizen participatory processes as follows: process 1: the co-design of urban interventions; process 2: collaborative legislation; process 3: participatory budgeting; and process 4: citizens proposals. In addition, the project suggests best practices for effective citizen participation considering other smart city projects, European Commission initiatives and other organizations. These have been defined as (1) define the community; (2) clear purpose and front loading; (3) continuous engagement: capacity building and feedback; (4) open process, open source, and open data; (5) co-design, co-create, and co-produce; and (6) privacy by design [20]. The MAtchUp project has focused on several aspects such as participation, education and co-creation, and the local strategy of social services and local energy offices to mitigate energy poverty at the district scale, as well as citizen feedback channel for traffic management [21].

POCITYF (POsitive Energy CITY Transformation Framework), a similar European funded project follows a "rapid prototyping" approach that enables simultaneous and almost real-time feedback by citizens. The project focuses on incentives for co-creating, co-delivering, and co-capturing value, and it also caters to disadvantaged communities while promoting sustainable tourism [22].

The city of Tampere developed the Tampere. Finland, application as part of the innovative partnership of the Enlighten Tampere Hackathon process by Geniem Oy. The application uses open data and application programming interfaces (APIs) available in the city, and collects user location data for input on the smart street lighting system in the Viinikka pilot area. The application now has 45,000 users [23].

The mySMARTLife project focuses on the energy retrofitting of houses in Helsinki, Nantes, and Hamburg. Evidence has shown that single-family house owners in Finland are most concerned with energy savings [24], and it is usually very hard for housing cooperatives to buy comprehensive energy retrofits despite potential profitability [25]. For this reason, energy advisory workshops held in Helsinki offered information on the most profitable energy retrofits, and both industry and city experts were invited to learn about the concerns, provide peer support, and initiate common projects [26].

CIVITAS ELAN "Mobilising Citizens for Vital Cities", launched by the European Commission as part of CIVITAS (City VITAlity and Sustainability), introduced transport measures and policies related to sustainable urban mobility and has the approach of 'putting the citizen first.' In Ljubljana, where citizen opinions were considered a burden and excluded from decision-making, the project team established a cycling platform to improve cycling conditions for residents and include the needs of disabled groups while also ensuring citizen participation through the new spatial plan. The city of Gent already had an active citizen population that was further enhanced by the use of dialogue cafes and social media to engage all age groups. Porto went through a change, as citizens who previously believed their suggestions would not be taken into consideration were pushed to voice their ideas through user interviews, face-to-face surveys, flyers, and brochures. In Brno, citizens were not used to discussions with transport operators, but after encouragement through public debates, public opinion research, and working groups, the traffic situation and mobility was greatly improved. In Zagreb, the lack of public participation was resolved through mobility dialogues, the training of citizens, the continuous provision of information, and the establishment of the Zagreb forum for direct discussions with experts [27].

1.4. Purpose of This Paper

The authors of this paper focused on the cities of Espoo (Finland) and Leipzig (Germany). They assessed the status quo of citizen engagement in the two cities and determined whether the cities are prepared to implement PEDs. This was the first analysis of its kind that focused on the cities of Espoo and Leipzig, as well as their citizens. It is envisioned that the analysis will also support other European cities in evaluating and enhancing their citizen engagement efforts and consequently support the energy transition towards PEDs.

The paper is structured as follows: Section 1 is the introduction to citizen engagement. Section 2 presents the approach and method adopted for the paper. Section 3 presents the selected case studies for each city, and Section 4 is the discussion and conclusion of the paper.

2. Approach and Methodology

The cities of Espoo and Leipzig have very different starting points regarding citizen engagement practices. As the next section shows, the existing infrastructure and regulations greatly vary between cities, consequently affecting how local activities are performed. This led the authors to perform an analysis of the two cities rather than a comparison of the two. It should also be highlighted that while the authors aimed to collect as diverse a portfolio as possible, part of the study material belongs to a European Horizon Europe (H2020) project where Espoo and Leipzig are lighthouse cities. As stated earlier, the topic of citizen engagement in the two cities has not been addressed in earlier studies, and this paper is a stepping stone for understanding and enhancing citizen engagement in both Espoo and Leipzig. A similar methodology was utilized for both cities to analyze and present the status quo of citizen engagement. The aim of the analysis was (1) to find the best local practice examples to build upon, (2) identify relevant (local) networks to cooperate with, and (3) to recognize (local) success factors for engagement activities.

The first step was to choose diverse case studies per city that showcased citizen engagement in urban development and transformation. The authors chose three case studies per city (Table 1) with the following criteria to assure relevance and to give a diverse view:

(1) Identify case studies with relatable target groups.
(2) Identify case studies with relatable objectives.
(3) Identify case studies with relatable themes.

Table 1. Description of selected case studies.

	Espoo Case 1	Espoo Case 2	Espoo Case 3	Leipzig Case 1	Leipzig Case 2	Leipzig Case 3
Target groups	All citizens	Youngsters	Youngsters	Private building owners in the districts	Actors in urban politics, business, and science	All citizens
Objectives	Understand mobility needs and challenges	Understand future mobility from young people's perspective	Safety of youngsters when visiting shopping centers	Convince building owners to invest in energy efficient retrofitting	Develop a future strategy for the next 10–15 years	Development and planning of green and blue infrastructure
Themes	Urban mobility and sustainable lifestyle	Urban mobility	New construction and renovation projects	Reduce emissions and climate change	Sustainable future	Healthy environment for better wellbeing

The selection of the case studies to represent the city of Espoo was based on discussions with the city representatives and those responsible for conducting citizen activities within the H2020 project. In addition, the authors conducted a broad literature review to understand the city processes and to know what initiatives and resources for the citizens already exist. The city of Espoo already has a broad history of engaging residents in both small and large-scale activities (Section 3), such as through schools and city libraries. However, in order to assure relevance to the paper, the authors selected case studies that had been conducted as part of the project as they were more recent and directly connected to PEDs. The first and second case studies addressed a variety of citizen types and age groups while focusing on mobility, while the third case study addressed a target group of youngsters and focused on creating a safe environment. Overall, the selected case studies present the recent efforts of Espoo and allow for the assessment of what could be the next steps for the city.

The selection of the Leipzig case studies was based on an in-depth desktop research and literature review, discussions within the local the project consortium, and interviews with representatives of the city, including relevant project partners of the case studies. The first case study focused on the behavioral change of tenants in apartment houses in the field of energy-efficiency in daily life. The second case study related to target groups on a district level and was connected to the development of PEDs. The third case study focused on citizens and civil society actors, which will be of interest for more general engagement activities and to raise awareness for the topic of PED to facilitate replication activities across Europe.

The selected case studies presented a wide range of experience with participation formats over a period of several years and variety of citizens. The presentation of the case studies from Espoo and Leipzig in Section 3 is structured according to the following questions:

General questions:

- What are the main responsibilities, structures and processes of implementation in the cities?
- What methods do Espoo and Leipzig use to include citizens?

Case specific questions:

- What was the aim of each city case?
- How was it done? What was the format?
- Who was the audience? How many people were reached? What was the response?
- What could be changed for replication and better results?

3. Case Study Analysis

3.1. Focus on Espoo

The Espoo Story is the city's strategy of the future. It has been diligently prepared with intensive cooperation together with residents, staff, and elected officials. The Espoo Story directs the city's operations toward common goals. The city council approved the strategy in 2017 for the current council term of 2017–2021 [28]. Espoo is to have municipal elections in April 2021, and preparations for the new strategy have been ongoing since summer 2020. Values in the Espoo Story include the active involvement of residents in the development of services and comprehensive co-operation with partners.

The four administrative development programs until the year 2021 are: (a) Participatory Espoo (Osallistuva Espoo), (b) Inspiring and dynamic Espoo, (c) Sustainable Espoo, and (d) Healthy Espoo. The city is aiming for carbon-neutral status by the year 2050, and, in parallel, Espoo will attempt to reduce resident-specific emissions by 60% by the year 2030 compared to the levels of 1990 [29]. In Espoo, resident participation that has extensively utilized new methods and tools has existed and been promoted for a long time. In addition, the municipal democracy has also been continuously revised over the years. The city believes that Espoo residents are much happier than the residents in other parts of Finland [30]. In Finland, the basis for citizen involvement in city development is set as laws. The Participatory Espoo program developed a participation model, with cornerstones including a handbook. A new position of a development manager for citizen engagement has also been established and work started in August 2020. In order to bring Espoo employees located in different parts of city organization together, there is a participatory network to share ideas and information.

The Espoo Voluntary Local Review (VLR) is the framework to evaluate and communicate economic, ecological, social, and cultural sustainability, and it was developed with regard to the 17 Sustainable Development Goals (SDGs) of the UN's 2030 Agenda. The review was performed together with hundreds of people (Espoo employees, residents, customers etc.). It was based on three themes: (1) leave no one behind, (2) let us do it together, and (3) accelerated action [31]. Within the city structure, there are different councils that cater to the need of different age groups such as the youth council (nuva), which has been active for 20 years and comprises forty 13–18-year-old youths to contribute in the decision making and planning. The Espoo youth council is the largest in Finland, and it has representatives in the city council and committees with the right to attend and speak. In 2020, it was also agreed that the youth council representatives have the right to attend and speak at city board meetings. The Espoo Elderly Council serves as an advocate for the elderly in the municipal decision making. In parallel, there is also a council for the disabled. All councils together influence the planning and preparation of the city's activities, as well as monitoring issues relevant to well-being, health, inclusion, work life, living environment, housing, mobility, day-to-day activities, and services [31].

Service development in Espoo is based on experimental culture and co-creation. The city as a service model has the following key aspects: the engagement of residents, the accessibility of services by means such as digitalization, and the creation and deployment of new business and operating models. In Espoo, this means that the city invites all stakeholders of municipal services from companies to associations, research institutions, and residents to refine old services and innovate new ones together. The joint "6Aika" strategy of the six largest cities in Finland aims to develop more transparent and intelligent services. It directs its focus on promoting transparent operating models that help the entire urban community, research and development actors, and authorities [31].

An overview of the residents in Espoo showed that the number of foreign-speaking residents in Espoo is gradually increasing. It is estimated that 18% of the Espoo residents speak a different language, and there are 150 different nationalities currently residing in Espoo [31]. Espoo citizens are able to give feedback to the municipal decision makers in a number of ways, either through an electronic feedback system [32] (https://easiointi.espoo.fi/eFeedback/en (accessed on 15 November 2020)) or traditional methods of communication. As per the Local Government Act, the resident of a municipality has the right to propose an initiative, e.g., the improvement of street or traffic. The submission of a proposal is an official and regulated procedure that is handled more carefully than unofficial feedback messages. In addition, the city of Espoo arranges residents' evenings, where it is possible to influence projects and plans. The residents of Espoo may also influence matters that concern them through residents' forums. Each residents' forum also has a preparation team [33].

The three successful citizen engagement projects of the city of Espoo are presented as follows. These have been conducted as part of the ongoing European H2020 project as mentioned earlier and include several local partners (City of Espoo, VTT, KONE Corporation (KONE), and Citycon).

Case 1—citizen user studies (2020): This case study was led by KONE, which is a global leader in the elevator and escalator industry, as well as a solution provider for the maintenance and modernization to add value to buildings throughout their life cycle. KONE's current development efforts focus on urban flow and user experience in smart cities. The case study aimed to gain insights into Espoo citizens' mobility behaviors, i.e., understand the ways and reasons for people to transition from one place to another, as well as understanding the challenges and opportunities for designing behavioral interventions for more sustainable mobility behaviors. The user study focused on two diverse districts of Espoo: Espoonlahti and Leppävaara. The Espoonlahti district is a developing area where a new shopping center (Lippulaiva) is currently being built to replace the old shopping center and the construction of an underground metro is to be completed by 2023. The Leppävaara district is a fairly developed area, characterized by a shopping center (Sello) and good public transportation train connections (both local and long-distance train, bus, and tram to start operation in 2024). Thus, the mobility infrastructure and the opportunities for (behavioral) interventions between these two places greatly vary.

With a focus on these two districts, citizens' mobility behaviors were studied through mobile probing, interviews, and co-creation workshops. The citizen insights were supported with mobility and citizen engagement expert interviews in order to tie the findings from the user studies into a more systemic understanding. Citizen needs were put into the core of innovation, while relevant stakeholders such as city representatives and service providers in mobility were involved in the research and development process from the early stages of the project. The ultimate goal of the applied citizen engagement process was to develop and test behavioral interventions and smart ecosystems that support citizens' sustainable mobility behaviors and habit formation, as well as more sustainable lifestyles in Espoo by the end of 2022.

The citizens were selected through an application process advertised on local social media channels (Facebook). The application was open to all citizens in Espoo, and 41 respondents answered a preliminary survey. The citizens were asked to provide background information about their mobility modes (such as walking, biking, private car, public transportation, and micro mobility solutions), attitudes towards sustainability, and life situation. The final sample consisted of 10 diverse Espoo citizens, representing varying demographics (Table 2). Five of the participants were selected from the Espoonlahti district, and the other five were selected from the Leppävaara district to achieve a full understanding of mobility needs of Espoo citizens. The participants documented their daily mobility behaviors, as well as related experiences and thoughts through a mobile probing method in which images, text, videos, and voice messages were sent on WhatsApp for a period of eight weeks (Figure 1). The mobile probing method captured and communicated people's daily mobility behaviors on a holistic level, enabling a deeper understanding of citizens' mobility needs. The method enabled an active role for the citizens in the research and design process of future mobility [34]. KONE provided the participants phone devices to document their daily journeys. The mobile probing activity was supported with weekly questions and tasks sent on WhatsApp to prompt reflection on the participants' current mobility decisions and alternative mobility modes. After the eight weeks, participants were interviewed to gain a deeper understanding based on the documented contents. In the final workshop, the participants co-ideated mobility concepts related to, for example, autonomous cars and material logistics based on the identified challenges and opportunities deriving from the user research. Due to COVID-19, all research activities during the study were held online and facilitated through digital platforms. Some challenges, such as limited digital capabilities and skills, were faced in the final workshop, as the participants did not have a prior experience of using digital co-creation platforms.

Table 2. Demographic information about the final research participants.

Gender	Age	Life Situation
Male	58	Spouse and two children (one living on their own)
Female	29	Spouse and a newborn baby
Male	41	Spouse and two small children
Female	20	Parents and sister
Male	44	Spouse and three children with active hobbies
Female	67	Spouse (using wheelchair)
Male	39	Single with three children (kindergarten/school)
Female	27	Living alone
Male	37	Living alone
Female	54	Teenage child every second week

Figure 1. Two screenshots from the WhatsApp probing contents sent by the participants—driving daughter and her bike to the ice rink (**left**) and travelling with a wheelchair (**right**).

At the end of the final workshop, participants were asked for feedback on the quality of citizen engagement achieved through the user study. Through the described comprehensive citizen engagement process that lasted for four months in total, a holistic understanding of various Espoo citizens' mobility challenges, needs, and desires was gained. It must be highlighted that informed consent was signed by the participants to maintain their privacy and data protection during the study period.

Case 2—testing a participatory design research workshop concept with youth (2020): This study was also led by KONE and arranged as part of a high school entrepreneurship course. A group of nine students (aged 17 years) were invited to a participatory design research workshop with the aim to investigate desirable future (sustainable) urban mobility from young people's perspective. Participatory design research workshops (1.5–2 h) were conducted on five themes: pedestrian, bicycling, public transportation, shared mobility, and mobility inside hybrid buildings. The workshop was formed around a case of an imaginary new living area currently developed in Espoo.

The workshop was facilitated face-to-face. In the workshop, the youngsters were asked two questions: "What kind of challenges do you experience related to the above-mentioned themes?" and "What would be your dream user experience (based on the five themes)? The ideas were organized through post-its, e.g., blue post-it-notes illustrated the challenges and yellow post-it-notes showed the desires (Figure 2).

Figure 2. Youth discussing the workshop results related to their future mobility needs and wishes.

The preliminary findings of the workshop indicated that the youth experience many challenges related to their current mobility. These challenges were related to matters such as weather and dressing, road safety, bike storing, noise pollution, and crowding. In addition, many desires for an improved city infrastructure that supports walking, biking, and public transport were communicated in the workshop. According the participants, this could be achieved by creating a mobility infrastructure with reduced noise pollution and shorter distances to places. All the students received a sticker for voting on the change that they would prefer in the close future of urban mobility developments. The workshop also helped build empathy for this sometimes-marginalized target group and allowed the experts to understand the criteria for future urban development efforts. The next step would be to plan and organize collaborative workshops for ideating more tangible ideas for tackling the identified challenges and high-level desires.

Case 3—co-creating shopping centers together with youngsters (2012 and ongoing): Shopping centers can be characterized as quasi-public spaces that may be defined as being open to all but are under private ownership [35]. Citycon is a leading owner and developer of mixed-use centers for urban living located in the Nordic region [36]. In Finland, Citycon owns and manages 12 centers, and the largest one called Iso Omena is located in Espoo. It attracts approximately 19.9 million visitors yearly and is a great example of urban mixed-use center with retail, public services (such as library, health care, and social insurance institution office), office spaces, and housing. This makes Citycon shopping centers even closer to public spaces where citizens handle multiple activities.

The aim of the workshop was to ensure that young citizens feel safe and comfortable while visiting shopping centers and that they are aware of the common rules to behave in shopping centers. The three main citizen engagement highlights for young citizens implemented the shopping centers are: (1) co-creation workshops, (2) making common shopping center rules together, and (3) working with shopping centers' guards (known as NOJA-guards) devoted and educated to work with youngsters (Table 3). In Finnish, the NOJA Guards is derived from nuorten oma järjestyksenvalvoja® and translates to youth's own security guard. All these activities were developed and held together with Nuorten Palvelu ry [37] an organization that works with and for the youth and whose support was crucial in order to reach as many young people as possible.

Table 3. Targets and timelines of different youth citizen engagement actions developed and held in Citycon together with Nuorten Palvelu ry.

	Co-Creation Workshops	Making Common Rules	NOJA-Guards
Target	To involve youngsters and to idea together	To make common rules of behavior in shopping center	To have known and safe adults to help youngsters in shopping centers and to decrease the amount of disorder
Timeline	During renovation or construction project	Before opening or when needed	While operation

The co-creation workshops are organized for youngsters when larger renovation or construction projects are ongoing. For example, the Iso Omena shopping center in Espoo was expanded in 2013–2017, so several co-creation workshops were held during that time. Approximately 20–30 youngsters (aged 12–16 years) participated in the workshop and worked together with architects, youth workers from Nuorten Palvelu ry, representatives from Espoo library, and security guards. The youngsters were invited to directly participate in the workshops by the NOJA guards who were present in the shopping center. In addition, Nuorten Palvelu also invited youngsters through their own networks. These methods ensured diversity amongst the group.

The workshops were conducted in an interactive manner where the construction project was first described to the youth, and later the youngsters were divided into groups to work on different topics. A few of the most popular ideas from the workshop were developed further (Figure 3). At the end of the workshops, the youngsters were rewarded with movie tickets for their valuable contribution.

Figure 3. Ideas from youngsters that were realized in Iso Omena—large aquariums (**left**) and a large clock in Iso Omena (**right**). These ideas came from youngsters, and they were planned together with Citycon employees.

By making rules for the shopping center together with youngsters, the designers and building owners had the opportunity to understand and solve problems that may have been overlooked. The youngsters invented possible rules applicable to all the visitors in the shopping centers. It was assumed that rules that are made together are usually obeyed better compared to rules given from above (managers). NOJA-guards (security guards) are still active, and they are educated and devoted to work with young customers. Observation showed that shopping centers with NOJA-guards have decreased the amount of disorder. Their presence is also a great way for shopping center owners to pay attention to young customers, implement social responsibility, and increase the amount of positive publicity.

3.2. Focus on Leipzig

Leipzig is the initiator and namesake of the Leipzig Charter on Sustainable European Cities [38], which was signed by 27 EU member states in 2007. The Leipzig Charter was the basis for a new urban policy in Europe, in particular for the concept of integrated urban development, which emphasizes the involvement of residents and an improved dialogue between representatives from politics, residents, and economic actors. The current framework for urban development in Leipzig is formed by the strategic goals of municipal policy, manifested in the integrated urban development concept "INSEK Leipzig 2030" of 2018 [39]. The main four target areas for Leipzig are: quality of life, social stability, national and international importance, and competition between cities and regions. These target areas are concretized by defining specific focus areas and technical concepts. One of the main goals of Leipzig is to be climate-neutral by 2050, and the city administration itself plans to be climate-neutral by 2030 [40]. Citizen and stakeholder engagement is explicitly listed as one of the principles of the INSEK implementation strategy.

In Germany, the law requires various forms of citizen engagement such as citizen petitions, citizen meetings, participation in city district and local councils, and formal public engagement in urban land use planning. However, many municipalities have established local guidelines that go beyond these legal requirements in order to include the perspectives of their citizens, to identify barriers early, and to improve the acceptance of their actions. In 2009, Leipzig started to develop a framework for an informal citizen participation resolution involving various actors from politics, science, administration, and citizenship. In 2012, the guidelines for citizen engagement [41] were manifested as the mayor's instruction to the city administration. They included the involvement of the citizens, city council, and administration in municipal processes in all areas of responsibility (trialogue principle), and they defined citizen engagement as an integrative part of these processes. Key points related to transparency and cooperation in decision-making. The basic principle is the "early and continuous involvement of citizens in planning and decision-making processes." In addition to the strategic goals, the guidelines contained recommendations for implementation, as well as a set of instruments.

Depending on the character of the projects, various offices are responsible for planning and implementing urban participation processes. Since 2014, they have been supported by the "Leipzig Weiterdenken" coordination center ("Thinking Leipzig Ahead") [42]. Its area of responsibility can be divided into three fields: Coaching and consulting within the city administration, testing innovative processes, and finally evaluating the implemented projects. In 2015, the strengths and weaknesses of the participation culture in Leipzig were analyzed. As a result, the coordinating office has since then maintained a central communication platform that provides transparent reports on completed, ongoing, and planned participation projects. The central participation structure also includes the city office [43], as a platform for citizen participation and commitment, and the youth parliament [44].

Successful formats of decentralized participation management include regular district forums, Leipzig neighborhood management, and district offices. District forums are organized by the city in the context of ongoing participation projects to inform the residents of a district about urban projects and to stimulate discussion. By making a representative selection of participants and embedding smaller discussion groups, the citizens' opinions can be analyzed in a representative way. In some key areas of urban development (the districts of Grünau, Leipzig East, Leipzig West, Paunsdorf, and Schönefeld), neighborhood or community management acts on behalf of the office of residential construction and urban renewal (AWS), implemented by private agencies. Neighborhood management is always connected to national or European urban development funding programs (e.g., Soziale Stadt, Stadtumbau Ost, and URBAN (European initiatives)). Its task is to initiate communication between administration, politics, and local stakeholders, as well as to support local associations, initiatives, and projects, thus strengthening identification and the culture of participation in a neighborhood. The specific design of a management structure is based on the needs of the respective area. Neighborhood councils have been formed in two of the priority areas (the districts of Grünau and Leipzig West) as points of contact within the neighborhood, as well as for city administration and politics. These councils consist of citizens or representatives of citizen committees and representatives from other thematic areas (child and youth work, culture, education, housing companies, local economy, church, social affairs, etc.). In the context of neighborhood management, the city operates district offices in some districts as points of contact for residents.

In addition to the experience gained from local projects, Leipzig also has experience with European smart city projects. From 2015 to 2020, Leipzig was one of the follower cities in the project Triangulum [45], focusing on cutting-edge concepts for smart district development. Within this project, the city developed an implementation strategy for smart city solutions in the sectors of energy, mobility, information, and communications technology for the Leipzig West area. Moreover, the findings of the Triangulum project have been considered in the development of the 2030 Leipzig INSEK Integrated Urban Development Concept.

The three successful citizen engagement projects of the city of Leipzig are presented as follows.

Case 1—Energy-efficient Retrofitting Management (ESM) (2015–2018): The ESM [46] established in two districts (Alt-Schönefeld, Lindenau-Plagwitz) in 2015 was part of the Leipzig Energy and Climate Protection Program 2014–2020, developed for European Energy Award certification. The objective was to convince building owners to invest in energy-efficient retrofitting, thereby improving the energy balance of the district. For this purpose, the seecon Ingenieure GmbH (Leipzig, Germany) and the Deutsche Stadt- und Grundstücksentwicklungsgesellschaft mbH & Co. KG, Regionalbüro Leipzig (DSK, Leipzig, Germany) established a desktop support system for measures to reduce emissions and to adapt to climate change, especially regarding integrated neighborhood concepts, as well as to improve the energy efficiency of buildings and infrastructures. The main target group comprised private building owners in the districts. The different formats have reached between several dozen (consulting) and several thousand people (campaign) and are as follows [46,47]:

- Postal and telephone contact with residential building owners (2015–2016).
- Energy and retrofitting consulting for private owners and tenants (2015–2017).
- Public event on the occasion of the cooperation with the Consumer Advice Center combined with exhibition (2016).
- Energy and retrofitting consulting in cooperation with the Consumer Advice Center Saxony (2016–2018).
- Joint campaign "Leipzig is Climate Conscious" with the Department for Environmental Protection (AfU) and the Department for Urban Development and Construction (ASW)—inserts in the district journals combined with vouchers for free energy consulting (2017).
- Thermography tour: guided thematic walk (2017).
- MobilityChange stakeholder workshop in the neighborhood: common elaboration of practical project ideas and suggestions (2015).
- Talks on cooperation for environmental education projects with local education actors and schools in the neighborhood, as well as the presentation of coordinated project outlines (2015).

During the three years of this project, it became clear that the function of a municipal advisor for energy issues with the option to advise key stakeholders and interested private owners is an important task for future district management. In the case of private owners, a high proportion of owner-occupiers is advantageous, as investment-oriented property owners of apartment houses have little interest in energy-efficient, socially-responsible renovation measures due to the lack of incentives on the rental market in Leipzig. Cooperation with the municipal or cooperative owners of apartment houses is also promising. In these cases, the tenants are to be involved in order to gain their acceptance and support. In addition, community or commercial facilities showed clear interest in the energy optimization of the building fabric and management. Here, the commercial tenants should be involved as well.

One of the most important conclusions to be drawn from this ESM project is that the cooperation with key stakeholders in the district, who take over the function of multipliers, is essential for the successful involvement of local residents. If no stakeholder networks exist, a network and trust building phase has to be scheduled.

Case 2—2030 Leipzig INSEK Integrated Urban Development Concept (2015–2017): One of the biggest and most successful participation projects in Leipzig was the participative development of the 2030 Leipzig INSEK Integrated Urban Development Concept [48,49]. Based upon the previous integrated urban development concept, SEKo 2020, the city planning office initiated a participative process to mutually design the approach for realizing the strategic vision for Leipzig 2030. This participative process was managed by the "Thinking Leipzig Ahead" coordination center and an external agency. The objective of the project was to develop a future strategy for the next 10–15 years to ensure economic power, finances, understanding of democracy, and natural resources. The different formats have reached between several dozen (workshops, forums) and several hundred people (presentations) [50,51]:

- Public kick-off event (2015).
- Four thematic workshops: moderated discussions in small and big groups, partly combined with presentations, exhibitions, and speed dating with city administration staff (2016–2017).
- Four district forums: moderated discussions in small groups combined with presentations (2016).
- Three working groups: discussion at the LivingLab Leipzig West (2016).
- Public presentation and discussion of the draft (2017).
- Presentation of the draft in five district forums: Grünau, Leipzig North, Schönfeld/Mockau, Leipzig East, and Paunsdorf (2017).
- Exhibition on the concept supported by Leipzig Lego model makers (2017).

The main topics of the participative discussions were defined as follows: challenges for living together in a growing city, future growth of the medium-sized economy, further spatial development, digital city, sustainable mobility, and active urban society. The contents of the discussion were based on sectoral plans, such as urban development plans and specific field-related plans. City administration, several city departments, and various interdepartmental working groups were involved in the planning and development process. The target groups of the participation formats were either the urban society per se or specific actors in urban politics, urban society, business, and science. The findings and results of the workshop and forum discussions, and thus the opinions and wishes of the citizens of Leipzig, were integrated into the concept.

The city of Leipzig gained a lot of experience in involving different stakeholders using various participation formats dependent on the specific objectives, the size of the participant group, and the level of involvement. One of the key aspects throughout the process was transparency—all steps and results were documented and published on the city's website.

Case 3—Green Masterplan: Leipzig green–blue 2030 (2018–2020): The Green Masterplan [52] is a development concept for the green–blue infrastructure of Leipzig (green space and waters). It is meant to become the politically accepted, socially accepted, and application-oriented basis for upcoming decisions on the spatial development of the city. Results from the Open Space Strategy 2017 are part of the 2030 Leipzig INSEK Integrated Urban Development Concept. The Green Masterplan shall define specific implementation concepts for green space and water design. The participation process is managed by the Office of Green Space and Waters (ASG).

The objective of the project was to involve citizens in the development and planning process of the green and blue infrastructure for the purpose of healthy environmental conditions, individually usable recreation and exercise opportunities in open spaces that are accessible to all, and promoting health, as well as a biotope network for species and habitats. The different formats have reached between several dozen (for consulting and workshops/green walks) and several thousand people (for campaign and survey) [53]:

- Online survey on attitudes and expectations (2019).
- Two citizen forums: discussion in groups with different questions combined with presentations and exhibition of online survey results (2019–2020).
- Four thematic workshops: discussion at round table combined with presentations (2019–2020).
- Green labs: various activities that turn urban green spaces and water bodies into a cultural experience and field of experimentation.
- Green walks: regular moderated walks on different topics combined with the presentation of the Green Masterplan; participants wear green vests and there is a record of observations/suggestions on cards/posters (2019–2020).

The main topics of the participative discussions were: environmental protection, health, climate adaptation, biodiversity, environmentally-friendly mobility, and urban gardening. The main target group of the participation process was the urban society—citizens and civil society actors. Points of critiques and proposals arising from these forums and workshops are to be included in the Green Masterplan. In addition to conventional formats, the city uses creative formats such as green labs and green walks to strengthen the citizens' relationship to their city and to raise awareness for their environment.

4. Discussion and Conclusions

The united nations intergovernmental panel on climate change's claims that there are 11 years left to save the planet before an irreversible climate catastrophe occurs. The climate urgency is to be resolved not only on the macro level but also at the local level where municipalities are required to engage in the transformation of the use and production of energy for the wellbeing of citizens [54]. While PEDs are low-carbon solutions for cities, they must provide co-benefits to citizens and local authorities (such as better wellbeing and health, job creation, increased gross domestic product (GDP), and tourism) [11]. In addition, success with a PED requires that all stakeholders join forces to develop the best-fit solution for and by the citizens to tackle the global climate crisis. Furthermore, the new Green Deal released by the European Commission demands new and more effective ways of including city leaders, authorities, and citizens to establish political dialogue across various levels of government [55]. In parallel, the concept of 'renewable energy communities' is also being discussed and holds great importance when planning the implementation of PEDs. Looking at the definition of PEDs provided in the report by the European Commission, there is a clear and strong emphasis on ownership by citizens and change being led by citizens [11].

This paper performed an analysis of the status quo of citizen engagement in the two emerging cities of Espoo and Leipzig, where the local authorities, together with the local partners, have planned several actions to initiate the uptake of PEDs for a carbon-neutral sustainable future. It has to be recognized that these cities have had very different starting points and varying experiences, so it is not feasible to make a comparison; rather, it was feasible to study the current situation and provide support in progressing towards the successful implementation of PEDs.

Through this paper, it can be seen that there is no one definition for citizen engagement in cities. Citizens play an active role in defining issues, finding solutions, and identifying priorities for action. However, involving citizens is voluntary for cities, and such an inclusion process varies between cities and may be even lacking completely in certain cities.

When discussing how the city of Espoo plans to create a smooth transition to PEDs together with citizens, the authors looked at some of the planned actions, such as engaging different age groups. In parallel, the city is determined to encourage a sustainable lifestyle with a special emphasis on urban mobility behaviors and citizens' daily journeys by following a design thinking process and utilizing methods from co-design. The described case studies provide evidence that the city and its expert partners can effectively engage a variety of actors, and the participation formats may be readily adapted to suit current requirements. With the existing participation infrastructure within the city, spanning over decades, building up and replicating the described citizen engagement activities will be easy to do. As citizen engagement is also part of Finnish law, over the years, there has been a strong emphasis on co-creation and encouragement of participation through the local council meetings, forums and through local networks across Finland. In addition, Espoo frequently conducts city-wide online surveys on different topics, such as 'Mun Espoo' (My Espoo) where residents indicate their satisfaction with the city services and provide suggestions.

The Espoo case studies illustrated how Espoo has worked with different age groups on various topics. The recruitment of citizens could be a challenge, but it may be resolved through rigorous promotion and reaching out to residents through various channels, such as social media, and providing motivation for participation. The youth are considered a marginalized group of society, but the experts have reached the youth through various methods to gather as diverse a group as possible and to ensure that all opinions are heard. Based on the success of the youth workshops, the city of Espoo has now begun work with two junior high schools and its students to plan a new shopping center in Espoo. Considering the global pandemic, this activity has been adapted to an online format, and co-creation with the students continues. It must also be highlighted that because the purpose of the Espoo workshops was to hold discussions on a general level and to hear citizen concerns regarding city development, the actual development of ideas that might include intellectual property was not realized and there were no issues related to ownership of ideas.

For the city of Leipzig, the former participation projects were found to be a good basis for further developing its participatory approach in urban development and energy transformation. The objective is to involve and to enable residents and local stakeholders in the planning and implementation processes of positive energy districts in urban society. This includes both convincing stakeholders who are directly involved and raising general awareness for the topic itself. Networking and cooperation with existing structures that already form a central point of contact in a district have proven to be advantageous. They can provide a first assessment of relevant stakeholders and their needs, and they can arrange for relevant contacts. Established structures and local key actors are perceived as trustworthy and can successfully promote project activities as multipliers. Similarly, the establishment of a permanent contact point for stakeholders has also proven useful. One of the promising formats with transfer potential for energy topics is ESM energy consulting in cooperation with the consumer center in Saxony, which could conceivably accompany the introduction of certain products or activities. The consumer centers in Germany are non-governmental organizations representing consumers' interests. They enjoy great trust among the population as advisory bodies for energy issues, among other things, and can be valuable partners in energy-related projects. As a third point, the participation format differs for each target group. One of the lessons learned from the ESM project is that there are no adequate incentives for investment-oriented property owners of apartment houses in Leipzig to improve the energy performance of their buildings. On the other hand, local tenants have little influence on the owners' decisions because they have no levers under the current circumstances. The better target group for energy-retrofitting are the owner-occupiers who directly profit from this kind of investment in the end (socially and economically). Tenants, in turn, are limited in their action and are the right target group for promoting energy-saving activities in everyday life. In the end, the city has to define different starting points and goals for different target groups with different frameworks of action.

This paper aimed to answer the question whether the two cities are ready to implement PEDs; considering the background of each city and the diverse experiences that each one has in engaging actors, it may be said that both Espoo and Leipzig are prepared to effectively move towards the implementation of PEDs in a citizen-inclusive manner. A variety of citizens have been invited to participate and provide input on services and development projects in both cities. The inclusion of the young generation, children, teenagers, and vulnerable young people is crucial in urban development because it leads to awareness on energy transition from an early stage, builds up or strengthens the emotional bond with their own neighborhood, and creates ownership. Moreover, as early adopters, children and young people can exert a motivating influence on their local environment, including their family. For this reason, it is important to include special engagement formats for this target group, as observed in the Espoo case studies. The cooperation with local schools, education centers, and training centers in the district, as well as with associations and sponsors of children's and youth work on site, should be part of projects dealing with urban development. Citizens must also feel that their opinions and feedback are heard and considered by the experts. Therefore, it becomes essential that interaction with citizens is not carried out for the sake of interaction. Instead, experts have to be ready to modify and adapt the plans based on citizen needs. In parallel, it must also be highlighted that common formats such as presentations, discussions, workshops, and working groups, sometimes combined with informational formats such as exhibitions, are indispensable parts of the participation strategy according to the project phase, the defined target group, and the purpose of the activity. Depending on the circumstances, the influence of participants ranges from information to the development of proposed solutions and co-creative formats. Additionally, it is advisable to include more event-like and creative formats to raise awareness, create emotional impact and get into discussion with local residents. The 2017 exhibition on INSEK Leipzig 2030 was supported, e.g., by Leipzig Lego model makers. The Green Masterplan, and the ESM project provided guided thematic walks or tours illustrating real-life implications of political topics.

With the global pandemic, the key question is the adaptation of participation plans. The crucial points are not only the different levels of digital literacy, access to digital infrastructure, and of willingness to use digital technologies but also the mode of interaction itself that differs of course from that of face-to-face events. Recent developments have forced us to rethink our concepts and to include much more digital activities than intended. Certain activities, however, have the capability to be immediately adapted the online space, e.g., the user mobility study was conducted entirely remotely through phone devices. Workshops held by the city, such as inviting residents to develop new participation ways, have already been switched online. In the coming months, the cities of Espoo and Leipzig will investigate inclusive digital and hybrid participation formats, and they will adapt them further to create innovative use cases for the development of PEDs for replication across Europe. Moreover, the cities will look into possible key performance indicators (KPIs) that could be deployed to measure success with citizens through different participation formants and to assess how these could be adapted to suit the different contexts across Europe.

Author Contributions: Conceptualization, Z.F.; methodology, Z.F. and U.P.; writing—original draft preparation, Z.F., U.P., S.-S.S., K.K. and M.T.; writing—review and editing, Z.F., U.P., S.-S.S., K.K. and M.T. All authors have read and agreed to the published version of the manuscript.

Funding: This research was funded by Sustainable energy Positive & zero cARbon CommunitieS (SPARCS). Grant agreement ID: 864242 and the APC was waived.

Institutional Review Board Statement: Not applicable.

Informed Consent Statement: Informed consent was obtained from all subjects involved in the study. This is only applicable to Espoo case studies and not applicable to the Leipzig cases.

Data Availability Statement: Not applicable.

Acknowledgments: The authors are grateful to Elina Wanne, Reetta Turtiainen and Merja Ryöppy for their valuable contribution regarding the Espoo case studies.

Conflicts of Interest: The authors declare no conflict of interest.

References

1. Carpini, M.X.D.; Cook, F.L.; Jacobs, L.R. Public deliberation, discursive participation, and citizen engagement: A Review of the Empirical Literature. *Annu. Rev. Politi. Sci.* **2004**, *7*, 315–344. [CrossRef]
2. Warren, A.M.; Sulaiman, A.; Jaafar, N.I. Social media effects on fostering online civic engagement and building citizen trust and trust in institutions. *Gov. Inf. Q.* **2014**, *31*, 291–301. [CrossRef]
3. Gil De Zúñiga, H.; Jung, N.; Valenzuela, S. Social Media Use for News and Individuals' Social Capital, Civic Engagement and Political Participation. *J. Comput. Commun.* **2012**, *17*, 319–336. [CrossRef]
4. Zukin, C.; Keeter, S.; Andolina, M.; Jenkins, K.; Carpini, M.X.D. *A New Engagement?: Political Participation, Civic Life, and the Changing American Citizen*; Oxford University Press: Oxford, UK; New York, NY, USA, 2006; pp. 1–270. [CrossRef]
5. Ekman, J.; Amnå, E. Political participation and civic engagement: Towards a new typology. *Hum. Aff.* **2012**, *22*, 283–300. [CrossRef]
6. Bonsón, E.; Perea, D.; Bednárová, M. Twitter as a tool for citizen engagement: An empirical study of the Andalusian municipalities. *Gov. Inf. Q.* **2019**, *36*, 480–489. [CrossRef]
7. Smart Cities Information System. In *Citizen Engagement Solution Booklet*; European Union: Brussels, Belgium, 2019.
8. Canadian Institutes of Health Research (CIHR), CIHR'S Citizen Engagement Handbook. 2012. Available online: http://www.cihr-irsc.gc.ca/e/42196.html (accessed on 1 July 2020).
9. The Association of Netherlands Municipalities VNG. Participatie. Available online: https://vng.nl/artikelen/participatie (accessed on 1 July 2020).
10. European Innovation Partnership on Smart Cities and Communities. Citizen Focus Action Cluster. Available online: https://eu-smartcities.eu/clusters/3/description (accessed on 1 July 2020).
11. Schütz, F.; Heidingsfelder, M.L.; Schraudner, M. Co-shaping the Future in Quadruple Helix Innovation Systems: Uncovering Public Preferences toward Participatory Research and Innovation. *She Ji J. Des. Econ. Innov.* **2019**, *5*, 128–146. [CrossRef]
12. Paci, S.; Bertoldi, D.; Shnapp, S. Enabling Positive Energy Districts across Europe: Energy efficiency couples renewable energy. *JRC Tech. Rep.* **2020**. [CrossRef]
13. European Commission. *Final Report of the High-Level Panel of the European Decarbonisation Pathways Initiative*; European Commission: Brussels, Belgium, 2018. [CrossRef]
14. European Commission. *Cities by 2030—By and for the Citizens. Interim Report of the Mission Board for Climate-Neutral and Smart Cities*; European Commission: Brussels, Belgium, 2020. [CrossRef]
15. European Commission. The European Green Deal. 2019. Available online: https://ec.europa.eu/info/strategy/priorities-2019-2024/european-green-deal_en (accessed on 1 July 2020).
16. Allen, B.; Tamindael, L.E.; Bickerton, S.H.; Wonhyuk, C. Does citizen coproduction lead to better urban services in smart cities projects? An empirical study on e-participation in a mobile big data platform. *Gov. Inf. Q.* **2020**, *37*, 101412. [CrossRef]
17. Choi, J.-C.; Song, C. Factors explaining why some citizens engage in E-participation, while others do not. *Gov. Inf. Q.* **2020**, *37*, 101524. [CrossRef]
18. Simonofski, A.; Vallé, T.; Serral, E.; Wautelet, Y. Investigating context factors in citizen participation strategies: A comparative analysis of Swedish and Belgian smart cities. *Int. J. Inf. Manag.* **2021**, *56*, 102011. [CrossRef]
19. Johnson, P.A.; Robinson, P.J.; Philpot, S. Type, tweet, tap, and pass: How smart city technology is creating a transactional citizen. *Gov. Inf. Q.* **2020**, *37*, 101414. [CrossRef]
20. Integrated and Replicable Solutions for Co-Creation in Sustainable Cities. In *H2020 Project, Deliverable 1.6*; European Commission: Brussels, Belgium, 2018.
21. García, J.B.; Sánchez Mora, M.; Webb, R.; Næss, K.S.; Perkins, A.; Fica, C.; Prokýšek, M.; Delgado Ortiz, L.C.; Petkova, E.; Hallimäe, T. *+CityxChange D3.2: Delivery of the Citizen Participation Playbook*; European Commission: Brussels, Belgium, 2020.
22. MAtchUp. MAtchUp EU Project—Citizens Solution. 2018. Available online: https://www.matchup-project.eu/solutions/citizens-engagement/ (accessed on 1 July 2020).
23. POCITYF. H2020 POCITYF. Available online: https://pocityf.eu/solutions/#citizen-driven-co-creation (accessed on 1 July 2020).
24. City of Tampere. Stardust. Available online: https://www.tampere.fi/smart-tampere/kestava-tampere-2030-ohjelma/stardust.html (accessed on 1 July 2020).
25. Hirvonen, J.; Vanhatalo, M. *Ympäristöasenteet ja kaupunkikehitys Helsingissä ja Vantaalla*; TUTKIMUKSIA: Helsinki, Finland, 2018.
26. Murto, P.; Jalas, M.; Juntunen, J.; Hyysalo, S. The difficult process of adopting a comprehensive energy retrofit in housing companies: Barriers posed by nascent markets and complicated calculability. *Energy Policy* **2019**, *132*, 955–964. [CrossRef]
27. Viitanen, M. *Mission Board Seminar. Energy Renaissance. MySmartLife Project*; European Union: Brussels, Belgium, 2020.
28. Marega, M.; van Aken, E.; Braun, M.; Kontić, V.; Delanghe, P.; Pavić-Rogošić, L.; Štěpnička, J.; Martinho, B.S.; Engels, D. *Civitas Elan Measure Leaders, Citizen Engagement in the Field Of Mobility Work and Lessons Learned*; European Commission: Brussels, Belgium, 2012.
29. City of Espoo. Espoo Story. 2017. Available online: https://www.espoo.fi/en-us/city_of_espoo/Decisionmaking/The_Espoo_Story (accessed on 1 July 2020).
30. City of Espoo. Goal: A Sustainable Future. 2020. Available online: https://www.espoo.fi/en-US/Housing_and_environment/Sustainable_development (accessed on 1 July 2020).

31. City of Espoo. Osallistuva Espoo. 2020. Available online: https://www.espoo.fi/fi-fi/espoon_kaupunki/paatoksenteko/espootarina/osallistuva_espoo (accessed on 1 July 2020).
32. City of Espoo. City of Espoo—Feedback system. 2020. Available online: https://easiointi.espoo.fi/eFeedback/en (accessed on 15 November 2020).
33. City of Espoo. Voluntary Local Review—Implementation of the United Nations' Sustainable Development Goals 2030 in the City. Available online: https://www.local2030.org/pdf/vlr/EspooVLR2020Web.pdf (accessed on 1 July 2020).
34. City of Espoo. Influence. Available online: https://www.espoo.fi/en-US/City_of_Espoo/Influence (accessed on 1 July 2020).
35. Koch, D.; Maaß, S. Digital Probes Kit. *i-com* **2018**, *17*, 169–178. [CrossRef]
36. Sahito, N.; Han, H.; Nguyen, T.V.T.; Kim, I.; Hwang, J.; Jameel, A. Examining the Quasi-Public Spaces in Commercial Complexes. *Sustainability* **2020**, *12*, 1830. [CrossRef]
37. Citycon. Available online: https://www.citycon.com/ (accessed on 1 November 2020).
38. European Commission. Leipzig Charter on Sustainable European Cities. 2007. Available online: https://ec.europa.eu/regional_policy/archive/themes/urban/leipzig_charter.pdf (accessed on 30 November 2020).
39. City of Leipzig. Integrated Urban Development Concept "INSEK Leipzig 2030". 2018. Available online: https://static.leipzig.de/fileadmin/mediendatenbank/leipzig-de/Stadt/02.6_Dez6_Stadtentwicklung_Bau/61_Stadtplanungsamt/Stadtentwicklung/Stadtentwicklungskonzept/Leipzig-2030_Beschluss_Gesamtfassung.pdf (accessed on 30 November 2020). (in German).
40. City of Leipzig. Emergency Climate Action Program 2020. Available online: https://static.leipzig.de/fileadmin/mediendatenbank/leipzig-de/Stadt/02.3_Dez3_Umwelt_Ordnung_Sport/36_Amt_fuer_Umweltschutz/Publikationen/200910_Sofortma-nahmenprogramm_Klimanotstand_2020_FINAL.pdf (accessed on 30 November 2020). (in German).
41. City of Leipzig. Public Submission to the City Council on the Guidelines. 2012. Available online: http://english.leipzig.de/fileadmin/mediendatenbank/leipzig-de/Stadt/02.6_Dez6_Stadtentwicklung_Bau/61_Stadtplanungsamt/Stadtentwicklung/Leipzig_weiter_denken/OEffentliche_Vorlage_an_den_Stadtrat_zu_den_Leitlinien.pdf (accessed on 30 November 2020). (in German).
42. City of Leipzig. City of Leipzig, Official Homepage of the Coordination Centre. 2020. Available online: https://english.leipzig.de/services-and-administration/opportunities-for-residents-to-get-involved-and-make-a-difference/thinking-leipzig-ahead/ (accessed on 30 November 2020). (in English).
43. City of Leipzig. Official Homepage of the City Office. 2020. Available online: https://www.leipzig.de/buergerservice-und-verwaltung/buergerbeteiligung-und-einflussnahme/stadtbuero/ (accessed on 30 November 2020). (in German).
44. City of Leipzig. Official Homepage of the Youth Parliament. 2020. Available online: https://jugendparlament.leipzig.de/ (accessed on 30 November 2020). (in German).
45. City of Leipzig. TRIANGULUM Local Project Activities. 2020. Available online: https://www.leipzig.de/bauen-und-wohnen/stadterneuerung-in-leipzig/stadterneuerungsprojekte/eu-projekt-triangulum/ (accessed on 30 November 2020). (in German).
46. City of Leipzig. Energy-efficient Retrofitting Management (ESM). 2020. Available online: https://www.leipzig.de/bauen-und-wohnen/foerdergebiete/klimaquartiere/ (accessed on 30 November 2020). (in German).
47. Seecon, DSK. *Energy-Efficient Retrofitting Management Lindenau-Plagwitz. Final Report—Energetisches Sanierungsmanagement Lindenau-Plagwitz. Endbericht*; Seecon: Leipzig, Germany, 2018; non-public. (in German)
48. Seecon, DSK. *Energy-Efficient Retrofitting Management Alt-Schönefeld. Final Report—Energetisches Sanierungsmanagement Alt-Schönefeld. Endbericht*; Seecon: Leipzig, Germany, 2018; non-public. (in German)
49. City of Leipzig. Content and Participative Development of INSEK Leipzig 2030, Summary. 2020. Available online: https://english.leipzig.de/construction-and-residence/urban-development/insek-leipzig-2030-integrated-urban-development-concept/content-and-process/ (accessed on 30 November 2020). (in English).
50. City of Leipzig. Participative Development of INSEK Leipzig 2030. 2020. Available online: https://www.leipzig.de/bauen-und-wohnen/stadtentwicklung/stadtentwicklungskonzept-insek/prozess-und-beteiligung/ (accessed on 30 November 2020). (in German).
51. City of Leipzig. Detailed Documentation of the INSEK Participation Process. 2020. Available online: https://www.leipzig.de/buergerservice-und-verwaltung/buergerbeteiligung-und-einflussnahme/leipzig-weiter-denken/beteiligen/leipzig-2030/ (accessed on 30 November 2020). (in German).
52. City of Leipzig. Official Homepage of Masterplan Green. 2020. Available online: https://www.leipzig.de/freizeit-kultur-und-tourismus/parks-waelder-und-friedhoefe/masterplan-gruen/ (accessed on 30 November 2020). (in German).
53. City of Leipzig. Detailed Documentation of the Masterplan Green Participation Process. Available online: https://www.leipzig.de/freizeit-kultur-und-tourismus/parks-waelder-und-friedhoefe/masterplan-gruen/beteiligungsprozess/ (accessed on 30 November 2020).
54. Saheb, Y.; Shnapp, S.; Paci, D. *From Nearly-Zero Energy Buildings to Net-Zero Energy Districts—Lessons Learned from Existing EU Projects*; EUR 29734 EN.; Publications Office of the European Union: Luxembourg, 2019. [CrossRef]
55. Eurocities. The European Green Deal: Delivering Results for Citizens with Europe's Cities. 2020. Available online: https://eurocities.eu/latest/the-european-green-deal-delivering-results-for-citizens-with-europes-cities/ (accessed on 1 December 2020).

buildings MDPI

Article

An Approach to Data Acquisition for Urban Building Energy Modeling Using a Gaussian Mixture Model and Expectation-Maximization Algorithm

Mengjie Han [1], Zhenwu Wang [2] and Xingxing Zhang [1,*]

1 School of Technology and Business Studies, Dalarna University, 79188 Falun, Sweden; mea@du.se
2 Department of Computer Science and Technology, China University of Mining and Technology, Beijing 100083, China; wzw@cumtb.edu.cn
* Correspondence: xza@du.se; Tel.: +46-(0)-23-77-87-89

Abstract: In recent years, a building's energy performance is becoming uncertain because of factors such as climate change, the Covid-19 pandemic, stochastic occupant behavior and inefficient building control systems. Sufficient measurement data is essential to predict and manage a building's performance levels. Assessing energy performance of buildings at an urban scale requires even larger data samples in order to perform an accurate analysis at an aggregated level. However, data are not only expensive, but it can also be a real challenge for communities to acquire large amounts of real energy data. This is despite the fact that inadequate knowledge of a full population will lead to biased learning and the failure to establish a data pipeline. Thus, this paper proposes a Gaussian mixture model (GMM) with an Expectation-Maximization (EM) algorithm that will produce synthetic building energy data. This method is tested on real datasets. The results show that the parameter estimates from the model are stable and close to the true values. The bivariate model gives better performance in classification accuracy. Synthetic data points generated by the models show a consistent representation of the real data. The approach developed here can be useful for building simulations and optimizations with spatio-temporal mapping.

Keywords: gaussian mixture model; Expectation-Maximization; urban building energy modeling; data acquisition

Citation: Han, M.; Wang, Z.; Zhang, X. An Approach to Data Acquisition for Urban Building Energy Modeling Using a Gaussian Mixture Model and Expectation-Maximization Algorithm. *Buildings* **2021**, *11*, 30. https://doi.org/10.3390/buildings11010030

Received: 29 November 2020
Accepted: 14 January 2021
Published: 16 January 2021

1. Introduction

Buildings account for 40% of global energy consumption [1]. Of this figure over 60% of the energy is consumed in the form of electricity, and only 23% of it is supplied by renewable sources [2]. Studies about nearly zero-energy building (NZEB) and positive energy districts (PEDs) have recently drawn much attention to possible ways to reduce this energy demand [3,4]. NZEB buildings have a very high energy performance level, with the nearly zero or very low amount of energy provided by significant renewable sources, and their energy performance is determined on the basis of calculated or actual annual energy usage [5–7]. PEDs are defined as energy-efficient and energy-flexible urban areas with a surplus renewable energy production and net zero greenhouse gas emissions. For both NZEB and PED, building energy performance is a crucial criteria for indicating their energy achievements [8].

Building energy modeling is an efficient way to predict the different possible performance levels of a building [9]. Among modeling methods, data-driven approaches have shown their advantages in building energy modeling, especially at an urban scale [10–13]. Basically, a data-driven approach is a systematic framework of data modeling techniques comprising model selection, parameter estimation and model validation that creates analytical decision-making. Most of the machine learning methods are data-driven since the machines or models are trained by learning a mapping function that operates between the input and output of the data. The more experience a machine has at learning, the better performance it will get. Thus, acquiring sufficient data is the basis to identify accurate energy use patterns and decide on the optimal actions to take in response. However, acquiring sufficient data in high quality for buildings at the urban level is a real challenge. Either random missing values or a large amount of incomplete information jeopardizes the model's validity. As data in urban building energy modeling (UBEM) collected from different sources, it can take significant effort to integrate datasets into a standardized format for interoperability [14].

By identifying such a research gap around the acquisition of data for urban building energy modeling, this paper aims to develop a novel approach. The specific contributions of this work are as follows: (1) it proposes to use a Gaussian mixture model (GMM), trained by Expectation-Maximization (EM) algorithm, as a generative model to discover the populations where the data can be drawn from; (2) it uses real datasets to validate parameter estimation and generative performance; (3) it suggests that the bivariate Gaussian model is more robust than the univariate model; and (4) it discusses the practical ways in which the initial values of the EM algorithm can be set.

The rest of the paper is structured as follows: based on an extensive literature review, the necessities and challenges of modeling with sufficient data are discussed in Section 2. Section 3 continues with a brief summary of the different ways to acquire more data. The philosophies of GMM and EM are presented in Section 4. In Section 5, the real datasets, parameter estimation details, and model evaluations are introduced. Section 6 discusses the spatio-temporal mapping of the synthetic data. It is followed by a conclusion in the final section.

2. Necessities and Challenges of Building Performance Modeling with Big Data

The analysis of building energy performance is switching from single buildings to district and urban levels. It yields new research domains that are associated with building energy performance, such as transportation, spatial correlations, grid operations, energy market actions and so on. Together with the factors, such as occupant behavior and climate change affecting single building modeling, a large amount of data are being produced in different domains, and the causal relationships between them are becoming complex. For instance, Zhang et al. reviewed a modeling technique for urban energy systems at the building cluster level that incorporated renewable-energy-source envelope solutions, in which they highlighted that a high-resolution energy profile, a spatio-temporal energy demand (both building and mobility) and detailed engineering/physical and statistical models are desirable for further development [15]. Salim et al. defined occupant-centric urban data and the pipeline to process it in a review paper that also outlined the different sources of urban data for modeling urban-scale occupant behavior: mobility and energy in buildings [16]. Perera et al. developed a stochastic optimization method to consider the impact of climate change and extreme climate events on urban energy systems across 30 cities in Sweden by considering 13 climate change scenarios [17]. Yang et al. started the data-driven urban energy benchmarking of buildings that uses recursive partitioning and stochastic frontier analysis in their study of a dataset of over 10,000 buildings in New York City [18]. By examining 3640 residential buildings, Ma and Cheng estimated building energy use intensity at an urban scale by integrating geographic information system (GIS) and big data technology [19]. Pasichnyi et al. proposed a data-driven building archetype for urban building energy modeling that uses rich datasets [10]. Risch et al. presented a level scheme to study the influence of data acquisition on the individual Bayesian calibration of archetype for UBEMs [20]. Ali et al. developed a data-driven approach to optimize urban-scale energy retrofit decisions for residential buildings while acknowledging that developing a building stock database is a time-intensive process that requires extensive data (both geometric and non-geometric), that can, in its uncollected form, be sparse, inconsistent, diverse and heterogeneous in nature [11].

When training a model, inadequacies in the data acquisition process generally mean that important information will be lacking, and the capture of underlying features is badly carried out. For example, energy policies may be misleading, and the data derived from them can be unclear or ambiguous. Securing sufficient data from different domains and in a finer resolution at the urban level will improve the quality of a model's decision-making. Understanding the impact of insufficient data on building energy modeling allows challenges to be identified, limits to the model's capacity to be broken and ways of improving the data situation to be found [21]. Modern machine learning techniques, especially deep learning (DL) methods, are a powerful way to model large amounts of urban energy data. The reason that DL outperforms other methods is that it uses millions of parameters to create sophisticated, nonlinear relationships that map the input data to the output data. The central goal is to train the bulk of the parameters so that they not only fit the training set well, but that they are also able to work on a dataset that the model has not seen. The ability to perform well on previously unobserved inputs is called generalization [22]. Small data sets do not provide enough epochs to update the parameters, which means that the model can be perfectly trained, and the output can be mapped to an extremely high accuracy, but only on an observed dataset. This leads to the problem of overfitting. Feeding sufficient data to the model is the equivalent of enabling it to discover more comprehensive mapping rules and enhancing, therefore, the model's generalization ability.

Data capture and storage, data transformation, data curation, data analysis and data visualization are all challenges when working with big data [23]. Establishing new systems that can gather the data required to estimate building energy performance requires multidisciplinary efforts and comes with high financial and time costs. Consequently, missing data at different levels in building energy modeling is a common problem. Most of the existing statistical and nonlinear machine learning methods can provide reliable interpolations when the missing rate is small, for example, lower than 30%, and the problem is considered to be random missing. When the missing rate is as large as 50–90% or if the sample information is completely missing from a large-scale dataset, it is unknown how well these methods will perform [24]. Alternatively, when a simplified building model handles incomplete data, its findings are usually fairly robust and efficient [25]. However, the assessment methodology has depended on the situation in each specific area, and it can be difficult to generalize [26].

3. Methods for Acquiring Building Energy Performance Data

Traditional methods for acquiring building energy performance data include direct collection from a meter, data augmentation and simulation. Combining these sources will enrich a dataset. Despite this, statistical methods, especially mixture models, from generative model point of view have not been seriously examined.

3.1. Collecting Energy Performance Data

Direct collection of building energy data from energy meters and sub-meters can be done in three different ways: energy readings, high-frequency energy logging and building management systems (BMS) [27]. Reading energy consumption, data are easy and cheap to do if meters are on-site, but readers can make mistakes and these are not easily discovered. One alternative is to apply computer vision techniques to automatically read the meter [28]. High-frequency energy data logging is the process of both collecting and storing data over a period of relatively short time intervals. The core device is a data logger comprising an electronic sensor, a processor and storage memory. The development of cloud and edge computing make the management of these data much smoother. These kinds of data are usually used for accurate forecasting in high time resolution and serves as the basis for a real-time bidding system. A BMS is a digital system that monitors, controls and records a building's operations. However, its automated processes, analysis capabilities and its integration with other systems are still not well developed. The data that BMS systems provide have been shown to contain errors [27]. These challenges can be tackled with the aid of the modeling approaches and integration capabilities provided by building information modeling (BIM) [29].

Direct data collection from meters is fast and precise. Where there is uncertainty about energy performance, meter data can act as a system benchmark. With the help of computers, the handling of data is becoming increasingly efficient. However, errors, including missing and incorrect values, can still be made by humans and machines. Thus, investing in data infrastructures such as sensors, meters, sub-meters and data archiving systems and linking these to data-driven building operation applications are still essential [30].

3.2. *Data Augmentation*

Data augmentation is a particular technique in computer vision that allows images to be flipped, rotated, scaled and translated in order to produce more samples. New features can also be generated to aid model generalization. Although building energy data are not directly stored in the form of images, there have been a few studies that explore the different ways that data could be augmented.

3.2.1. Energy Consumption

Unlike face recognition, intelligent transportation, precision medicine and other AI industries where computer vision has been comprehensively developed, an efficient, image-based approach to analyzing building energy is still in its early stages [31]. Traditional data augmentation methods have seldom been considered, although a deep convolutional neural network is able to detect equipment usage and predict the energy heat gains in office buildings [32]. In other research, building energy consumption data incorporating equipment use, lighting and air conditioning systems were augmented and enabled to capture more sequences by imposing a sliding sample window on the original training data [33]. With this technique, the training sample was enlarged to the length of original sequence minus the window length.

3.2.2. Short-Term Load Forecasting

Short-term load forecasting (STLF) for buildings, which secures operations in a smart grid, focuses on predicting the energy load of a building for a certain number of hours or days ahead. Deep learning has become the principal method for securing accurate predictions within this process from the demand side [34,35]. The inclusion of multi-source data such as user profiles, efficiency and weather data has been shown to improve the performance of loading forecasting [36]. Without a sufficient supply of data, deep learning modeling may fail to cope with a huge number of parameters. However, a large historical load dataset is very likely unavailable. One of the reasons is the development of distribution networks where newly built distribution transformers are growing. The other reason is that the responsible party for data management frequently restricts access to it [37].

One recent study proposed a method that merged the three-phase dataset to form a larger dataset so that the load forecasting of a particular phase could also be based on other phases' information. The new dataset was then thrown into a rolling forest to generate a training and testing set [37]. Another study proposed that, where a method that concatenates the series to generate larger amounts of data was viable, for a single building, the loading series should have less uncertain and more homogeneous information [38]. The size of the data was enlarged $(K^2 + K)/2$ times by figuring out 1, 2, …, K centroid load profiles and the corresponding residuals. In order to improve the concatenation method, instead of aggregating all of the historical data from previous years, a historical data augmentation method inserted one feature factor, which adopts adjacent loads as new feature, into the original input [39].

3.2.3. Learning-Based Methods

Generative Adversarial Network (GAN) has been developed in many applications [40,41]. In GAN, a generator generates random objects that are mixed with real objects for a discriminator to discriminate. The discriminator learns to assign high scores to real objects and low scores to generated objects. The generator is then updated so that it generates new objects that are likely to be assigned a high score by the fixed discriminator. The procedure stops when the discriminator is not able to discriminate whether the objects are generated or real. In a recent work [42], GAN was applied to one year of hourly whole building electrical meter data from 156 office buildings so that the individual variations of each building were eliminated. The generated load profiles were close to the real ones suggesting that GAN can be further used to anonymize data, generate load profiles and verify other generation models. A recurrent GAN preceded by core features pre-processing was also used to test the same dataset. The model trained with the synthetic data achieved a similar accuracy as the one trained with real data [43]. In another work, a conditional variational auto-encoder was developed to detect electricity theft in buildings. The method considered the shapes and distribution characteristics of samples at the same time, with the training set improving detection performance in comparison with other classifiers [44].

3.2.4. Simulation

Building simulation is an economical method for evaluating building performance and comparing it with real data [45,46]. With a pre-defined physical environment and generally acceptable levels of accuracy, simulation tools can rapidly generate sufficient amounts of analyzable data. Stochastic factors that incur uncertainties, such as occupant behaviors, have been integrated into building performance simulations and have improved their accuracy [47]. Combined with Geographical Information System (GIS) programs, urban energy simulations are already demonstrating that they are likely to further reduce input data uncertainty and simplify city district modeling [48]. Some of challenges of building simulation, such as model calibration, efficient technology adoption and integrated modeling and simulation have also been addressed in various modeling scales, from the single building to the national level, and at different stages in the building life cycle, from initial design to retrofit [49]. For building simulation, the applicability of integrated tools, not only during the early design but also throughout the building operation, management and retrofit stages should be improved to make the most effective decisions [50].

4. The Gaussian Mixture Model and Expectation-Maximization Method

4.1. *Gaussian Mixture Model*

Building energy data are often a mixture of samples from different populations where the parameters differ from each other. A natural strategy is to identify these populations and generate new data points using respective populations where a Gaussian mixture model (GMM) is built. GMM is a simple linear superposition of Gaussian components, aimed at providing a richer class of density models than the single Gaussian [51]. GMM is also a generative model, wherein arbitrary amounts of data can be generated. While *K*-means, a special form of GMM, assigns an individual to the nearest center, GMM gives a soft allocation for all data points. GMM is an unsupervised method where a data point is assumed to belong to a component with a certain probability. A categorical latent variable Z is adopted to determine a specific component by letting $Z_k = 1$ and $Z_{-k} = 0$, where Z_{-k} is the elements other than Z_k. The marginal distribution of Z is denoted as $P(Z_k = 1) = \pi_k$ with $\sum \pi_k = 1$ for $k = 1, 2, \ldots, K$. Thus, in Equation (1), the marginal distribution for the observable variable $\boldsymbol{X}$ will be

$$p(\boldsymbol{x}) = \sum_{k=1}^{K} \pi_k \mathcal{N}(\boldsymbol{x}|\boldsymbol{\mu}_k, \boldsymbol{\Sigma}_k), \tag{1}$$

where $\mathcal{N}(x|\mu_k, \Sigma_k)$ is the Gaussian density. A posterior probability for Z_k when x is given indicates how much each component contributes to the realization of x:

$$\gamma(k|\ ) = p(Z_k = 1|x,\ \mu,\ \Sigma) = \frac{\pi_k \mathcal{N}(x|\mu_k, \Sigma_k)}{\sum_{j=1}^{K} \pi_j \mathcal{N}(x|\mu_j, \Sigma_j)} \tag{2}$$

$\gamma(k|)$ in Equation (2) is also known as the responsibility probability allowing us to partition an observation into K components. For a given set of independent observations $_1,\ _2, \ldots,\ _N$ with sample size N, we assume that z_n, $n = 1, 2, \ldots, N$, is the latent variable for each observation. In addition to the location and shape parameters μ_k and Σ_k, the marginal probabilities $\pi_1, \pi_2, \ldots, \pi_k$ also contribute the parameter space of the log-likelihood function in Equation (3):

$$\mathcal{L}(\pi,\ \mu,\ \Sigma) = lnp(x|\pi_k, \mu_k, \Sigma_k) = \sum_{n=1}^{N} ln\left[\sum_{k=1}^{K} \pi_k \mathcal{N}(_n|\mu_k, \Sigma_k)\right] \tag{3}$$

The graphical representation for an observation $_n$ can be illustrated in Figure 1. The explicit form of the derivatives to Equation (3) is not available due to the summation term in the logarithm operation. The EM algorithm introduced in Sections 4.2 and 4.3 will address this difficulty and evaluate its likelihood in a tractable way.

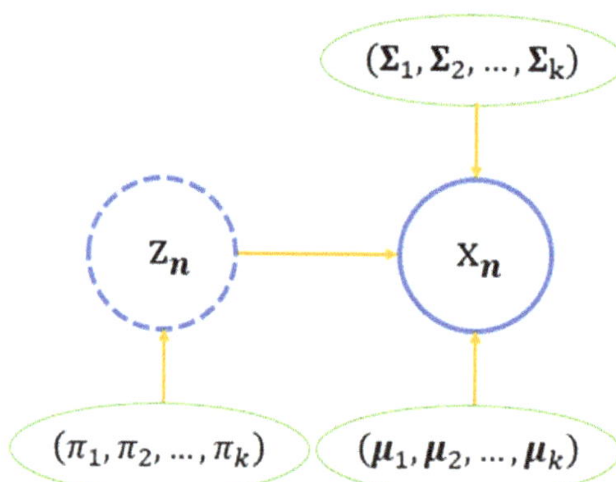

Figure 1. Graphical representation of a set of data points.

4.2. The Expectation-Maximization (EM) Algorithm

It has been proposed that an EM algorithm might be used to iteratively compute maximum-likelihood with incomplete information, where many applications such as filling in missing data, grouping and censoring, variance components, hyperparameter estimation, reweighted least-squares and factor analysis are explicitly addressed [52]. EM is considered one of the top ten algorithms in data mining and simplifies the maximization procedures for GMM [53]. Density estimation for GMM via EM can also handle high-dimensional data [54]. Thus, for a parametrized probability model $P(X|\theta)$, the joint distribution $P(X, Z|\theta)$ is introduced to rewrite the log-likelihood as

$$\mathcal{L}(\theta) = lnP(X|\theta) = lnP(X,\ Z|\theta) - lnP(Z|X, \theta), \tag{4}$$

where X is the observable variable and Z is the hidden variable. It should be noted that $lnP(X|\theta)$ is given in the form of a random variable. It will be equivalent to $\sum_{i=1}^{N} lnP(_i|\theta)$ when the sample $_1,\ _2, \ldots,\ _N$ is obtained. We denote a density function of Z as $q(Z)$ with $q(Z) > 0$. Since $q(Z)$ is irrelevant to θ, taking an expectation to $lnP(X|\theta)$ with regard to the distribution of Z will not affect the value of $\mathcal{L}(\boldsymbol{\theta})$. On the other hand, taking the expectation to the right-hand side in Equation (4) results in

$$\mathcal{L}(\boldsymbol{\theta}) = \int_Z q(Z) ln \frac{P(X, Z|\boldsymbol{\theta})}{q(Z)} dZ - \int_Z q(Z) ln \frac{P(Z|X, \boldsymbol{\theta})}{q(Z)} dZ. \quad (5)$$

In Equation (5), $-\int_Z q(Z) ln \frac{P(Z|X,\boldsymbol{\theta})}{q(Z)} dZ$, known as the Kullback–Leibler divergence and denoted as $KL(q \parallel P)$, is used to measure the distance between $q(Z)$ and $P(Z|X, \boldsymbol{\theta})$ [55]. $KL(q \parallel P)$ takes the value of 0 when $q(Z) = P(Z|X, \boldsymbol{\theta})$, otherwise it is greater than 0. If we denote $\int_Z q(Z) ln \frac{P(X,Z|\boldsymbol{\theta})}{q(Z)} dZ$ as the evidence lower bound (ELBO), $\mathcal{L}(\boldsymbol{\theta})$ can be represented in Equation (6) as

$$\mathcal{L}(\boldsymbol{\theta}) = ELBO + KL(q \parallel P). \quad (6)$$

For fixed $\boldsymbol{\theta}$ and $_1, _2, \ldots, _N$, $\mathcal{L}(\boldsymbol{\theta}) \geq ELBO$. $\mathcal{L}(\boldsymbol{\theta})$ takes the value of its lower bound ELBO only when $KL(q \parallel P) = 0$. Thus, the task becomes to find the estimate of $\boldsymbol{\theta}$ such that

$$\begin{aligned} \hat{\boldsymbol{\theta}}^{(t+1)} &= arg \max_{\boldsymbol{\theta}} ELBO \\ &= \operatorname{argmax}_{\boldsymbol{\theta}} \int_Z P\left(Z|X, \boldsymbol{\theta}^{(t)}\right) \left[lnP(X, Z|\boldsymbol{\theta}) - lnP\left(Z|X, \boldsymbol{\theta}^{(t)}\right)\right] dZ \\ &= \operatorname{argmax}_{\boldsymbol{\theta}} \mathbb{E}_{P(Z|X,\boldsymbol{\theta}^{(t)})}[lnP(X, Z|\boldsymbol{\theta})]. \end{aligned} \quad (7)$$

$\boldsymbol{\theta}^{(t)}$ is fixed for $P\left(Z|X, \boldsymbol{\theta}^{(t)}\right)$ that is one of the specific options for $q(Z)$ in Equation (7). The superscript (t) indicates the values obtained from the last iteration of the EM algorithm. Hence, $\int_Z P\left(Z|X, \boldsymbol{\theta}^{(t)}\right) lnP\left(Z|X, \boldsymbol{\theta}^{(t)}\right) dZ$ is independent of $\boldsymbol{\theta}$ and can be treated as constant in the estimation. It should be noted that $P(Z|X, \boldsymbol{\theta})$ is the ideal representation of the form of $q(Z)$ in some specific models. For implicit $q(Z)$ that is hard to obtain, an approximation method is usually applied to find the best $q(Z)$.

Two recursive steps for updating $\hat{\boldsymbol{\theta}}^{(t+1)}$ in Equation (7) form the EM algorithm. In the E-step, we use old $\boldsymbol{\theta}^{(t)}$ to find the posterior distribution for the latent variable, which is used to calculate the expectation of complete-data likelihood:

$$\mathcal{H}\left(\boldsymbol{\theta}; \boldsymbol{\theta}^{(t)}\right) = \mathbb{E}_{P(Z|X,\boldsymbol{\theta}^{(t)})}[lnP(X, Z|\boldsymbol{\theta})] \quad (8)$$

In the M-step, $\boldsymbol{\theta}^{(t+1)}$ is updated by maximizing Equation (8):

$$\hat{\boldsymbol{\theta}}^{(t+1)} = arg \max_{\boldsymbol{\theta}} \mathcal{H}\left(\boldsymbol{\theta}; \boldsymbol{\theta}^{(t)}\right). \quad (9)$$

An iterative evaluation of Equations (8) and (9) guarantees the convergence of $\mathcal{L}(\boldsymbol{\theta})$ [56]. An illustration of this process can be seen in Figure 2. The M-step searches the new parameter to increase the value of $\mathcal{H}(\boldsymbol{\theta}; \bullet)$ that is further increased by plugging to $\mathcal{L}(\boldsymbol{\theta})$ because the property $\mathcal{L}(\boldsymbol{\theta}) \geq ELBO$ always holds. The convergence will be found until $\boldsymbol{\theta}$ does not significantly update its value or $\mathcal{L}(\boldsymbol{\theta})$ does not improve.

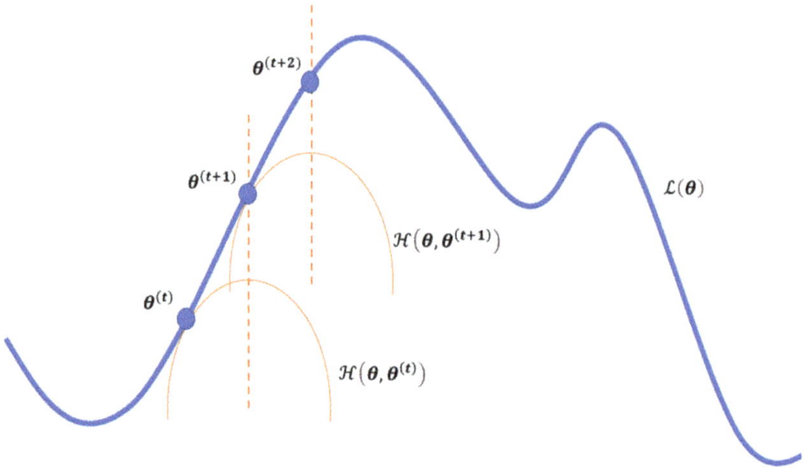

Figure 2. Parameter update in the EM algorithm.

4.3. Parameter Estimation for GMM

$\mathcal{H}\left(\boldsymbol{\theta};\, \boldsymbol{\theta}^{(t)}\right)$ is evaluated on the joint log-likelihood with complete data, which differs from the incomplete log-likelihood described in Equation (3). If we let the distribution of latent variable Z be $P(Z) = \prod_{k=1}^{K} \pi_k^{z_k}$ and the conditional distribution be $P(X|Z) = \prod_{k=1}^{K} \mathcal{N}(\boldsymbol{x}|\boldsymbol{\mu}_k, \boldsymbol{\Sigma}_k)^{z_k}$, the likelihood for complete data is a product of the two distributions:

$$P(\ _n, \mathbf{z}_n|\boldsymbol{\pi},\ \boldsymbol{\mu},\ \boldsymbol{\Sigma}) = \prod_{n=1}^{N}\prod_{k=1}^{K} \pi_k^{z_{nk}} \mathcal{N}(\ _n|\boldsymbol{\mu}_k, \boldsymbol{\Sigma}_k)^{z_{nk}}, \tag{10}$$

where z_{nk} indicates the kth component for $_n$. Compared with Equation (3), the benefit for evaluating Equation (10) is that the logarithm part of $lnP(\ _n, \mathbf{z}_n|\boldsymbol{\pi},\ \boldsymbol{\mu},\ \boldsymbol{\Sigma})$ will not be calculated on any summation terms, and there will only be a linear relationship between the latent variable Z and the observed variable, namely

$$lnP(\ _n, \mathbf{z}_n|\boldsymbol{\pi},\ \boldsymbol{\mu},\ \boldsymbol{\Sigma}) = \sum_{n=1}^{N}\sum_{k=1}^{K} z_{nk}[ln\pi_k + ln\mathcal{N}(\ _n|\boldsymbol{\mu}_k, \boldsymbol{\Sigma}_k)]. \tag{11}$$

Based on Equation (11), $\mathcal{H}\left(\boldsymbol{\theta},\ \boldsymbol{\theta}^{(t)}\right)$ can be easily specified as

$$\begin{aligned} &\mathcal{H}\left(\boldsymbol{\pi},\ \boldsymbol{\mu},\ \boldsymbol{\Sigma};\ \boldsymbol{\pi}^{(t)},\ \boldsymbol{\mu}^{(t)},\ \boldsymbol{\Sigma}^{(t)}\right) = \mathbb{E}_{P(\mathbf{z}_n|\ _n, \boldsymbol{\pi}^{(t)},\ \boldsymbol{\mu}^{(t)},\ \boldsymbol{\Sigma}^{(t)})}[lnP(\ _n, \mathbf{z}_n|\boldsymbol{\pi},\ \boldsymbol{\mu},\ \boldsymbol{\Sigma})] \\ &= \sum_{n=1}^{N}\sum_{k=1}^{K} \gamma_{nk}[ln\pi_k + ln\mathcal{N}(\ _n|\boldsymbol{\mu}_k, \boldsymbol{\Sigma}_k)], \end{aligned} \tag{12}$$

where $\gamma_{nk} = \mathbb{E}[z_{nk}]$ is the responsibility probability for a given $_n$ to be partitioned into component k. Thus, the specification of $\gamma(k|)$ in Equation (2), γ_{nk}, can be evaluated by

$$\gamma_{nk} = \gamma(k|_n) = \frac{\pi_k \mathcal{N}(_n|\boldsymbol{\mu}_k, \boldsymbol{\Sigma}_k)}{\sum_{j=1}^{K} \pi_j \mathcal{N}(_n|\boldsymbol{\mu}_j, \boldsymbol{\Sigma}_j)}. \tag{13}$$

The maximization of Equation (12) is in a tractable form if taking derivatives to the parameters. Due to the constraint $\sum \pi_k = 1$, a Lagrange multiplier λ is introduced for π_k.

Finally, the E-step for GMM is to evaluate the log-likelihood given $\boldsymbol{\theta}^{(t)} = \{\boldsymbol{\pi}^{(t)}, \boldsymbol{\mu}^{(t)}, \boldsymbol{\Sigma}^{(t)}\}$ and the M-step updates the parameters:

$$\begin{aligned}
\boldsymbol{\mu_k}^{(t+1)} &= \frac{1}{\sum_{n=1}^{N} \gamma_{nk}} \sum_{n=1}^{N} \gamma_{nkn}; \\
\boldsymbol{\Sigma_k}^{(t+1)} &= \frac{1}{\sum_{n=1}^{N} \gamma_{nk}} \sum_{n=1}^{N} \gamma_{nk}(\boldsymbol{n} - \boldsymbol{\mu_k}^{(t+1)})(\boldsymbol{n} - \boldsymbol{\mu}_k^{(t+1)})^T; \\
\pi_k^{(t+1)} &= \frac{\sum_{n=1}^{N} \gamma_{nk}}{N}.
\end{aligned}$$

When all the parameters are updated, we return to the E-step and evaluate Equation (13). The terminal of the algorithm, as introduced in Section 4.2, is reached either by observing stable $\boldsymbol{\theta}$ or $\mathcal{L}(\boldsymbol{\theta})$.

5. Data and Performance

5.1. Test Datasets

This paper considers two different datasets to validate the proposed method. Both datasets are well organized and free to use. The first one is the public building energy and water use data from Boston in the United States (the data are available at https://data.boston.gov/dataset/building-energy-reporting-and-disclosure-ordinance). The dataset was collected according to the Building Energy Reporting and Disclosure Ordinance (BERDO), which allows different types of building stakeholders to track their energy usage and greenhouse gas emissions and provides an assessment source for local government to implement energy-saving policies. In the original file from 2019 there were 28 variables in total. This includes general variables related to categories such as building name, location, physical information and energy-related variables. The variable *Energy Use Intensity* (EUI) reflects total annual energy consumption divided by the gross floor area. EUI is an effective measurement of energy performance. By eliminating the missing values and outliers from the Boston dataset, we identified 659 buildings labeled as multifamily housing.

The second dataset was collected in Aarhus (Århus in Denish), Denmark in 2020 in order to examine the district heating (DH) efficiency of different Danish building types (the data are available at https://data.mendeley.com/datasets/v8mwvy7p6r/1) [57]. From this dataset we extracted the EUI data for multifamily housing built between the 1960s and the 1980s and identified 183 buildings. The mean values of EUIs for Boston and Aarhus were 66.68 and 130.46 respectively. Given the two samples now comprised homogeneous information, we merged them into one and assumed the number of populations to be two. Thus, univariate Gaussian distribution was considered.

In addition, our calculations took the variable age group from the Danish dataset, representing the period when the building was built, as a new population indicator to illustrate the bivariate case. The segmentation was determined by shifts in Danish building traditions and the tightening of energy requirements in Danish Building Regulations. Two specific age groups were chosen to constitute the populations: '1951–1960' with 3461 buildings and 'After 2015' with 927. We then selected a secondary variable, *Ga*, to measure the daily heat load variation of these buildings since we postulated that the load variation may form a representative feature for both populations. Ga is calculated as the accumulated positive difference between the hourly and daily average heat loads during a year divided by both the annual average heat load and the number of hours (8760) in the year. Most of the values of Ga were around 0.2 indicating 20% load deviations on average. Thus, two populations (two age groups) and two variables (EUI and Ga) were constructed only from the Danish dataset for the bivariate case.

5.2. The Performance of EM Algorithm

5.2.1. The Univariate Case

The histograms are plotted to present the mixed and grouped distribution of EUIs for both cities. In the left panel of Figure 3, the mixed distribution created two peaks, although there was no obvious separation in the overlapped part. The true distributions can, however, be seen quite clearly in the grouped (separated) distribution displayed in the right panel of Figure 3. One limitation when using a Gaussian distribution is that the value of EUI cannot be negative. A truncated Gaussian may be a more appropriate representation. However, since all the truncated data points belonged to the Boston population, using GMM will not affect the responsibility probability γ_{nk} when conducting the E-step. Thus, we still follow the Gaussian assumption. If Gaussian property is severely violated, for example, because of the heat load variation in the bivariate variables, a Box-Cox transformation is required before implementing EM.

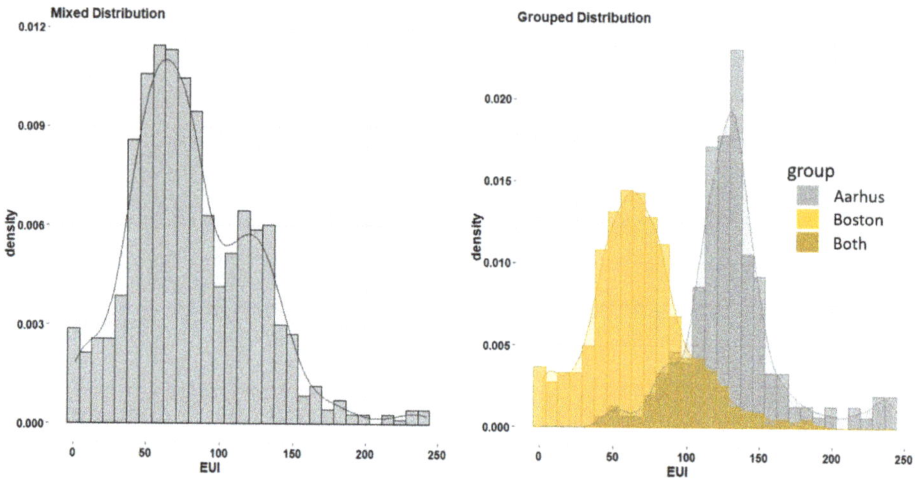

Figure 3. Mixed and grouped distribution of EUI.

Another issue for the parameter estimation is to determine the initial values of $\boldsymbol{\theta}$. It is not a difficult task to compare several combinations for the univariate case, but it will become a problem when the parameter size is large. Thus, we tested a number of possible combinations by varying the initial choice of $\pi_1, \pi_2, \mu_1, \mu_2, \sigma_1, \sigma_2$ and applied the same scheme to the bivariate case. More details on this process can be found in previous work on the setting of initial values [58]. The choice was determined by considering whether the final estimation could well represent both populations. The empirical results showed that the choices of $\pi_1, \pi_2, \sigma_1, \sigma_2$ did not seem to be dominant for the convergence, and we simply took $\pi_1 = \pi_2 = 0.5$ and σ_1, σ_2 to be the sample standard deviations. On the other hand, close values for μ_1 and μ_2 failed to separate the populations. In most of the experiments, μ_1 and μ_2 converged to a single value. Thus, we initialized μ_1 and μ_2 by letting them be constrained in the upper and lower $1/3$ quantile of the mixed population, respectively. Further, one hundred initial values for μ_1 and one hundred for μ_2 were randomly generated to obtain ten thousand combinations in which we randomly opted ten for evaluating the performance.

The ten sets of parameters are summarized in Table 1. For every parameter, there was no significant variation, and the mean values can be used to represent $\hat{\mu}_1$ and $\hat{\mu}_2$. We also computed the absolute errors in percentage terms between the mean and true values. Most of them were within 5%, while the overestimation of σ_2 for Aarhus might be due to the slightly smaller estimation for μ_2. Unlike fixed initial values, we also allowed for random variations up to 25% for σ_1 and σ_2 to validate our argument. As Table 2 shows, the result resembled Table 1. The same conclusion could be drawn for π_1 and π_2, which are not shown here. It is also observed that the performance of the log-likelihood values in Figure 4 is uniform. All of the experiments stopped within 15 updates and seemed to converge at the same point. In other words, it is enough to make inferences based on current estimations.

Table 1. Summary of the parameter estimation for fixed variance.

Population	Parameter	Min	Max	Mean	True Value	Error
Boston	μ_1	66.03	67.07	66.71	66.68	0.04%
	σ_1	30.99	31.60	31.39	32.62	3.77%
	π_1	0.76	0.78	0.77	0.78	1.28%
Aarhus	μ_2	126.84	129.23	128.33	130.46	1.63%
	σ_2	39.33	39.48	39.38	34.00	15.8%
	π_2	0.22	0.24	0.23	0.22	4.55%

Table 2. Summary of the parameter estimation for random variance.

Population	Parameter	Min	Max	Mean	True Value	Error
Boston	μ_1	66.00	67.42	66.59	66.68	0.13%
	σ_1	30.98	31.80	31.32	32.62	3.98%
	π_1	0.76	0.79	0.77	0.78	1.28%
Aarhus	μ_2	126.65	129.97	128.00	130.46	1.88%
	σ_2	39.35	39.48	39.42	34.00	15.9%
	π_2	0.21	0.24	0.23	0.22	4.55%

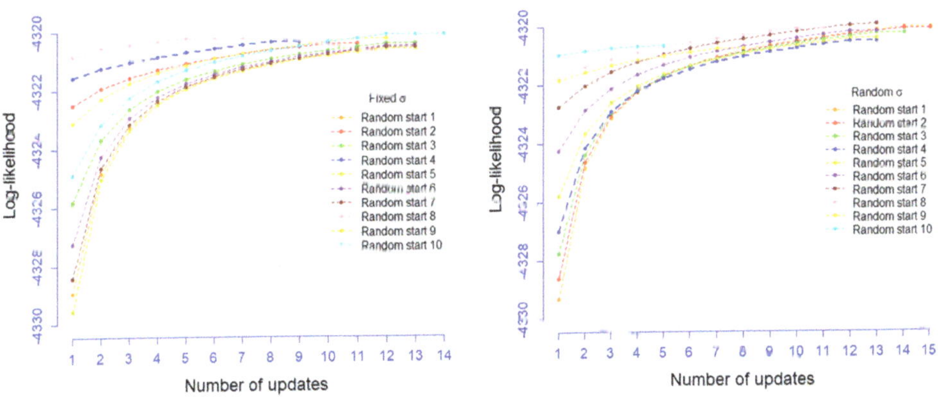

Figure 4. Log-likelihood performance for the univariate case.

We created two scenarios and assigned the sample points to the population with the larger density for classification. Two corresponding confusion matrices are presented in Table 3. We divided all the data points into four categories: true Boston, false Boston, true Aarhus and false Aarhus. The accuracy is the sum of both true classifications that is close to 90%.

Table 3. Classification accuracy for the univariate case.

Population (Predicted)	True Population, Fixed σ		True Population, Random σ	
	Boston	Aarhus	Boston	Aarhus
Boston	72.92%	7.13%	73.40%	7.72%
Aarhus	5.34%	14.61%	4.87%	14.01%

We then demonstrated the fitness between theoretical and empirical proportions. Proportion here refers to the quotient for which Boston's EUI should theoretically and empirically account. Theoretical proportion is made by

$$\mathcal{P}_{th}(Boston) = \frac{0.77\mathcal{N}(\tilde{x}|66.71,\ 31.39)}{0.77\mathcal{N}(\tilde{x}|66.71,\ 31.39) + 0.23\mathcal{N}(\tilde{x}|128.33,\ 39.38)},$$

where $\tilde{x}$ corresponds to the probability quantile segmentation on the x-axis in Figure 5 for the mixed distribution. Similarly, the empirical proportion, $\mathcal{P}_{em}(Boston)$, counts the number of data points from Boston divided by all data points between two adjacent values of $\tilde{x}$. For example, if 100% of the observations are taken from Boston, $\mathcal{P}_{th}(Boston)$ should be extremely close to 1 at the quantile 0. Both $\mathcal{P}_{th}(Boston)$ and $\mathcal{P}_{em}(Boston)$ are supposed to decrease because $\hat{\mu}_2$ is greater than $\hat{\mu}_1$. Thus, the results showed a good fit except for the quantiles close to 1 on the bottom right corner. This is because of the long tail of Boston's EUI. However, there were only a total of eight data points in this area, which is a minor deviation compared with $\mathcal{P}_{th}(Boston)$. The performance for Aarhus would look exactly the same just by vertically reversing Figure 5.

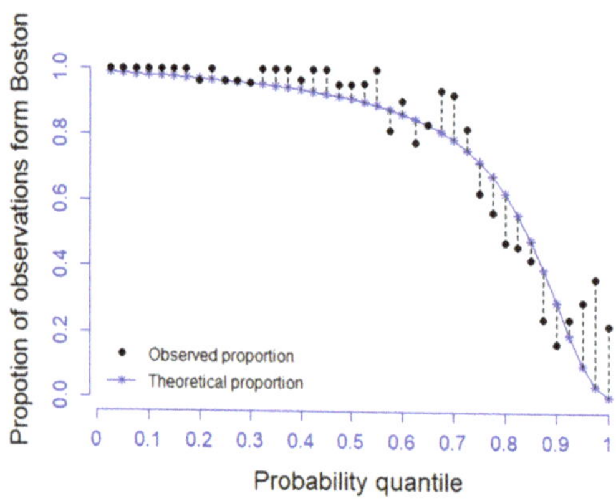

Figure 5. Theoretical and empirical proportions.

Given the results, we have drawn an additional two-step random sampling from the distributions to examine the generated EUIs. In step 1, we created a vector to store a random sequence of 0s and 1s that imply to which population a generated point belongs. The probabilities are taken as $P(I = 0) = 0.77$ and $P(I = 1) = 0.23$, where I is the indicator. The length of the vector is the same as the one of the observed sample, namely 842. In step 2, Gaussian random samples were drawn for each population to constitute the generated data. The quantile–quantile plot is shown for both fixed and random initial σs. As seen in Figure 6, the quantile values were taken every 5%. It is not surprising that neither setting for initial σ differed a great deal, and almost all of the quantile values were located on the 45-degree line, which means that the quantile values matched each other. In this sense, the generated sample under GMM presented a reliable representation of the populations. Here we only used the true sample to validate a generated sample of the same size. When the method is used in an authentic context, the sample size will depend on the total number of buildings in the original cohort.

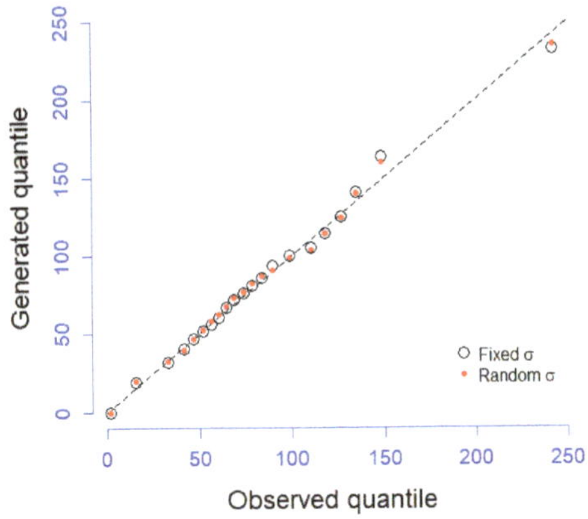

Figure 6. Quantile–quantile plot for observed and generated data.

5.2.2. The Bivariate Case

Testing bivariate case requires more parameters to be estimated. The selection of the initial values followed the same paradigms that were adopted for the univariate case. In order to keep the Gaussian property, as mentioned in Section 5.2.1, the daily heat load variation (Ga) was treated by Box-Cox transformation [59]. Since the estimations then became complex, we also increased the number of experiments for determining the estimates. We picked 20 initial values from each of the population means: μ_{EUI1}, μ_{Ga1}, μ_{EUI2} and μ_{Ga2}. The number of combinations became $20^4 = 160,000$. In all the experiments, we highlighted the combinations with a log-likelihood in the top 10%. We present the resulting pattern in Figure 7 where 400 combinations were located on the x-axis for the population '1951–1960', while 400 for the population 'After 2015' were located on the y-axis. The combinations were arranged from {minimum EUI, minimum Ga} to {minimum EUI, maximum Ga} and then to {maximum EUI, maximum Ga}. The figure shows that there were slight periodic patterns among the bigger 20 × 20 grids. Higher log-likelihood was slightly denser in the top left part. In almost all the bigger 20 × 20 grids, however, high log likelihood could always be found. Thus, the selection of initial values for the EM algorithm in the bivariate case appears to be somewhat isotropic at finding estimates with high log-likelihood values.

Something similar happened when we summarized the results of the 10% experiments in Table 4. In the bivariate case the overall errors decreased significantly compared with the univariate case. The majority were now below 3%. The reason for the bivariate model's success might be that it is able to use more of the energy performance features to separate the populations more correctly. It should be noted that the estimated $\hat{\mu}_{Ga}$ was not equal to the transformed mean value of Ga because the Box-Cox transformation is nonlinear. Thus, we only show the results for the transformed values here. We present both transformed and non-transformed Ga in the generative model evaluation later on. The classification accuracy is further computed in concordance with the univariate case in Table 5. Given the better parameter estimations, the true accuracy was over 99%.

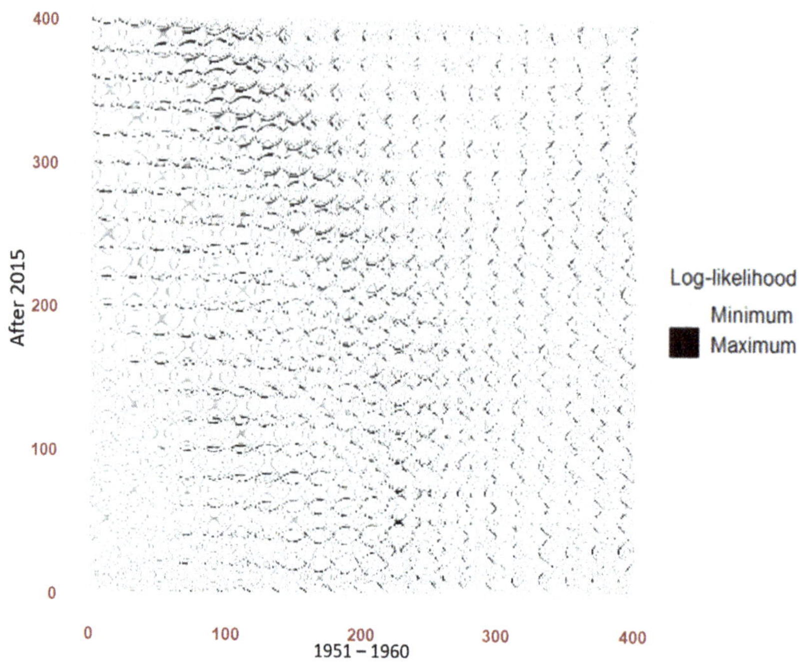

Figure 7. Distribution of top 10% log-likelihood for the combination of the initial values.

Table 4. Summary of the parameter estimation in the bivariate case.

Building Age Group	Parameter	Min	Max	Mean	True Value	Error
1951–1960	μ_{EUI1}	142.43	142.51	142.46	142.36	0.07%
	μ_{Ga1}(transformed)	−2.31	−2.31	−2.31	−2.31	0.00%
	σ^2_{EUI1}	1706.44	1713.20	1710.78	1728.90	1.05%
	σ^2_{Ga1}	0.26	0.26	0.26	0.26	0.00%
	$cov(EUI1, Ga1)$	−8.03	−7.81	−7.98	−8.17	2.33%
	$\pi_{1951-1960}$	0.79	0.79	0.79	0.79	0.00%
After 2015	μ_{EUI2}	54.12	54.24	54.19	54.86	1.22%
	μ_{Ga2}(transformed)	−0.59	−0.59	−0.59	−0.59	0.00%
	σ^2_{EUI2}	213.28	217.99	215.87	248.95	13.29%
	σ^2_{Ga2}	0.02	0.02	0.02	0.02	0.00%
	$cov(EUI2, Ga2)$	−0.62	−0.58	−0.60	−0.67	10.45%
	$\pi_{After\ 2015}$	0.21	0.21	0.21	0.21	0.00%

Table 5. Classification accuracy for the bivariate case.

Age Group (Predicted)	True Population	
	1951–1960	After 2015
1951–1960	78.56%	0.20%
After 2015	0.32%	20.92%

The estimates obtained in Table 4 were used to generate density contours, with dense and sparse areas distinguished in the two-dimensional surface. We compared these with the true distributions because they disclose the real scales of the densities. The contours are displayed in the left panel of Figure 8, and show the comparison that is made for the transformed heat load variation, while the result of the non-transformed distributions is shown in the right panel. The generative models are supposed to characterize the distributions in both dense and sparse areas. Both panels had obvious and observable centers. Both of these centers converged at the densest part of the real data. In other words, the generative models represented the real data to an acceptable degree. As discussed in the univariate case, the actual number of generated samples depends on the city's capacity to evaluate its energy performance with insufficient data.

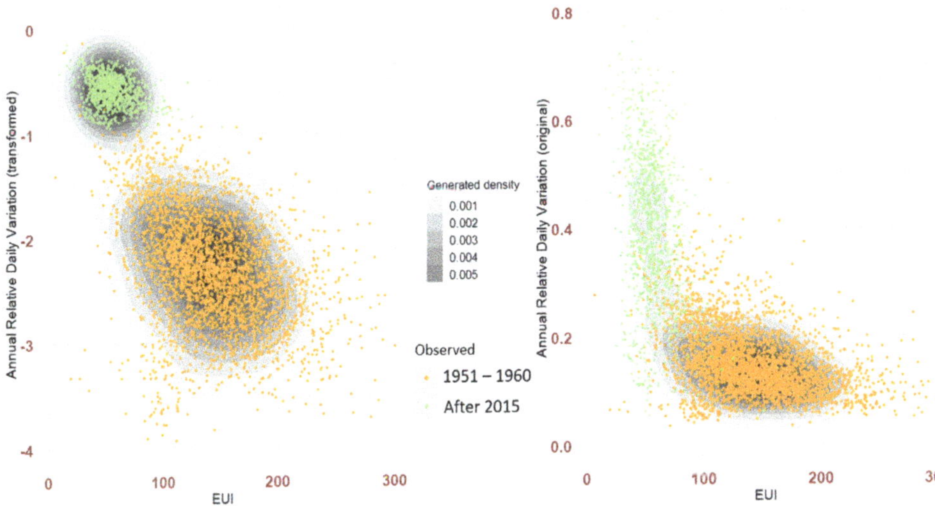

Figure 8. Density comparisons for the bivariate generative model.

6. Discussion

Using our proposed method to generate synthetic data is based on a distributional overview of the energy performance across all of the buildings in a district or urban area. However, our method does not take any spatial or temporal information into account. Imposing spatio-temporal mapping onto the synthetic data will help to draw an even better picture of building energy performance at the urban scale. As shown in Figure 9, spatial mapping takes the geolocations of buildings into consideration. Practically, the size of unlabeled data is far larger than labeled data. The performance of supervised learning models varies with limited labeled data. By including physical features, synthetic data are used to select and validate the supervised learning models for each population in a much more robust way.

The GMM method discussed in this work handles temporally aggregated data. With temporal mapping, it is possible to create higher time resolution data (for example hourly data) by including additional variables such as building class and hourly weather data. A set of buildings with known hourly energy consumption and building class could be taken as a set of reference buildings. By weighting from the reference buildings and sorting out a convex optimization problem, the synthetic data could then generate a set of energy performance profiles [60].

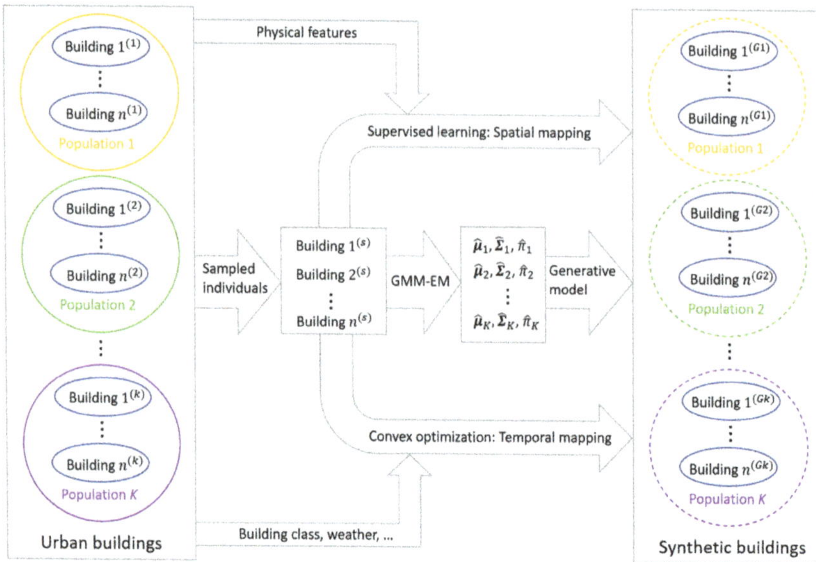

Figure 9. Spatio-temporal mapping of urban buildings from GMM synthetic data.

7. Conclusions

Big data are costly to collect and use. Even when this process is automated, many data collection systems operate with incomplete data or, in the case of building energy performance, are only able to collect data on a limited scale. From a statistical point of view, data from different groups are mixed together with little attempt made to distinguish between their representative populations. This hinders the efficient modeling of energy performance data, particularly at a large scale, and the construction of synthetic data for target groups.

In this paper, we proposed a GMM with an EM algorithm that can model building energy performance data for both univariate and bivariate cases. The energy performance indicators of a sample of buildings from Boston and Aarhus were adopted to segment mixed populations. For the univariate case, the Energy Use Intensity data from the two different cities were analyzed, and the updates were shown to have quick convergence. The derived models were able to capture the distributional features and to reflect the true population. The classification rate was almost 90%, and the generated data matched in quantile to the observed data. For the bivariate case, we showed that the inclusion of the new variable daily heat load variation further increased the power in parameter estimation, thus making the classification rate higher than in the univariate case. These data not only generate reliable density representations, but they also can be adjusted according to the real building capacity of a city with a spatio-temporal mapping.

Moreover, there are a number of topics in connection with these findings that would be interesting to explore in future studies.

- Firstly, in this paper we assumed that the number of populations was known. An interesting investigation would be to detect the optimal number of populations by introducing an objective function. Since Akaike information criterion (AIC) or Bayesian information criterion (BIC) reduces the effect on penalizing the model with small number of parameters, it is still an unsolved issue when the number of populations is small.
- Secondly, we suggest that other indicators of building energy use need to be considered because overall building performance is usually affected by multiple factors.

- Finally, it is also appealing that more probability distributions could be studied instead of using data transformation.

Author Contributions: Conceptualization, M.H., Z.W. and X.Z.; methodology, M.H. and Z.W.; software, M.H.; validation, M.H.; formal analysis, M.H. and Z.W.; data curation, M.H.; writing—original draft preparation, M.H. and X.Z.; writing—review and editing, M.H. and X.Z.; funding acquisition, M.H. and X.Z. All authors have read and agreed to the published version of the manuscript.

Funding: This research was funded by the IMMA research network between Microdata Analysis and Energy and Built Environments in the School of Technology and Business Studies at Dalarna University; and the UBMEM project at the Swedish Energy Agency (Grant no. 46068).

Institutional Review Board Statement: Not applicable.

Informed Consent Statement: Not applicable.

Conflicts of Interest: The authors declare no conflict of interest.

Abbreviations

DL	deep learning
ELBO	evidence lower bound
EM	Expectation-Maximization
EUI	Energy Use Intensity
Ga	daily heat load variation
GAN	Generative Adversarial Network
GMM	Gaussian mixture model
UBEM	urban building energy modeling
Notations	
$\mathbb{E}$	expectation
$\mathcal{H}$	objective function of $\boldsymbol{\theta}$
I	indicator of population
K	a category of latent variable
$KL(q \parallel P)$	Kullback–Leibler divergence
$\mathcal{L}$	log-likelihood function
N	sample size
$\mathcal{N}$	Gaussian density function
$q(Z)$	a density function of Z
$(t), (t+1)$	iteration state
	an observation of $\boldsymbol{X}$
r	a quantile of $\boldsymbol{X}$
$\boldsymbol{X}$	observable variable
$\widetilde{x}$	quantile segmentation
z_n	latent variable for each
Z	latent variable
γ	responsibility probability
$\boldsymbol{\theta}$	set of unknown parameters
$\boldsymbol{\mu}$	mean vector of $\boldsymbol{x}$
$\hat{\mu}_1, \hat{\mu}_2$	estimates of mean parameter
π_k	marginal distribution of Z
$\boldsymbol{\pi}$	vector of marginal probabilities
$\boldsymbol{\Sigma}$	covariance matrix of $\boldsymbol{x}$

References

1. Cao, X.; Dai, X.; Liu, J. Building Energy-Consumption Status Worldwide and the State-of-the-Art Technologies for Zero-Energy Buildings during the Past Decade. *Energy Build.* **2016**, *128*, 198–213. [CrossRef]
2. Castillo-Calzadilla, T.; Macarulla, A.M.; Kamara-Esteban, O.; Borges, C.E. A Case Study Comparison between Photovoltaic and Fossil Generation Based on Direct Current Hybrid Microgrids to Power a Service Building. *J. Clean. Prod.* **2020**, *244*, 118870. [CrossRef]

3. D'Agostino, D.; Mazzarella, L. What Is a Nearly Zero Energy Building? Overview, Implementation and Comparison of Definitions. *J. Build. Eng.* **2019**, *21*, 200–212. [CrossRef]
4. Tabar, V.S.; Hagh, M.T.; Jirdehi, M.A. Achieving a Nearly Zero Energy Structure by a Novel Framework Including Energy Recovery and Conversion, Carbon Capture and Demand Response. *Energy Build.* **2021**, *230*, 110563. [CrossRef]
5. Hermelink, A.; Schimschar, S.; Boermans, T.; Pagliano, L.; Zangheri, P.; Armani, R.; Voss, K.; Musall, E. *Towards Nearly Zero-Energy Buildings*; European Commission: Köln, Germany, 2012.
6. Magrini, A.; Lentini, G.; Cuman, S.; Bodrato, A.; Marenco, L. From Nearly Zero Energy Buildings (NZEB) to Positive Energy Buildings (PEB): The next Challenge—The Most Recent European Trends with Some Notes on the Energy Analysis of a Forerunner PEB Example. *Dev. Built Environ.* **2020**, *3*, 100019. [CrossRef]
7. De Luca, G.; Ballarini, I.; Lorenzati, A.; Corrado, V. Renovation of a Social House into a NZEB: Use of Renewable Energy Sources and Economic Implications. *Renew. Energy* **2020**, *159*, 356–370. [CrossRef]
8. Kurnitski, J.; Saari, A.; Kalamees, T.; Vuolle, M.; Niemelä, J.; Tark, T. Cost Optimal and Nearly Zero (NZEB) Energy Performance Calculations for Residential Buildings with REHVA Definition for NZEB National Implementation. *Energy Build.* **2011**, *43*, 3279–3288. [CrossRef]
9. Li, W.; Zhou, Y.; Cetin, K.; Eom, J.; Wang, Y.; Chen, G.; Zhang, X. Modeling Urban Building Energy Use: A Review of Modeling Approaches and Procedures. *Energy* **2017**, *141*, 2445–2457. [CrossRef]
10. Pasichnyi, O.; Wallin, J.; Kordas, O. Data-Driven Building Archetypes for Urban Building Energy Modelling. *Energy* **2019**, *181*, 360–377. [CrossRef]
11. Ali, U.; Shamsi, M.H.; Bohacek, M.; Hoare, C.; Purcell, K.; Mangina, E.; O'Donnell, J. A Data-Driven Approach to Optimize Urban Scale Energy Retrofit Decisions for Residential Buildings. *Appl. Energy* **2020**, *267*, 114861. [CrossRef]
12. Fathi, S.; Srinivasan, R.; Fenner, A.; Fathi, S. Machine Learning Applications in Urban Building Energy Performance Forecasting: A Systematic Review. *Renew. Sustain. Energy Rev.* **2020**, *133*, 110287. [CrossRef]
13. Nutkiewicz, A.; Yang, Z.; Jain, R.K. Data-Driven Urban Energy Simulation (DUE-S): Integrating Machine Learning into an Urban Building Energy Simulation Workflow. *Energy Procedia* **2017**, *142*, 2114–2119. [CrossRef]
14. Hong, T.; Chen, Y.; Luo, X.; Luo, N.; Lee, S.H. Ten Questions on Urban Building Energy Modeling. *Build. Environ.* **2020**, *168*, 106508. [CrossRef]
15. Zhang, X.; Lovati, M.; Vigna, I.; Widén, J.; Han, M.; Gal, C.; Feng, T. A Review of Urban Energy Systems at Building Cluster Level Incorporating Renewable-Energy-Source (RES) Envelope Solutions. *Appl. Energy* **2018**, *230*, 1034–1056. [CrossRef]
16. Salim, F.D.; Dong, B.; Ouf, M.; Wang, Q.; Pigliautile, I.; Kang, X.; Hong, T.; Wu, W.; Liu, Y.; Rumi, S.K.; et al. Modelling Urban-Scale Occupant Behaviour, Mobility, and Energy in Buildings: A Survey. *Build. Environ.* **2020**, *183*, 106964. [CrossRef]
17. Perera, A.T.D.; Nik, V.M.; Chen, D.; Scartezzini, J.-L.; Hong, T. Quantifying the Impacts of Climate Change and Extreme Climate Events on Energy Systems. *Nat. Energy* **2020**, *5*, 150–159. [CrossRef]
18. Yang, Z.; Roth, J.; Jain, R.K. DUE-B: Data-Driven Urban Energy Benchmarking of Buildings Using Recursive Partitioning and Stochastic Frontier Analysis. *Energy Build.* **2018**, *163*, 58–69. [CrossRef]
19. Ma, J.; Cheng, J.C.P. Estimation of the Building Energy Use Intensity in the Urban Scale by Integrating GIS and Big Data Technology. *Appl. Energy* **2016**, *183*, 182–192. [CrossRef]
20. Risch, S.; Remmen, P.; Müller, D. Influence of Data Acquisition on the Bayesian Calibration of Urban Building Energy Models. *Energy Build.* **2021**, *230*, 110512. [CrossRef]
21. Goy, S.; Maréchal, F.; Finn, D. Data for Urban Scale Building Energy Modelling: Assessing Impacts and Overcoming Availability Challenges. *Energies* **2020**, *13*, 4244. [CrossRef]
22. Goodfellow, I.; Bengio, Y.; Courville, A. *Deep Learning*; Adaptive Computation and Machine Learning; MIT Press Ltd: Cambridge, MA, USA, 2016; ISBN 978-0-262-03561-3.
23. Chen, C.L.P.; Zhang, C. Data-Intensive Applications, Challenges, Techniques and Technologies: A Survey on Big Data. *Inf. Sci.* **2014**, *275*, 314–347. [CrossRef]
24. Ma, J.; Cheng, J.C.P.; Jiang, F.; Chen, W.; Wang, M.; Zhai, C. A Bi-Directional Missing Data Imputation Scheme Based on LSTM and Transfer Learning for Building Energy Data. *Energy Build.* **2020**, *216*, 109941. [CrossRef]
25. Marta, M.; Belinda, L. Simplified Model to Determine the Energy Demand of Existing Buildings. Case Study of Social Housing in Zaragoza, Spain. *Energy Build.* **2017**, *149*, 483–493. [CrossRef]
26. Cho, K.; Kim, S. Energy Performance Assessment According to Data Acquisition Levels of Existing Buildings. *Energies* **2019**, *12*, 1149. [CrossRef]
27. Guerra-Santin, O.; Tweed, C.A. In-Use Monitoring of Buildings: An Overview of Data Collection Methods. *Energy Build.* **2015**, *93*, 189–207. [CrossRef]
28. Laroca, R.; Barroso, V.; Diniz, M.A.; Goncalves, G.R.; Schwartz, W.R.; Menotti, D. Convolutional Neural Networks for Automatic Meter Reading. *J. Electron. Imaging* **2019**, *28*, 013023. [CrossRef]
29. Oti, A.H.; Kurul, E.; Cheung, F.; Tah, J.H.M. A Framework for the Utilization of Building Management System Data in Building Information Models for Building Design and Operation. *Autom. Constr.* **2016**, *72*, 195–210. [CrossRef]
30. Afroz, Z.; Burak Gunay, H.; O'Brien, W. A Review of Data Collection and Analysis Requirements for Certified Green Buildings. *Energy Build.* **2020**, *226*, 110367. [CrossRef]

31. Despotovic, M.; Koch, D.; Leiber, S.; Döller, M.; Sakeena, M.; Zeppelzauer, M. Prediction and Analysis of Heating Energy Demand for Detached Houses by Computer Vision. *Energy Build.* **2019**, *193*, 29–35. [CrossRef]
32. Wei, S.; Tien, P.W.; Calautit, J.K.; Wu, Y.; Boukhanouf, R. Vision-Based Detection and Prediction of Equipment Heat Gains in Commercial Office Buildings Using a Deep Learning Method. *Appl. Energy* **2020**, *277*, 115506. [CrossRef]
33. Gao, Y.; Ruan, Y.; Fang, C.; Yin, S. Deep Learning and Transfer Learning Models of Energy Consumption Forecasting for a Building with Poor Information Data. *Energy Build.* **2020**, *223*, 110156. [CrossRef]
34. Ryu, S.; Noh, J.; Kim, H. Deep Neural Network Based Demand Side Short Term Load Forecasting. *Energies* **2017**, *10*, 3. [CrossRef]
35. Lu, J.; Qian, J.; Zhang, Q.; Liu, S.; Xie, F.; Xu, H. Best Practices in China Southern Power Grid Competition of AI Short-Term Load Forecasting. In Proceedings of the 2020 12th IEEE PES Asia-Pacific Power and Energy Engineering Conference (APPEEC), Nanjing, China, 20–23 September 2020; pp. 1–5.
36. Wang, Y.; Liu, M.; Bao, Z.; Zhang, S. Short-Term Load Forecasting with Multi-Source Data Using Gated Recurrent Unit Neural Networks. *Energies* **2018**, *11*, 1138. [CrossRef]
37. Zhang, Y.; Ai, Q.; Li, Z.; Yin, S.; Huang, K.; Yousif, M.; Lu, T. Data Augmentation Strategy for Small Sample Short-term Load Forecasting of Distribution Transformer. *Int. Trans. Electr. Energ. Syst.* **2020**, *30*, 1–18. [CrossRef]
38. Acharya, S.K.; Wi, Y.; Lee, J. Short-Term Load Forecasting for a Single Household Based on Convolution Neural Networks Using Data Augmentation. *Energies* **2019**, *12*, 3560. [CrossRef]
39. Lai, C.S.; Mo, Z.; Wang, T.; Yuan, H.; Ng, W.W.Y.; Lai, L.L. Load Forecasting Based on Deep Neural Network and Historical Data Augmentation. *IET Gener. Transm. Distrib.* **2020**. [CrossRef]
40. Goodfellow, I.J.; Pouget-Abadie, J.; Mirza, M.; Xu, B.; Warde-Farley, D.; Ozair, S.; Courville, A.; Bengio, Y. Generative Adversarial Nets. In Proceedings of the Advances in Neural Information Processing Systems 27 (NIPS), Montreal, QC, Canada, 8–13 December 2014.
41. Gui, J.; Sun, Z.; Wen, Y.; Tao, D.; Ye, J. A Review on Generative Adversarial Networks: Algorithms, Theory, and Applications. *arXiv* **2020**, arXiv:2001.06937.
42. Wang, Z.; Hong, T. Generating Realistic Building Electrical Load Profiles through the Generative Adversarial Network (GAN). *Energy Build.* **2020**, *224*, 110299. [CrossRef]
43. Fekri, M.N.; Ghosh, A.M.; Grolinger, K. Generating Energy Data for Machine Learning with Recurrent Generative Adversarial Networks. *Energies* **2019**, *13*, 130. [CrossRef]
44. Gong, X.; Tang, B.; Zhu, R.; Liao, W.; Song, L. Data Augmentation for Electricity Theft Detection Using Conditional Variational Auto-Encoder. *Energies* **2020**, *13*, 4291. [CrossRef]
45. Hong, T.; Chou, S.K.; Bong, T.Y. Building Simulation: An Overview of Developments and Information Sources. *Building Environ.* **2000**, *35*, 347–361. [CrossRef]
46. Maile, T.; Bazjanac, V.; Fischer, M. A Method to Compare Simulated and Measured Data to Assess Building Energy Performance. *Building Environ.* **2012**, *56*, 241–251. [CrossRef]
47. Abuimara, T.; O'Brien, W.; Gunay, B.; Carrizo, J.S. Towards Occupant-Centric Simulation-Aided Building Design: A Case Study. *Build. Res. Inf.* **2019**, *47*, 866–882. [CrossRef]
48. Schiefelbein, J.; Rudnick, J.; Scholl, A.; Remmen, P.; Fuchs, M.; Müller, D. Automated Urban Energy System Modeling and Thermal Building Simulation Based on OpenStreetMap Data Sets. *Building Environ.* **2019**, *149*, 630–639. [CrossRef]
49. Hong, T.; Langevin, J.; Sun, K. Building Simulation: Ten Challenges. *Build. Simul.* **2018**, *11*, 871–898. [CrossRef]
50. Solmaz, A.S. A Critical Review on Building Performance Simulation Tools. *Int. J. Sustain. Trop. Des. Res. Pract. UPM* **2019**, *12*, 7–20.
51. Bishop, C.M. *Pattern Recognition and Machine Learning*; Springer Science+Business Media, LLC: New York, NY, USA, 2006.
52. Dempster, A.P.; Laird, N.M.; Rubin, D.B. Maximum Likelihood from Incomplete Data via the EM Algorithm. *J. R. Stat. Soc. Ser. B (Methodol.)* **1977**, *39*, 1–38.
53. Wu, X.; Kumar, V.; Quinlan, J.R.; Ghosh, J.; Yang, Q.; Motoda, H.; McLachlan, G.J.; Ng, A.; Liu, B.; Yu, P.S.; et al. Top 10 Algorithms in Data Mining. *Knowl. Inf. Syst.* **2008**, *14*, 1–37. [CrossRef]
54. Ghahramani, Z.; Jordan, M.I. *Supervised Learning from Incomplete Data via an EM Approach*; Morgan Kaufmann: San Francisco, CA, USA, 1994.
55. Kullback, S.; Leibler, R.A. On Information and Sufficiency. *Ann. Math. Stat.* **1951**, *22*, 79–96. [CrossRef]
56. Wu, C.F.J. On the Convergence Properties of the EM Algorithm. *Ann. Stat.* **1983**, *11*, 95–103. [CrossRef]
57. Kristensen, M.H.; Petersen, S. District Heating Energy Efficiency of Danish Building Typologies. *Energy Build.* **2021**, *231*, 110602. [CrossRef]
58. Blömer, J.; Bujna, K. Simple Methods for Initializing the EM Algorithm for Gaussian Mixture Models. *arXiv* **2013**, arXiv:1312.5946v3.
59. Box, G.E.P.; Cox, D.R. An Analysis of Transformations. *J. R. Stat. Soc. Ser. B (Methodol.)* **1964**, *26*, 211–252. [CrossRef]
60. Roth, J.; Martin, A.; Miller, C.; Jain, R.K. SynCity: Using Open Data to Create a Synthetic City of Hourly Building Energy Estimates by Integrating Data-Driven and Physics-Based Methods. *Appl. Energy* **2020**, *280*, 115981. [CrossRef]

MDPI
St. Alban-Anlage 66
4052 Basel
Switzerland
Tel. +41 61 683 77 34
Fax +41 61 302 89 18
www.mdpi.com

Buildings Editorial Office
E-mail: buildings@mdpi.com
www.mdpi.com/journal/buildings